EXPERIMENTAL TECHNIQUES in MATERIALS and MECHANICS

EXPERIMENTAL TECHNIQUES in MATERIALS and MECHANICS

C. Suryanarayana

CRC Press
Taylor & Francis Group
Boca Raton London New York

CRC Press is an imprint of the
Taylor & Francis Group, an **informa** business

CRC Press
Taylor & Francis Group
6000 Broken Sound Parkway NW, Suite 300
Boca Raton, FL 33487-2742

CRC Press is an imprint of Taylor & Francis Group, an Informa business

Version Date: 2011908

International Standard Book Number: 978-1-4398-1904-3 (Hardback)

Library of Congress Cataloging-in-Publication Data

Suryanarayana, C.
Experimental techniques in materials and mechanics / C. Suryanarayana.
p. cm.
Includes bibliographical references and index.
ISBN 978-1-4398-1904-3 (hardcover : alk. paper)
1. Materials--Mechanical properties--Research--Methodology. 2. Materials--Experiments. 3. Materials--Testing--Technique. I. Title.

TA404.8.S87 2011
620.1'12--dc23 2011026917

Visit the Taylor & Francis Web site at
http://www.taylorandfrancis.com

and the CRC Press Web site at
http://www.crcpress.com

Contents

Preface

Materials are pervasive and everyone has to deal with some type of material or other. A majority of the undergraduate students of engineering (aerospace, chemical, electrical, materials, mechanical, etc.) go through a first course on materials, usually titled *Structure and Properties of Materials*, to enable them to appreciate the basic principles and applications of different types of materials—metals, ceramics, polymers, and composites. The market is currently flooded with a large number of textbooks dealing with different aspects of materials and at different levels. However, the emphasis in most of these textbooks has been to "load" the student with theoretical information. Very rarely, if at all, a laboratory component is included as part of this course, wherein it will be possible to experimentally determine the properties and relate them to the structure and processing behavior of materials. Also, the course content for this elementary course is so heavy that it is almost impossible to cover the full text, let alone having a laboratory component.

The materials engineer is often called upon to design, select, and recommend usage of materials for diverse applications. One may also be asked to investigate the reasons for failure of materials. For all these purposes, a thorough knowledge of the basic principles of materials science and engineering is essential. Even though materials engineers and, for that matter some mechanical engineers and aerospace engineers, go through additional courses on materials; a practical knowledge of the correlations between crystal structure, microstructure, and properties is deemed very much necessary.

This book has evolved from teaching a lecture-cum-laboratory course, *Experimental Techniques in Mechanics and Materials*, to undergraduate mechanical and materials engineering students at the University of Central Florida. The topics chosen for inclusion in the book are based on the possibility of completing them in a one-semester course. Emphasis is placed on practical aspects of crystal structure determination, microstructural observation, heat treatment, and mechanical testing. As an experienced instructor would realize, the microstructure, heat treatment, and mechanical properties are completely interrelated. Therefore, samples used in one experiment would be useful in another experiment. For example, specimens prepared for microscopic observation could be used for hardness testing. Similarly, samples used for tensile or impact testing could be used for scanning electron microscopic examination to determine the type of failure.

Although a large number of textbooks are available on structure and properties of materials and these explain the basic phenomena in some detail, they do not explain the practical aspects in sufficient detail to enable the student to go into the laboratory and perform the experiment. Thus, for example, even though one may know the theoretical background, one may have little understanding of how to prepare a specimen in the laboratory for optical microscopic examination and how to derive useful information from the microstructures. The primary aim of this book is to enable students to understand the practical aspects of the different techniques discussed here and derive the maximum possible information from the results obtained. The book has placed emphasis on techniques that are commonly available in most laboratories and those that are accessible to undergraduate students.

The book contains 12 chapters that deal with crystal structure determination, optical and scanning electron microscopy, phase diagrams and heat treatment, and different types of mechanical testing methods. The format of the chapters is fairly uniform, but not identical. Some variation has been brought in depending on the contents of the chapter. The importance of the technique is first introduced. This is followed by reasonably detailed background information to provide the student with the necessary theory to fully understand the experiment. In this sense, the book is self-contained and the student need not go to another book for the required theoretical background. A number of examples have been worked out in every chapter to further clarify the concepts explained. This is followed by a detailed description of the experiment to be conducted and how the data could be tabulated and interpreted. A special feature of the book is that a large number of illustrations, figures, and micrographs, wherever possible, have been included in each chapter. At the end of the chapter, a number of exercises have been provided; by answering, the student will realize the extent to which the subject matter was clear. If any student is interested in knowing much more about the topics covered in any chapter, references for further reading are also given.

This book is primarily intended for undergraduate students majoring in materials science and engineering or mechanical engineering. All the topics covered in the book can be covered in a one-semester course. It is realized that not *all* the techniques used by materials scientists and engineers are included in the book. For example, techniques such as differential thermal analysis or differential scanning calorimetry, temperature measurement, vacuum systems, and so on are not included. Even though these are all useful techniques, a choice had to be made, depending on the time available in one semester.

The book should also be useful for practicing engineers or technicians involved in microscopy, crystal structure determination, and mechanical testing. Preliminary knowledge of materials science and engineering will be useful but not essential because the concepts have been explained in easy-to-understand terms in the background information.

Sufficient care has been taken to ensure that the information presented is correct, clear, and easy to understand for the students; accuracy has not been sacrificed. But if the discerning reader notices any mistakes, I would be most grateful if these are communicated to me at Challapalli.Suryanarayana@ucf.edu

C. Suryanarayana
Orlando, Florida

Acknowledgments

During the preparation of this book I had the good fortune of receiving help from many people and I take pleasure in acknowledging their assistance. First of all I would like to acknowledge the large number of students who used the draft copy of the book and acted as involuntary reviewers during the lecturing and laboratory sessions over the years.

A number of friends and students have read parts of the book and provided helpful comments to improve the clarity and/or readability. I am very grateful to them for sparing their time to read the chapters and to comment on the contents of the book. I would like to particularly mention Professor Thomas Klassen of Helmut-Schmidt University, Hamburg, Germany, Dr. C. L. Mendis of National Institute for Materials Science, Tsukuba, Japan, and Dr. Raj Vaidyanathan of the University of Central Florida, Orlando, for several constructive comments on the early versions of the chapters. Some chapters of the book were also informally read by others and I am grateful to all of them for their comments. Particular mention should be made of Professor J. J. Moore of Colorado School of Mines, Golden and Professor K. L. Murty of North Carolina State University, Raleigh for their encouragement.

Parts of this book were written when I was a visiting professor at the King Fahd University of Petroleum and Minerals in Dhahran, Saudi Arabia and also at Vel Tech Dr. RR & Dr. SR Technical University, Chennai, India. I am very grateful to both these institutions for providing the necessary facilities for successfully completing some chapters. I am also grateful to the Chair of the Department of Mechanical, Materials and Aerospace Engineering at the University of Central Florida in Orlando for providing the right environment for completing the book.

I am grateful to several organizations that have provided me with original figures used in this book. A number of figures used in this book were drawn meticulously by Dr. U. M. R. Seelam. I sincerely thank him for his help and his patience in revising the figures until they reached the final stage for inclusion in the book. The aesthetics of the figures is all because of him. But if there are errors in any of them, the responsibility is mine. I thank the staff of CRC Press for their patience and for the high level of cooperation and interest in the production of this book.

Finally, I owe a huge debt of gratitude to my wife Meena for her patience with me during the course of preparation of the book.

Author

C. Suryanarayana is professor of materials science and engineering in the Department of Mechanical, Materials and Aerospace Engineering of the University of Central Florida in Orlando, Florida. He has conducted research investigations in the areas of rapid solidification processing, mechanical alloying, innovative synthesis/processing techniques, metallic glasses, superconductivity, quasicrystals, and nanostructured materials for over 40 years. He has published more than 330 technical papers in archival journals and authored/edited 19 books and conference proceedings. He has extensive teaching experience and has taught at several universities in India, United States, and other countries. Earlier he was a professor of metallurgy at Banaras Hindu University in Varanasi, India. He held visiting assignments at the University of Oxford, United Kingdom; the Atomic Energy Center in Mol, Belgium; the Wright-Patterson Air Force Base in Dayton, Ohio; the University of Idaho in Moscow, Idaho; the Colorado School of Mines in Golden, Colorado; the GKSS Research Center in Geesthacht, Germany; the Helmut Schmidt University in Hamburg, Germany; the Tohoku University in Sendai, Japan; the National Institute for Materials Science in Tsukuba, Japan; the Chungnam National University in Taejon, Korea; and the Shenyang National Laboratory for Materials Science in Shenyang, China.

Professor Suryanarayana is on the editorial boards of several prestigious materials science journals. He has received several awards for his research contributions to nonequilibrium processing of materials including the Young Scientist Medal of the Indian National Science Academy, Pandya Silver Medal of the Indian Institute of Metals, National Metallurgists Day Award of the Government of India, Distinguished Alumnus Award of Banaras Hindu University, and Lee Hsun Research Award from the Chinese Academy of Sciences. Thomson Reuters has just announced that Professor Suryanarayana is one of the top 40 materials scientists who achieved the highest citation impact scores for their papers published since 2000. He is a fellow of ASM International and also of the Institute of Materials, Minerals and Mining, London, United Kingdom. He received the BE degree in metallurgy from the Indian Institute of Science, Bangalore, and MS and PhD degrees in physical metallurgy from Banaras Hindu University, Varanasi, India.

1 Introduction

1.1 MATERIALS SCIENCE AND ENGINEERING

Materials science and engineering is a continuously evolving and developing field. Materials are pervasive and all scientists (whether physicists, chemists, or geologists) and engineers (whether civil, mechanical, electrical, chemical, or materials) deal with some type of material or other. Every aspect of our day-to-day life revolves around materials. In fact, the different periods of human civilization have been named after materials—stone, copper, bronze, iron, and so on. The present age can be most aptly named the silicon age.

The discipline of materials science and engineering, which has come into existence since the 1960s, involves establishing interrelationships among the structure, processing, and properties of materials. Even prior to this, engineers were doing similar things independently for metals, ceramics, or polymers. The structure and properties of materials will eventually determine the performance of the material. Thus, the four components—structure, properties, processing, and performance—are typically represented as the four corners of a tetrahedron, emphasizing the interrelationship among these. Figure 1.1 represents the tetrahedron of materials science and engineering with the four corners representing the structure, properties, processing, and performance. It is important to realize that the structure of the material determines the properties. The properties can also be altered by changing the composition and/or processing. In other words, the structure, composition, and processing will determine the final properties of materials. Similarly, the other components are also dependent on each other.

The types of materials of common interest include metals, ceramics, polymers, and composites. As it will become clear in the following chapters, the crystal structure and microstructure, to a great extent, determine the mechanical properties of materials. For example, materials with a face-centered cubic (FCC) structure are more ductile than those with a body-centered cubic (BCC) structure. Further, fine-grained materials are stronger than coarse-grained materials of the same composition. Therefore, determination of the crystal structure and microstructure assumes great importance in selection of materials for applications. The microstructure of materials is determined by the processing stages through which the material passes. It also becomes necessary to determine the properties of materials to enable these materials to be used for appropriate applications. Alternately, if one has to choose and/or design a material for a given application, then the engineer needs knowledge of the crystal structure, microstructure, and properties of the material. The performance of the material will depend on the type of mechanical loads it is subjected to, the type of environment it is exposed to, and the different combinations of

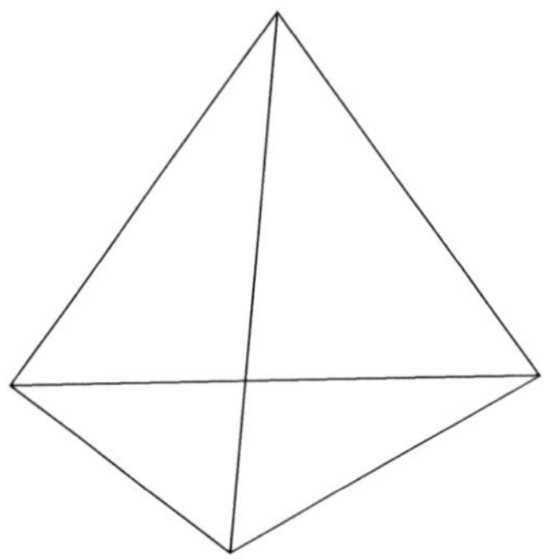

FIGURE 1.1 The four elements of materials science and engineering are interdependent and are equally important in determining the usefulness of a material.

temperature, stresses, and environmental effects the material experiences. The material could degrade under unfavorable or inappropriate conditions. Finally, economics also plays a major role in determining the selection of a material.

An engineer needs to conduct tests to obtain information about the structure and properties of materials. The purpose of this book is to make the undergraduate engineering students familiar with the different techniques available in the fields of mechanical engineering and materials science and engineering for determining the structure and properties of materials. However, we will focus here only on techniques for determining the crystal structures of cubic materials by x-ray diffraction (XRD) and on the conventional mechanical properties of materials. Further, we will concentrate on techniques that are commonly available in most undergraduate laboratories of mechanical engineering and materials science and engineering programs.

1.2 STRUCTURE

As mentioned above, the structure and properties of materials are very important. But the structure of a material can be determined at different levels. When we talk of structure we can consider the structure at macroscopic, microscopic, atomic/molecular, or subatomic levels. However, it is the structure at the atomic and microscopic levels that significantly affects the properties of engineering materials. Therefore, we will consider the crystal structure (i.e., atomic level structure) and microstructure (i.e., microscopic structure) of materials.

1.2.1 Crystal Structure

The different ways in which atoms are arranged in a crystalline material is referred to as crystal structure. Remember that materials can be classified as crystalline or amorphous (glassy and noncrystalline are the other terms used). A majority of the materials we use are in the crystalline state. The hallmark of a crystal is that the constituent atoms are arranged in a regular and periodic manner in three dimensions. On the other hand, the atoms are arranged in a random fashion in noncrystalline materials. Therefore, to characterize a crystalline solid, we need to define the size and shape of the unit cell (the smallest building block of the space lattice) and the distribution of atoms in the unit cell. While the size and shape of the unit cell determine the crystal system to which the material belongs, the positions of atoms in the unit cell determine the space lattice. Complete determination of the crystal structure of materials is a very involved process and can take from a few minutes (for simple structures with a few atoms) to a few years (for complex structures containing a very large number—hundreds and thousands—of atoms). It is very rare that an engineer is called upon to determine the complete crystal structures of materials, especially the complex ones. But an engineer should be familiar with the basic principles and how to use them for determining at least the size and shape of the unit cell. This is what we will discuss in Chapter 2. Chapter 2 will deal with the basic principles of crystallography including the concepts of space lattice, crystal structure, and indexing of planes and directions. We will then consider how x-rays are produced, how diffraction of x-rays occurs from crystals, and the conditions under which diffraction takes place from different types of structures. It will be shown that the size and shape of the unit cell determine the positions of the XRD peaks in a diffraction pattern, while the positions of atoms in the unit cell determine the intensities of the diffracted beams. Therefore, we will use the positions of the diffraction peaks in the pattern to determine the size and shape of the unit cell (i.e., the structure).

Since XRD is the most popular technique for structure determination, we will concentrate our discussion on this method to determine the crystal structure of materials. But, because of the simplicity, we will describe the methods for determining the structure of only materials with cubic structures. The student will be able to use the information and the worked out example presented in the chapter, and go ahead to determine the lattice parameters of other unknown materials if they have the cubic structure. Knowing the basic principles of structure determination, it should also be possible to extend the methods to determine the structures of noncubic materials. Other books on XRD (e.g., Suryanarayana and Norton, 1998; Cullity and Stock, 2001) could be consulted for further details of structure determination of cubic and noncubic materials and also additional information about how to obtain the fine structure of materials.

In addition to the technique of XRD, crystal structures of materials can also be determined using electron diffraction and neutron diffraction methods. Techniques of crystal structure determination using these techniques and the relative advantages and disadvantages of the different methods have been only briefly outlined.

1.2.2 Microstructure

The structure of the material at the microscopic level is referred to as microstructure, since this can be observed only with the help of a microscope. The microstructure can be described as the size, shape, and distribution of the phases present in the material in addition to its defect structure. Even though we had just mentioned that a microscope is used to describe and understand the microstructure, the scientist has a variety of microscopes for this purpose. These include the optical microscope (OM), scanning electron microscope (SEM), and transmission electron microscope (TEM), which are routinely used in most laboratories. In each of these microscopy fields, there are specialized techniques to enable one to obtain the required special information. Thus, one can have polarized light microscopy, phase-contrast microscopy, Nomarski contrast microscopy, and so on in optical microscopy; secondary electron imaging, backscattered electron imaging, orientation imaging microscopy, electron channeling patterns, and energy-dispersive spectroscopy techniques in scanning electron microscopy; and bright-field electron microscopy, dark-field electron microscopy, weak-beam microscopy, convergent beam electron microscopy, high-resolution electron microscopy, electron diffraction, energy-dispersive spectroscopy, and so on in transmission electron microscopy. An undergraduate student is not normally expected to use all these techniques. At the most, a student will be using some of these techniques if he or she is pursuing a research program. For these reasons, we have included a detailed account of optical microscopy and a brief account of scanning electron microscopy techniques in the book.

In addition to describing the principles of optical and scanning electron microscopy and how an image is formed in these cases, attempts have also been made to describe the interpretation of images and methods for obtaining quantitative information about the microstructure. Methods of preparing specimens suitable for microscopic examination in both the optical and scanning electron microscopy techniques have also been described.

The interested student should consult the several books that describe these techniques in great detail (see, e.g., Gifkins, 1970; Bradbury, 1989; Goldstein et al., 2003; Williams and Carter, 2009).

The microstructure of materials can be significantly changed by subjecting the materials to either cold working or different types of heat treatment. This assumes added significance in the case of steels, which are the most used structural materials. Therefore, a couple of chapters in this book have been devoted to explaining the basic principles of phase diagrams, the special features of the Fe–C phase diagram, and the types of heat treatments and the resultant microstructures that can be obtained. Interested students could also consult many good textbooks devoted to the physical metallurgy of steels. Three important books that provide the most useful information for students are Sinha, 1989; Krauss, 1990; Bhadeshia and Honeycombe, 2006.

1.3 PROPERTIES

In the selection and design of materials, an engineer is expected to be familiar with the properties of materials. In books on materials science and engineering, the mechanical, electrical, magnetic, thermal, optical, and deteriorative properties of materials are normally discussed at some length. In addition, physical properties such as density, viscosity, and others also become important. But among all these properties the mechanical properties assume the most importance for structural applications. Therefore, techniques for determining the mechanical properties of materials have been emphasized here.

Mechanical properties of materials can be determined in different ways. In principle, this is done by the application of load on a specimen and studying the specimen's response. The mechanical properties can be determined either under static conditions as in a conventional tensile testing or under dynamic conditions as in an impact test. The time-dependent deformation in materials could also be studied as in creep methods. Accordingly, the book describes hardness, tensile testing, impact, fatigue, and creep methods. Again, like in the previous cases of crystal structure and microstructure, the student is advised to consult standard textbooks that deal with the mechanical behavior of materials to get a detailed knowledge of the topics discussed (see, e.g., Dieter, 1996; Hertzberg, 1996; Courtney, 2000; Meyers and Chawla, 2008).

1.4 OUTLINE OF THE BOOK

In each of the chapters, the scientific background and the necessary theory for interpreting the results of the experiment and the basic principles of the experiment are first described. The way of conducting the test and the method of interpreting the results are then described. The method of conducting the test is described, keeping in mind of course that the details could be different depending on the type of equipment available in the laboratory.

A special feature of the book is that a large number of figures and examples have been provided in each chapter. For example, several XRD patterns have been provided in Chapter 2, and a variety of optical micrographs depicting different microstructural features have been presented in Chapter 3. A large number of micrographs showing the variation of microstructure with composition and heat treatment have been presented in Chapters 5 and 6. Another important feature of the book is that a number of example problems have been worked out in each chapter. Additionally, a number of questions have been presented at the end of each chapter, by answering which the students will be able to determine how much they have understood the subject matter.

Since the background information required for understanding the experiments described later in the chapter has been provided at the beginning, the book is self-contained and a student need not refer to any other books.

All the experiments described in this book can be completed in a one-semester course. Additional experiments have been suggested in each chapter if the students wish to conduct other experiments. The course instructor can also use these additional experiments to allot different experiments to different groups of students.

REFERENCES

Bhadeshia, H. K. D. H. and R. W. K. Honeycombe. 2006. *Steels: Microstructure and Properties*, third edition, London, UK: Butterworth-Heinemann.

Bradbury, S. 1989. *An Introduction to the Optical Microscope*, Oxford, UK: Oxford University Press.

Courtney, T. H. 2000. *Mechanical Behavior of Materials*, second edition, New York, NY: McGraw-Hill.

Cullity, B. D. and S. R. Stock. 2001. *Elements of X-Ray Diffraction*, third edition, Upper Saddle River, NJ: Prentice-Hall.

Dieter, G. E. 1996. *Mechanical Metallurgy*, third edition, New York, NY: McGraw-Hill.

Gifkins, R. C. 1970. *Optical Microscopy of Metals*, New York, NY: Elsevier Publishing.

Goldstein, J. I., D. E. Newbury, D. C. Joy, C. E. Lyman, P. Echlin, E. Lifshin, L. Sawyer, and J. R. Michael. 2003. *Scanning Electron Microscopy and X-ray Microanalysis*, third edition, New York, NY: Springer.

Hertzberg, R. W. 1996. *Deformation and Fracture of Engineering Materials*, fourth edition, New York, NY: John Wiley & Sons.

Krauss, G. 1990. *Steels: Heat Treatment and Processing Principles*, Materials Park, OH: ASM International.

Meyers, M. A. and K. K. Chawla. 2008. *Mechanical Behavior of Materials*, second edition, Cambridge, UK: Cambridge University Press.

Sinha, A. K. 1989. *Ferrous Physical Metallurgy*, Stoneham, MA: Butterworth.

Suryanarayana, C. and M. G. Norton. 1998. *X-Ray Diffraction: A Practical Approach*, New York, NY: Springer.

Williams, D. B. and C. B. Carter. 2009. *Transmission Electron Microscopy: A Textbook for Materials Science* (four volume set), New York, NY: Springer.

2 X-Ray Diffraction

2.1 INTRODUCTION

X-ray diffraction (XRD) is a powerful technique for determining many structural and microstructural parameters of materials. The most common application of the XRD technique is in determining the *crystal structure* of materials—metals, alloys, compounds, intermetallics, quasicrystals, semiconductors, and ceramic materials. It could also be used to determine the "structure" of liquids and amorphous (noncrystalline) materials. But, in this case, we will determine only the distances and the number of nearest neighbors. However, this analysis involves tedious and time-consuming techniques. The other applications of XRD to crystalline materials include determination of (1) solid-state phase boundaries in phase diagrams, (2) crystallite size and lattice strain in heavily cold-worked materials, (3) long-range order parameter in ordered alloys, (4) phase proportions in multiphase alloys and powder mixtures, (5) textures, and (6) magnitude of residual stresses, among others.

The crystal structure is an important attribute of a material since alloy formation, phase transformations, some of the physical properties (e.g., coefficient of thermal expansion), and mechanical behavior (plastic deformation, e.g., the number of slip systems and therefore the ductility) are determined by its crystal structure. As an example, metals with an FCC structure are more ductile than metals with a BCC or hexagonal close-packed (HCP) structure. Materials with an FCC crystal structure (e.g., copper or aluminum) are ductile since they have 12 slip systems (four slip planes and three slip directions in each plane) while materials with an HCP structure (e.g., magnesium or zinc) are typically brittle since they have only three slip systems (one slip plane and three slip directions in that plane).

Before we go into the details of structure determination using XRD techniques, it is important to have a clear understanding of the basics of crystallography and information needed to interpret the XRD patterns.

2.2 CRYSTAL STRUCTURE

The crystal structure of a material can be conveniently represented by a *unit cell*, which can be considered as the smallest building block of a *space lattice*. A space lattice represents regular arrangement of points in space, also called a point lattice. *The basic requirement of the space lattice is that each point has identical surroundings.* The unit cell in the space lattice is characterized by the six parameters of the unit cell, namely, a, b, and c representing unit repeat distances along the X, Y, and Z axes of the unit cell and α, β, and γ, the interaxial angles between Y and Z, Z and X, and X and Y axes, respectively. These six parameters defining the space lattice are referred

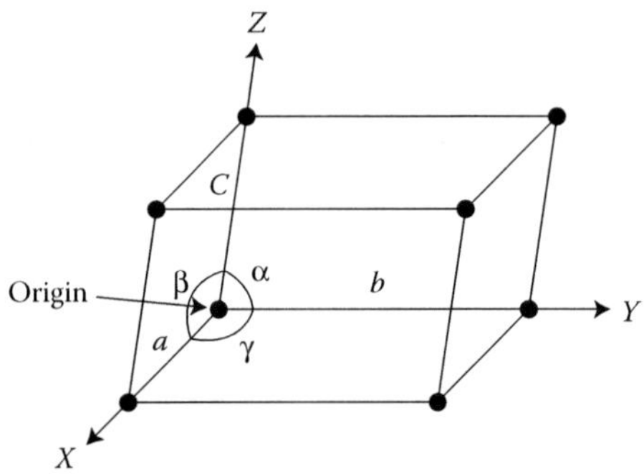

FIGURE 2.1 Schematic of a typical unit cell indicating the lattice parameters. a, b, and c are the unit repeat distances along the X, Y, and Z axes and α, β, and γ are the interaxial angles between Y and Z, Z and X, and X and Y axes, respectively.

to as *lattice parameters*. The schematic of a typical unit cell is shown in Figure 2.1 and the lattice parameters are also indicated. Depending on the interrelationship among these six lattice parameters, all crystalline materials can be classified as belonging to any one of the seven crystal systems listed in Table 2.1.

If the angle included between two axes is equal to 90°, then it is not specifically mentioned. That is, the interaxial angle is listed only if it is different from 90°. If nothing is mentioned, then the interaxial angle is assumed to be 90°. Further, if the

TABLE 2.1
The Axial and Angular Relationships in the Seven Crystal Systems

Crystal System	Axial Relationship	Angular Relationship	Space (Bravais) Lattice
Cubic	$a = b = c$	$\alpha = \beta = \gamma = 90°$	• Primitive • Body-centered • Face-centered
Rhombohedral (or trigonal)	$a = b = c$	$\alpha = \beta = \gamma \neq 90°$	• Primitive
Hexagonal	$a = b \neq c$	$\alpha = \beta = 90°; \gamma = 120°$	• Primitive
Tetragonal	$a = b \neq c$	$\alpha = \beta = \gamma = 90°$	• Primitive • Body-centered
Orthorhombic	$a \neq b \neq c$	$\alpha = \beta = \gamma = 90°$	• Primitive • Body-centered • Face-centered • Base-centered
Monoclinic	$a \neq b \neq c$	$\alpha = \gamma = 90°; \beta \neq 90°$	• Primitive • Base-centered
Triclinic	$a \neq b \neq c$	$\alpha \neq \beta \neq \gamma \neq 90°$	• Primitive

The crystal systems have been arranged in increasing order of asymmetry. The cubic system is the most symmetric and the triclinic system is the most asymmetric.

repeat distances along the three axes are the same, then only one of them is listed. If two of the repeat distances are the same and the third is different, then only two of them are listed. However, in a hexagonal system the included angle of $\gamma = 120°$, between the X and Y axes, is also not listed and is assumed as 120°. Thus, in a cubic system, since $a = b = c$, and $\alpha = \beta = \gamma = 90°$, listing of only the value of the lattice parameter a completely describes the lattice. But, in a hexagonal system, the two lattice parameters a and c are listed. In the orthorhombic system, the three parameters a, b, and c, are listed, and in the triclinic system, all the six parameters a, b, c, α, β, and γ are listed.

2.2.1 Bravais Lattices

The property of a space lattice, namely, that each point must have identical surroundings, can also be satisfied, within a crystal system, by placing additional points in the unit cell. This can be done by placing points at the center of the unit cell, at the face centers, or at the base centers (i.e., centers of any two parallel faces). Therefore, in any crystal system, one can, in principle, have four different arrangements of points in space such that the criterion of identical surroundings for each point is satisfied. These four possibilities are:

1. Points only at the corners of the unit cell, that is, *primitive lattice*
2. Points at all the corners of the unit cell and additional points at the body (volume) center, that is, *body-centered lattice*
3. Points at all the corners of the unit cell and additional points at all the face centers, that is, *face-centered lattice*
4. Points at all the corners of the unit cell and additional points at the base centers, that is, *base-centered lattice*

In the literature on crystallography, these four lattices are referred to as P (primitive), I (*innenzentrierte* or body-centered), F (face-centered), and C (centered on the C-faces only). In the last case, it is also possible to have atoms centered only on the A- or B-faces.

Figure 2.2 shows the unit cells of these four types of lattices. Since there can be four different arrangements in each of the seven crystal systems, one can, in principle, visualize a total of 28 possible space lattices in the seven crystal systems. But, some of these 28 possibilities are identical to others, and therefore, the maximum number of unique space lattices one can have is only 14 and these are also listed in Table 2.1. These 14 different space lattices are commonly referred to as *Bravais lattices*, in honor of Auguste Bravais.*

* Auguste Bravais (1811–1863), was a French crystallographer born on August 23, 1811 in Annonay, France. He completed his classical education at Collège Stanislas in Paris and obtained a PhD degree from Lyon in 1837. His interest in exploration prompted him to join the Navy. Bravais taught a course in applied mathematics for astronomy in the Faculty of Sciences in Lyon and later became Chair of Physics at École Polytechnique, Paris between 1845 and 1856. In 1854 he was admitted to the Académie des Sciences, Paris. Bravais was responsible for revival of interest in the study of external forms of crystals and their internal structures. After intensive study of lattice properties, he pointed out in 1845 that there are 14 unique lattices in three-dimensional crystalline systems, adjusting the previously existent result (15 lattices) by Frankenheim, obtained three years before. In his honor, these 14 space lattices are referred to as Bravais lattices. He died on March 30, 1863 at Le Chesnay in France.

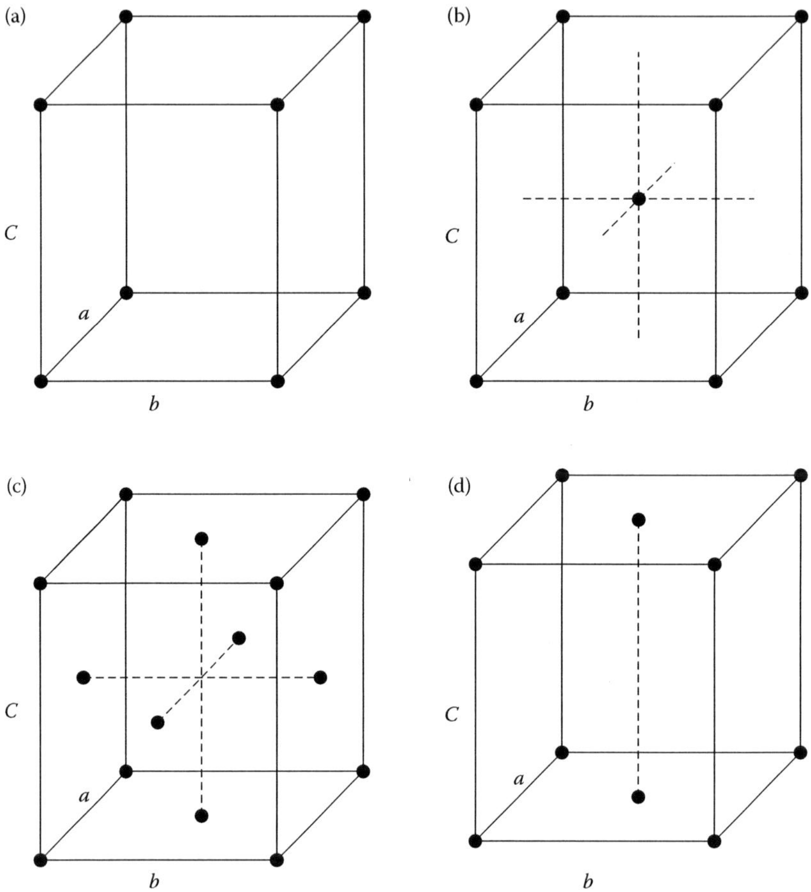

FIGURE 2.2 Schematic representation of the unit cells of (a) primitive, (b) body-centered, (c) face-centered, and (d) base-centered lattices. The unit cell chosen represents the orthorhombic system where $a \neq b \neq c$ and $\alpha = \beta = \gamma = 90°$.

In actual crystals, it is the atoms that are arranged in a regular and periodic manner and not the points. Therefore, the crystal structure can now be obtained by placing one (or more) atom(s) at each lattice point. This is called the basis. The relationship between Bravais lattices and crystal structures involves the basis. This relationship can be expressed as

$$\text{Bravais lattice} + \text{Basis} \rightarrow \text{Crystal structure} \tag{2.1}$$

Therefore, different crystal structures are obtained from the framework of the 14 Bravais lattices and contain a basis consisting of one or more atoms per lattice point.

Note, however, that when atoms are placed at the lattice points, these atoms belong to more than one unit cell. For example, the atom at the corner of the unit cell (whether it is cubic or noncubic) is shared by eight unit cells. Similarly, an atom placed at the face center is shared by two unit cells. But, an atom placed at the center of the unit cell belongs entirely to that unit cell and is not shared by any other unit cell. Accordingly, the *effective* number of atoms (just referred to as the number of atoms) in a unit cell can be calculated by considering $\frac{1}{8}$ of all the unit cell corner atoms, $\frac{1}{2}$ of all the face-centered atoms, and all the body-centered atoms. Thus,

$$\begin{aligned}\text{Number of atoms in a unit cell} &= \frac{1}{8}\times(\text{number of corner atoms})\\ &+\frac{1}{2}\times(\text{number of face-centered atoms})\\ &+1\times(\text{number of body-centered atoms})\end{aligned} \tag{2.2}$$

A BCC unit cell will have $(\frac{1}{8}\times 8)+(\frac{1}{2}\times 0)+(1\times 1)=2$ atoms per unit cell while an FCC unit cell will have $(\frac{1}{8}\times 8)+(\frac{1}{2}\times 6)+(1\times 0)=4$ atoms per unit cell. We can also consider that these unit cells have two and four lattice points, respectively.

The fact that a face-centered unit cell has four atoms (or points) per unit cell can also be understood as follows. Let us first determine the coordinates of all the atoms constituting the isolated unit cell. These are 0,0,0; 1,0,0; 0,1,0; 0,0,1; 1,0,1; 1,1,0; 0,1,1; 1,1,1; $\frac{1}{2}$,0,$\frac{1}{2}$; 0,$\frac{1}{2}$,$\frac{1}{2}$; $\frac{1}{2}$,$\frac{1}{2}$,0; $\frac{1}{2}$,$\frac{1}{2}$,1; $\frac{1}{2}$,1,$\frac{1}{2}$; and 1,$\frac{1}{2}$,$\frac{1}{2}$. Since the space lattice is obtained by repeating the atom coordinates along the *X*, *Y*, and *Z* directions by unit repeat distances, all coordinates involving 1 can be ignored. That is, 0,0,0; 1,0,0; 0,1,0; 0,0,1; 1,0,1; 1,1,0; 0,1,1; and 1,1,1 are equivalent and can be represented by 0,0,0. Similarly, $\frac{1}{2}$,0,$\frac{1}{2}$ and $\frac{1}{2}$,1,$\frac{1}{2}$ are equivalent and can be represented by $\frac{1}{2}$,0,$\frac{1}{2}$. A similar argument applies to the other co-ordinates as well. Therefore, we will end up with only four unique coordinates, namely, 0,0,0; $\frac{1}{2}$,$\frac{1}{2}$,0; $\frac{1}{2}$,0,$\frac{1}{2}$; and 0,$\frac{1}{2}$,$\frac{1}{2}$.

Let us consider an FCC lattice, which contains four lattice points in the unit cell. If we place one atom at each of the lattice points, we get the typical FCC structure adopted by metals like copper, aluminum, or silver. On the other hand, if we place two atoms of the same kind at each of the four lattice points, we get another structure adopted by semiconducting metals such as silicon and germanium. We can also place two atoms of a different kind at each of the four lattice points. We will then get a structure adopted by semiconducting compounds such as ZnS or GaAs. But, if we place three atoms at each lattice point, the structure is again different; it is the fluorite structure, adopted by materials like UO_2, an important nuclear fuel element. If we place 60 carbon atoms (a C_{60} molecule) at each lattice point, we get the Buckminsterfullerene or the "Buckyball" structure. Thus, it is possible to have many different crystal structures based on a single lattice. Therefore, we can say that while the number of crystal systems is limited to 7, and the number of space lattices to 14, the number of crystal structures can be infinite.

2.2.2 Lattice Planes

In XRD we will be discussing about diffraction occurring from crystal lattice planes. Therefore, it is essential that we have a way of uniquely describing the different crystal planes in a lattice and the *Miller indices* provide us such a way. The Miller indices are determined as follows:

1. Identify the points at which the plane intersects the *X*, *Y*, and *Z* axes. The intercept on each axis is measured in terms of fractions or multiples of the unit repeat distances (lattice parameters). If the plane moves through the origin, then the origin must be shifted.
2. Take reciprocals of the intercepts. This is done because if the plane never intersects one of the axes the intercept will be at ∞. We do not want to have ∞ in the indices so we take reciprocals.
3. Clear the fractions. We only want whole numbers in the indices, but do not reduce the numbers to the lowest integers.*
4. Enclose the resulting numbers in parentheses ( ). Negative numbers are written with a bar over the number ($\bar{1}$); and pronounced, for example, as bar one.

These steps give us the Miller indices of any plane in any crystal system—whether cubic or orthorhombic or triclinic. You should also note that the interaxial angles are not important in this case—the A face has the indices (100) and the C face has the indices (001) in all crystal systems. The actual values of the lattice parameters *a*, *b*, and *c* are also unimportant, because we take the intercepts in terms of fractions or multiples of the repeat distances along the *X*, *Y*, and *Z* axes. Thus, whether the repeat distance is 0.4 or 4.0 nm, if the plane intersects the axis at half the repeat distance (0.2 nm in the first case and 2.0 nm in the second case), the intercept is $\frac{1}{2}$.

Let us now work out a couple of examples of how to determine the Miller indices of planes. Consider the plane A shown shaded in Figure 2.3a. Following the procedure listed above, we obtain, for plane A:

	X	*Y*	*Z*
Intercepts	1/3	1/2	3/4
Reciprocals	3	2	4/3
Clear fractions	9	6	4

Enclose the integers in parentheses. So, plane A has the Miller indices (964).

Now let us consider the plane B, shown in Figure 2.3b. Following the above procedure, we obtain, for plane B:

* In some books, you will notice that these whole numbers are reduced to the smallest integers. But, this may not be the proper indexing of the plane. For example, if the intercepts made by a plane on the *X*, *Y*, and *Z* axes are at $\frac{1}{2}$, $\frac{1}{2}$, and $\frac{1}{2}$, respectively, the Miller indices of the plane are (222) according to the above procedure. If these whole numbers are reduced to the smallest integers, the Miller indices of the plane will be (111). On the other hand, if the intercepts made by another plane are 1, 1, and 1 on the *X*, *Y*, and *Z* axes, respectively, then also the Miller indices of this new plane are (111). A distinction between these two situations is possible only when the whole numbers are not reduced to the lowest integers. Even though the (111) and (222) planes are parallel to each other, their interplanar spacings are very different. The magnitude of interplanar spacing is very important since the angle at which diffraction occurs is dependent on the interplanar spacing (see Bragg equation, Section 2.5).

	X	*Y*	*Z*
Intercepts	1	−1	∞
Reciprocals	1	−1	0
Clear fractions	1	−1	0 (it was not necessary in this case)

Enclose the integers in parentheses. So, plane B has the Miller indices $(1\bar{1}0)$.

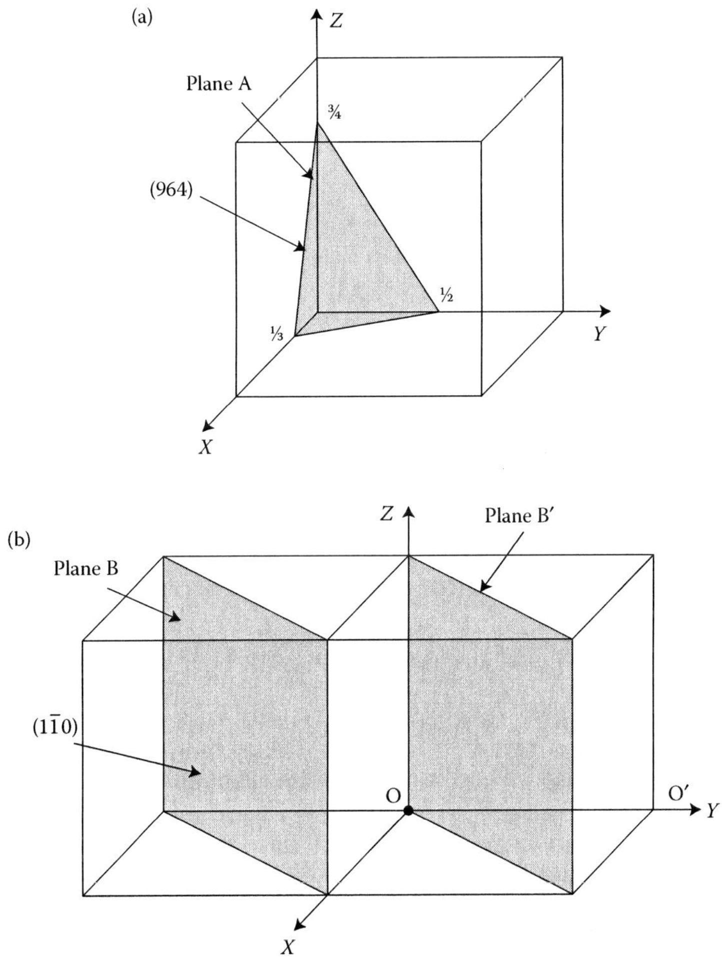

FIGURE 2.3 Indexing of planes. (a) The plane labeled A has the Miller indices (964). (b) The plane labeled as B has the Miller indices $(1\bar{1}0)$. Note that the indices of plane B′, derived by shifting the origin by one unit repeat distance along the *Y* direction are also $(1\bar{1}0)$. The planes B and B′ are parallel to each other (and have the same interplanar spacing) and so they are expected to have the same Miller indices.

We had mentioned earlier that the plane whose Miller indices have to be determined should not pass through the origin. If it does, the intercept is zero and the Miller index will be ∞. One of the ways to avoid having ∞ in the Miller indices is to shift the origin of the unit cell. (Remember that the choice of origin is arbitrary. Since each point in the space lattice has identical surroundings, the origin can be chosen to be at any point.) Let us assume that we wish to determine the Miller indices of the plane B′, which is parallel to plane B (in Figure 2.3b). Since this plane passes through the origin, we can shift the origin to point O′, that is, it is shifted one repeat unit along the positive direction of the Y axis. If we now wish to determine the Miller indices of this plane, the intercepts on the X, Y, and Z axes are 1, −1, and ∞, respectively. Consequently, the Miller indices, following the above procedure, are $(1\bar{1}0)$. The result is identical to what we had for the plane B! This is not surprising since planes B and B′ are parallel to each other.

In general, the Miller indices of any plane are specified by the three Miller indices $(hk\ell)$ enclosed in parentheses. A family of planes (all those planes that are crystallographically equivalent and have the same interplanar spacing) is designated by $\{hk\ell\}$ enclosed in braces. Figure 2.4 shows the (100), (110), (111), and (112) planes drawn in a cubic crystal. Note, however, that the way of assigning the Miller indices does not depend on the crystal system to which the crystal belongs. The procedure is identical whether we are considering cubic, tetragonal, monoclinic or any other structure.

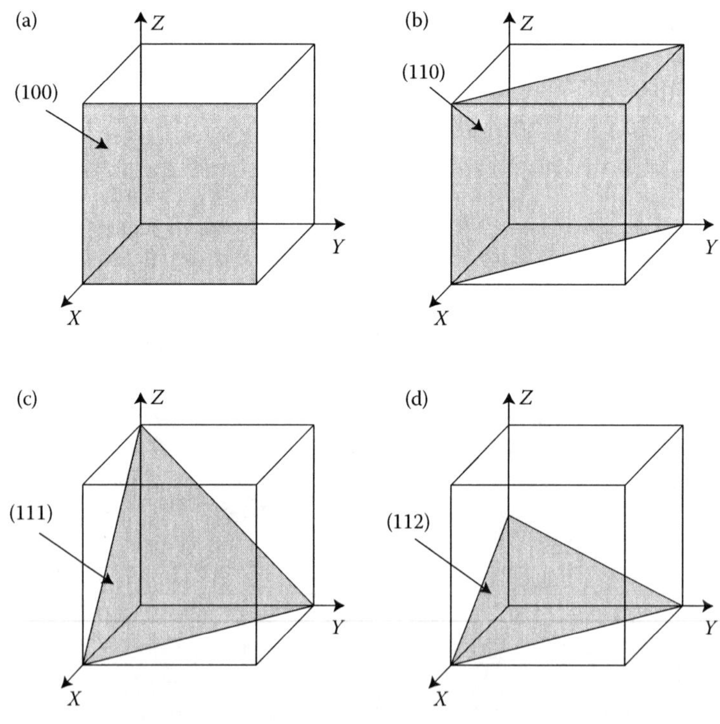

FIGURE 2.4 Representation of some common planes in a cubic lattice: (a) (100), (b) (110), (c) (111), and (d) (112).

2.2.2.1 Indexing of Planes in Hexagonal Crystals

A process similar to the above could be used to index planes in a hexagonal crystal also. But, by using this three-index system, planes which are crystallographically equivalent to each other will have different indices and this causes confusion. To avoid this confusion, indexing of planes in a hexagonal crystal is done using a four-index system to bring about the symmetry of the different planes (i.e., similar planes will have similar indices).

The method of indexing the crystal planes in the hexagonal system is similar to the method described above. The only difference, however, is that we now use four axes—X_1, X_2, and X_3 axes in the basal plane at an angle of 120° to each other and the Z axis normal to the basal plane. The intercepts made by the plane whose indices are to be determined are measured on these four axes and their reciprocals taken. The fractions are then cleared, but the whole numbers are not reduced to the lowest integers. The resulting numbers are then enclosed in parentheses. The indices of a plane in the four-index system are listed as $(hkil)$ and are known as the *Miller–Bravais indices.* Note that since the intercepts of a plane on the X_1 and X_2 axes determine the intercept on the X_3 axis, the value of i depends on the values of both h and k, according to the relationship:

$$h + k = -i \tag{2.3}$$

A family of planes in the hexagonal system is represented by $\{hkil\}$ enclosed in braces.

2.2.3 Lattice Directions

Sometimes, it also becomes important to identify the directions in a crystal. This is much easier than indexing the planes in a crystal. Directions in a crystal can be determined in a very straightforward way if the coordinates u,v,w of two points lying along that direction are known. If we have a line connecting two points A and B, then the direction of that line from A to B is $(u,v,w)_B - (u,v,w)_A$. For example, if a line points from the origin 0,0,0 (point A) to a point B located at (1,1,1), then the directional indices of that line are [1–0, 1–0, 1–0] ≡ [111]. In general, one can go through the following procedure to obtain the directional indices:

1. Choose a vector of convenient length such that it starts from the origin of the unit cell.
2. Determine the coordinates of any point on this vector in terms of the fraction or multiples of the three repeat distances along the three axes.
3. Clear the fractions. The indices are always integers. Reduce these numbers to the smallest integers. Remember that the direction does not depend on the distance you travel in that direction.
4. Enclose the resulting smallest integers in square brackets [ ]. Negative numbers are written with a bar over the number.

Using the above procedure, let us determine the indices of directions A and B in Figure 2.5. Let us consider the point M lying in the direction A. It has the coordinates

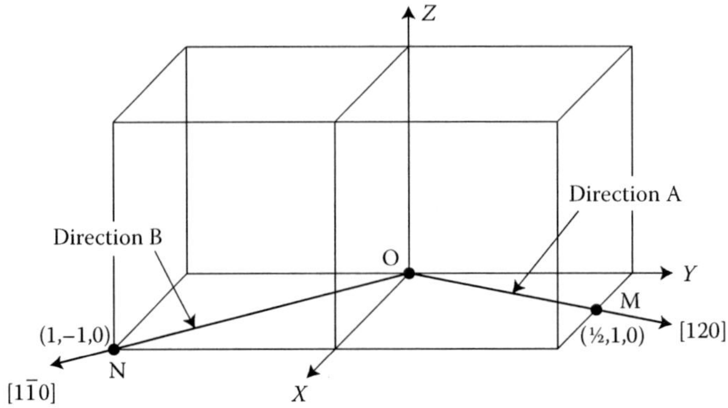

FIGURE 2.5 Determining the indices of the directions A and B. They are [120] for A and $[1\bar{1}0]$ for B.

($\frac{1}{2}$,1,0). Clearing the fractions by multiplying all the values with 2, we get the integers as 1,2,0. Thus, the direction A has the indices [120]. The same indices can also be obtained by using the procedure described above. The origin has the coordinates (0,0,0) and the point M has the coordinates ($\frac{1}{2}$,1,0). Thus, the directional indices of line A are ($\frac{1}{2}$ – 0, 1 – 0, 0 – 0) or ($\frac{1}{2}$,1,0). Clearing the fractions, we get 1,2,0. Thus, the directional indices of the line A are [120].

Similarly, for direction B, the point N along this direction has the coordinates (1,–1,0). Since all these are integers, the indices of direction B are $[1\bar{1}0]$. Note that we can visualize points along this direction with the coordinates (2,–2,0), (3,–3,0), and so on and therefore the directional indices could be written as $[2\bar{2}0]$, $[3\bar{3}0]$, and so on. But, since the direction is the same, the indices are reduced to the smallest integers. Figure 2.6 shows some common directions—[100], [110], [111], and [221]—marked in a unit cell.

A specific direction is designated by enclosing the indices (no commas between them) in square brackets [ ]. A generic direction is represented as [*uvw*]. A family of directions is indicated by enclosing the indices in angle brackets as <*uvw*>. The directions in a hexagonal crystal are represented by four indices, generally represented as [*uvtw*].

2.2.4 Conversion from Three-Index to Four-Index System

It is easy to convert indices of planes from the four-index Miller–Bravais system to the three-index Miller system and vice versa in a hexagonal crystal. To convert from the three-index into the four-index system, add the *h* and *k* indices, put a negative sign before that sum and use it as the third index in the four-index system. For example, the Miller indices (110) become $(11\bar{2}0)$ in the Miller–Bravais system. Similarly, to convert from the four-index into the three-index system, simply drop the third index in the Miller–Bravais indices. Thus, the indices $(10\bar{1}2)$ in the Miller–Bravais system become (102) in the Miller system.

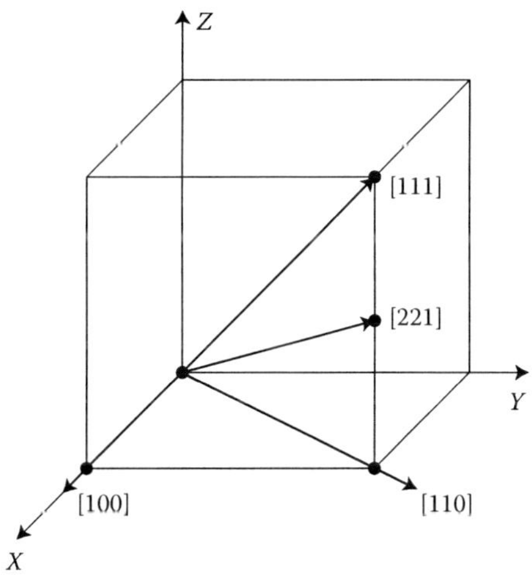

FIGURE 2.6 Representation of [100], [110], [111], and [221] directions in a cubic lattice.

Similar to the planes, the directions in the hexagonal system can also be converted from the three-index into the four-index system and vice versa. However, the conversion from the three-index system into the four-index system for directions in a hexagonal system is a little more involved and we can use the following relationships to convert the direction [*UVW*] in the three-index system into [*uvtw*] in the four-index system:

$$U = u - t, \quad V = v - t, \quad W = w \tag{2.4}$$

$$u = \frac{2U - V}{3}, \quad v = \frac{2V - U}{3}, \quad t = -(u + v) = -\frac{U + V}{3}, \text{ and } w = W \tag{2.5}$$

Thus, the direction [100] in the three-index system becomes $[2\bar{1}\bar{1}0]$ in the four-index system. Similarly, [110] becomes $[11\bar{2}0]$, and [111] becomes $[11\bar{2}3]$. While [100] and [110] appear to represent completely different directions in the three-index system, they are in fact equivalent directions in the hexagonal system. This is clearly brought out by adopting the four-index system when these two directions are translated into $[2\bar{1}\bar{1}0]$ and $[11\bar{2}0]$, respectively.

2.2.5 Structure Determination

Determination of crystal structures of metals and alloys is possible when x-rays, electrons, or neutrons are allowed to diffract from crystals. Since the diffracted beams possess characteristic diffraction angles and intensities determined by the crystal structure, one could use these beam positions and intensities in the diffraction patterns to determine the crystal structures of materials. XRD is the most common method used by scientists to determine the crystal structures of materials. This is because

1. One needs only a small sample volume
2. Specimen preparation is easy
3. Investment in capital equipment required for obtaining diffraction patterns is relatively small

But, before we proceed to do this, let us see how x-rays are produced, what their properties are, and how they are diffracted by crystals.

2.3 PRODUCTION OF X-RAYS

X-rays were discovered in 1895 by the German physicist Wilhelm Conrad Röntgen.* X-rays are high-energy electromagnetic radiation, with their energies ranging from about 200 eV to 1 MeV, which puts them between γ-rays and ultraviolet (UV) radiation in the electromagnetic spectrum. Each quantum of electromagnetic radiation, or photon, has an energy E, which is proportional to its frequency ν, that is, $E = h\nu$, where h is Planck's constant (4.136×10^{-15} eV s or 6.626×10^{-34} J s). Since the frequency of the radiation is related to its wavelength λ, through the speed of light, c ($\nu = c/\lambda$), the wavelength of the radiation can be written as

$$\lambda = \frac{hc}{E} \tag{2.6}$$

where c is the velocity of light (2.998×10^8 m s^{-1}). Thus, the x-ray wavelengths vary from about 10 nm to 1 pm. (The shorter wavelength corresponds to the higher energy of the radiation.) The useful range of wavelengths for XRD study is, however, between 0.05 and 0.25 nm.

* Wilhelm Conrad Röntgen (1845–1923) was born on March 27, 1845 in Lennep (which is today a borough of Remscheid), Prussia. He had his early education in the Netherlands and later on he joined the Federal Polytechnic Institute in Zurich (the present Eidgenössische Technische Hochschule, ETH) and obtained a PhD in mechanical engineering in 1869. He then moved to the University of Strassburg in 1873 and later to several other places. In 1990, he was offered the physics chair at the University of Munich, and he remained there for the rest of his career. While investigating the external effects of passing an electrical discharge through different vacuum tube equipment in 1895, Röntgen noticed that the invisible cathode rays caused a fluorescent effect on a small cardboard screen painted with barium platinocyanide. He discovered the new kind of rays, which he temporarily termed x-rays, using the mathematical designation for something unknown, on November 8, 1895. Nearly two weeks after his discovery, he took the very first x-ray picture of his wife's hand, Anna Bertha. Although the new rays came to bear his name and became known as Röntgen rays, he did not want the rays to be named after him, and therefore the term "x-rays" continues to be used. Röntgen was awarded the very first Nobel Prize in Physics in 1901 "in recognition of the extraordinary services he has rendered by the discovery of the remarkable rays subsequently named after him." He was also awarded the Rumford Medal and the Matteucci Medal. In November 2004, IUPAC named element 111 Roentgenium (Rg) in his honor. He died on February 10, 1923 in Munich, Germany.

Example Problem 2.1

Calculate the energy of the x-ray photon if it has a wavelength of 0.15 nm.

Solution 2.1

According to Equation 2.6,

$$E = \frac{hc}{\lambda}$$

Substituting the values of h, c, and λ in the above equation, we get

$$E = \frac{(4.136 \times 10^{-15}\text{eV s}) \times (2.998 \times 10^{8}\,\text{m s}^{-1})}{0.15 \times 10^{-9}\,\text{m}}$$

$$= 8.2 \times 10^{3}\ \text{eV} \quad \text{or} \quad 8.2\ \text{keV}$$

Example Problem 2.2

Calculate the wavelength of the x-ray photon if its energy is 6 keV.

Solution 2.2

According to Equation 2.6,

$$\lambda = \frac{hc}{E}$$

Substituting the values of h, c, and E in the above equation, we get

$$\lambda = \frac{4.136 \times 10^{-15}\text{eV s} \times 2.998 \times 10^{8}\ \text{ms}^{-1}}{6 \times 10^{3}\,\text{eV}}$$

$$= 2.067 \times 10^{-10}\ \text{m} \quad \text{or} \quad 0.2067\ \text{nm}$$

It is clear from the above two example problems that the energy of the x-ray beam and its wavelength are inversely proportional to each other. That is, when the wavelength is shorter the energy is higher.

X-rays are produced when fast-moving electrons are suddenly stopped by a metal target. This is achieved in an x-ray tube consisting of two metal electrodes enclosed in a vacuum chamber. Electrons are produced by heating a tungsten filament cathode, maintained at a high negative potential (usually about 30–50 kV). The electrons produced are accelerated toward the anode, which is normally at ground potential. The high-energy electrons lose their energy by collision with the water-cooled anode and this loss of energy is manifested as x-rays. Actually, less than 1% of the energy of the electrons is converted into x-rays, and the rest is dissipated as heat.

2.3.1 Continuous Radiation

A typical x-ray spectrum (intensity vs. wavelength) usually consists of characteristic radiation (with a specific wavelength) superimposed over the continuous (or white) radiation, which is made up of many different wavelengths (Figure 2.7). The *continuous radiation* is obtained due to the loss of energy of the electrons in a series of collisions with the atoms making up the target. X-rays with the shortest wavelength are produced when all the energy is lost in one impact. Longer wavelengths are obtained when the energy is lost in more than one collision. The continuous radiation is not frequently used in diffraction studies, unless one requires a number of different wavelengths in a single experiment, as for example, in the Laue method.

2.3.2 Characteristic Radiation

The origin of the *characteristic radiation* is, however, different. If the incident electron has sufficient energy to eject an inner-shell electron, the atom will be left in an excited state with a hole in the electron shell (Figure 2.8). When this hole is filled by an electron from an outer shell, an x-ray photon is produced, which has an energy equal to the difference in the electron energies of the two shells. Thus, if the incident electron has sufficient energy to knock out an electron from the K shell of the target

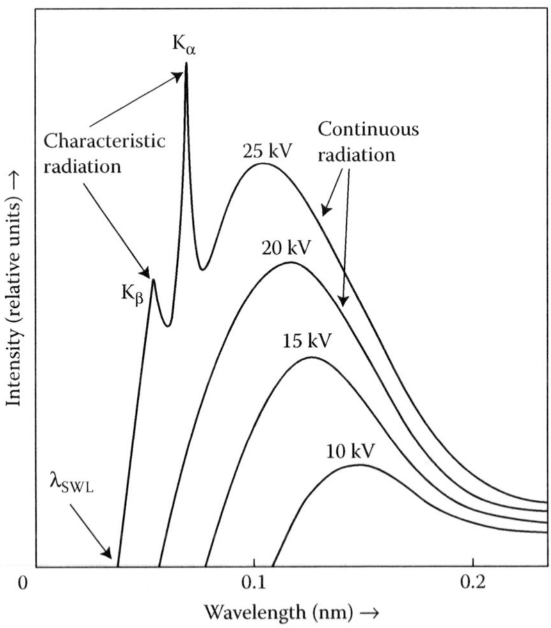

FIGURE 2.7 X-ray spectrum of molybdenum at different voltages. The voltages refer to those applied between the anode and the cathode. The line widths of the characteristic radiation are not to scale.

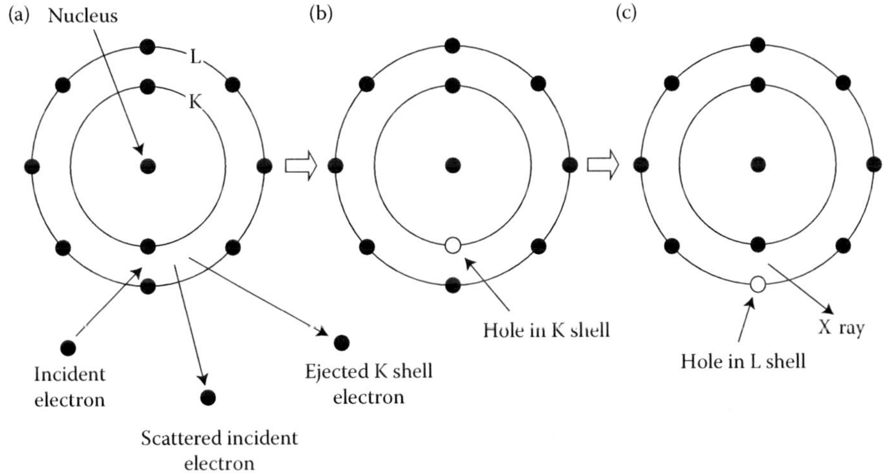

FIGURE 2.8 Illustration of the inner shell ionization and the subsequent emission of a characteristic x-ray. (a) An incident electron ejects a K shell electron from an atom. (b) A hole is left in the K shell. (c) Electron rearrangement occurs resulting in the emission of an x-ray photon.

atom, and if the hole so created in the K shell is filled by an electron from the L shell, the x-ray photon produced is referred to as *Kα radiation*. These Kα x-rays will have the energy which corresponds to the difference in energies between the K and L shells of the atom. That is,

$$E_{K\alpha} = E_L - E_K \tag{2.7}$$

But, remember that L shell has three subshells—L_I, L_{II}, and L_{III}. Therefore, if the hole in the K shell is filled by an electron from the L_{III} subshell, it will be referred to as Kα_1 and if the K-shell hole is filled by an electron from the L_{II} subshell, then it is referred to as Kα_2. Thus, the energies of the $K_{\alpha 1}$ and $K_{\alpha 2}$ x-rays are given by

$$E_{K\alpha 1} = E_{LIII} - E_K \tag{2.8}$$

and

$$E_{K\alpha 2} = E_{LII} - E_K \tag{2.9}$$

(The transition from the L_I subshell to K shell is not allowed on the basis of rules governing electron transitions.) Since the L_{III} subshell has a slightly higher energy than the L_{II} subshell, the energy of the Kα_1 x-rays is higher than the Kα_2 x-rays. Consequently, the wavelength of the Kα_1 x-rays is slightly shorter than the Kα_2 x-rays.

It is also possible that the hole in the K shell is filled by an electron from the M shell, N shell, and so on. Thus, when the hole in the K shell is filled by an electron

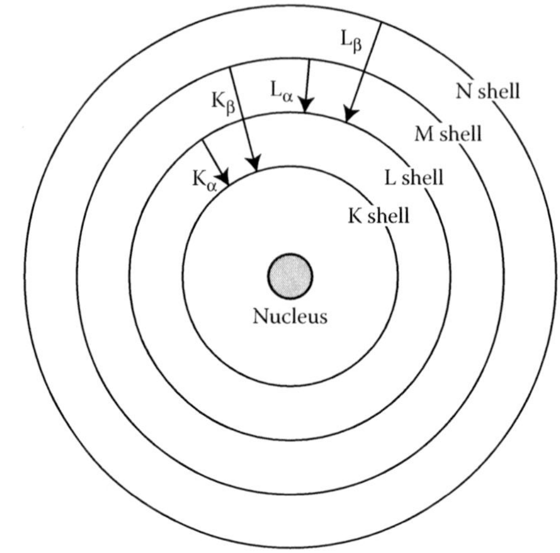

FIGURE 2.9 Electron transitions in an atom which produce the K_α, K_β, L_α, ... characteristic x-rays.

from the M shell, then the x-rays produced are referred to as *Kβ radiation*. But, the probability of filling the K-shell hole by an electron from the M shell is much smaller than the probability of it being filled by an electron from the L shell. Therefore, the intensity of the Kα radiation is higher than that of the Kβ radiation. As a corollary, the probability of a hole in the K shell being filled from an electron from the N shell (and obtaining the Kγ radiation) is almost insignificant.

Similarly, Lα x-rays are produced when a hole produced in the L shell is filled by an electron from the M shell, and Lβ x-rays are produced when the hole in the L shell is filled by an electron from the N shell. These are schematically shown in Figure 2.9.

Example Problem 2.3

The energies of the different shells in a molybdenum atom are given below. Calculate the wavelengths of the $K\alpha_1$ and $K\alpha_2$ radiations.

$$K: -20.00 \text{ keV},\ L_{II}: -2.63 \text{ keV},\ L_{III}: -2.52 \text{ keV}$$

Solution 2.3

Since the energies of the x-rays are equal to the difference in the energies of the two (sub)shells involved, the energy of the $K\alpha_1$ x-rays, according to Equation 2.8 is

$$E_{K\alpha 1} = E_{LIII} - E_K = -2.52 \text{ keV} - (-20.00 \text{ keV}) = 17.48 \text{ keV}$$

According to Equation 2.6, $\lambda = hc/E$. Substituting the appropriate values into this equation, we get

$$\lambda_{K\alpha_1} = \frac{\left(4.136 \times 10^{-15} \text{ eV s}\right)\left(2.998 \times 10^{8} \text{ m s}^{-1}\right)}{17.48 \times 10^{3} \text{ eV}}$$
$$= 0.70937 \times 10^{-10} \text{ m}$$
$$= 0.070937 \text{ nm}$$

Similarly, the wavelength of the Kα_2 radiation can be calculated, using Equation 2.9 as

$$E_{K\alpha 2} = E_{LII} - E_K = -2.63 \text{ keV} - \left(-20.00 \text{ keV}\right) = 17.37 \text{ keV}$$

and therefore,

$$\lambda_{K\alpha 2} = \frac{\left(4.136 \times 10^{-15} \text{ eV s}\right)\left(2.998 \times 10^{8} \text{ m s}^{-1}\right)}{17.37 \times 10^{3} \text{ eV}}$$
$$= 0.71386 \times 10^{-10} \text{ m}$$
$$= 0.071386 \text{ nm}$$

In reality, because of the small difference in the wavelengths, it may not always be possible (especially at small diffraction angles) to see the Kα_1 and Kα_2 components of the K_α radiation resolved, that is, as two separate peaks. In such a situation, it is common practice to use the wavelength of the Kα radiation as the weighted average of the two components, using the relationship:

$$\lambda_{K\alpha} = \frac{1}{3}\left(2\,\lambda_{K\alpha_1} + \lambda_{K\alpha_2}\right) \tag{2.10}$$

Since the Kα_1 line is twice as strong as the Kα_2 line, it is given twice the weightage. Thus, the *weighted average wavelength* of the Mo Kα radiation is (1/3) × (2 × 0.0 70937 nm + 0.071386 nm) = 0.071087 nm.

Table 2.2 lists the wavelengths of some of the common target metals used in XRD studies.

You would notice that the wavelengths of the x-ray beams are given to very high accuracy. Even though the accuracy in the above calculations depends on the accuracy with which the energy values are known, methods are available to measure the wavelengths of the x-ray beams very accurately from spectroscopic studies.

2.4 ABSORPTION OF X-RAYS

Like any other electromagnetic radiation, when x-rays encounter any form of matter, they are partially absorbed and partially transmitted. The intensity of the transmitted x-ray beam depends on x, the thickness of the material through which the x-ray

TABLE 2.2
Wavelengths (in Units of nm) of Some of the Commonly Used K X-rays

Target Element	$K\alpha_1$ (Very Strong)	$K\alpha_2$ (Strong)	$K\alpha$ (Weighted Average)	$K\beta$ (Weak)
Chromium (Cr)	0.228970	0.229361	0.229100	0.208487
Iron (Fe)	0.193604	0.193998	0.193736	0.175661
Cobalt (Co)	0.178897	0.179285	0.179026	0.162079
Copper (Cu)	0.154056	0.154439	0.154184	0.139222
Molybdenum (Mo)	0.070937	0.071386	0.071087	0.063229

beam is passing and μ, the *linear absorption coefficient* of the material for x-rays, according to the equation

$$I_x = I_o e^{-\mu x} \tag{2.11}$$

where I_x is the intensity of the transmitted x-ray beam after passing through a material of thickness, x, and I_o is the intensity of the incident x-ray beam. That is, $I_o - I_x$ is the amount of the x-ray intensity that is absorbed by the specimen. The linear absorption coefficient, μ has units of cm^{-1}, and is a function of the material, its density, and also the wavelength of the x-ray beam. But, μ is also proportional to ρ, the density of the material and therefore, μ/ρ is a constant of the material and is independent of the state of the material (solid, liquid, or gas). μ/ρ has the units of $cm^2\ g^{-1}$ and is called the *mass absorption coefficient.* This is the quantity which is used in practice, and standard books on XRD list the values of the mass absorption coefficient. Equation 2.11 may be rewritten in terms of the mass absorption coefficient as

$$I_x = I_o e^{-(\mu/\rho)\rho x} \tag{2.12}$$

Example Problem 2.4

Lead screens are used to protect personnel from exposure to x-rays. If a lead sheet of 1 mm thickness is used, calculate the "transmission factor," that is ($I_{transmitted}/I_{incident}$) of the lead sheet when Cu Kα and Mo Kα x-rays are used.

The mass absorption coefficients of lead for Cu Kα and Mo Kα radiations are 232.1 and 122.8 $cm^2\ g^{-1}$, respectively, and the density of lead is 11.34 $g\ cm^{-3}$.

Solution 2.4

We can use Equation 2.12 to calculate the intensity of the transmitted beam.

For Cu Kα x-rays,

$$\begin{aligned} I_{transmitted} &= I_o e^{-232.1 cm^2 g^{-1} \times 11.34\, g\, cm^{-3} \times 0.1 cm} \\ &= I_o e^{-263.2} \end{aligned}$$

Therefore,

$$\frac{I_{\text{transmitted}}}{I_{\text{incident}}} = e^{-263.2}$$
$$= 4.9 \times 10^{-115}$$

Similarly, for Mo Kα x-rays,

$$I_{\text{transmitted}} = I_o e^{-122.8 \text{ cm}^2 \text{ g}^{-1} \times 11.34 \text{ g cm}^{-3} \times 0.1 \text{cm}}$$
$$= I_o e^{-139.3}$$

Therefore,

$$\frac{I_{\text{transmitted}}}{I_{\text{incident}}} = e^{-139.3}$$
$$= 3.2 \times 10^{-61}$$

From the above example, you may notice, that the transmission factor, that is, the fraction of x-rays that have passed through the lead screen, is higher for the higher-energy Mo Kα x-rays than for the lower-energy Cu Kα x-rays. That is, the lower-energy x-rays get more absorbed by the lead sheets. In both the cases, however, the intensity of the x-ray beam is negligibly low.

In XRD studies, we use not only pure metals but also solid solutions, compounds, and mechanical mixtures of different components. The mass absorption coefficient of such materials (whether a solid solution, compound, or a mixture and irrespective of whether it is in the solid, liquid, or gaseous state), can be calculated using the equation:

$$\frac{\mu}{\rho} = w_1\left(\frac{\mu}{\rho}\right)_1 + w_2\left(\frac{\mu}{\rho}\right)_2 + w_3\left(\frac{\mu}{\rho}\right)_3 + \cdots \tag{2.13}$$

where w_1, w_2, ... are the weight fractions and $(\mu/\rho)_1$, $(\mu/\rho)_2$, ... represent the mass absorption coefficients of the individual components 1, 2, ..., respectively.

Example Problem 2.5

A copper-30 wt% nickel solid solution alloy sheet of 1 mm thickness is placed in the path of Cu Kα x-rays. Calculate the mass absorption coefficient of the alloy and the transmission factor for the x-ray beam. The mass absorption coefficients of copper and nickel for Cu Kα x-rays are 51.54 and 48.83 cm^2 g^{-1}, respectively. Assume the density of the Cu-30 wt% Ni alloy to be 8.924 g cm^{-3}.

Solution 2.5

The mass absorption coefficient of the solid solution alloy can be calculated using Equation 2.13 as

$$\begin{aligned}\frac{\mu}{\rho} &= w_{Cu}\left(\frac{\mu}{\rho}\right)_{Cu} + w_{Ni}\left(\frac{\mu}{\rho}\right)_{Ni} \\ &= 0.7 \times 51.54\ \text{cm}^2\ \text{g}^{-1} + 0.3 \times 48.83\ \text{cm}^2\ \text{g}^{-1} \\ &= 50.73\ \text{cm}^2\ \text{g}^{-1}\end{aligned}$$

The transmission factor can now be calculated in a manner similar to Example Problem 2.4 as

$$\begin{aligned}I_{transmitted} &= I_o e^{-50.73\ \text{cm}^2\ \text{g}^{-1} \times 8.924\ \text{g cm}^{-3} \times 0.1\text{cm}} \\ &= I_o e^{-45.27}\end{aligned}$$

Therefore,

$$\begin{aligned}\frac{I_{transmitted}}{I_{incident}} &= e^{-45.27} \\ &= 2.2 \times 10^{-20}\end{aligned}$$

It is important to note that x-ray beams travel in all directions and that they are not visible. Further, they can penetrate matter easily as shown above (they penetrate much deeper if the x-rays have a high energy). X-rays are damaging because they penetrate the human body and break up molecules of DNA in our cells. If these molecules do not repair themselves, they die. On the other hand, if the molecules are broken and they rejoin in a wrong way, then a mutant cell is produced and this can lead to cancer. Therefore, it is imperative that the user does not get exposed to x-rays. Although we cannot completely avoid exposure to radiation, we wish to keep the exposure *as low as reasonably achievable* (*alara*). Most x-ray diffractometers have a safety screen (or door) that encloses the instrument. The screen is opened to load or unload the specimen. The x-ray source cannot be turned on while the screen is open. The equipment is also provided with interlock mechanisms, and the x-ray machines automatically shut off if the safety screen or window is open.

2.5 BRAGG EQUATION

Having considered how x-rays are produced and what their absorption characteristics are, let us now discuss how they get diffracted by atoms in the crystals. When an x-ray beam is incident on an atom, the electrons in the atom oscillate about their mean positions. It was mentioned earlier that deceleration of electrons (sudden stopping of fast-moving electrons) produces an electromagnetic radiation. This process of absorption and reemission of electromagnetic radiation is known as *scattering*. If scattering takes place from many atoms, the scattered waves from the different atoms can interfere. That is, the amplitude and wavelength of the resultant beam could be different from the sum (or difference) of the component wavelengths.

Let us consider three different cases. First let us consider waves 1 and 2 of the same wavelength, but of slightly different amplitude, as shown in Figure 2.10a. Since these two waves are in phase, then the amplitudes get added up and therefore the resultant wave (3) will have larger amplitude. That is, the waves interfere in a constructive manner. On the other hand, if these two waves 1 and 2 are of the same wavelength and same amplitude, but are completely out of phase as shown in Figure 2.10b, then completely destructive interference takes place and the resultant wave will have zero amplitude. But, it is also possible that the two waves have the same wavelength and same amplitude, but are slightly out of phase (Figure 2.10c), then only partial constructive interference takes place and therefore the amplitude of the resultant wave will be less than what it is in Figure 2.10a.

If all the scattered waves reinforce one another, then the beam is considered as diffracted. Therefore, a diffracted beam may be defined as a beam consisting of a large number of scattered beams reinforcing one another.

The *diffraction directions* are governed by the wavelength (λ) of the incident radiation and the nature of the crystalline sample. Von Laue* had derived the conditions for diffraction to occur from crystals. He worked out the mathematical formulation and established the fact that x-ray waves are electromagnetic in nature and this opened the way to the later work of Bragg, who expressed the necessary conditions for diffraction to occur in a much simpler way. This equation, formulated by William Henry Bragg†

* Max Theodore Felix von Laue (1879–1960) was born on October 9, 1879 at Pfaffendorf, near Koblenz, Germany. He obtained his PhD degree in physics from the University of Berlin in 1903 working with Max Planck. In 1919, he was appointed professor of physics at the University of Berlin, a position he occupied till 1943. Von Laue had worked in different areas of physics—superconductivity, relativity, optics, quantum theory, crystallography, and x-ray diffraction. His best-known work, however, for which he received the Nobel Prize in Physics in 1914, was "for his discovery of the diffraction of X-rays by crystals." As related in his Nobel lecture, he argued that passage of short electromagnetic radiation, which x-rays were supposed to be, would cause some kind of diffraction or interference phenomena in crystals, and this was experimentally proven to be correct by Walter Friedrich and Paul Knipping. Von Laue worked out the mathematical formulation of it and the discovery was published in 1912. He won several awards for his research work, including the Ladenburg Medal, the Max-Planck Medal, and the Bimala-Churn-Law Medal. He was instrumental in reestablishing and organizing German science after World War II. He died on April 24, 1960 in Berlin, Germany as a result of injuries due to an auto accident.

† William Henry Bragg (1862–1942) was born on July 2, 1862 in Westward, Cumberland, England and graduated in mathematics from Cambridge University in 1885. He was appointed Elder Professor of Mathematics and Experimental Physics at the University of Adelaide in Australia. Subsequently he became the Cavendish Professor of Physics at Leeds (1909–1915), Quain Professor of Physics at University College London (1915–1925), and Fullerian Professor of Chemistry in the Royal Institution. At the University of Leeds, he invented the x-ray spectrometer, and with his son, Lawrence Bragg, founded the new science of x-ray analysis of crystal structure. If the fundamental discovery of the wave aspect of x-rays, as evidenced by their diffraction in crystals, was due to von Laue and his collaborators, it is equally true that the use of x-rays as an instrument for the systematic revelation of the way in which crystals are built was entirely due to the Braggs. W. H. Bragg received several awards for his work including the Matteucci Medal, Rumford Medal, Copley Medal, and Franklin Medal in addition to the Nobel Prize (which he shared with his son William Lawrence Bragg). He was President of the Royal Society of London from 1935 to 1940. He died on March 10, 1942 in London, England.

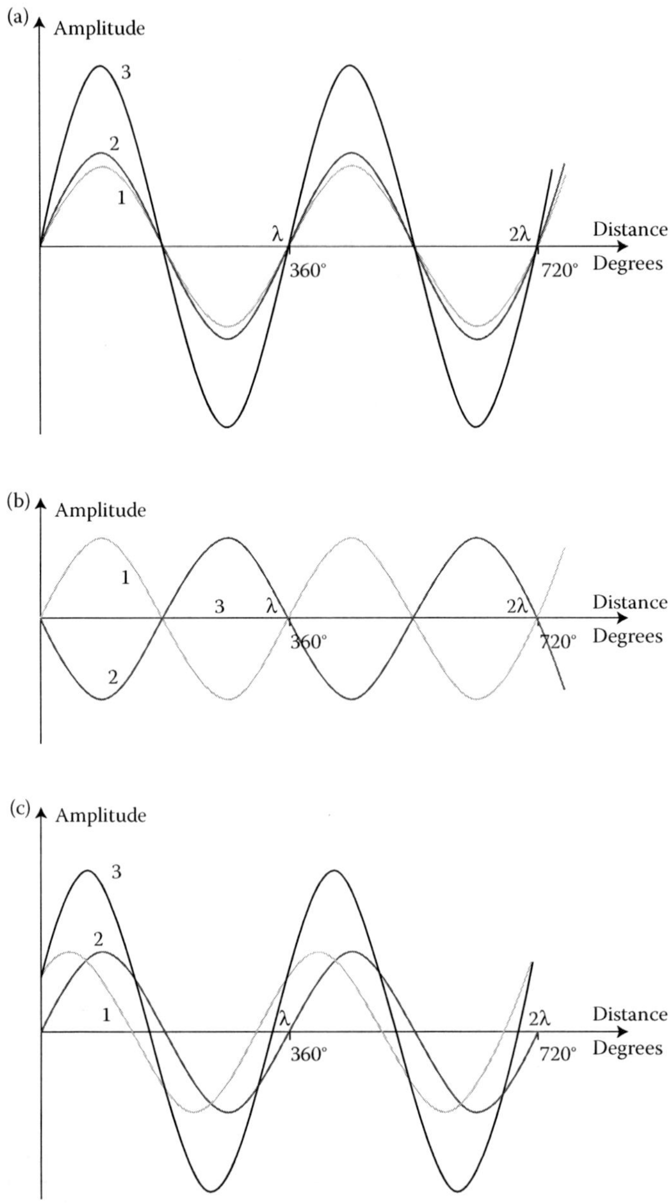

FIGURE 2.10 Constructive and destructive interference between two waves having the same wavelength. (a) Waves of the same wavelength, but of slightly different amplitude are completely in phase. Therefore, completely constructive interference takes place. The amplitudes of the two component waves add up. (b) Waves of exactly the same wavelength and same amplitude. When the two waves are completely out of phase, completely destructive interference occurs. The net amplitude of the resultant wave is zero. (c) When the two component waves are partly out of phase, partially destructive interference occurs. The amplitude of the resultant wave is smaller than what it is in (a).

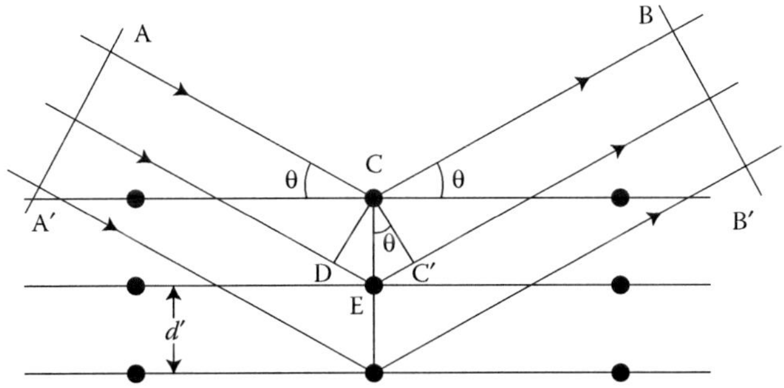

FIGURE 2.11 Schematic figure used to derive the Bragg equation. θ is the angle of incidence and d' is the interplanar spacing of the planes on which the x-ray beam is incident.

and William Lawrence Bragg*, the father and son duo, in 1913, relates the wavelength of the x-rays to the spacing of the atomic planes, d, in a crystal. It is this simpler expression, known as *Bragg equation*, which most researchers use. Therefore, let us now derive this equation from the basic principles of scattering.

The *Bragg equation* can be derived with reference to Figure 2.11, which represents a set of lattice planes ($hk\ell$) with an interplanar spacing of d'. Let us assume that the x-ray beam is incident on these planes at an angle θ, measured as the angle between the x-ray beam direction and the lattice planes. (Note that in conventional reflection phenomena, the angle of incidence is measured as the angle between the direction of the incident beam and the normal to the plane on which the radiation is incident.) For diffraction to occur, all the scattered beams should be in phase across a wavefront such as BB′. For this to be so, the path lengths between wavefronts AA′ and BB′ must differ by exactly an integral number of wavelengths, $n\lambda$. Therefore, for diffraction to occur, the path difference, δ, must be

$$\delta = n\lambda \tag{2.14}$$

* William Lawrence Bragg (1890–1971) was born on March 31, 1890 in Adelaide, Australia and was the son of William Henry Bragg. He moved to Cambridge, England in 1909 and started working on the von Laue phenomenon in 1912 and published his first paper on the subject in the *Proceedings of the Cambridge Philosophical Society*. From 1912 to 1914 he had been working with his father and it was this work that earned them jointly the Nobel Prize for physics in 1915 "for their services in the analysis of crystal structure by means of x-rays." At the age of 25, he was the youngest person ever to receive a Nobel Prize. The father–son pair winning the Nobel Prize the same time is still a unique accomplishment, because no other parent–child combination has yet shared a Nobel Prize in any field the same year. The very rare opportunity of celebrating a golden jubilee as a Nobel laureate was given special attention during the December ceremonies at Stockholm in 1965, when Sir Lawrence, at the invitation of the Nobel Foundation, delivered a lecture—the first and only Nobel Guest Lecture—on developments in his field of interest during the past 50 years. Lawrence Bragg was awarded the Matteucci Medal, Hughes Medal, Royal Medal, and Copley Medal. He was also director, National Physical Laboratory, Teddington; Cavendish Professor of Experimental Physics, Cambridge; and director of the Royal Institution. He was knighted in 1941 and made a Companion of Honour in 1967. Lawrence Bragg died on July 1, 1971 in Waldringfield, Ipswich, Suffolk, England.

where n is an integer. Now, since lines CD and CC′ in Figure 2.11 are also wavefronts, we can write the path difference between the two waves as

$$\delta = \mathrm{DE} + \mathrm{EC}' \tag{2.15}$$

From elementary trigonometry,

$$\mathrm{EC}' = \mathrm{CE} \sin\theta$$

Since CE = d', the interplanar spacing of the planes under consideration,

$$\mathrm{EC}' = d' \sin\theta.$$

Similarly, DE = $d' \sin\theta$.
Therefore, the path difference, δ is given by

$$\begin{aligned}\delta &= \mathrm{DE} + \mathrm{EC}' \\ &= d' \sin\theta + d' \sin\theta \\ &= 2\, d' \sin\theta \end{aligned} \tag{2.16}$$

By combining Equations 2.14 and 2.16, we get

$$n\lambda = 2\, d' \sin\theta \tag{2.17}$$

This equation is known as the Bragg equation (sometimes referred to as Bragg law), and is extremely important in indexing diffraction patterns and, therefore, for determining the crystal structures of materials. The Bragg equation is also equally applicable for *electron diffraction* and *neutron diffraction* studies.

The Bragg equation is, however, more customarily written as

$$\lambda = 2\, d \sin\theta \tag{2.18}$$

Equations 2.17 and 2.18 are not basically different, but they only differ in the way they are written. Note that n is the order of reflection. That is, a reflection is called first-order reflection when $n = 1$, second order when $n = 2$, and so on. Thus, for a first-order 100 reflection, according to Equation 2.17, $\lambda = 2d_{100} \sin\theta$, which is the same as Equation 2.18. Similarly, for a second-order 100 reflection, $2\lambda = 2d_{100} \sin\theta$, and for a first-order 200 reflection, $\lambda = 2d_{200} \sin\theta$. Since $d_{200} = \left(\frac{1}{2}\right)d_{100}$, and in general, $d_{nh\,nk\,n\ell} = (1/n) \times d_{hk\ell}$ the second-order 100 reflection could be considered as a first-order 200 reflection. Extending this argument further, an nth-order reflection from $hk\ell$ can be considered as a first-order reflection from planes spaced at a distance $1/n$ of the spacing of $hk\ell$. As an example, the 400 reflection can be considered a fourth-order reflection from the (100) planes of spacing d', or a second-order reflection from the (200) planes of spacing $d'/2$, or even as a first-order reflection from the (400) planes with the spacing $d = d'/4$. Thus, Equation 2.17 can be rewritten as

$$\lambda = 2\frac{d'}{n}\sin\theta \tag{2.19}$$

By setting $d = d'/n$, Equation 2.19 can be rewritten as $\lambda = 2d \sin\theta$, which is the same as Equation 2.18. And we will be using this form of the Bragg equation throughout this chapter. Recall that λ is the wavelength of the x-ray beam, d is the interplanar spacing, and θ is the Bragg angle.

The Bragg equation may be used in two different ways. Since it has three variables, namely, λ, d, and θ, knowing two of these parameters, the third can be determined. First, for a given XRD pattern recorded with a given wavelength, λ, it is possible to calculate the interplanar spacings, d of different planes by measuring the diffraction angles, θ. The lattice parameters can then be determined from these interplanar spacings. This will be referred to as *structure analysis* and is the main function of XRD studies, we will consider here. Second, it is also possible that by using a single crystal of known lattice spacing, d, the value of θ can be measured enabling measurement of the wavelength, λ of the radiation used. This is the field of *x-ray spectroscopy.*

It was mentioned earlier that diffraction generally occurs only when the wavelength of the radiation is of the same order of magnitude as the spacing between the scattering centers. Thus, from the Bragg equation, for a first-order reflection (the reflection that can be observed at the smallest angle),

$$\frac{\lambda}{2d} = \sin\theta \tag{2.20}$$

Since the highest value of $\sin\theta = 1$, $\lambda < 2d$. For a large majority of metals, alloys, and compounds the largest interplanar spacing, d is of the order of 0.25–0.3 nm. Therefore, the longest wavelength of the radiation we could use for diffraction purposes is less than about 0.5–0.6 nm. That is why the useful wavelengths of radiation for XRD purposes are about 0.05–0.25 nm. The use of wavelengths shorter than these values is also possible. But, the difficulty is that the diffraction angles are so small that it is not easy to measure them. (As we will see later in Section 2.13.1, we will use very short wavelengths in electron diffraction studies.)

The above equation also suggests that the minimum interplanar spacing that can be measured depends on the wavelength used. Thus, the minimum interplanar spacing that can be measured using Cu Kα radiation is $\lambda/2 = 0.077$ nm, while it is 0.1145 nm for Cr Kα and 0.0355 nm for Mo Kα radiations. That is why it is useful to use a radiation of short wavelength to obtain a diffraction pattern with a large number of reflections.

2.6 DIFFRACTION ANGLES

The usefulness of the Bragg equation is in calculating the possible angles at which diffraction can occur from different planes in the crystal. (Later, we will see that some of these possible reflections will be absent in the XRD patterns, based on the crystal structure of the material.) For calculating the possible angles at which the diffracted beams occur, Equation 2.18 can be used if we know the interplanar spacings, d, of the planes in the crystals. The values of d can be calculated by knowing the crystal system to which the material belongs and the lattice parameters of the material. For example, for a cubic crystal, the equation is

$$\frac{1}{d^2} = \frac{h^2 + k^2 + l^2}{a^2} \tag{2.21}$$

where d is the interplanar spacing, $(hk\ell)$ are the Miller indices of the plane, and a is the lattice parameter. Similarly, for a hexagonal structure with the lattice parameters a and c, the equation for interplanar spacings is

$$\frac{1}{d^2} = \frac{4}{3}\left(\frac{h^2 + hk + k^2}{a^2}\right) + \frac{l^2}{c^2} \tag{2.22}$$

Equations for other crystal systems may be obtained from standard XRD and/or Crystallography books.

Example Problem 2.6

The metal copper has an FCC structure with the lattice parameter $a = 0.3615$ nm. Calculate the angle at which Cu Kα x-rays with a wavelength $\lambda = 0.15406$ nm will be diffracted from the (111) planes.

Solution 2.6

The interplanar spacing of the (111) planes can be calculated from Equation 2.21 as

$$\frac{1}{d^2} = \frac{1^2 + 1^2 + 1^2}{(0.3615\,\text{nm})^2} = \frac{3}{(0.3615\,\text{nm})^2}$$

or

$$d = \frac{0.3615\text{nm}}{\sqrt{3}} = 0.2087\,\text{nm}$$

The diffraction angle can be calculated using Equation 2.18 and noting that the wavelength of the x-radiation is 0.15406 nm. Thus,

$$\lambda = 2\,d \sin\theta$$
$$0.15406\text{ nm} = 2 \times 0.2087\text{ nm} \times \sin\theta$$

or

$$\sin\theta = \frac{0.15406\text{ nm}}{2 \times 0.2087\text{ nm}} = 0.3691$$

or $\theta = \sin^{-1}(0.3691) = 21.66°$.

Example Problem 2.7

The metal tungsten has a BCC structure with the lattice parameter $a = 0.3165$ nm. Calculate the angle at which Cu Kα x-rays with a wavelength $\lambda = 0.15406$ nm will be diffracted from the (110) planes.

Solution 2.7

The interplanar spacing of the (110) planes can be calculated from Equation 2.21 as

$$\frac{1}{d^2} = \frac{1^2 + 1^2 + 0^2}{(0.3165\,\text{nm})^2} = \frac{2}{(0.3165\,\text{nm})^2}$$

or

$$d = \frac{0.3165\,\text{nm}}{\sqrt{2}} = 0.2238\,\text{nm}$$

Like in the previous example problem, the diffraction angle can be calculated using Equation 2.18, noting that the wavelength of the x-radiation is 0.15406 nm. Therefore,

$$\lambda = 2\,d \sin\theta$$
$$0.15406\text{ nm} = 2 \times 0.2238\text{ nm} \times \sin\theta$$

or

$$\sin\theta = \frac{0.15406\,\text{nm}}{2 \times 0.2238\,\text{nm}} = 0.3442$$

or $\theta = \sin^{-1}(0.3442) = 20.13°$. Even though we calculate the value of θ using the Bragg equation, in reality we will be measuring the angle 2θ, due to the popular geometry in which XRD patterns are recorded.

Note that the nature of the Bravais lattice (body-centered for tungsten and face-centered for copper) did not enter into the above calculations for the interplanar spacing or the diffraction angle. These values only depend on the crystal system to which the material belongs. But, we will see later in Section 2.7.2 that the Bravais lattice will decide which reflections will be present and which reflections will be absent in the diffraction patterns.

Example Problem 2.8

Calculate the interplanar spacings for the (100), (110), (111), (200), (220), and (300) planes in copper, which has the FCC structure and the lattice parameter $a = 0.3615$ nm. Show that the relationship $d_{nh\,nk\,n\ell} = (1/n) \times d_{hk\ell}$ is valid.

Solution 2.8

The interplanar spacings in a cubic structure can be calculated using Equation 2.21 as

$$\left(\frac{1}{d^2}\right) = \frac{h^2 + k^2 + \ell^2}{a^2}$$

or

$$d = (a/\sqrt{h^2 + k^2 + \ell^2})$$

Therefore, for (100) plane,

$$d_{100} = \frac{0.3615\text{ nm}}{\sqrt{1^2 + 0^2 + 0^2}} = 0.3615\text{nm}$$

For the (110) plane,

$$d_{110} = \frac{0.3615\text{ nm}}{\sqrt{1^2 + 1^2 + 0^2}} = 0.2557\text{nm}$$

Similar calculations yield the following results:

$hk\ell$	100	110	111	200	220	300
d (nm)	0.3615	0.2557	0.2087	0.1807	0.1278	0.1205

From the above values it can be clearly seen that d_{200} is exactly one-half of d_{100} and that d_{300} is $\frac{1}{3}$ of d_{100}. Similarly, d_{220} is exactly one-half of d_{110}. Thus, it is clear that the relationship $d_{nh\,nk\,n\ell} = (1/n) \times d_{hk\ell}$ is valid.

Although we will see which diffracted beams will be actually observed in materials with different crystal structures, similar to the above example, one can calculate the angles at which the Cu Kα x-rays ($\lambda = 0.15406$ nm) are diffracted from different planes in the copper lattice. This can be done by first calculating the interplanar spacings for the possible reflections (using Equation 2.21) and then choosing the allowed reflections (this will be discussed later in Section 2.7.2 of this chapter) and calculating the angles (using Equation 2.18) at which the reflections appear. For a copper specimen, the results will be

Miller indices of plane ($hk\ell$)	111	200	220	311	222	400	331
Interplanar spacing, d (nm)	0.2087	0.1807	0.1278	0.1090	0.1044	0.0904	0.0829
Diffraction angle (θ°)	21.66	25.23	37.07	44.97	47.55	58.44	68.31

As additional exercises, calculate the diffraction angles for an aluminum specimen (FCC, $a = 0.4049$ nm) and an α-iron specimen (BCC, $a = 0.2866$ nm), when Cu Kα x-rays are incident on the specimens.

TABLE 2.3
The 2θ Values at which the (111) Reflection from a Copper Specimen Appears with Radiations of Different Wavelength

Radiation	Wavelength, $\lambda_{K\alpha1}$ (nm)	2θ (°)
Mo Kα	0.070937	19.57
Cu Kα	0.154056	43.32
Co Kα	0.178897	50.76
Fe Kα	0.193604	55.27
Cr Kα	0.228970	66.54

Equation 2.18 can be used to calculate the 2θ value for the (111) reflection, once the value of d is calculated using Equation 2.21. Since $a = 0.3615$ nm for copper, $d_{111} = a/\sqrt{3} = 0.3615\ \text{nm}/\sqrt{3} = 0.2087$ nm.

Note that, according to Equation 2.18, the angle θ at which diffraction occurs from a given ($hk\ell$) plane is dependent upon the value of λ chosen. Thus, the smaller the value of λ, the smaller the angle at which diffraction occurs. As an example, Table 2.3 lists the 2θ values at which the (111) diffraction peak appears for a copper crystal, when different x-radiations (of different λ) are used. Figure 2.12 shows two XRD patterns of the pure metal copper—one recorded using the Cu Kα radiation and the other with the Co Kα radiation. Note that the sequence of reflections is the same in both the patterns. But, because of the shorter wavelength of the Cu Kα radiation, the number of reflections observed in the XRD pattern is larger than when the pattern is recorded with Co Kα (having a longer wavelength). Further, also note that the first reflection [from the (111) planes] is at a lower angle in the pattern with Cu Kα radiation than with the Co Kα radiation. In general, it may be noted that the reflections are shifted to higher angles when the pattern is recorded with Co Kα radiation since the wavelength of Co Kα radiation (0.178897 nm) is longer than that of Cu Kα (0.154056 nm).

From Equation 2.21 it is also clear that the larger the lattice parameter of the metal, the larger is the interplanar spacing for a given ($hk\ell$) plane. Consequently, in XRD patterns recorded with a given radiation of wavelength λ, the θ values will be smaller, the larger the lattice parameter of the metal (Table 2.4).

The Bragg equation 2.18 allows us to calculate the angle, θ at which diffraction occurs from a plane with the interplanar spacing, d. Note, however, that when an XRD pattern is recorded in a machine, the plot will be the intensity of the x-ray diffracted beam as a function of 2θ. This is because of the design of the diffractometer. Consequently, it is the 2θ values that are frequently calculated rather than the θ values.

2.7 INTENSITIES OF DIFFRACTED BEAMS

Having known how to calculate the angles at which diffraction occurs from the different lattice planes in a crystal, let us now calculate the intensity of the x-ray beams diffracted by all atoms making up the crystal. For this, we need to first consider scattering of x-rays by an electron, then an atom (containing all the electrons in it), and finally all the atoms in the unit cell. Even though the crystal contains a large number

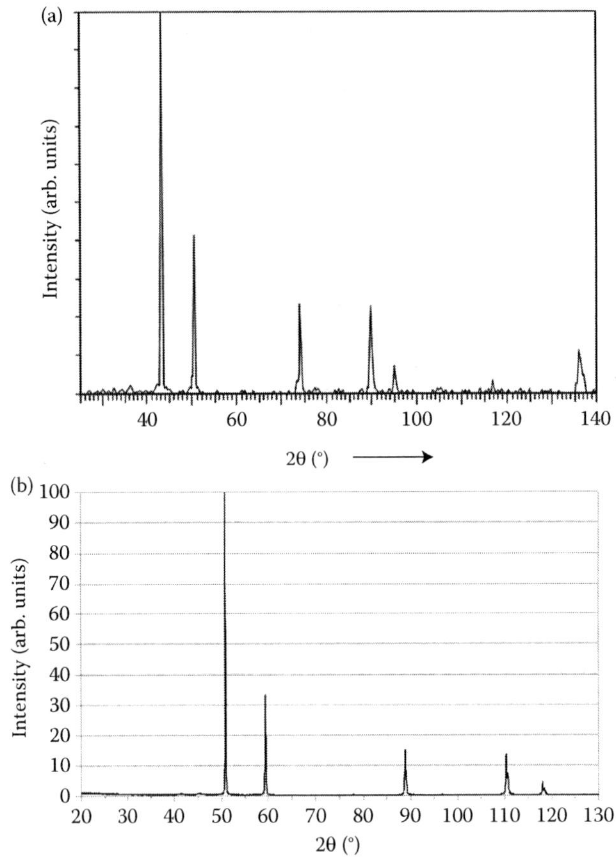

FIGURE 2.12 XRD patterns of pure metal copper recorded with (a) Cu Kα radiation and (b) Co Kα radiation.

TABLE 2.4
The 2θ Values at which the (111) Reflection from Different Metals Appears Using Cu Kα Radiation

Metal	Lattice Parameter, a (nm)	d_{111} (nm)	2θ (°)
Copper	0.3615	0.2087	43.32
Palladium	0.3891	0.2246	40.12
Aluminum	0.4049	0.2338	38.47
Silver	0.4086	0.2359	38.12
Lead	0.4951	0.2858	31.27

Equation 2.18 can be used to calculate the 2θ value for the (111) reflection, once the value of d is calculated using Equation 2.21. The wavelength of the Cu $K\alpha_1$ radiation is 0.154056 nm.

of atoms, we know that the atoms in the crystal are arranged in a regular and periodic manner. Since the crystal structure is represented by the unit cell, it is sufficient to calculate the amplitude of the diffracted beam as a function of the arrangement of the atoms in one unit cell. The total intensity of the diffracted beam is then obtained by squaring the amplitude of the diffracted beam and also by incorporating other factors such as multiplicity, Lorentz polarization, and temperature factors.

We should appreciate that the x-ray beam is incident on a small volume of the crystal, which still contains a very large number of atoms. Recall that one gram atomic weight (atomic weight expressed in grams) of a metal contains the Avogadro number of atoms, that is, 6.023×10^{23}.

Example Problem 2.9

Calculate the number of atoms present in a 0.5 mm-thick sheet of copper whose dimensions are 1 cm × 1 cm. The density of copper is 8.94 g cm^{-3}

Solution 2.9

The volume of the copper sheet is

$$\text{Volume} = 1\ \text{cm} \times 1\ \text{cm} \times 0.05\ \text{cm} = 0.05\ \text{cm}^3.$$

Since the density of the copper metal is 8.94 g cm^{-3}, the weight of the copper sheet is

$$\begin{aligned}\text{Weight} &= \text{Volume} \times \text{Density}\\ &= 0.05\ \text{cm}^3 \times 8.94\ \text{g cm}^{-3}\\ &= 0.447\ \text{g}\end{aligned}$$

One gram atomic weight of copper (63.55 g) contains 6.023×10^{23} atoms. Therefore, 0.447 g of copper contains (0.447 g× 6.023×10^{23} atoms)/63.55 g = 4.236×10^{21} atoms.

The intensity of the diffracted beam from the specimen will be due to the reinforcement of the diffracted beams from all the atoms in the crystal. Thus, the larger the number of atoms present in the sample, the more intense will be the diffracted beam. Also, the heavier the atom is, the higher is the intensity of the diffracted beam.

2.7.1 Atomic Scattering Factor

It was mentioned earlier that scattering of x-ray beams occurs when the incident x-ray beam interacts with the electron and is reemitted. If an atom contains more than one electron, then we have to consider scattering of x-rays by all the electrons in the atom. The amplitude of the scattered x-ray beam along a particular direction can be calculated using the concept of *atomic scattering factor*, f, defined as

$$f = \frac{\text{Amplitude of wave scattered by an atom}}{\text{Amplitude of wave scattered by one electron}} \tag{2.23}$$

The value of f depends on the scattering angle, θ. If scattering takes place in the forward direction ($\theta = 0°$), then the amplitudes of the waves scattered by all the

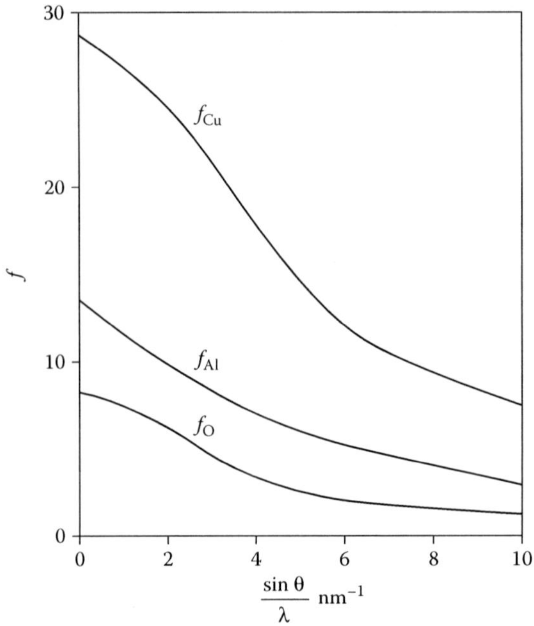

FIGURE 2.13 Variation of the atomic scattering factor of copper, aluminum, and oxygen with $(\sin\theta)/\lambda$.

electrons in the atom add up (since they are all in phase) and then $f = Z$, the atomic number, that is, the total number of electrons in the atom. If scattering takes place in any other direction, for example, when $\theta > 0°$, then the waves become more and more out of phase, and so the effective amplitude of the scattered wave is less than when $\theta = 0°$. The atomic scattering factor also depends on the wavelength of the incident x-rays. For a fixed value of θ, f is smaller for shorter wavelength radiation. In fact, it is usual to plot f as a function of $(\sin\theta)/\lambda$ (units are nm^{-1}) and a typical plot is shown in Figure 2.13. The maximum value of f is the atomic number (Z) for $(\sin\theta)/\lambda = 0$ and it starts to decrease with increasing value of $(\sin\theta)/\lambda$. The rate of decrease of f with increasing values of $(\sin\theta)/\lambda$ is different for different elements. As we will see later (Section 2.7.2), these values of f are useful in calculating the intensities of the diffracted x-ray beams.

2.7.2 Structure Factor

Let us now see how the amplitude of the diffracted wave is modified by the positions of atoms in the unit cell. For this we need to use the concept of *structure factor*, *F* defined as

$$F = \frac{\text{Amplitude of the wave scattered by all the atoms of a unit cell}}{\text{Amplitude of the wave scattered by one electron}} \quad (2.24)$$

Without going into the actual derivation of the equation for the structure factor, we can state that

$$F_{hk\ell} = \sum_{n=1}^{n=N} f_n e^{2\pi i(hu_n + kv_n + \ell w_n)} \tag{2.25}$$

where $F_{hk\ell}$ represents the structure factor (i.e., amplitude) of the diffracted beam from the $(hk\ell)$ plane, f_n represents the atomic scattering factor for the nth atom, $(hk\ell)$ are the Miller indices of the diffracting plane, and u_n,v_n, and w_n are the fractional coordinates of the nth atom in the unit cell. The summation extends over all the N atoms in the unit cell. The intensity of the diffracted beam is proportional to the square of the structure factor. Therefore, if the structure factor calculated is zero, the intensity of the diffracted beam will be zero, that is, the reflection is not observed in the XRD pattern. Therefore, the main objective in calculating the structure factor will be in determining whether a reflection with the indices $hk\ell$ will be present in the XRD pattern or not.

The value of F can be calculated for any plane $(hk\ell)$ if we know f, the atomic scattering factor for the diffracting atom(s) and its (their) fractional coordinates in the unit cell.

Since the equation for the structure factor contains exponential terms, it is important to know some of the useful relationships. These are

$$\begin{aligned} &e^{n\pi i} = (-1)^n \text{ where } n \text{ is any integer,} \\ &e^{n\pi i} = e^{-n\pi i} \text{ where } n \text{ is any integer, and} \\ &e^{ix} + e^{-ix} = 2\cos x \end{aligned} \tag{2.26}$$

Example Problem 2.10

A primitive cubic structure contains one atom in the unit cell whose coordinates are 0,0,0. Calculate the structure factor and list the rules governing the presence or absence of reflections.

Solution 2.10

The structure factor equation is

$$F_{hk\ell} = \sum_{n=1}^{n=N} f_n e^{2\pi i(hu_n + kv_n + \ell w_n)}$$

The primitive cubic unit cell contains one atom, and its coordinates are 0,0,0. Therefore, substituting this value in the above equation, we get

$$\begin{aligned} F_{hkl} &= f.e^{2\pi i(h\times 0 + k\times 0 + l\times 0)} \\ &= f\left[e^0\right] \\ &= f \end{aligned}$$

Therefore, $F = f$ and this is independent of the $hk\ell$ values. Thus, for any $(hk\ell)$ plane in the primitive cubic structure, there is a finite value of diffracted amplitude F, and therefore, some intensity. That is, reflections from all the possible planes are present and no diffraction peak is absent. Hence, for a material with the primitive cubic structure we will observe reflections corresponding to all $(hk\ell)$ values, namely, (100), (110), (111), (200), (210), and so on.

Example Problem 2.11

A BCC structure contains two atoms of the same kind in the unit cell with the fractional coordinates 0,0,0 and $\frac{1}{2},\frac{1}{2},\frac{1}{2}$. Calculate the structure factor and list the rules governing the presence or absence of reflections.

Solution 2.11

The structure factor equation is

$$F_{hk\ell} = \sum_{n=1}^{n=N} f_n e^{2\pi i(hu_n + kv_n + \ell w_n)}$$

Since the BCC unit cell has two atoms of the same kind, f is the same for both of the atoms. Further, since the fractional coordinates of the atoms are listed as 0,0,0 and $\frac{1}{2},\frac{1}{2},\frac{1}{2}$, substituting these values in the above equation, we get

$$\begin{aligned} F_{hkl} &= f.e^{2\pi i(h\times 0 + k\times 0 + l\times 0)} + f.e^{2\pi i\left(h\times\frac{1}{2} + k\times\frac{1}{2} + l\times\frac{1}{2}\right)} \\ &= f\left[e^0 + e^{\pi i(h+k+l)}\right] \\ &= f\left[1 + e^{\pi i(h+k+l)}\right] \end{aligned}$$

Thus, $F = 2f$, when $(h + k + \ell)$ is even, and $F = 0$, when $(h + k + \ell)$ is odd. (Recall that $e^{n\pi i} = +1$ for even integers of n and -1 for odd integers of n.) That is, the amplitude (and therefore the intensity) of the scattered wave is zero, when the sum of the indices $(h + k + l)$ is odd. In such a situation, it is said that the diffraction peak is absent.

Hence, for a material with the BCC structure we will observe reflections corresponding to planes such as (110), (200), and (211), but we will not observe reflections for planes with the indices (100), (111), (210), (221), and so on.

Example Problem 2.12

An FCC structure contains four atoms of the same kind in the unit cell with the fractional coordinates 0,0,0; $\frac{1}{2},\frac{1}{2}$,0; $\frac{1}{2}$,0,$\frac{1}{2}$; and 0,$\frac{1}{2},\frac{1}{2}$. Calculate the structure factor and list the rules governing the presence or absence of reflections.

Solution 2.12

The structure factor equation is

$$F_{hk\ell} = \sum_{n=1}^{n=N} f_n e^{2\pi i(hu_n + kv_n + \ell w_n)}$$

Since the FCC unit cell has four atoms of the same kind, f is the same for all of the atoms. Further, since the fractional coordinates of the atoms are given as: 0,0,0; $\frac{1}{2},\frac{1}{2}$,0; $\frac{1}{2}$,0,$\frac{1}{2}$; and 0,$\frac{1}{2}$, $\frac{1}{2}$ substituting these values in the above equation, we obtain:

$$\begin{aligned} F_{hkl} &= f.e^{2\pi i(h\times 0 + k\times 0 + l\times 0)} + f.e^{2\pi i\left(h\times\frac{1}{2} + k\times\frac{1}{2} + l\times 0\right)} + f.e^{2\pi i\left(h\times\frac{1}{2} + k\times 0 + l\times\frac{1}{2}\right)} + f.e^{2\pi i\left(h\times 0 + k\times\frac{1}{2} + l\times\frac{1}{2}\right)} \\ &= f\left[e^0 + e^{\pi i(h+k)} + e^{\pi i(h+l)} + e^{\pi i(k+l)}\right] \\ &= f\left[1 + e^{\pi i(h+k)} + e^{\pi i(h+l)} + e^{\pi i(k+l)}\right] \end{aligned}$$

If h, k, and ℓ are all even or all odd (we say they are "unmixed"), then the sums $h + k$, $h + \ell$, and $k + \ell$ are even integers, and each term in the above equation has the value +1. Therefore, for the FCC structure, $F = 4f$, for unmixed indices. If h, k, and ℓ are mixed (i.e., some are odd and some are even), then the sum of the three exponentials is −1. Let us suppose that $h = 0$, $k = 1$, and $\ell = 2$, that is, the plane has indices (012), then $F = f\,(1–1 + 1–1) = 0$ and there is no reflection. In other words, $F = 0$ for mixed indices.

Hence, for a material with the FCC structure we will observe reflections corresponding to planes such as (111), (200), and (220), but we will not observe reflections for planes with the Miller indices (100), (110), (210), (211), and so on.

An important observation from the above results is that we did not use anywhere the fact that the structure was cubic. Nor did we use the lattice constant of the material (it was not provided also). Therefore, it is important to realize that *the structure factor is independent of the shape and size of the unit cell* and that it depends only on the positions of the atoms in the unit cell. Thus, the same result as above will be obtained for other noncubic lattices also, provided that the number and positions of the atoms are the same as above. All primitive lattices will have the same rules. Similarly, all body-centered lattices will have the same rules. Table 2.5 summarizes the conditions under which the diffraction peaks will be present or absent in different lattices. These are sometimes referred to as *extinction rules*.

The rules listed in Table 2.5 are for lattices which contain only one atom per lattice point. If the lattice contains more than one atom per lattice point (e.g., the diamond cubic (DC) structure contains two atoms per lattice point in the FCC unit cell), that is, the basis is different, then some of the peaks allowed in the FCC structure (with one atom per lattice point) may be absent. In such cases, one needs to calculate the structure factor by knowing the fractional coordinates of all the atoms in the unit cell.

As an exercise, you may derive the selection rules (to determine which reflections will be present and which will be absent) using the structure factor equation for

TABLE 2.5
Conditions under which Diffraction Peaks will be Present or Absent in Different Lattices

Bravais Lattice	Reflections Possibly Present	Reflections Necessarily Absent
Primitive	All	None
Body-centered	$(h + k + \ell)$ even	$(h + k + \ell)$ odd
Face-centered	h, k, and ℓ unmixed	h, k, and ℓ mixed
Base-centered	h and k unmixed[a]	h and k mixed[a]

[a] These relations are for a cell centered on the C face, and will be different for other situations. For example, if the cell is B centered, then reflections will possibly be present for h and ℓ unmixed and absent for h and ℓ mixed. Similarly, for the A centered lattice, reflections will possibly be present for k and ℓ unmixed and absent for k and ℓ mixed.

diamond (that has a DC structure) that contains eight carbon atoms in the unit cell at the following positions:

$$\begin{array}{cccc} 0,0,0 & \frac{1}{2},\frac{1}{2},0 & \frac{1}{2},0,\frac{1}{2} & 0,\frac{1}{2},\frac{1}{2} \\ \frac{1}{4},\frac{1}{4},\frac{1}{4} & \frac{3}{4},\frac{3}{4},\frac{1}{4} & \frac{3}{4},\frac{1}{4},\frac{3}{4} & \frac{1}{4},\frac{3}{4},\frac{3}{4} \end{array}$$

The rules for the presence and absence of reflections in this case are

$$\begin{aligned} F^2 &= 64 f_C^2 && \text{for } h + k + \ell = 4N \\ F^2 &= 32 f_C^2 && \text{for } h + k + \ell \text{ odd} \\ F^2 &= 0 && \text{for } hk\ell \text{ mixed...} \\ F^2 &= 0 && \text{for } h + k + \ell = 4N \pm 2 \end{aligned} \tag{2.27}$$

where f_C represents the atomic scattering factor of the C atom in the DC structure and N is an integer. Note that the DC structure is based on the FCC Bravais lattice. Therefore, the reflections possibly present in the diffraction pattern are only those, for which the indices are unmixed, that is, the indices are all even or all odd. But, due to the fact that additional atoms are present in the lattice, some of the reflections allowed in the FCC lattice, for example, (200), (222), and so on are absent in diffraction patterns from the DC structures.

The intensity of a diffracted beam depends not only on F^2 but also on other factors. The integrated intensity, I (the area under the diffraction peak) of a diffracted beam is given by the expression

$$I = |F^2| p \left(\frac{1 + \cos^2 2\theta}{\sin^2 \theta \cos \theta} \right) e^{-2M} \tag{2.28}$$

TABLE 2.6
Multiplicity Factors for Different Planes in the Cubic and Hexagonal Systems

			Cubic System				
Miller indices	*hkℓ*	*hhℓ*	0*kℓ*		0*kk*	hhh	00ℓ
Example	345	112	012		011	111	002
Multiplicity	48	24	24		12	8	6
			Hexagonal System				
Miller indices	*hkℓ*	*hhℓ*	0*kℓ*	*hk*0	*hh*0	0k0	00ℓ
Example	123	110	010	120	110	010	002
Multiplicity	24	12	12	12	6	6	2

where p is the multiplicity factor (the number of planes in a family that have the same interplanar spacing), the term in parentheses is the Lorentz-polarization factor, and e^{-2M} is the temperature factor. As an example, p for a (100) plane in a cubic structure is 6, since the indices of (100) can be expressed as (100), (010), (001), ($\bar{1}$00), (0$\bar{1}$0), and (00$\bar{1}$). Values for multiplicity and Lorentz-polarization factors are listed in standard text books on XRD. Table 2.6 gives the multiplicity values for the cubic and hexagonal systems.

2.8 XRD EQUIPMENT

The three basic components of an x-ray diffractometer are:

- X-ray source
- Specimen
- X-ray detector

and they all lie on the circumference of a circle, known as the focusing circle (Figure 2.14). The angle between the plane of the specimen and the x-ray source is θ, the

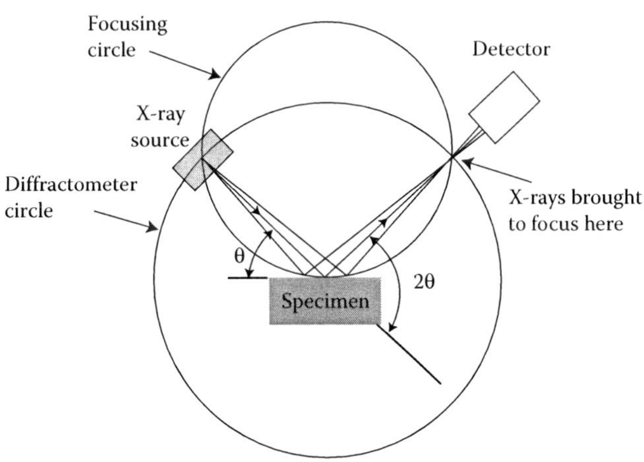

FIGURE 2.14 Schematic of an x-ray diffractometer.

Bragg angle. The angle between the projection of the x-ray source and the detector is 2θ. For this reason the XRD patterns which are produced using this geometry are often known as θ–2θ (theta–two theta) scans. In the θ–2θ geometry, the x-ray source is fixed and the detector moves through a range of angles. The radius of the focusing circle is not constant, it increases as the angle 2θ decreases. The 2θ measurement range is typically from 0° to about 170°. In an actual experiment you need not necessarily scan over the whole range of detector angles. A 2θ range from 20° to 120° would be an example of a typical scan. The choice of range depends on the crystal structure of the material (if known) and the time you want to spend obtaining the diffraction pattern. For an unknown specimen a large range of angles is often used because the positions of the reflections are not known—at least not yet!

The θ–2θ geometry is the most common but there are other geometries which you may come across. In the $\theta - \theta$ (theta–theta) geometry both the x-ray source and the detector move in the vertical plane in opposite directions above the center of the specimen. In some forms of XRD analysis the sample is tilted about an axis ψ (psi). This scan, often known as an ω (omega) scan, provides a measure of the strain (change in length divided by original length, $\Delta\ell/\ell$) in the specimen from which it is then possible to determine the stress (force per unit area, F/A). The specimen can be rotated about the ψ axis and also rotated about an orthogonal axis called the ϕ (phi) axis. This type of scan, sometimes called a ϕ scan, is particularly useful for characterizing oriented polycrystalline thin films where the rotation axis of the grain boundaries is often predominantly aligned perpendicular to the surface of the film. These two techniques, the ω scan and the ϕ scan, are very specialized and, although very useful, are not widely used. The most common mode of operation of the x-ray diffractometer is the θ–2θ geometry.

The *diffractometer circle*, also shown in Figure 2.14, is different from the focusing circle. The diffractometer circle is centered at the specimen and both the x-ray source and the detector lie on the circumference of the circle. The radius of the diffractometer circle is fixed. The diffractometer circle is also referred to as the *goniometer circle*. The goniometer is the central component of an x-ray diffractometer and contains the specimen holder and has arms to which the x-ray source and the detector are mounted. In most powder diffractometers the goniometer is vertical but in some other diffractometers, for example, those used to study thin films, the goniometer may be horizontal.

Each manufacturer's instrument has its unique features, and for specific details one should refer to the instrument operating manual that came with the diffractometer. Figure 2.15 shows a photograph of a commercially available x-ray diffractometer and details of the goniometer stage.

2.8.1 X-Ray Source

As mentioned earlier, x-rays are generated by directing an electron beam of high voltage at a metal target anode inside an evacuated x-ray tube. The voltage and current are both variables and suitable operating voltages depend on the target metal, for example, 50–55 kV for Mo and 25–40 kV for Cu. Copper is the most frequently used target and typical operating conditions are 40 kV and 40 mA. One should choose the

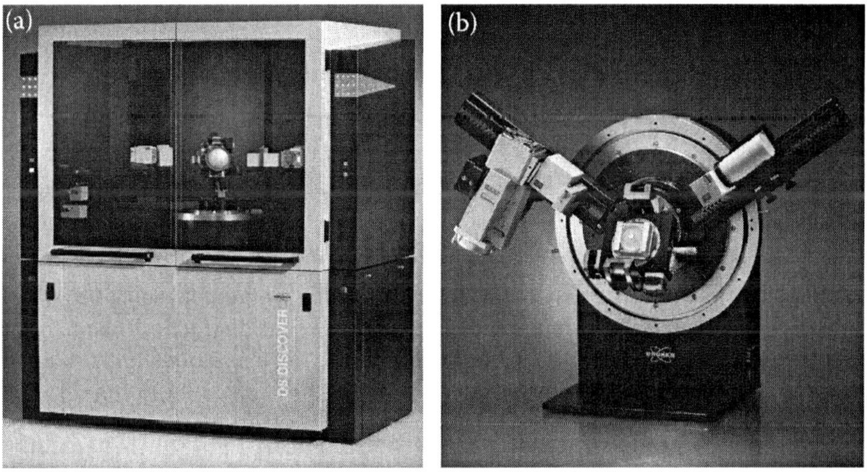

FIGURE 2.15 (a) Photograph of the commercially available x-ray diffractometer D8 DISCOVER with Davinci and (b) details of the goniometer stage. (Photos courtesy of Bruker AXS Inc. With permission.)

appropriate voltage and current values by checking with either the operating manual or the person-in-charge of the instrument. Even though the intensity of the characteristic lines in the x-ray spectrum increases with an increase in the applied voltage, *the values of current and voltage will not affect the positions (2θ values) of the reflections in the diffraction pattern.* The load (product of voltage and current) is usually kept just under the maximum rating (typically around 2 kW) for the x-ray tube in order to make it last longer. The choice of the radiation depends on several factors including how penetrating it is.

2.8.2 The Specimen

In a typical XRD experiment, a thin layer of crystalline powder is spread onto a planar substrate, which is often a nondiffracting material such as a glass microscope slide, and exposed to the x-ray beam. The quantity of powder used for each experiment is quite small, typically on the order of a few milligrams. Ideally, the specimen should contain a large number of small, equiaxed, and randomly oriented grains. The grain size of the powders should be ≤50 μm, that is, the powder will pass through a U.S. standard 325 mesh sieve. It is also critical that the grain size of the specimen is not too small; in that case the diffraction peaks get broadened. The peak breadth increases with a decrease in the grain size of the material. It is important that the specimen you choose for analysis is representative of the entire sample. For example, if you are going to characterize a mixture of two powders, then it is necessary to make sure that the two powders are thoroughly mixed, otherwise the small amount of powder taken for XRD analysis may not contain the two powders in the same proportion as in the global mixture.

2.8.3 X-Ray Detector

There are three main types of x-ray detector—proportional, scintillation, and solid state—that are used on x-ray diffractometers. Majority of the powder diffractometers use a proportional detector. The solid-state detector offers many advantages, but its high cost is often a factor against its use. A discussion on the principles of operation of different types of detector is beyond the scope of the present chapter and the interested reader should consult standard books on XRD. But, note that the type of detector will not affect either the position or the intensity of the diffracted beams in the XRD pattern.

2.9 EXAMINATION OF A TYPICAL XRD PATTERN

An example of a typical XRD pattern recorded from a polycrystalline pure aluminum specimen using Cu Kα radiation ($\lambda = 0.15406$ nm) is shown in Figure 2.16. The pattern consists of a series of peaks. The peak intensity is plotted on the ordinate (*Y*-axis) and the measured diffraction angle, 2θ, along the abscissa (*X*-axis). These peaks are also sometimes referred to as reflections. Each peak, or reflection, in the diffraction pattern corresponds to x-rays diffracted from a specific set of planes in the specimen. These peaks are of different heights (intensities) and the intensity is proportional to the number of x-ray photons of a particular energy which have been counted by the detector for each angle 2θ. The intensity is usually expressed in arbitrary units because it is difficult to measure the absolute intensity. It is fortunate that we are generally not interested in the absolute intensity of the peak but rather the relative intensities of the peaks and the relative differences in their integrated intensity (the area under the peak). The intensities of the reflections depend on several factors including the structure factor, multiplicity factor, Lorentz-polarization factor, intensity of the incident x-ray beam, slit width and values of current and voltage used in the x-ray source, and so on.

The as-recorded XRD patterns generally have a background. The background arises in the XRD patterns because of several reasons. The important among these

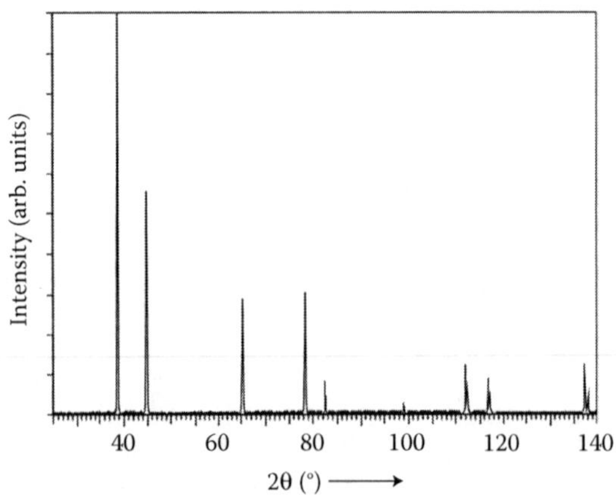

FIGURE 2.16 XRD pattern of aluminum metal powder recorded using Cu Kα radiation.

are: fluorescent radiation emitted by the specimen, diffraction of the continuous spectrum, incoherent scattering from the specimen (this becomes more intense when the specimen contains low-atomic number elements), temperature-diffuse scattering, and diffraction and scattering from materials other than the specimen. The background is usually subtracted and the peaks smoothened. Further, since the most intense peak may have a much higher intensity than the other peaks in the diffraction pattern, the intensity scale is often adjusted so as to have a reasonable intensity for all the peaks. (In this situation, one cannot use the relative intensities of the peaks to calculate the phase proportions in a multiphase mixture or degree of ordering in a partially ordered alloy. The original XRD pattern, without adjusting the intensities, needs to be used for such a purposes.)

The position of the peaks in an XRD pattern depends on the crystal structure (more specifically, the shape and size of the unit cell) of the material and the wavelength of the x-rays used. Recall that we had calculated the values of 2θ for the diffracted beam from the (111) planes of copper for different λ values (Table 2.3), and for the (111) planes of different metals using Cu Kα radiation (Table 2.4). The angular positions of the diffracted beams enable us to determine the size and shape of the unit cell, that is, the Bravais lattice and lattice parameters of the material.

As the symmetry of the crystal structure *decreases* there is an *increase* in the number of peaks. For example, diffraction patterns from materials with cubic structures have only a few peaks. On the other hand, materials with hexagonal (and other less symmetric) structures show many more peaks in their diffraction patterns.

The intensities of the reflections in a single-phase material provide us with information about the atom positions (including their ordering) in the crystal. From the intensities in a multiphase material, it is possible to obtain the proportions of the different amounts of phases. In this chapter, we will not go into the exact determination of atom positions from peak intensities. It is beyond what would normally be covered in an undergraduate materials science course. The width of the peaks, often defined as the full-width at half the maximum height (FWHM), can be used to determine crystallite size and the presence of lattice distortions (strain) in the crystal.

For small values of 2θ each reflection appears as a single sharp peak. For larger values of 2θ, each reflection consists of a pair of peaks, and these correspond to diffraction of the $K\alpha_1$ and $K\alpha_2$ wavelengths. As you will recall from the Bragg equation ($\lambda = 2d \sin \theta$), a peak will be produced at a particular Bragg angle, θ, and for a specific value of *d*. Even if *d* is the same, we will have one θ value for $K\alpha_1$ and another for $K\alpha_2$. Also, note that the separation between the α_1 and α_2 components at any angle will be larger when we expand the 2θ scale and therefore, we may see the α_1 and α_2 separated even at smaller values (2θ less than 80°). It is useful to remember that a peak *always* contains the α_1 and α_2 components. Whether we see it resolved or not depends on the 2θ values, the scale on the *X*-axis, and the resolution of the diffractometer. At low values of 2θ the separation of the peaks is quite small, but increases at larger 2θ values. For example, the separation of the Cu $K\alpha_1$ and Cu $K\alpha_2$ peaks increases from 0.05° at 20° (2θ) to 1.08° at 150° (2θ).

One parameter that you may very well wish to vary during recording of the XRD patterns is the length of time that it takes to obtain the diffraction pattern. The length of time depends on two factors:

1. The range of 2θ values that you want to incorporate into the pattern
2. The length of time that you want to spend collecting data at each 2θ value

It was mentioned earlier that we normally cover a 2θ range of 20–120°; but for an unknown specimen it may be useful to cover the full range of angles in order to avoid missing any of the important reflections. The longer you spend collecting data at each angle the better your counting statistics will be—you will have a higher signal-to-noise ratio and better resolution. On the other hand, the amount of time (and budget!) you have available will provide a practical limit to the length of each scan. Typically, XRD patterns are recorded using a step size of 0.05° (i.e., the counts are measured at every 0.05° for sufficient length of time) and a scanning rate of about 1°/min. The time spent recording an XRD pattern should be typically between 15 and 60 min.

2.10 CRYSTAL STRUCTURE DETERMINATION

As mentioned earlier, full structure determination consists of three steps:

1. Calculation of the size and shape of the unit cell from the angular positions of the diffraction peaks, that is, determination of the six lattice parameters of the unit cell. But, before we could do this, we need to identify the planes from which diffraction is occurring and this step is known as "indexing" of the XRD pattern.
2. Computation of the number of atoms per unit cell from the volume of the unit cell, chemical composition, and its measured density.
3. Determination of the atom positions within the unit cell from the relative intensities of the diffraction peaks.

In this chapter, however, we will consider only step (1). That is, we will index the XRD pattern and determine the shape (Bravais lattice) and size (lattice parameters) of the unit cell. Further, instead of a complex structure, we will choose a cubic structure since it is much easier to do the analysis. Indexing of XRD patterns from more complex structures can also be done; the reader is referred to standard textbooks on x-ray diffraction for this purpose.

2.11 INDEXING THE XRD PATTERN

Indexing the XRD pattern involves assigning the correct Miller indices to each peak in the diffraction pattern. Correct indexing is done only when all the peaks in the diffraction pattern are accounted for and no peaks expected for the structure are missing from the diffraction pattern. We will now describe the indexing of a diffraction pattern recorded from a material (metal, ceramic or semiconductor) that has a cubic structure. We will determine the type of space lattice (primitive, body-centered, or face-centered) and also calculate the lattice parameter.

Even though we had earlier mentioned that there are six lattice parameters that need to be specified for a unit cell, the situation in the case of a cubic structure is simple. Since the three axial lengths are equal to each other, we need to specify only one of them. Further, the three interaxial angles are equal and equal to 90°. If the interaxial angles are not specified, it is assumed that they are 90°. Therefore, we need not specify the interaxial angles also. Consequently, it is necessary to specify only one lattice parameter, namely, the cube edge length, a.

2.11.1 Comparison with a Calculated or Standard XRD Pattern

This method is perhaps the easiest. If you know the crystal structure and lattice parameter of the material (i.e., the identity of the material), then one can determine the allowed reflections (hkl) using the extinction rules listed in Table 2.5, calculate their interplanar spacings (d) using Equation 2.21, and then the diffraction angles from Equation 2.18. You may then compare the 2θ values from the experimental diffraction pattern with the calculated values. If both of them match, then it is possible to assign the hkl values to the different peaks. Since the identity of the material is already known, you also know the Bravais lattice and the lattice parameter.

Another possibility is to look up the standard XRD pattern from the Powder Diffraction Files (PDF) maintained by the International Center for Diffraction Data (ICDD). These data contain the 2θ values, $hk\ell$ values, and the relative intensities of the different reflections for every known structure. If the peak positions (2θ) and relative intensities of all the peaks in the experimental diffraction pattern match with those in the standard, then the identification is unambiguous. This comparison can be done even if you do not know the identity of the material from which the experimental diffraction pattern is recorded. Since the PDF contains the identity (space lattice, lattice parameter, and nature of the metal or alloy) of the material, it may be assumed that the material from which the experimental diffraction pattern is produced is the same as the standard material. The exact way of doing this is described in standard textbooks on XRD.

2.11.2 Relationship between $\sin^2\theta$ and $hk\ell$ Values

The interplanar spacing for any set of hkl planes in a cubic structure can be calculated using

$$\frac{1}{d^2} = \frac{h^2 + k^2 + \ell^2}{a^2} \tag{2.21}$$

Since the Bragg equation is written down as $\lambda = 2d \sin\theta$ (Equation 2.18), then

$$\frac{1}{d^2} = \frac{4\sin^2\theta}{\lambda^2} \tag{2.29}$$

Relating the above two Equations 2.21 and 2.29, we get

$$\frac{1}{d^2} = \frac{h^2 + k^2 + \ell^2}{a^2} = \frac{4\sin^2\theta}{\lambda^2} \tag{2.30}$$

By rearranging the above equation, we obtain

$$\sin^2\theta = \left(\frac{\lambda^2}{4a^2}\right)(h^2 + k^2 + \ell^2) \tag{2.31}$$

Realizing that $\lambda^2/4a^2$ is a constant for any one XRD pattern (since λ, the wavelength of the radiation and a, the lattice parameter of the material are constant), we note that $\sin^2\theta$ is proportional to $(h^2 + k^2 + \ell^2)$; that is, as θ increases, planes with higher Miller indices will diffract.

Example Problem 2.13

Calculate the 2θ values for the first six reflections from a material having a primitive cubic structure with the lattice parameter, $a = 0.296$ nm. Assume that Cu Kα radiation ($\lambda = 0.15406$ nm) was used to record the XRD pattern.

Solution 2.13

The relationship between interplanar spacing, d, Miller indices ($hk\ell$), lattice parameter, a, wavelength of the incident radiation, λ, and the angle, θ, at which the diffraction peak appears for a cubic material are related through Equation 2.31, namely,

$$\sin^2\theta = \left(\frac{\lambda^2}{4a^2}\right)(h^2 + k^2 + \ell^2)$$

Since $\lambda = 0.15406$ nm and $a = 0.296$ nm, the above equation can be simplified as

$$\sin^2\theta = \left(\frac{(0.15406\text{ nm})^2}{4 \times (0.296\text{ nm})^2}\right)(h^2 + k^2 + \ell^2)$$
$$= 0.06772 \times (h^2 + k^2 + \ell^2)$$

Since the first six reflections in the primitive cubic structure have the Miller indices, ($hk\ell$)–(100), (110), (111), (200), (210), and (211), the $h^2 + k^2 + \ell^2$ values will be 1, 2, 3, 4, 5, and 6, respectively. Substituting these values in the above equation, we can calculate the θ values. For example, for (100), $h^2 + k^2 + \ell^2 = 1$. Thus,

$$\sin^2\theta_{100} = 0.06772 \times 1 = 0.06772, \text{ or}$$
$$\sin\theta_{100} = 0.2602, \text{ or}$$
$$\theta_{100} = 15.08°$$

Performing similar calculations for the other reflections, we get

$hk\ell$	100	110	111	200	210	211
θ (°)	15.08	21.59	26.79	31.36	35.58	39.60
2θ (°)	30.16	43.18	53.58	62.72	71.16	79.20

The 2θ values for any structure with a given lattice parameter can be calculated in a manner similar to the above example. As an exercise, calculate the 2θ values for the first six reflections in a diffraction pattern of (a) a BCC metal with $a = 0.3306$ nm and (b) an FCC metal with $a - 0.4086$ nm, recorded with Cu K⊠ radiation ($\lambda = 0.15406$ nm).

It is also critical to realize that the diffraction angles (2θ) at which the allowed reflections appear in any diffraction pattern also depend on the wavelength of the radiation used. The shorter the wavelength of the radiation used, the smaller is the diffraction angle. Note, however, that the interplanar spacing for a given ($hk\ell$) plane is fixed and does not depend on the wavelength of the x-radiation used. Accordingly, Table 2.3 lists the 2θ values at which the (111) reflection of copper appears when the diffraction pattern is recorded with different x-ray wavelengths. Further, maintaining the wavelength of the radiation and the ($hk\ell$) reflection constant, the angle at which the reflection appears is different for metals with different lattice parameters. Thus, the smaller the lattice parameter, the larger is the 2θ value (Table 2.4).

We can now write Equation 2.31 for two different planes with the Miller indices $(h_1k_1l_1)$ and $(h_2k_2l_2)$ as

$$\sin^2\theta_2 = \left(\frac{\lambda^2}{4a^2}\right)(h_2^2 + k_2^2 + \ell_2^2) \tag{2.32}$$

$$\sin^2\theta_1 = \left(\frac{\lambda^2}{4a^2}\right)(h_1^2 + k_1^2 + \ell_1^2) \tag{2.33}$$

Dividing Equation 2.32 with Equation 2.33 we get

$$\frac{\sin^2\theta_2}{\sin^2\theta_1} = \frac{h_2^2 + k_2^2 + \ell_2^2}{h_1^2 + k_1^2 + \ell_1^2} \tag{2.34}$$

From the structure factor calculations (Example Problems 2.10 through 2.12), we had noted that the first peak in the diffraction pattern from a cubic material is due to diffraction from planes with the Miller indices (100) for primitive, (110) for body-centered, and (111) for face-centered lattices. (These are the planes that have the highest planar density in each case.) Therefore, the $h^2 + k^2 + \ell^2$ values are equal to 1, 2, and 3, respectively, for primitive cubic, BCC, and FCC lattices. Since the values of h, k, and ℓ are always integers, the sum of their squares is also an integer. Further, according to Equation 2.34, the ratio of $\sin^2\theta$ values scales with the ratio of $h^2 + k^2 + \ell^2$ values for different planes, and so it should be possible to obtain the $h^2 + k^2 + \ell^2$ values by dividing the $\sin^2\theta$ values of different reflections with the minimum value (i.e., $\sin^2\theta$ value of the first reflection) and multiplying the ratios so obtained by an appropriate integer. Table 2.7 lists the ratios of the allowed $h^2 + k^2 + \ell^2$ values for primitive cubic, BCC, FCC, and DC structures.

Therefore, to index the diffraction pattern, the $\sin^2\theta$ values calculated for all the peaks in the diffraction pattern are divided by the smallest value (i.e., $\sin^2\theta$ value of

TABLE 2.7
Ratios of $\sin^2\theta$ Values for Different Cubic Structures

		Simple Cubic (SC)	BCC	FCC	DC
$h^2 + k^2 + \ell^2$	$hk\ell$	$\sin^2\theta/\sin^2\theta_{100}$	$\sin^2\theta/\sin^2\theta_{110}$	$\sin^2\theta/\sin^2\theta_{111}$	$\sin^2\theta/\sin^2\theta_{111}$
1	100	1	—	—	—
2	110	2	1	—	—
3	111	3	—	1	1
4	200	4	2	1.333	—
5	210	5	—	—	—
6	211	6	3	—	—
7		—	—	—	—
8	220	8	4	2.667	2.667
9	221, 300	9	—	—	—
10	310	10	5	—	—
11	311	11	—	3.667	3.667
12	222	12	6	4.000	—
13	320	13	—	—	—
14	321	14	7	—	—
15		—	—	—	—
16	400	16	8	5.333	5.333
17	410, 322	17	—	—	—
18	411, 330	18	9	—	—
19	331	19	—	6.333	6.333
20	420	20	10	6.667	—
21	421	21	—	—	—
22	332	22	11	—	—
23		—	—	—	—
24	422	24	12	8.000	8.000
25	500, 430	25	—	—	—
26	510, 431	26	13	—	—
27	511, 333	27	—	9.000	9.000
28		—	14	—	—
29	520, 432	29	—	—	—
30	521	30	15	—	—
31		31	—	—	—
32	440	32	16	10.667	10.667
33	522, 441	33	—	—	—
34	530, 433	34	17	—	—
35	531	35	—	11.667	11.667
36	600, 442	36	18	12.000	—

Note: "—" Indicates that the reflection is absent.

the first reflection). These ratios (if they are not already integers), when multiplied by 2 or 3, yield integers. These subsequent integers represent the $h^2 + k^2 + \ell^2$ values of the planes from which that diffraction peak has formed. The Miller indices [$(hk\ell)$ values] of the planes can then be easily identified from the quadratic forms listed in Table 2.8.

The diffraction patterns may also be indexed using the ratios of $1/d^2$ values (instead of the $\sin^2\theta$ values), using the equation:

$$\frac{d_1^2}{d_2^2} = \frac{h_2^2 + k_2^2 + \ell_2^2}{h_1^2 + k_1^2 + \ell_1^2} \tag{2.35}$$

However, since θ is a directly measurable quantity, $\sin^2\theta$ values are more commonly used. But, the results will not be any different whether one uses $1/d^2$ or $\sin^2\theta$ values. Both are related to each other through equation:

$$\frac{1}{d^2} = \frac{4\sin^2\theta}{\lambda^2} \tag{2.29}$$

Figure 2.16 shows the XRD pattern recorded from a pure metal aluminum sample using Cu Kα radiation ($\lambda = 0.15406$ nm). Note that the peaks with $2\theta > 80°$ are resolved into the α_1 and α_2 components. But, here we will only take the α_1 component. The details of indexing the XRD pattern, using the above procedure, are shown in Table 2.9.

From the indices of the observed diffraction peaks, it is clear that only reflections with unmixed $hk\ell$ values (all odd or all even) are present. Therefore, it can be concluded that the Bravais lattice of aluminum is FCC.

Once the space lattice is identified, then the next step is to calculate the lattice parameter of the material. This can be done using Equation 2.31:

$$\sin^2\theta = \left(\frac{\lambda^2}{4a^2}\right)(h^2 + k^2 + \ell^2)$$

Rearranging this equation gives

$$a^2 = \left(\frac{\lambda^2}{4\sin^2\theta}\right)(h^2 + k^2 + \ell^2) \tag{2.36}$$

or

$$a = \frac{\lambda}{2\sin\theta}\sqrt{h^2 + k^2 + \ell^2} \tag{2.37}$$

Using the above equation, the lattice parameter is calculated as 0.4049 nm. Thus, the space lattice and lattice parameter (i.e., shape and size of the unit cell) can be conveniently determined from XRD patterns.

TABLE 2.8
Quadratic forms of Miller Indices and Selection Rules for Cubic Crystal Structures

		Space Lattice			
$h^2 + k^2 + \ell^2$	$hk\ell$	Primitive Cubic	BCC	FCC	DC[a]
1	100	Present	—	—	—
2	110	Present	Present	—	—
3	111	Present	—	Present	Present
4	200	Present	Present	Present	—
5	210	Present	—	—	—
6	211	Present	Present	—	—
7		—	—	—	—
8	220	Present	Present	Present	Present
9	221, 300	Present	—	—	—
10	310	Present	Present	—	—
11	311	Present	—	Present	Present
12	222	Present	Present	Present	—
13	320	Present	—	—	—
14	321	Present	Present	—	—
15		—	—	—	—
16	400	Present	Present	Present	Present
17	410, 322	Present	—	—	—
18	411, 330	Present	Present	—	—
19	331	Present	—	Present	Present
20	420	Present	Present	Present	—
21	421	Present	—	—	—
22	332	Present	Present	—	—
23		—	—	—	—
24	422	Present	Present	Present	Present
25	500, 430	Present	—	—	—
26	510, 431	Present	Present	—	—
27	511, 333	Present	—	Present	Present
28			—	—	—
29	520, 432	Present	—	—	—
30	521	Present	Present	—	—
31		—	—	—	—
32	440	Present	Present	Present	Present
33	522, 441	Present	—	—	—
34	530, 433	Present	Present	—	—
35	531	Present	—	Present	Present
36	600, 442	Present	Present	Present	—
37	610	Present	—	—	—
38	611, 532	Present	Present	—	—
39		—	—	—	—
40	620	Present	Present	Present	Present

[a] Even though DC is not one of the 14 Bravais lattices, important elements such as Si and Ge, and many of the semiconductor materials adopt this structure. That is why we have included it here separately for convenience.

Note: "—" Indicates that the reflection is absent.

TABLE 2.9
Indexing the XRD Pattern of Aluminum

Peak #	2θ (°)	θ (°)	$\sin^2\theta$	$\sin^2\theta / \sin^2\theta_{min}$	$(\sin^2\theta / \sin^2\theta_{min}) \times 3$	$h^2 + k^2 + \ell^2$	$hk\ell$	a (nm)
1	38.52	19.26	0.1088	1.000	3.000	3	111	0.40448
2	44.76	22.38	0.1450	1.333	3.999	4	200	0.40457
3	65.14	32.57	0.2898	2.664	7.992	8	220	0.40471
4	78.26	39.13	0.3983	3.661	10.983	11	311	0.40480
5	82.47	41.235	0.4345	3.994	11.982	12	222	0.40480
6	99.11	49.555	0.5792	5.324	15.972	16	400	0.40485
7	112.03	56.015	0.6876	6.320	18.960	19	331	0.40491
8	116.60	58.30	0.7238	6.653	19.958	20	420	0.40491
9	137.47	68.735	0.8684	7.982	23.945	24	422	0.40494

Material: Aluminum
Radiation: Cu Kα
$\lambda_{K\alpha 1} = 0.154056$ nm

From Table 2.9, one can see that the lattice parameter increases with increasing 2θ values. Since the systematic error in $\sin^2\theta$ decreases as θ increases, one should select the value of the lattice parameter for the peak at the highest 2θ value in the diffraction pattern as the most accurate.

It was mentioned above that an XRD pattern could also be indexed using the interplanar spacings of the different planes instead of the diffraction angles. Table 2.10 provides the different stages involved in indexing the XRD pattern using the interplanar spacings. In brief, the $1/d^2$ values of the different reflections are first calculated. The $1/d^2$ values of all the reflections are divided by the smallest value (i.e., of the first reflection) and these ratios correspond to the ratios of the sum of $h^2 + k^2 + \ell^2$ values. From these ratios, the $hk\ell$ values of the diffracting planes and the lattice parameter can be calculated. As may be noted, there is no difference in the result obtained whether one uses the $\sin^2\theta$ values or the $1/d^2$ values since these two quantities are related to each other through Equation 2.29.

2.11.3 Analytical Approach

An analytical approach can also be used to index diffraction patterns from materials with a cubic structure. It was stated earlier that

$$\sin^2\theta = \left(\frac{\lambda^2}{4a^2}\right)(h^2 + k^2 + \ell^2) \tag{2.31}$$

Since $\lambda^2/4a^2$ is constant for any pattern, which we call A, we can rewrite the above equation as

$$\sin^2\theta = A(h^2 + k^2 + \ell^2) \tag{2.38}$$

The possible values for $(h^2 + k^2 + \ell^2)$ in a cubic system are 1, 2, 3, 4, 5, 6, 8,…. (even though all the reflections may not be present in every type of cubic lattice). The observed $\sin^2\theta$ values for all peaks in the diffraction pattern are therefore divided by the integers 1, 2, 3, 4, 5,…. to obtain a common quotient, which is the value of A corresponding to $h^2 + k^2 + \ell^2 = 1$. Dividing the $\sin^2\theta$ values of the different reflections by A gives the $h^2 + k^2 + \ell^2$ values, from which the $hk\ell$ values of each reflection can be obtained using the quadratic forms listed in Table 2.8.

We can then calculate the lattice parameter from the value of A using the relationship

$$A = \frac{\lambda^2}{4a^2} \quad \text{or} \quad a = \frac{\lambda}{2\sqrt{A}} \tag{2.39}$$

The aluminum XRD pattern indexed earlier is now indexed using the analytical approach and the results are presented in Table 2.11. Note that the smallest common $\sin^2\theta$ value in Table 2.11 is 0.0363 (indicated in bold), and so we can conclude that this value represents A in Equation 2.38. Since Equation 2.38 suggests that division of the $\sin^2\theta$ values by A yields the $h^2 + k^2 + \ell^2$ values from which

TABLE 2.10
Indexing the XRD Pattern of Aluminum

Peak #	2θ (°)	θ (°)	d (nm)	$1/d^2$ (nm^{-2})	$(1/d^2) / (1/d^2_{min})$	$(1/d^2)/(1/d^2_{min}) \times 3$	$h^2 + k^2 + \ell^2$	$hk\ell$	a (nm)
1	38.52	19.26	0.2335	18.341	1.000	3.000	3	111	0.4044
2	44.76	22.38	0.2023	24.435	1.332	3.996	4	200	0.4046
3	65.14	32.57	0.1431	48.834	2.663	7.989	8	220	0.4047
4	78.26	39.13	0.1221	67.076	3.657	10.971	11	311	0.4049
5	82.47	41.235	0.1167	73.427	4.003	12.009	12	222	0.4043
6	99.11	49.555	0.1012	97.643	5.323	15.969	16	400	0.4048
7	112.03	56.015	0.0929	115.869	6.317	18.951	19	331	0.4049
8	116.60	58.30	0.0905	122.096	6.657	19.971	20	420	0.4047
9	137.47	68.735	0.0827	146.214	7.972	23.916	24	422	0.4051

Material: Aluminum
Radiation: Cu Kα
$\lambda_{K\alpha 1} = 0.154056$ nm

TABLE 2.11
Calculation of the Common Quotient

Peak #	2θ (°)	$\sin^2\theta$	$(\sin^2\theta/2)$	$(\sin^2\theta/3)$	$(\sin^2\theta/4)$	$(\sin^2\theta/5)$	$(\sin^2\theta/6)$	$(\sin^2\theta/8)$
1	38.52	0.1088	0.0544	**0.0363**	0.0272	0.0218	0.0181	0.0136
2	44.76	0.1450	0.0725	0.0483	**0.0363**	0.0290	0.0242	0.0181
3	65.14	0.2898	0.1449	0.0966	0.0725	0.0580	0.0483	**0.0362**
4	78.26	0.3983	0.1992	0.1328	0.0996	0.0797	0.0664	0.0498
5	82.47	0.4345	0.2173	0.1448	0.1086	0.0869	0.0724	0.0543
6	99.11	0.5792	0.2896	0.1931	0.1448	0.1158	0.0965	0.0724
7	112.03	0.6876	0.3438	0.2292	0.1719	0.1375	0.1146	0.0860
8	116.60	0.7238	0.3619	0.2413	0.1810	0.1448	0.1206	0.0905
9	137.47	0.8684	0.4342	0.2895	0.2171	0.1737	0.1447	0.1086

Material: Aluminum
Radiation: Cu Kα
$\lambda_{K\alpha 1} = 0.154056$ nm

TABLE 2.12
Indexing of the Diffraction Pattern of Aluminum using the Analytical Approach

Peak #	$\sin^2\theta$	$\sin^2\theta/A$	$h^2 + k^2 + \ell^2$	$hk\ell$
1	0.1088	2.997	3	111
2	0.1450	3.995	4	200
3	0.2898	7.983	8	220
4	0.3983	10.972	11	311
5	0.4345	11.970	12	222
6	0.5792	15.956	16	400
7	0.6876	18.942	19	331
8	0.7238	19.939	20	420
9	0.8684	23.923	24	422

Material: Aluminum
Radiation: Cu Kα
A = 0.0363

the $hk\ell$ values can be obtained, the calculations can be tabulated as shown in Table 2.12.

Note that the common quotient of 0.0363 appears only in the columns of $(\sin^2\theta)/3$, $(\sin^2\theta)/4$, and $(\sin^2\theta)/8$, in Table 2.11. This suggests that the sequence of $h^2 + k^2 + \ell^2$ values in the XRD pattern will be in the ratio 3, 4, 8, ... This sequence of $h^2 + k^2 + \ell^2$ values suggests that the Bravais lattice of aluminum is FCC. This is also proved when we divided the $\sin^2\theta$ values by A. The other Bravais lattices can also be identified in a similar way.

You will notice from Table 2.12 that the Miller indices of the reflections again are unmixed (all odd or all even) and so the Bravais lattice is FCC. The lattice parameter is calculated using Equation 2.37. Since $A = 0.0363$ and $a = \lambda/2\sqrt{A}$, for aluminum

$$a = \frac{0.154056 \text{ nm}}{2\sqrt{0.0363}} = 0.4049 \text{ nm} \tag{2.40}$$

As you would expect, the results obtained by this analytical method are the same as those obtained by the earlier method.

2.11.4 Identification of the Bravais Lattice

The Bravais lattice can be identified by noting the systematic presence (or absence) of reflections in the diffraction pattern. Table 2.5 had earlier summarized the selection rules (or extinction conditions as they are also known) for different lattices.

According to these selection rules, the $h^2 + k^2 + \ell^2$ values for the different *cubic* lattices follow the sequence:

Primitive (or simple) cubic (SC): 1, 2, 3, 4, 5, 6, 8, 9, 10, 11, 12, 13, 14, 16, ...

Body-centered cubic (BCC): 2, 4, 6, 8, 10, 12, 14, 16, 18, 20, 22, 24, 26, ...

Face-centered cubic (FCC): 3, 4, 8, 11, 12, 16, 19, 20, 24, 27, 32, 35, 36, ...

Diamond cubic (DC): 3, 8, 11, 16, 19, 24, 27, 32, 35, 40, ...

The crystal structures of materials with cubic structures are relatively easy to identify in the XRD patterns. From the above sequence of $h^2 + k^2 + \ell^2$ values, one can notice that the reflections in the primitive cubic and BCC lattices are spaced at approximately equal distances (with the exceptions when a reflection will not be present corresponding to $h^2 + k^2 + \ell^2 = 7$, 15, 23, 28, 31, ..., which cannot be expressed as the sum of three squares. These numbers correspond to $h^2 + k^2 + \ell^2 = 8N - 1$, where N is an integer which is ≥ 1). In the case of FCC structures, two peaks are closely spaced, then the next peak is farther, and the next pair is close, and this sequence of closely spaced pairs followed by a widely spaced single peak gets repeated. In the case of the DC structure, the spacing (gap) alternates between large and small differences between the peaks. These are shown schematically in Figure 2.17 for the four different cubic structures discussed above. Therefore, it should be possible to identify the crystal structures of these metals by careful observation of the gap between the reflections (and without actually performing the analysis!).

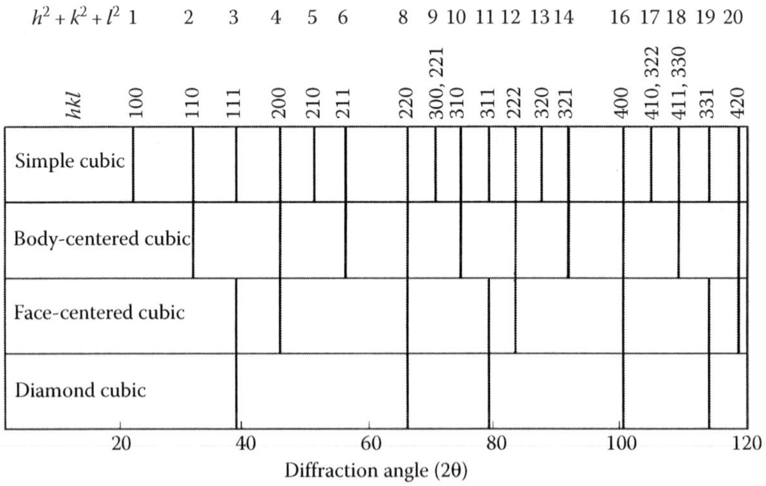

FIGURE 2.17 Calculated diffraction patterns of materials with primitive cubic, body-centered cubic, face-centered cubic, and diamond cubic structures. It is assumed that all of the patterns are recorded with Cu Kα radiation ($\lambda = 0.15406$ nm) and that the lattice parameter is the same in every case ($a = 0.4$ nm).

The Bravais lattice and the crystal structure can be thus unambiguously identified from the sequence of $h^2 + k^2 + \ell^2$ values in diffraction patterns from materials with a cubic structure.

2.12 DIFFERENTIATION BETWEEN SC AND BCC LATTICES

If the diffraction pattern contains only six peaks and if the ratio of the $\sin^2 \theta$ values is 1, 2, 3, 4, 5, and 6 for these peaks, then the Bravais lattice may be either primitive cubic (also known as simple cubic, SC) or BCC (if these ratios are multiplied by 2 to get the sequence as 2, 4, 6, 8, 10, 12). That means, it is not possible to unambiguously distinguish between the two, on the basis of the diffraction pattern alone. Two points are, however, worth noting here. First, SC structures are not very common. Therefore, you will probably be right if you had indexed the pattern as belonging to a material with a BCC structure. Second, if the material has truly an SC structure, its lattice parameter will be small and since all reflections are permitted by the structure factor, the number of peaks in the diffraction pattern will be large. (But, in this example, you are told that only six peaks are present in the XRD pattern.) However, if the structure is actually SC and the pattern is wrongly indexed as BCC, the first peak will be indexed as 110 instead of 100. (Note that while the Miller indices of planes are enclosed in parentheses, such as (110), diffraction peaks from the planes are indicated without parentheses, e.g., 110.) Consequently, the lattice parameter of the unit cell will be $\sqrt{2}$ times larger than the actual value.

This confusion can be avoided if the diffraction pattern has at least 7 peaks. In that case, the seventh peak has Miller indices (321) ($h^2 + k^2 + \ell^2 = 14 = 7 \times 2$) for the BCC lattice and (220) ($h^2 + k^2 + \ell^2 = 8$) for the SC lattice. Note that 7 cannot be expressed as the sum of three squares, and therefore the seventh peak in the diffraction pattern from a primitive cubic lattice will be much more separated from the sixth peak than the earlier sets (see also Figure 2.17). One way of getting more number of peaks in the diffraction pattern is to use a radiation with a shorter wavelength. For example, use Mo Kα ($\lambda = 0.071087$ nm) instead of Cu Kα ($\lambda = 0.154184$ nm).

Note that the sequence and spacing of reflections in the XRD pattern mentioned above (with reference to Figure 2.17) are applicable *only* to cubic materials. These are *not* applicable to noncubic materials, since the repeat distances along the three axes in noncubic materials are not equal to each other; the three axes are also not necessarily orthogonal to each other.

Table 2.13 lists the crystal structures and lattice parameters of some metals that have the three most common crystal structures—BCC, FCC, and HCP. Data for some metals with the DC structure is also included.

2.13 COMPARISON WITH ELECTRON AND NEUTRON DIFFRACTION

It was mentioned earlier that electrons and neutrons, in the appropriate energy range, can be made to diffract from crystalline materials and that these diffraction patterns can also be used to determine the structural features of materials. In this sense, the type of information one can obtain from electron and neutron diffraction patterns is

TABLE 2.13
Crystal Structures and Lattice Parameters of Some Metallic Elements

Element	Crystal Structure	Lattice Parameters		
		a (nm)	c (nm)	c/a
Polonium (α)	Primitive cubic	0.3359	—	—
Chromium	BCC	0.2884	—	—
Iron (α)	BCC	0.2866	—	—
Lithium	BCC	0.3510	—	—
Molybdenum	BCC	0.3147	—	—
Niobium	BCC	0.3307	—	—
Potassium	BCC	0.5247		
Sodium	BCC	0.4291	—	—
Tantalum	BCC	0.3306	—	—
Titanium(β)	BCC	0.3307	—	—
Tungsten	BCC	0.3165	—	—
Vanadium	BCC	0.3023	—	—
Zirconium (β)	BCC	0.3609	—	
Aluminum	FCC	0.4049	—	—
Cobalt (β)	FCC	0.3544	—	—
Copper	FCC	0.3615	—	—
Gold	FCC	0.4079	—	—
Iridium	FCC	0.3839	—	—
Iron (γ)	FCC	0.3647	—	—
Lead	FCC	0.4951	—	—
Nickel	FCC	0.3524	—	—
Palladium	FCC	0.3891	—	—
Platinum	FCC	0.3924	—	—
Rhodium	FCC	0.3805	—	—
Silver	FCC	0.4086	—	—
Carbon (diamond)	DC	0.3567	—	—
Germanium	DC	0.5658	—	—
Silicon	DC	0.5431	—	—
Tin (α-Gray)	DC	0.6489	—	—
Beryllium	HCP	0.2286	0.3584	1.5678
Cadmium	HCP	0.2979	0.5618	1.8857
Cobalt (α)	HCP	0.2507	0.4070	1.6235
Magnesium	HCP	0.3209	0.5211	1.6238
Titanium (α)	HCP	0.2951	0.4683	1.5872
Zinc	HCP	0.2665	0.4947	1.8563
Zirconium (α)	HCP	0.3231	0.5148	1.5933

BCC: body-centered cubic; FCC: face-centered cubic; HCP: hexagonal close packed.

similar to what one obtains from the XRD patterns. Even though the basic diffraction phenomenon is the same, whether one uses x-rays, electrons, or neutrons, there are significant differences in the requirements for the specimen, equipment, recording of diffraction patterns, and interpretations of the patterns. It is also possible to obtain different types of additional information depending on the technique used. We will briefly mention about electron diffraction and neutron diffraction here for the sake of completion. A detailed description of these techniques will be outside the scope of this chapter. Further, it is only rarely that an undergraduate student is called upon to record and interpret electron and neutron diffraction patterns.

2.13.1 Electron Diffraction

The diffraction of electrons was demonstrated by Davisson and Germer in 1927. A beam of electrons is produced in an electron gun when a hair-pin filament of tungsten is heated to very high temperatures. Instead of the tungsten filament, present-day instruments use a lanthanum hexaboride (LaB_6) single crystal or a cold cathode (appropriately oriented tungsten single crystal); these have the advantages of higher electron intensity and longer life. The diameter of the electron beam (e-beam) is 1–10 μm, and electromagnetic lenses decrease the divergence to about 10^{-3}–10^{-6} rad. These electrons are accelerated through a potential difference of V volts. The wavelength of the electron beam, λ can be calculated using the equation:

$$\lambda = \frac{h}{2m_o eV\left(\frac{1+eV}{2m_o c^2}\right)^{1/2}} \tag{2.41}$$

In the above equation, h is Planck's constant (4.136×10^{-15} eV s), m_o is the rest mass of the electron (9.109×10^{-31} kg), e is the charge on the electron (1.602×10^{-19} C), V is the applied voltage in volts, and c is the velocity of light (2.998×10^8 m s^{-1}). By substituting the appropriate values for the constants in the above equation, we can approximate the above equation to

$$\lambda = \frac{12.247}{(V + 10^{-6}V^2)^{1/2}} \tag{2.42}$$

Thus, it can be seen that the wavelength of the electron beam decreases as the acceleration voltage increases. Recall that the wavelength of the x-ray beam decreases as the applied voltage between the cathode and the anode (and therefore the energy) increases. Further, the wavelength of the electron beam is much shorter than what we had seen for x-ray beams. For example, the electron wavelength will be 0.00370 nm for an acceleration voltage of 100 kV and it will be only 0.00087 nm for an acceleration voltage of 1 MV. Table 2.14 lists the wavelengths of the electron beams for different accelerating voltages.

Electron diffraction patterns are recorded in a transmission electron microscope. Since the electron beam has to go through the specimen thickness (transmission), a reasonably strong electron diffraction pattern can be obtained only when the specimen thickness is a few tens of nanometers; a typical specimen thickness is

TABLE 2.14
Electron Wavelength as a Function of the Accelerating Voltage

Accelerating Voltage (kV)	Wavelength (nm)
100	0.0037
120	0.00335
200	0.00251
300	0.00197
400	0.00164
1000	0.00087

about 100 nm. Electron scattering is caused by the interaction of the incident electrons with the electrostatic field of the atoms. The scattering will be referred to as *elastic scattering* when there is no loss in the energy of the electrons after scattering. This can occur when the incident electrons are scattered by the Coulombic potential due to the nucleus, where the proton mass is much larger than that of the electron. The scattering will be referred to as *inelastic scattering*, when there is a loss of energy of the incident electron beam after scattering; this occurs when the incident electrons interact with the electrons of the atom. Consequently, electron scattering is much more efficient than x-ray scattering and, therefore, diffraction effects are easily detected even from volumes much smaller than those required for XRD studies. Figure 2.18 shows the typical scattering curves for electrons, x-rays, and neutrons. Note that the atomic scattering factor, f, for electrons is much larger for electrons than for x-rays, especially at small values of θ. Since the electron wavelength is short, electron diffraction from crystals takes place at small angles and therefore the increased atomic scattering factor becomes useful.

In an XRD experiment from polycrystalline materials or powders, we will always get a diffraction pattern that is commonly referred to as a powder diffraction pattern.

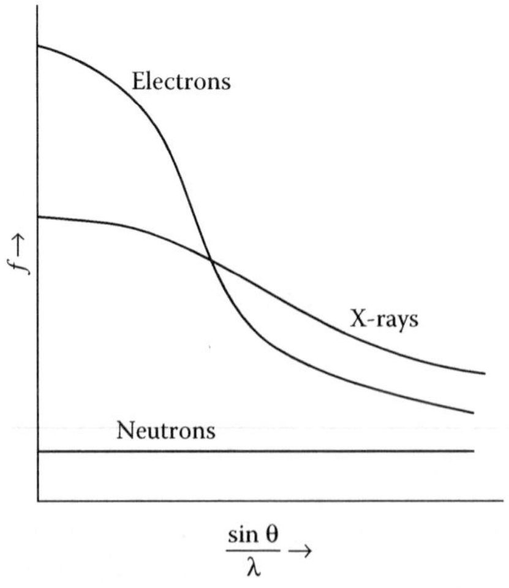

FIGURE 2.18 Typical scattering curves for electrons, x-rays, and neutrons.

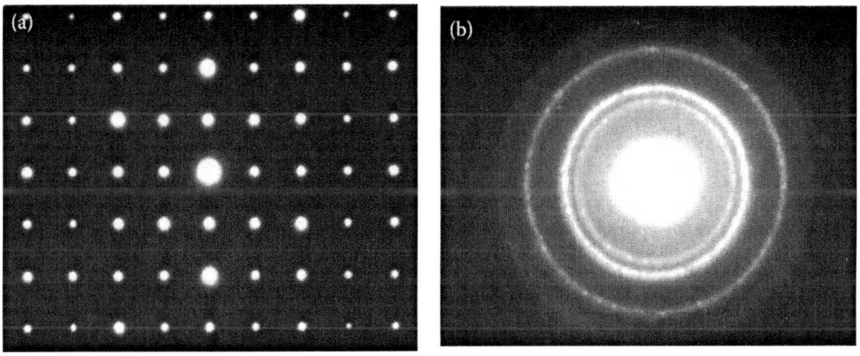

FIGURE 2.19 Typical electron diffraction patterns from (a) single-crystal and (b) polycrystalline specimens. Each of the spots in the single-crystal pattern represents a plane from which diffraction has occurred and the concentric rings in the polycrystalline pattern represent the family of planes from which diffraction has occurred.

We will obtain a single-crystal diffraction pattern only from a single-crystal specimen. On the other hand, in electron diffraction experiments, it is possible to obtain either a single-crystal diffraction pattern or a polycrystalline (or powder) diffraction pattern from the same specimen, depending on the grain size of the material and the size of the aperture used. A single-crystal electron diffraction pattern is possible from a polycrystalline specimen as long as the grain size is not too fine. This is because the electron beam can be focused to a fine spot and therefore it is possible to obtain diffraction from one grain—or a single crystal. Figure 2.19 shows a single-crystal (spot) and a polycrystalline (ring) diffraction pattern recorded in a transmission electron microscope. While the single-crystal diffraction pattern shows a set of spots, each representing a crystal plane from which diffraction has taken place, the diffraction pattern from a polycrystalline material shows a series of concentric rings. Each of these rings represents a family of planes (which are differently oriented, but have the same interplanar spacing) from which diffraction has occurred. Recall that a family of planes consists of a set of crystallographically identical planes which have the same interplanar spacing.

These electron diffraction patterns can also be indexed to obtain crystal structure information (plus many other additional features) from the material. But, a large number of single-crystal diffraction patterns in different orientations are required to obtain the full crystal structure information. More commonly, single-crystal diffraction patterns are used to obtain orientation of the specimen and orientation relationships between phases in a multiphase material. Additionally, it is possible to determine the microstructural features (microscopy) and chemical composition (analysis) of the different phases present in the sample. Therefore, the technique of transmission electron microscopy is very useful in material characterization.

2.13.2 Neutron Diffraction

A neutron is a heavy particle (mass of a neutron is 1.67×10^{-27} kg, and therefore a neutron is significantly heavier than an electron) with spin 1/2 and magnetic moment of 1.9132 nuclear magnetrons. Neutron diffraction experiments require high fluxes provided nowadays by modern reactors. They produce fast neutrons (having high energy) whose energy is reduced by collisions with a moderator of heavy water or graphite. These neutrons, whose energy is reduced, are referred to as thermal neutrons. A monochromator selects the desired wavelength of this neutron beam. Neutrons can also be produced in a pulsed manner by spallation, at a repetition rate between 24 and 50 Hz. High-energy protons (about 1 GeV) in short pulses at the appropriate pulse frequency, strike a target such as uranium or tungsten releasing several tens of neutrons per proton. The neutron flux is very high. High intensity at short wavelengths is a very significant characteristic of pulsed neutron sources.

The neutron–atom interaction comprises interaction with the nucleus and interaction of the magnetic momentum associated with the spin of the neutron with the magnetic momentum of the atom. This effect mainly occurs for atoms with incompletely filled outer electron shells, for example, transition metals.

The neutron–nucleus interaction is governed by very short range nuclear forces. Since the radius of the nucleus is of the order of 10^{-15} cm, that is, several orders of magnitude less than the wavelength of the incident neutron beam, the nucleus will behave like a point and its scattering factor b_o will be isotropic and not dependent on $\sin\theta/\lambda$ (Figure 2.18). When the neutron is very close to the nucleus, a metastable system—nucleus + neutron—is created which decays by reemitting the neutron. For appropriate energy a resonance effect can occur; then the scattering factor can be given as $b = b_o - \Delta b'$. Since $\Delta b'$ can be greater than b, it is possible to have negative scattering factors for some nuclei (e.g., ^{48}Ti, ^{62}Ni, ^{55}Mn).

Neutron scattering differs from electron and x-ray scattering in the following ways:

1. The interaction of neutrons with matter is weaker than that of x-rays or electrons. Generally speaking, scattering amplitudes are about 10^{-12}–10^{-11} cm for x-rays, 10^{-8} cm for electrons, and 10^{-12} cm for neutrons. Therefore, high neutron fluxes and crystals with dimensions of several millimeters are needed for measuring appreciable scattering intensities with neutron beams.
2. Unlike for x-rays and electrons, the b values for neutrons vary nonmonotonically with the atomic number Z; isotopes of the same element can have very different values of b. This allows one to distinguish between atoms having very close values of Z (i.e., neighbors in the Periodic Table), since they may have very different values of b. As an example, $b_{Mn} = -0.36$, $b_{Fe} = 0.96$ and $b_{Co} = 0.25$. This will also allow localizing the positions of light atoms (especially hydrogen) in the presence of heavy atoms.

Further, when ordered phases form between elements which are neighbors in the periodic table (e.g., the CuZn (B2) phase in the Cu–Zn system), the superlattice reflections in the XRD patterns will be extremely weak. In fact, they are so weak that they cannot be detected. This is due to the fact that the x-ray atomic scattering factors of Cu and Zn are so close to each other. But, it should be possible to detect the presence of superlattice reflections in this system using neutron diffraction methods since the atomic scattering factors of Cu and Zn are 0.76 and 0.59×10^{-12} cm, respectively.

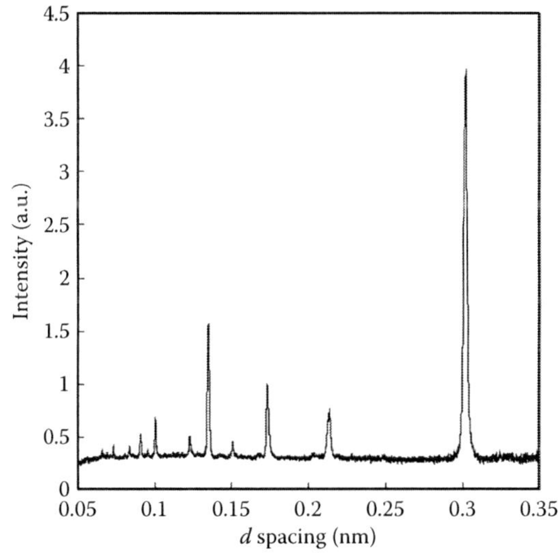

FIGURE 2.20 A typical neutron diffraction pattern. Note that this is very similar to the XRD patterns we had considered earlier, except that the intensity of the diffracted neutrons is plotted against the interplanar spacing, rather than 2θ.

Figure 2.20 shows a typical neutron diffraction pattern. Note that it is very similar to the XRD patterns we have seen earlier in the chapter, except that the intensity of the diffracted neutrons is plotted against the interplanar spacing instead of the diffraction angle. But, the diffraction angle and interplanar spacings are related to each other through the Bragg equation 2.18. Therefore, indexing of these patterns can be done in a way similar to what we have used for XRD patterns.

2.14 EXPERIMENTAL PROCEDURE

Get familiar with the operation of the x-ray diffractometer available in your laboratory. Take fine powders (–325 mesh, <45 μm in size) of some metals with a cubic structure, say iron, nickel, and germanium. Load each powder separately into the x-ray holder and record the XRD patterns using Cu Kα radiation in the 2θ angular range of 20–90°. The XRD pattern is a record of the intensity of the diffracted peaks (on the *Y*-axis) and 2θ on the *X*-axis. Using the cursor, locate the maxima of all the peaks in each diffraction pattern and list the 2θ values in increasing order. If the α_1 and α_2 peaks are resolved, you are advised to take the 2θ value of the α_1 peak. Using the worked example above for aluminum, index all the diffraction patterns (find the Miller indices of the planes giving rise to the diffraction peaks). From the systematic absence (or presence) of reflections in each pattern and from Table 2.5, identify the Bravais lattice. Calculate the lattice parameter in each case and tabulate your calculations in tables similar to that in Table 2.9. Look up the literature for the lattice parameter values of the metals under consideration and compare your values with the standard values. Once you have finished the calculations for all the patterns, summarize your results in the following table:

Summary of Crystal Structure Determination

Material	Bravais Lattice	Crystal Structure	Lattice Parameter	
			Calculated (nm)	Literature Value (nm)
Aluminum	FCC	FCC	0.4049	0.4049
Iron				
Nickel				
Germanium				

Some additional experiments can be carried out based on the above discussion of calculation of lattice parameters. For example, one can determine how the lattice parameter of a metal changes with solute content when the solute atom has an atomic size different from the solvent atom. For example, take the pure metal copper, which has an FCC structure, and determine the lattice parameter. Then take alloys of copper containing small amounts of nickel (say 5, 10, and 15 wt%) and measure the lattice parameter of the solid solution. You will notice that the lattice parameter of the Cu(Ni) solid solution decreases with increasing nickel content. Similarly, take alloys of copper containing gold (say 2, 5, and 10 wt%) and measure the lattice parameter of the solid solution. You will notice that the lattice parameter of the Cu(Au) solid solution increases with increasing gold content. Try to relate the variation of the lattice parameter with solute content to the atomic sizes of the solvent and solute atoms.

EXERCISES

2.1 Sketch the (100), $(1\bar{1}0)$, $(11\bar{1})$, and (123) planes in a cubic unit cell.

2.2 Sketch the $[\bar{1}10]$, $[1\bar{1}1]$, $[12\bar{1}]$, and [123] directions in a cubic unit cell.

2.3 Transform the Miler indices of the (110) plane in a hexagonal system into the Miller–Bravais system.

2.4 Transform the [112] direction of a hexagonal crystal from the three-index into the four-index system.

2.5 How are x-rays produced?

2.6 Distinguish between continuous and characteristic x-radiation.

2.7 How is the characteristic x-radiation produced?

2.8 Calculate the wavelength of the x-ray photon corresponding to energy of 8.04 keV.

2.9 Calculate the energy of the x-ray photon corresponding to a wavelength of 0.071 nm.

2.10 How are the K_α and K_β radiations produced?

2.11 Calculate the mass absorption coefficient of austenitic stainless steel containing Fe-18 wt% Cr-8 wt% Ni for Cu Kα radiation. The mass absorption coefficients of Fe, Cr, and Ni are 304.4, 252.3, and 48.83 cm^2 g^{-1}, respectively.

2.12 Calculate the transmission factors, that is, ratio of the intensity of the transmitted beam to the intensity of the incident beam, for Cu Kα radiation when it passes through a 1-mm-thick sheet of Al, Cu, Pb, or

W. The densities of these metals are 2.71, 8.94, 11.35, and 19.3 g cm^{-3}, respectively, and the mass absorption coefficients are 50.23, 51.54, 232.1, and 170.5 cm^2 g^{-1}, respectively.

2.13 State the Bragg equation and define the terms involved.

2.14 What can we use Bragg equation for?

2.15 What is the smallest interplanar spacing that can be measured using Cu K_α x-radiation ($\lambda = 0.154$ nm)?

2.16 Calculate the interplanar spacings for the (111), (200), and (220) planes of silver metal, which has the FCC crystal structure and a lattice parameter of 0.4086 nm.

2.17 Calculate the interplanar spacings for the (110), (200), and (211) planes of iron metal, which has the BCC crystal structure and a lattice parameter of 0.2866 nm.

2.18 If the $(hk\ell)$ planes in a crystal diffract x-rays of $\lambda = 0.154$ nm at a diffraction angle (2θ) of 42.5°, calculate the interplanar spacing of the $(hk\ell)$ planes.

2.19 An XRD pattern recorded using Cu Kα radiation ($\lambda = 0.154056$ nm) from a material with an FCC structure showed that $2\theta = 36.71°$ for the {111} planes. Calculate the lattice parameter of this metal.

2.20 Calculate the angle at which Cu Kα x-rays with a wavelength $\lambda = 0.15406$ nm will be diffracted from the {110} planes of α-iron which has a BCC structure with $a = 0.2866$ nm.

2.21 Calculate the angle at which Co Kα x-rays with a wavelength $\lambda = 0.178897$ nm will be diffracted from the {110} planes of α-iron which has a BCC structure with $a = 0.2866$ nm.

2.22 Calculate the angle at which Cu Kα x-rays with a wavelength $\lambda = 0.15406$ nm will be diffracted from the {111} planes of γ-iron which has an FCC structure with $a = 0.3647$ nm.

2.23 Calculate the angle at which Mo Kα x-rays with a wavelength $\lambda = 0.071$ nm will be diffracted from the {110} planes of α-iron which has a BCC structure with $a = 0.2866$ nm.

2.24 X-rays are diffracted at an angle of $2\theta = 38.52°$ from the {111} planes of aluminum ($a = 0.4049$ nm). Assuming that this is a first-order reflection, calculate the wavelength of the x-ray beam used.

2.25 Calculate the 2θ angle at which diffraction occurs when a Mo Kα radiation ($\lambda = 0.071$ nm) is incident on the {110} planes of tungsten ($a = 0.3165$ nm).

2.26 Define atomic scattering factor.

2.27 Define structure factor.

2.28 Derive simplified expressions for the structure factor and the rules for the observance of reflections for a NaCl crystal which has a cubic structure and contains 8 atoms per unit cell at the following positions:

$$\text{Na} \quad 0,0,0 \quad \tfrac{1}{2},\tfrac{1}{2},0 \quad \tfrac{1}{2},0,\tfrac{1}{2} \quad 0,\tfrac{1}{2},\tfrac{1}{2}$$

$$\text{Cl} \quad \tfrac{1}{2},\tfrac{1}{2},\tfrac{1}{2} \quad 0,0,\tfrac{1}{2} \quad 0,\tfrac{1}{2},0 \quad \tfrac{1}{2},0,0$$

2.29 Is it possible to identify the space lattice from x-ray diffraction patterns?

2.30 How can you identify the space lattice from x-ray diffraction patterns?

2.31 An XRD pattern recorded from a cubic metal showed diffraction peaks at the following 2θ values: 44.41°, 64.59°, 81.77°, 98.26°, 115.34°, and 135.47°. Assuming that Cu Kα radiation (λ = 0.15406 nm) was used to record this diffraction pattern determine (a) the crystal structure and (b) lattice parameter of the metal.

2.32 An XRD pattern recorded from a cubic metal showed diffraction peaks at the following 2θ values: 43.16°, 50.30°, 73.99°, 89.85°, 95.03°, 116.92°, and 136.59°. Assuming that Cu Kα radiation (λ = 0.15406 nm) was used to record this diffraction pattern, determine (a) the crystal structure and (b) lattice parameter of the metal.

2.33 An XRD pattern recorded from a cubic metal showed diffraction peaks at the following 2θ values: 50.75°, 59.35°, 88.85°, 110.35°, and 118°. Assuming that Co Kα radiation (λ = 0.178897 nm) was used to record this diffraction pattern, determine (a) the crystal structure and (b) lattice parameter of the metal.

2.34 Calculate the 2θ values for the first six reflections from tantalum having a BCC structure and a lattice parameter of 0.3306 nm. Assume that Cu Kα radiation (λ = 0.15406 nm) was used to record the diffraction pattern.

2.35 Calculate the 2θ values for the first six reflections from silver having an FCC structure and a lattice parameter of 0.4086 nm. Assume that Cu Kα radiation (λ = 0.15406 nm) was used to record the diffraction pattern.

2.36 Distinguish between the XRD patterns of BCC and FCC metals.

2.37 What are the typical applications of XRD?

2.38 How do the x-ray and electron diffraction patterns differ from each other?

2.39 What are the important differences between x-ray and neutron diffraction phenomena?

2.40 List, in increasing order, the atomic scattering factor of any element for x-ray, electron, and neutron scattering at low diffraction angles.

FURTHER READING

Bacon, G. E. 1975. *Neutron Diffraction.* Oxford: Oxford University Press.

Bacon, G. E. (ed.) 1987. *Fifty Years of Neutron Diffraction: The Advent of Neutron Scattering.* Boca Raton, FL: Taylor & Francis.

Brandon, D. G. and W. D. Kaplan. 1999. *Microstructural Characterization of Materials.* Chichester, West Sussex, England: John Wiley & Sons, Inc.

Cullity, B. D. and S. R. Stock. 2001. *Elements of X-Ray Diffraction*, 3rd edition. Upper Saddle River, NJ: Prentice-Hall.

Krawitz, A. D. 2001. *Introduction to Diffraction in Materials Science and Engineering.* New York, NY: John Wiley & Sons, Inc.

Suryanarayana, C. and M. Grant Norton. 1998. *X-Ray Diffraction: A Practical Approach.* New York, NY: Springer.

Williams, D. B. and C. B. Carter. 2009. *Transmission Electron Microscopy: A Textbook for Materials Science.* New York, NY: Springer.

Zhang, S., L. Li, and A. Kumar. 2009. *Materials Characterization Techniques.* Boca Raton, FL: CRC Press.

3 Optical Microscopy

3.1 INTRODUCTION

We will see in Chapters 6 to 12 that the internal structure of materials (i.e., the microstructure) plays a very important role in determining the properties of materials. For example, the strength and hardness of materials are determined by the number of phases and their (grain or particle) sizes. The electrical and magnetic properties and also the chemical behavior (corrosion) are determined by the grain size and defects (vacancies, dislocations, grain boundaries, etc.) present in the material. These structural features, which are often on a microscopic level, constitute part of the *microstructure* of a material. A complete description of the microstructure involves describing the size, shape, and distribution of grains and second-phase particles and their composition along with the defect structure of the specimen.

The microstructure of a specimen is not unique; it can be modified by different methods. For example, a pure metal will exist as a single-phase material. On adding a solute element, depending on the nature and amount of alloying element added, the material may exist as a solid solution or as a mixture of two phases or as an entirely different phase. These phases may undergo further changes as a function of temperature and/or pressure. The microstructure of the material will also be different after the alloy undergoes a phase transformation. In Chapter 6, we will see that the microstructure of steels can be significantly modified starting from the austenitic phase field. One could get coarse pearlite, fine pearlite, feathery bainite, acicular bainite, martensite, spheroidite, and so on, depending on the heat treatment given to the specimen. The microstructure could also be changed by the way the material is processed; the grains could be elongated if the material is cold worked. Stress-free equiaxed grains could be again obtained from a cold-worked material by annealing it. The cast structure of materials containing dendritic features or segregation patterns from a solidified or cast material could be changed to obtain a homogeneous structure by a combination of cold working and annealing treatments. Thus, a great variety of microstructures could be obtained by changing the composition, processing, and/or heat treatment procedures. Since microstructural features mainly decide the mechanical behavior of materials, it becomes important to obtain at least a basic understanding of the microstructural behavior of materials.

Microstructural features of interest are the size, shape, and distribution of grains in the case of a single-phase material. In a two-phase or multiphase material the size, shape, and distribution of the second-phase particles or grains are also important in addition to the microstructural parameters of the matrix (the major component). A majority of the microstructural parameters of interest in common engineering alloys are on the micrometer (μm) scale (1 μm = 10^{-6} m). In recent times, there has been a lot of interest in characterizing nanocrystalline materials that have a grain size of ≤100 nm (1 nm = 10^{-9} m). Since the human eye cannot observe details below about 0.07 mm (1 mm = 10^{-3} m), and the microstructural features of interest are 1–2 orders of magnitude finer than this value, we will have to magnify the microstructural features in the specimen to enable the human eye to observe them. For this, we use different types of microscopes capable of achieving different magnifications and resolutions.

A microscope is used mainly:

- To visualize fine details in the microstructure of a sample
- To provide a magnified image of an object
- To measure the sizes and shapes of the microstructural features (e.g., grain sizes, sizes and shapes of precipitates, volume fractions of different phases in a multiphase material, and sizes and shapes of precipitates)
- As an analytical tool for measuring the optical properties and chemical compositions of samples

The microstructures may be viewed directly, recorded using photographic methods, stored digitally, or received by a television camera for input into an analysis system.

A simple magnifying lens can magnify objects from about 5 times (expressed as 5×) to as much as 20×. However, these magnifications are too low for observing the microstructural features in engineering materials. Therefore, the most common microscope that is employed for microstructural observation is a light (or optical) microscope (OM), capable of magnifying objects up to about 1500–2000× , with a resolution of about 200–300 nm. High magnifications are obtained when a lens (or lens system) first magnifies a part of the specimen producing a real magnified image, which is further magnified by another lens (or lens system) to produce a final, magnified, virtual image for observation by the user's eye or a real image for recording on a photographic plate or digital storage. Such microscopes are referred to as compound microscopes. We will see later the difference between magnification and resolution. The light microscope is inexpensive, easy to use, and available in almost every laboratory. Further, it is easy to interpret the microstructures observed using a light microscope.

However, there are other types of more powerful and expensive microscopes that can magnify the microstructural features up to a few tens of thousands or a few hundred thousand times (10,000–500,000×). Associated with this high magnification, the resolution achieved in the microscopes is also high, up to about 0.1 nm. The scanning electron microscope (SEM) and the transmission electron microscope (TEM) fall under this category. An additional advantage of using these microscopes is that it is possible to conduct chemical analysis on a local scale to obtain information about the compositions of the different phases present in the sample. In the TEM, it is also possible to obtain crystal structure information from selected area electron diffraction patterns. One could also use the field ion microscope (FIM) or the atom probe FIM or atomic force microscope to observe atomic scale details and also to analyze individual atoms for their chemical identity. These microscopes are expensive, not easy to use (one needs specialized training to use these microscopes), and interpretation of the microstructural features is not straightforward. One needs to be familiar with the "contrast theory" to interpret the transmission electron microscopic images. Table 3.1 summarizes the capabilities of these different types of microscopes.

The science of observing microstructures of metals under a microscope is called metallography. (Sometimes people extend the term to "materialography" when one observes microstructures of all types of materials—metals, polymers, ceramics, or composites. However, the term "metallography" has come to stay in the literature and includes observation of microstructures of different types of materials.) Metallography is a valuable tool that can greatly aid in:

- Quality control for materials processing, material product, and hardware
- Determining reasons for material failure
- Establishing structure–property correlations in materials*

3.2 PRINCIPLE OF THE OPTICAL MICROSCOPE

The main purpose of a microscope is to reveal the microstructural details in a specimen that are too small to be seen with the naked or unaided eye. A *light microscope* (also referred to as an OM or a metallurgical microscope) consists of three important components—*illumination system*, *objective*, and *eyepiece*. The objective is either a single lens or a combination of lenses that is closest to the object (specimen). Similarly, the eyepiece also is either a single lens or a combination of lenses that is closest to the eye. That is, we see the final image through the eyepiece. In addition to these, the microscope will also contain condenser lenses, diaphragms, and reflectors. Using any microscope, we will be able to *observe* the microstructure. However, provision must also be made to *record* the observed image. A metallurgical microscope that

* Henry Clifton Sorby (1826–1908) was an English microscopist and geologist, who pioneered the use of optical microscopy for investigating the microstructures of a variety of materials in Sheffield, UK. He was born on May 10, 1826 at Woodbourne near Sheffield in Yorkshire and attended Sheffield Collegiate School. In 1849, he pioneered a new branch of geology known as microscopical petrography, wherein he investigated thin sections of rock using a microscope. He took up the study of rocks and minerals under the microscope, and published an important memoir "On the Microscopical Structure of Crystals" in 1858. He then moved on to study the microstructure of meteorites and meteoric iron and microscopical examination of modern manufactured iron and steel. In 1863, he again pioneered a new field of study—microscopic metallurgy—which is now an accepted part of modern materials science and engineering. Toward the end of his life Sorby commented with justifiable satisfaction "In those early days, if railway accident had occurred and I had suggested that the company should take up a rail and have it examined with the microscope, I should have been looked upon as a fit man to send to an asylum. But that is what is now being done ..." He was elected Fellow of the Royal Society of London at the age of 31 and was also president of the Royal Microscopical Society. Sorby also worked hard for the establishment of the University of Sheffield and had a university hall of residence named after him, Sorby Hall, which was built in the 1960s and demolished in August 2006. Both the International Association of Sedimentologists and the Yorkshire Geological Society have Sorby Medals named in honor of Sorby's achievements in geology. Dorsa Sorby on the Moon is named after him. The International Metallographic Society, Inc. presents annually The Sorby Award in recognition of lifetime achievement in the field of metallurgy. Sorby died on March 9, 1908.

TABLE 3.1
Capabilities of the Different Types of Microscopes Used for Microstructural Observations

Microscope	"Illumination" Source	Maximum Magnification	Limit of Resolution (nm)	Structural Information
Light microscope	Visible light	2000 ×	λ/2, typically 200–300 nm	• Microstructure
SEM	Electrons	50,000 ×	1–10	• Microstructure • Chemical analysis
TEM	Electrons	500,000 ×	0.2–0.3	• Microstructure • Crystal structure • Chemical analysis
FIM	Ions	>1,000,000 ×	Atomic	• Microstructure • Chemical analysis

Only typical values are given here; the actual values will be different depending on the type of instrument and the manufacturer.
Note that λ is the wavelength of the light source used.

also has a provision for photographic arrangement, or recording facility, is generally referred to as a *metallograph*.

A light microscope with which you are familiar is most likely of the transmission type. That is, the visible light from the illumination system will pass through the thin specimen and reach the objective lens. Such microscopes are used for observing microstructural features of biological samples. However, since metallic specimens are not transparent to light, however thin their sections may be, the microscopes for observing microstructural features of metallic specimens are of the reflecting type. In these microscopes, the light is reflected from the surface of the specimen before it reaches the objective lens. For this reason, it becomes necessary to prepare a specimen to a mirror finish so that it could efficiently reflect the light from its surface. We will see later about the different steps involved in preparing a specimen for microstructural observation under a light microscope. (The preparation of specimens for transmission electron microscopy investigations is more involved and time consuming because we need to thin the metallic specimens down to about 100 nm in thickness.)

A schematic diagram of the optical arrangement in a light microscope is shown in Figure 3.1. The light from the illumination source is condensed by the condenser and then collimated by the diaphragms. While the first diaphragm controls the intensity of light, the second diaphragm limits the field of view. The light beam will then be diverted with the help of a *plane glass reflector* or a *prism reflector* and made to pass through the objective lens and be incident on the surface of the sample. The specimen is kept just outside the focal point of the objective lens. A part of this light will then be reflected from the surface of the specimen, pass through the objective lens, and will form an enlarged image of the illuminated area. This *primary image* (a real

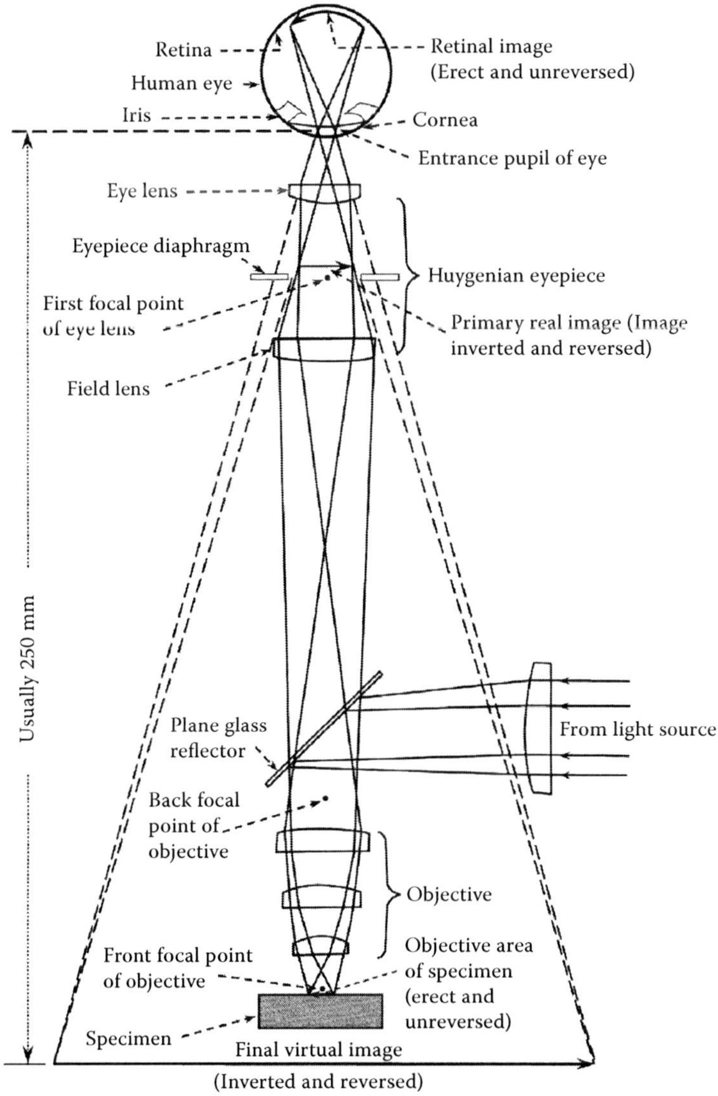

FIGURE 3.1 Principle illustrating image formation in a typical light microscope.

image) formed by the objective lens is located at or within the focal point of the eyepiece. This forms the object for the eyepiece and this is further magnified by the eyepiece to form a virtual image to be viewed by the eye. If the image is to be photographed or recorded, then the image formed by the eyepiece needs to be real. Such an image may be obtained if the eyepiece is moved slightly such that the real image formed by the objective lens lies just outside the focal point of the eyepiece. Therefore, the final magnification is obtained in two stages—first by the objective lens and later by the eyepiece.

3.3 COMPONENTS OF THE MICROSCOPE

We had mentioned earlier that there are three important components in a metallurgical microscope. These are the illumination system or *light source*, objective lens, and eyepiece. Let us now look at the different components of a metallurgical microscope.

3.3.1 Light Source

A number of different light sources are used in metallurgical microscopes. The light beam passes through several components of the microscope and it is also likely that some intensity is lost during this tortuous path. Therefore, it is necessary that the light source is of high intensity so that the final image formed is reasonably bright and sharp. Further, the light incident on the specimen surface should illuminate all the microstructural features uniformly. The choice of the light source depends on its stability, the final brightness and sharpness required in the image, and whether visual observation or photography is involved. To meet these criteria, the following are the more common light sources available.

3.3.1.1 Tungsten-Filament Lamp

Often a low-voltage (6 or 12 V and rated at 100 W) tungsten-filament bulb such as a projection lamp is widely used for visual examination of microstructural features. These bulbs are powered from the mains via a transformer, which is usually fitted with a voltage control. The intensity of the light from these lamps can be controlled with the help of a variable transformer. These bulbs emit a continuous spectrum of light, with most of the light in the longer wavelength region. You will find this type of light source in microscopes that you will use routinely to observe microstructures of samples in the laboratory.

The light output (intensity) from conventional bulbs deteriorates as they age, as tungsten evaporates from the filament and condenses on the inner surface of the glass bulb, thus blackening it.

3.3.1.2 Quartz–Halogen Lamp

This is similar to the normal filament lamp except that the lamp is filled with iodine gas. These are referred to as tungsten–halogen bulbs (sometimes called quartz–iodine or QI bulbs), which have greater brightness, when operated at their designated voltage, than conventional tungsten-filament bulbs. The filament in these bulbs can be heated to much higher temperatures (requiring the lamp bulb to be made of quartz glass), and this produces more intense and steady illumination. These lamps are very dependable, even though there may be some deterioration in intensity for any given voltage as the bulb ages.

3.3.1.3 Xenon Lamp

For special purposes the light source may be a gas discharge tube filled with mercury or xenon vapor under pressure. High-pressure mercury lamps also have intensities similar to the D.C. carbon arc lamps but are much easier to use and maintain. The mercury lamps do not, however, have a continuous spectrum and consequently are not suitable where color reproduction is important. Therefore, a green or yellowish-green filter is frequently used to obtain the best possible definition in the image.

Lamps filled with high-pressure xenon gas are more common. These lamps have a very high intensity (approaching that of daylight) extending well into the UV and infrared (IR) regions. The spectrum is continuous and therefore these lamps are very useful for color photography. Its visual spectrum is of daylight quality and therefore daylight-type color film could be used for photographic purposes. (It is rare nowadays that photographic film is used for recording the microstructures; instead it is mostly digital recording.)

The light sources have a finite life. Therefore, these need to be changed when the light source does not work any more. Changing a tungsten-filament lamp is easy and these lamps are also inexpensive. The mercury arc lamp has a very high operating pressure and therefore these lamps should be allowed to completely cool down to room temperature before they are handled. The xenon lamp also has a high internal pressure both when hot and cold. The manufacturer's instructions should be strictly followed while changing the lamps to avoid accidents.

3.3.1.4 D.C. Carbon Arc

This source provides a more intense illumination than the tungsten-filament lamp. In fact, this is perhaps the oldest source that used to provide a high intensity, but it is rarely used nowadays. An arc is struck between two carbon electrodes, and this produces a continuous spectrum of bright light well suited for photography. This is especially true for use with polarized light where characteristic color changes occur. Some of the disadvantages associated with this light source are the occasional flickering of light and the requirement of regular maintenance. Because of these difficulties, and the availability of alternate high-intensity sources, the carbon arc is no longer used.

The method of illuminating the specimen could be different depending on the information required. Most commonly, metallurgical microscopes use the vertical *bright-field illumination*; that is, the light is incident on the specimen surface at right angles to the specimen surface. In such a situation, features that are at an angle to the incident illumination may not reflect the light back into the objective. Accordingly, those features appear dark. In general, it may be mentioned that in bright-field illumination, the features of interest appear dark on a bright background. The other types of illumination used in metallurgical microscopy are *oblique bright-field illumination* (the light is incident obliquely, i.e., at an angle, on the specimen surface), *dark-field illumination* (the light from the source does not pass through the objective, but is incident on a ring-shaped mirror or lens located around the objective). In oblique bright-field illumination, the surface relief is enhanced, but there may be some loss of resolution. In dark-field illumination, the contrast will be exactly opposite to that of bright-field illumination; that is, the features of interest appear bright against a dark background. Dark-field microscopy helps in positive identification of angled surfaces (those of pits, cracks, etched grain boundaries, etc.). It is also possible to use polarized light. Additionally, phase contrast or interference contrast can be achieved with certain additional components.

3.3.2 Lens Aberrations

All lenses suffer from aberrations (or defects), due to which the image produced by the lens is not an exact copy of the object. A majority of these defects in lenses can be corrected in a compound microscope by combining lenses of different characteristics.

There are basically six lens aberrations: (1) *chromatic aberration*, (2) *spherical aberration*, (3) *coma*, (4) *astigmatism*, (5) *curvature of field*, and (6) *distortion*. Like all lenses, the objective lenses also suffer from these aberrations and these are related to the use of white light (light consisting of different wavelengths) and/or the curved surfaces of the lens elements. However, in objectives used in modern microscopes, all aberrations, except the spherical aberration, are eliminated or reduced by the manufacturer to such a small extent that they are of no consequence. However, the spherical aberration is the most significant, since it is introduced into the image by incorrect use of the microscope or by using unsatisfactory preparations.

Chromatic Aberration: When white light (consisting of violet, indigo, blue, green, yellow, orange, and red colors, with the wavelength increasing from violet to red) passes through the lens elements, beams of different wavelengths come to focus at different points on the emergent side of the lens (Figure 3.2a). This is because beams

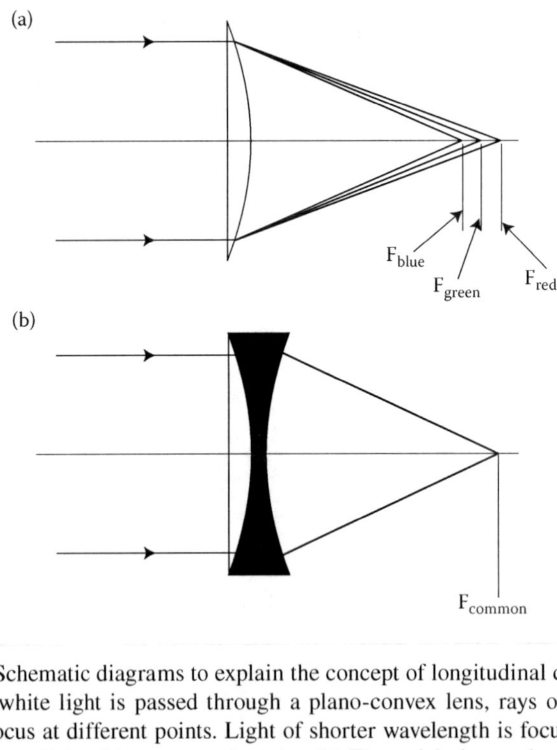

FIGURE 3.2 Schematic diagrams to explain the concept of longitudinal chromatic aberration. (a) When white light is passed through a plano-convex lens, rays of different colors (wavelengths) focus at different points. Light of shorter wavelength is focused nearer to the emergent side than light of longer wavelengths. (b) The axial chromatic aberration can be corrected by combing the plano-convex lens with a concave lens of higher dispersion glass. Rays of all wavelengths now come to focus at the same point.

of shorter wavelength have a greater refractive index than those of longer wavelength. Therefore, violet or blue radiations will focus closer to the lens (on the emergent side) than red radiation. This defect is known as axial *chromatic aberration*. When chromatic aberration is present in a lens, the image formed appears to be surrounded by fringes of varying color. Axial chromatic aberration can be corrected by combining a convex lens of crown glass (with low refractive index and low dispersion) with a concave lens of flint glass (with high refractive index and high dispersion) (Figure 3.2b).

Spherical Aberration: When a monochromatic light (light beam of a well-defined wavelength) is passed through a lens, the rays passing through the outermost part of the lens will be refracted to a greater extent than the rays passing through the axis. Consequently, the light passing through the outer regions will be focused at a point closer to the emergent side of the lens than those passing through the lens nearer the principal axis (Figure 3.3). This is referred to as *spherical aberration*. This aberration is attributable mainly to the curved surfaces characteristic of lens elements. It is impossible to achieve perfect correction for this aberration over the entire aperture of an objective. It is a general practice to use the central portion of the lens so that mostly the axial rays are used to form the image.

When white light is used, it is dispersed into its color components and this, together with the variation of refraction of light in the different zones of the lens, gives rise to the formation of a multiple series of successive color images along the principal axis. That is, both spherical aberration and chromatic aberration will be present simultaneously.

Coma is an aberration associated with points that lie off the axis and are imaged as a conical or comet-shaped blur. The more axial rays contribute to the image of the "head," while the more oblique rays form a series of images of decreasing intensity, but increasing size with their centers more and more displaced from the axis. When *astigmatism* is present in a lens element the image of an off-axis point is not reproduced as a point, but as a line, or an ellipse, or as a circle as the focus is moved away

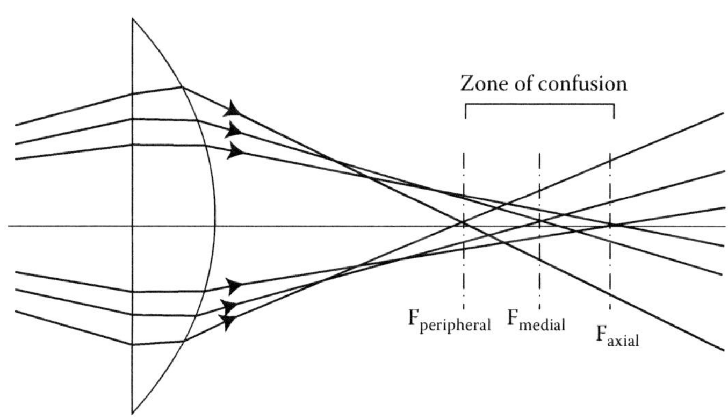

FIGURE 3.3 Schematic diagram illustrating the concept of spherical aberration in a simple lens. Axial rays are focused at a point farther from the lens than either medial or peripheral rays. Because of this, the image is not sharp and there is an extensive zone of confusion.

from the lens, the size of which is different depending on the focal plane chosen. When the defect of the *curvature of field* is present, the image of a linear object does not lie in a plane but on the surface of a sphere. When this defect is present in a lens, it is impossible to focus both the center of the field and the peripheral areas simultaneously. When *distortion* is present in a lens element, a truly rectangular object images either with barrel distortion (i.e., greater magnification at the center of the field) or with pincushion distortion (with a lower magnification at the center of the field).

For obtaining a "good" image (sharp and with all the details faithfully reproduced), it is necessary that the lenses are as free from the above aberrations as possible. It is also important to realize that a majority of these aberrations are eliminated in modern objectives; some of the above aberrations cannot be completely eliminated; they can only be minimized.

3.3.3 Objective Lens

The *objective lens* (or in short the objective) is undoubtedly the most important single optical component of the microscope. It

- Collects light reflected from the specimen surface and forms the primary real image
- Provides the resolution needed to observe details in the microstructure
- Provides the greater part of the magnification for the image

That is why the objective lens is considered the heart of the microscope. Objective lenses can nowadays be made with a wide array of optical glasses and other materials with specified refractive indices and dispersions, and of uniform composition, quality, and stability.

A modern metallurgical microscope can accommodate three or four objectives on a nosepiece. The objectives chosen cover the magnification and resolution range most commonly needed. Typically, the individual magnifications of the objectives used are 5×, 10×, 40×, and 100×, the last normally used with oil immersion. These objectives arranged on the nosepiece are *parfocal* in nature. That is, when one objective of the set mounted on the nosepiece is focused on the object, all the other objectives in the set will also produce a sharp image without any significant readjustment of the focus.

3.3.3.1 Types of Objective Lenses

Although there are different ways in which objective lenses may be classified, the most important classification is based on the extent of correction for axial chromatic aberration. If the lens is corrected for two wavelengths, usually red and green, then these objectives are referred to as *achromats*. That is, all the wavelengths between the red and green region are brought to focus at the same point. Such lenses are acceptable for visual observation and also for photography in black and white. They are relatively cheap and therefore used most commonly for routine type of microscopy work. *Apochromats*, on the other hand, are corrected for three wavelengths—red, green, and violet—and consequently axial chromatic aberration is almost completely eliminated from these lenses. Additionally, their spherical aberration is corrected for two wavelengths. As a result, the numerical aperture (NA) (to be discussed later) of these lenses is larger than those of the achromats. When used correctly, the apochromats produce the highest-quality images and are ideal for observation of images at high magnifications and also for photomicrography. The apochromats are very expensive.

If the residual color cannot be completely removed by combining elements constructed from crown and flint glasses, improvement can be obtained by using fluorspar or some new optical glasses. If the color correction is better than that of achromats, then they are referred to as *fluorites* or *semiapochromats*. Consequently, semiapochromats produce images with a higher degree of contrast than those from an achromat. Semiapochromats are probably the most popular class of microscope objectives used, because of their relative freedom from aberrations and their relatively low cost. Figure 3.4 shows a set of objective lenses used for optical microscopy work.

Objective lenses are generally used as dry; that is, only air is present between the front lens of the objective and the specimen. However, to improve the NA (and resolution), sometimes the objective lenses are used with an immersion fluid such as cedar oil or glycerol. Typically, an immersion objective is designed to use an oil of the same refractive index and dispersion as glass. In these cases, a drop of oil is placed on the specimen and the reflected light from the specimen will enter the objective through this drop of oil. These are then referred to as *oil immersion objectives*. The ability to obtain a better resolution depends on the increase in the effective NA of the objective, which depends on the refractive index of the oil. Cedar oil, for example, has a refractive index of 1.5.

FIGURE 3.4 A set of objective lenses used for optical microscopy. Note that the magnification and NA are engraved on the objectives. (Photo courtesy of Carl Zeiss Microimaging, LLC. With permission.)

For high-temperature microscopy, it is necessary that the lens elements are not too close to the hot specimen. In such a case, one uses *long working distance objectives*. Special objectives are also used with phase contrast and polarized light microscopes.

3.3.3.2 Properties of Objective Lenses

Some of the important and interrelated properties of a microscope objective lens are magnification, NA, resolving power, illuminating power (brightness), depth of focus, and curvature of the image field. Let us now look at the details of these properties.

Magnification: A characteristic feature of the objective lens is its *ability* to magnify the object a definite number of times without the aid of an eyepiece, referred to as *magnification* or *magnifying power*. This magnification is usually engraved on the objective mount as 5×, 10×, and so on. That is, the microstructural features observed are magnified 5 or 10 times by the objective lens. Notice that the objective lenses in Figure 3.4 show that the magnifications of the objectives are 20× and 50×. In some cases (in earlier years), the magnification used to be indicated by a letter, in which case one has to refer to the manufacturer's bulletins to ascertain the exact magnification value. An eyepiece is used to magnify this primary image further. Eyepieces, like objectives, also possess initial magnification, and these magnifying powers also are engraved on the eyepiece mounts. The final total magnification achieved in the microscope depends not only on the initial magnifications of the objective and eyepiece used, but also on the distance separating these in the optical system, which is typically 250 mm. The total magnification M is then obtained as

$$M = M_o \times M_e \tag{3.1}$$

where M_o is the initial magnification of the objective and M_e is the initial magnification of the eyepiece. Thus, if the individual magnification of the objective is 50× and that of the eyepiece is 10×, then the total magnification is 50 × 10 = 500×.

Sometimes it becomes necessary to project the image formed by the eyepiece. Such a situation arises when the image needs to be photographed or when the image has to be projected on to a screen so that a group of students can conveniently see the picture. The total magnification of a projected image then can be obtained from the equation

$$M = \frac{DM_oM_e}{250} \tag{3.2}$$

where D is the projection distance in mm, usually determined with sufficient accuracy by measuring the distance from the eye lens of the eyepiece to the focusing screen. The denominator of 250 in Equation 3.2 refers to the distance in millimeters for which the total magnification of the objective and the eyepiece was computed.

If the magnification is to be computed more accurately than can be obtained by using Equation 3.2, then one could use a stage micrometer. This is a metal plate, which is accurately engraved with lines spaced 0.01 mm apart over a distance of

1.0 mm. To accurately determine the magnification, the stage micrometer is used as an object, and the distance between the engraved lines in the projected image is measured. By dividing the distance between the projected engraved lines, measured in millimeters with 0.01 mm accuracy, the correct magnification can be determined.

Generally speaking, the magnifying power of the objectives is much higher than that of the eyepieces. The initial magnifying powers of the objectives are usually 5×, 10×, 20×, 40×, or 100× even though objectives with other magnifying powers also are available. The magnifying powers of the eyepieces are much smaller, usually 5×, 10×, or 20×.

Numerical aperture: The NA, or light-gathering capacity, of an objective is probably its most important property. It is this NA (which is a function of the design) that enables one to completely and clearly resolve fine details in an object. Other factors such as wavelength of illumination remaining constant, the resolving power (to be described later) of an objective is directly proportional to the value of the NA. That is, the larger the NA, the greater the resolution. This is because the resolution of fine details by an objective lens is determined by the amount of light entering the objective lens. The NA of the objective lens is also engraved on the objective mount. For example, note in Figure 3.4 that the combination of numbers 20 × /0.5 and 50 × /0.8 indicates that the magnifying powers are 20× and 50× and that the NA are 0.5 and 0.8, respectively.

The NA of an objective may be expressed numerically as the product of the sine of one-half the angle of light aperture (α in Figure 3.5a) and the index of refraction (μ) of the medium between the front lens of the objective and the object. That is,

$$\mathrm{NA} = \mu \sin \alpha \tag{3.3}$$

The maximum value of α is determined by the physical dimensions of the lens, and therefore it may not be possible to alter this value. Therefore, a convenient way

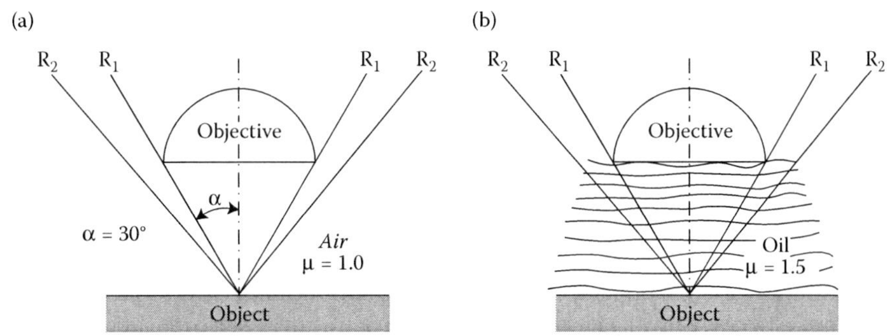

FIGURE 3.5 (a) Schematic diagram showing the concept of NA. NA is the light-gathering capacity of the lens and is the product of the refractive index, μ, of the medium between the objective lens and the specimen and sine of one-half the angle of the most oblique rays, α. That is, NA = μ sin α. (b) The NA of the objective lens can be increased by increasing μ in the case of the oil immersion objective.

of increasing the NA of an objective lens is to increase the value of μ. Thus, by replacing air ($\mu = 1.0$), which is the common medium between the objective lens and the specimen with oil, for example, cedar oil ($\mu = 1.5$), the NA of the lens can be increased by 50% (Figure 3.5b).

Example Problem 3.1

Calculate the NA when the value of α, the half-angle of the most oblique rays, is 30°, in the case of (a) a dry objective and (b) an oil immersion objective when cedar oil is used.

Solution 3.1

The NA is given by Equation 3.3, $\text{NA} = \mu \sin \alpha$. In this example, it is given that $\alpha = 30°$, and we know that μ for air is 1.0 and that for cedar oil is 1.5. Using these data, we can now calculate the NA in both the cases.

a. In the case of the dry objective,

$$\begin{aligned} \text{NA} &= \mu \sin \alpha \\ &= 1.0 \times \sin 30° \\ &= 1.0 \times 0.5 \\ &= 0.5 \end{aligned}$$

b. In the case of the oil immersion objective,

$$\begin{aligned} \text{NA} &= \mu \sin \alpha \\ &= 1.5 \times \sin 30° \\ &= 1.5 \times 0.5 \\ &= 0.75 \end{aligned}$$

It can be clearly seen from this example that NA can be significantly increased by increasing the refractive index of the medium between the objective lens and the specimen.

Even though we had mentioned earlier that resolution of the details in the image increases with increasing NA, it is important to realize that we need a minimum magnification of about 400–500 times the NA to achieve this. However, realize that NA and magnification are related to the brightness of the image according to the relationship:

$$\text{Brightness} = \frac{(\text{NA})^2}{(\text{Magnification})^2} \tag{3.4}$$

From the above equation it is clear that brightness is inversely proportional to the square of the magnification. Therefore, the image is brighter at lower magnifications. Further, the brightness is also higher when we use objectives of larger NA, since NA is nothing but its light-gathering capacity.

Resolving power: The *resolving power* (or *resolution* or *limit of resolution* or *fineness of detail*) is the ability of an objective to produce sharply defined separate images of closely spaced detail in the object. In other words, it is the least separation between two points at which they may be distinguished as separate. Many people (especially the beginners) believe (erroneously) that it is the magnification which is important to see the details in the microstructure. Even though high magnifications are needed to see the details, this is only part of the requirements.

Remember that the human eye can distinguish two points as separate only if they are separated by a distance of at least 0.07 mm. Thus, if two points in a microstructure are separated by a distance of 2.0 μm, we need to magnify the image at least by (0.07×10^{-3} m/2.0×10^{-6} m=) 35× to see the details. However, for comfortable viewing, we may observe the microstructure at a higher magnification, say 100×. That means we will not be able to see these details using a hand-held magnifying glass, since it can magnify only up to about 20×. We need a metallurgical microscope for this. Let us now assume that we observe the microstructure not at 100×, but at 1000×. Even though the separation between the details is now much larger than it was at 100×, we are not seeing any additional details by merely increasing the magnification. A magnification that does not provide any new information or detail is known as *empty magnification*.

On the other hand, let us assume that the points of interest are separated by 2 nm. Then, the minimum magnification required to see the details separately is (0.07×10^{-3} m/2×10^{-9} m=) 35,000×. Therefore, it will be impossible for us to see the details using a light microscope, since the maximum magnification we can obtain is only about 2000×. Then we say that the microscope does not have the required resolution to observe the details in the specimen.

The *limit of resolution* or fineness of detail in a specimen that can be observed using a microscope is given by

$$\text{Limit of resolution} = \frac{\lambda}{2\,\text{NA}} \tag{3.5}$$

where λ is the wavelength of the light beam used.

From Equation 3.5, it is clear that the limit of resolution can be smaller when the wavelength is shorter or NA is higher. Keep in mind that the lower the limit of resolution (value) the better it is, because we will be able to see objects more closely spaced. Therefore, for the same wavelength of the light beam used, the limit of resolution is smaller for objectives with a higher NA. We had seen above that the NA can be increased by using an oil immersion objective lens. On the other hand, for the same NA, the limit of resolution can be decreased by decreasing the wavelength of the light beam used to illuminate the object field. Thus, the limit of resolution will be smaller when we use blue light (λ is about 475 nm) than when we use red light (λ is about 650 nm).

Sometimes, we come across expressions such as the TEM has a better resolution than an OM. This simply means that the TEM has a lower limit of resolution. In other words, we will be able to see details separated by a smaller distance by using a TEM than by using an OM. Remember that the smaller the value of the limit of resolution, the more powerful the microscope. That is why we say that the electron microscope has a higher resolution than an OM.

Depth of focus: Depth of focus, also referred to as *vertical resolution*, is the ability of an objective to produce a sharply focused image when the surface of the object is not truly planar. If a metallurgical specimen is etched, then it is possible that certain constituents are depressed from the etched surface. Further, if a specimen is deeply etched, then some features in the specimen are more affected and thus some grooves may be formed. Again, if specimens have fractured either in service or after subjecting them to a mechanical test in a laboratory, then also the fractured surface can be very rough.

The depth of field is inversely proportional to NA and initial magnification of the objective lens. That is why you will note that if the specimen surface is not flat, then some parts of the specimen will be in focus while the other parts are not. This effect becomes more serious at higher magnifications. Unfortunately, there is not much that can be done to correct this, even though we will see later that an SEM has a larger depth of focus.

3.3.4 Eyepiece

The *eyepiece*, also referred to as an ocular, is used mainly to further magnify the primary image formed by the objective lens and renders it visible as a virtual image for observation, or to project the primary image as a real image for recording purposes. This magnified image helps in seeing the details that have been resolved by the objective. Additionally, the eyepiece may also be used to complete the correction of residual aberrations in the primary image, or to introduce reticules and pointers in a conjugate plane so that they may be seen in sharp focus at the same time as the image. This becomes very useful when one wishes to perform quantitative microscopy or to measure the sizes of microstructural features in any specimen. Like objective lenses, eyepieces also are not single lenses, but a combination of lenses. The lens closest to the observer's eye is called the "eye" lens, while that at the lower end of the eyepiece is called the "field" lens.

It is often necessary to measure the sizes of some specific microstructural features, dimensions of surface defects, or sizes of hardness indentations (see Chapter 8 for details). Eyepieces intended for measurement (known as *micrometer eyepieces* or *measuring eyepieces* or *reticules*) carry a reticule etched on to a circular glass disc in the plane of the field-limiting diaphragm of the eyepiece. The reticule could contain a reference mark, a micrometer scale, or other geometric markings. When this is superimposed on the microstructure, the size of the microstructural features of interest can be directly and accurately measured.

3.4 MICROSCOPIC OBSERVATION

The metallurgical microscopes can be very simple containing a light source, an objective, an eyepiece, and some apertures and condensers. Figure 3.6 shows a typical bench microscope available in most laboratories. It is very easy to operate. The power is turned on, and with the help of a transformer, the brightness of the light may be adjusted. The specimen that has been prepared for observation is placed on the stage

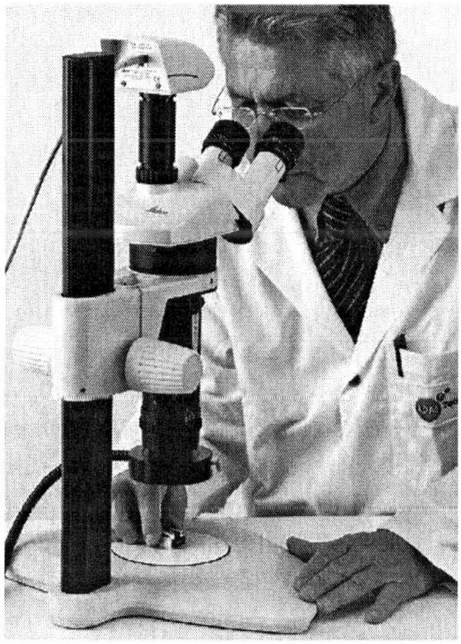

FIGURE 3.6 A typical bench-type metallurgical microscope, Leica MZ16. (Photo courtesy of Leica Microsystems, Inc. With permission.)

of the microscope with the polished and etched surface facing the objective lens. The specimen stage can be moved in the *X* and *Y* directions so as to choose the appropriate area in the specimen for magnifying and observing the image. Make sure that there is enough gap between the top (polished and etched) surface of the specimen and the objective lens. Sometimes, the microscopes are designed in such a way that very large specimens can be accommodated on the specimen stage (Figure 3.7).

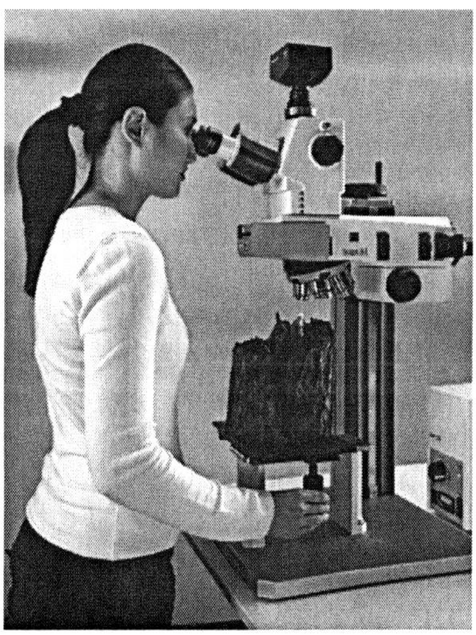

FIGURE 3.7 An optical microscope with a stage to accommodate very large specimens. (Photo courtesy of Carl Zeiss Microimaging, LLC. With permission.)

In order to ensure that the specimen (especially if it is small in size) stays in place and that the specimen surface is normal to the microscope axis, it is common practice to support the specimen from below using plasticine. For this, a small amount of plasticine is placed on a glass slide, the prepared specimen is placed on it (with the *unprepared* surface touching the plasticine), and a slight pressure is applied from the top so that the specimen is seated properly. Make sure that the prepared surface does not come directly in contact with the bottom side of the plunger used to apply the pressure. A thin sheet of lens paper could prevent direct contact between the plunger and the prepared surface of the specimen, and maintain the prepared surface intact.

Once the specimen is properly placed on the microscope stage, look through the eyepiece and raise the specimen stage slowly until the specimen surface comes into focus. At this stage, the specimen surface is exactly at the focal point of the objective lens. Realize that the gap between the surface of the specimen and the objective lens will be very small at high magnifications and reasonably large at low magnifications. As a matter of good practice, it is desirable that you *always* start observing the microstructure at a low magnification and slowly increase the magnification until you reach the desired value needed to look at the required detail in the microstructure. All the objectives in a microscope are arranged in a revolving nosepiece and are placed in such a way that once the specimen is in focus at one magnification, bringing in the other objective lenses by rotating the nosepiece does not require refocusing. That is, these are all parfocal objectives and therefore only very slight adjustments may be needed (Figure 3.8).

A microscope that has attachments to enable the image to be recorded is called a *metallograph*. Metallographs are more complex than the bench-type microscopes mentioned above and their operation (although similar to the description above) could be a little more elaborate. Figure 3.9 shows a typical optical metallograph. As mentioned earlier, recording of the microstructures nowadays is mostly digital in

FIGURE 3.8 Parfocal arrangement of four different objectives on a revolving turret. Once any of the four objectives is properly focused on the specimen, one need not adjust the focus when any other objective is brought into position. (Photo courtesy of Carl Zeiss Microimaging, LLC. With permission.)

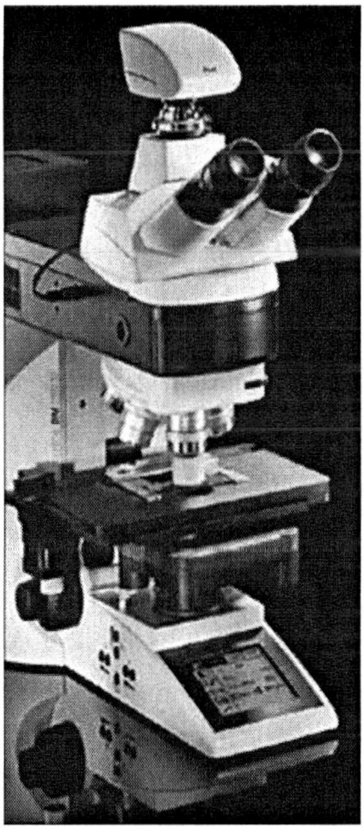

FIGURE 3.9 A typical optical metallograph, Leica DM6000 M. (Photo courtesy of Leica Microsystems, Inc. With permission.)

nature and therefore the microscope is interfaced with a computer so that image processing and analysis could be done. The recorded image of the microstructure of the sample is referred to as a *photomicrograph*. Sometimes, people use the expression "microphotograph," which is not the correct term to express a micrograph that has been recorded using a metallograph.

In the above discussion, we have considered a situation when the specimen is below the objective lens. This is the normal situation and here it is necessary that the specimen is mounted in plasticine to ensure that the top prepared surface is perfectly horizontal and that the light is incident normal to the surface of the sample. However, we also have inverted microscopes, where the specimen is kept on a stage above the objective lens (Figure 3.10). That is, the light from the source goes through the objective lens from bottom to top. In this case, the prepared specimen surface rests on an annular support and no mounting is necessary. Such an arrangement will be convenient for large-size specimens and it is also not necessary to have a flat bottom; that is, the two surfaces need not be parallel to each other. However, one should be careful not to scratch the surface of the prepared specimen during its placement and/or movement on the specimen stage.

The operation of a light microscope is very simple except that the finer details of operation and placement of different knobs for controlling the focus and adjusting the apertures may be slightly different from manufacturer to manufacturer. Therefore, it is always a good practice to check with the operation manual supplied with the instrument for details of operation.

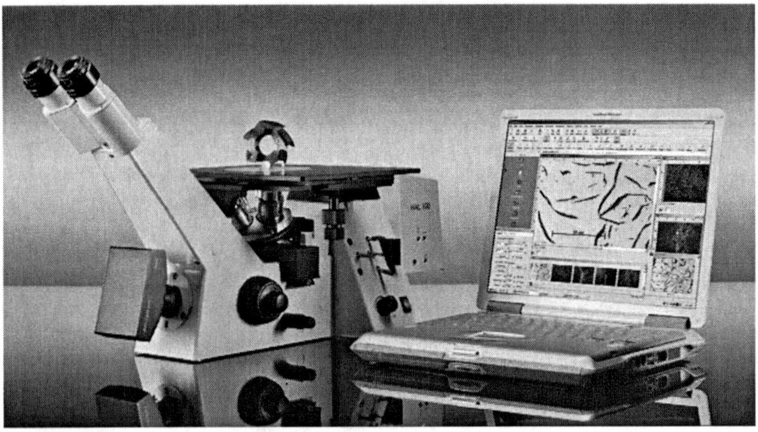

FIGURE 3.10 An inverted metallograph. (Photo courtesy of Carl Zeiss Microimaging, LLC. With permission.)

In addition to the microscopes described above, one may also have microscopes for some specialized investigations. These may include high-temperature microscopes (to study the changes in microstructure with temperature, e.g., to study grain growth, precipitation reactions, phase changes, sintering, diffusion, and surface reactions), polarized light microscopy (to study anisotropic materials and reveal grain structure and twinning in anisotropic metals and alloys), phase contrast microscopy (to study height differences in specimens produced during crystal growth), and interferometry (to accurately measure the microtopography of surfaces).

3.5 INFORMATION DERIVABLE FROM THE MICROSTRUCTURE

As mentioned at the beginning of this chapter, the microstructure of materials plays a very important role in determining the mechanical properties of materials. Therefore, let us now see what type of information can be obtained from the microstructures of materials. By observing the microstructure, we should be able to determine the number of phases present in the sample. In the case of single-phase materials, we can determine the grain size and shape. Additionally, by looking at the shape of the grains, it should also be possible to determine or at least guess the prior thermal and mechanical history the specimen has gone through. In the case of two-phase (or multiphase) materials, we can determine the size, shape, and distribution of the second-phase particles and volume fractions of the phases. In some special cases, for example, plain carbon steels (steels containing only carbon as the alloying element), it should also be possible to estimate the chemical composition of the steel sample by determining the proportions of the phases (ferrite or cementite and pearlite) in an annealed sample.

3.5.1 Number of Phases

Each phase present in the sample reacts differently to the action of an etching reagent. Consequently, the way the phase appears in the microstructure is different and therefore it should be possible to determine the number of phases present in the sample by observing the etched sample. For example, if one observes a single-phase polycrystalline material under an OM, the grain structure will appear bright with the grain boundaries appearing dark. The surfaces along which two grains meet are known as grain boundaries. Since we normally observe the microstructure in a direction parallel to the grain boundary surface (and the intersection of a grain boundary, a two-dimensional feature, with the surface of the specimen will be a line), the grain boundary in a microstructure appears as a line. The grain boundaries in any specimen are chemically more active and, consequently, they are etched to a greater extent than the grains. Therefore, while the grain surfaces are flat, grooves form along the grain boundaries. As a result, when light from the microscope is incident on the specimen, it gets reflected back into the objective from the grains and so they appear bright. On the other hand, light incident on the (grooved) inclined grain boundaries gets scattered at an angle different from that of the grains and so it will not reach the objective lens. That is why, the grain boundaries appear dark. Figure 3.11 shows schematically how this can happen. Typical optical micrographs of two materials—Fe–0.02 wt%C and an Fe–0.2 wt%C steel—are presented in Figure 3.12. Note that the Fe–0.02 wt%C specimen which is almost pure iron

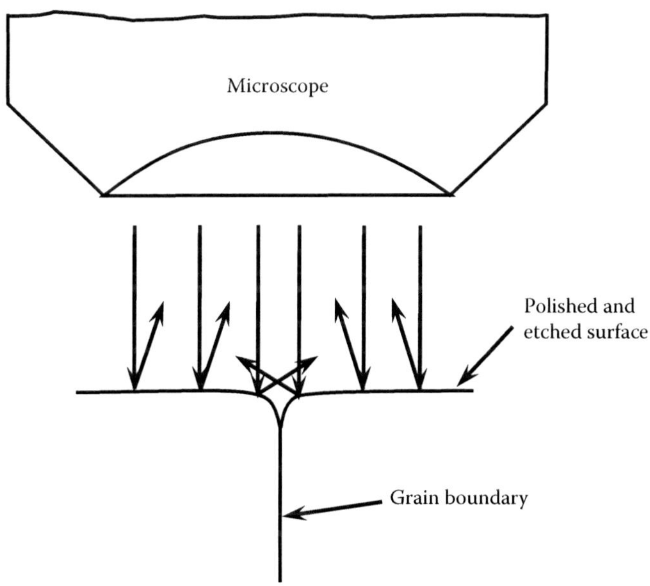

FIGURE 3.11 Etching characteristics of a single-phase material. Since grain boundaries are chemically more active than grains, the grain boundaries get more attacked by the etchant. Therefore, while the grain surfaces are flat, the grain boundaries generate grooves. The picture shows a section of a grain boundary and its surface groove produced by etching. Since the light incident on the grain surface reflects back into the objective lens of the microscope, the grains appear bright. The light from the grain boundaries is not reflected back into the objective and therefore they appear dark.

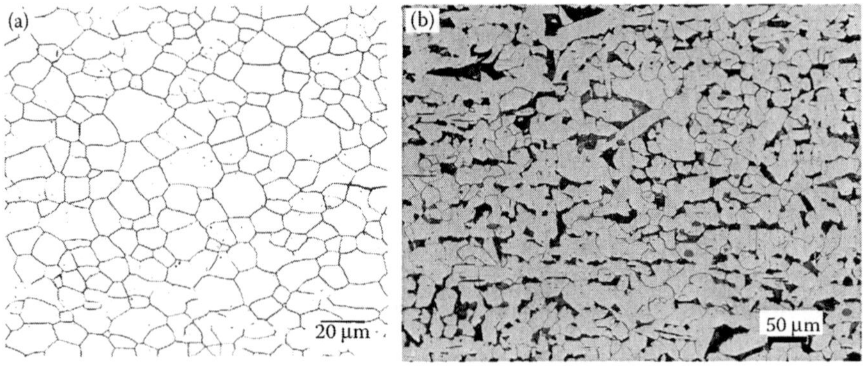

FIGURE 3.12 Photomicrographs of (a) Fe–0.02 wt% C and (b) Fe–0.2 wt% C steel specimens. While the Fe–0.02 wt% C sample is a single-phase material, the Fe–0.2 wt% C steel is a two-phase material. (Photos courtesy of ASM International, Materials Park, OH. With permission.)

(Figure 3.12a) is a single-phase material and so we will see only equiaxed grains of iron throughout the microstructure. The dark lines appearing in the microstructure are the grain boundaries. Figure 3.12b, on the other hand, shows that the microstructure now consists of equiaxed grains of one phase and another (dark-etching) feature present along the grain boundaries of the major phase. That is, the specimen now contains two phases. Since the specimen is Fe–0.2 wt% C steel, it now contains ferrite (as the matrix) and pearlite (the dark-etching feature). (We will see later in Chapter 6 that pearlite is not a single phase, but a mixture of two phases, namely, ferrite and cementite.)

When a second phase is present in small quantities, it can be present as small particles along the grain boundaries of the major (or matrix) phase. Alternately, it can also be present dispersed inside the grains of the matrix phase. Occasionally, the two phases may also be uniformly distributed throughout the matrix, especially if the volume fraction of both the phases is comparable.

3.5.2 Grain Shape

By observing the microstructure, it should be easy to determine the shape of the grains. The grains in a well-annealed material are equiaxed. That is, if one starts from the center of the grain and moves in different directions, one will be able to reach the grain boundary by traveling approximately the same distance in any direction. This does not, of course, mean that the grains are spherical in shape. Grains tend to take up a shape so as to fill the space effectively and completely and spheres cannot do this. The three-dimensional shape of grains is very complex, but a simple shape that fulfills the space-filling criterion is a tetrakaidecahedron (with 14 faces, 36 edges, and 24 corners). It is also interesting to note that the surface area of a tetrakaidecahedron is very close to that of a sphere. The ratio of their surface areas is actually only 1.099. (The sphere has the lower of the surface areas.)

Remember that we are observing only a two-dimensional section of a three-dimensional specimen when we are observing the microstructure in the microscope or observing a recorded photograph. Therefore, even if we notice that the grains appear equiaxed in two dimensions, it is possible that the length in the third dimension is quite different. Thus, the grains can vary through a very wide range of sizes.

The shape of the grains is dependent on whether the material is in the cast or annealed condition. The shape of the grains also changes depending on the mechanical and thermal history experienced by the specimen. For example, if the specimen is observed in the as-cast condition, one observes three clear microstructural features. These are (1) the chill zone at places where the hot molten metal came in direct contact with the cold mold wall surface, (2) the columnar zone where the crystals grow longitudinally, and (3) the central equiaxed zone, where the grains are equiaxed, but the grain size is larger than in the chill zone. One can also see clear dendritic (tree-like) structures in cast materials.

If a specimen is cold worked to a very small extent, slip lines may be observed inside the grains. These could be along one direction or multiple directions depending on the number of active slip systems in the crystals. If only one slip system is active, then the slip lines will be along one direction; if more slip systems are active, the slip lines will be along multiple directions. Since different grains have different orientations, the direction in which the slip lines form is different in different grains. If a specimen is only slightly cold worked, one can note the presence of slip lines inside the equiaxed grains. On the other hand, if the specimen is cold worked to a larger extent, then the grain shape will significantly change. The grains will get elongated in the direction of cold working. These will then be referred to as elongated grains. Figure 3.13 shows the microstructure showing the presence of heavy slip deformation and elongated grains in a cold-worked Cu–30 wt% Zn (α-brass) specimen. One could also observe internal structure inside these elongated grains. But it is so complex that optical microscopy will not be of much use in discerning the details; a TEM is required for such investigations.

FIGURE 3.13 Microstructure showing the presence of heavy slip deformation and elongated grains in a cold-worked Cu–30 wt% Zn specimen. (Photo courtesy of ASM International, Materials Park, OH. With permission.)

3.5.3 Grain Size

The *grain size* in a single-phase material (pure metal or solid solution alloy) is a very important microstructural parameter, especially in determining the mechanical behavior of a material. Roughly speaking, one can say that the largest distance from one point on the grain boundary to the other end along a straight line through the grain center may be considered as the grain size. The grain size is expected to be the same throughout the microstructure. However, it is a common observation that the grain size varies slightly from one area to another. This is because we are observing only a section of the three-dimensional aggregate, and therefore, depending on the portion at which the section is considered, the grain sizes can be different. The grain size also depends on the prior history of the specimen. For example, the grain size will be very small if the specimen has gone through a process of recrystallization. On the other hand, the grain size can be very large if the recrystallized material is annealed at a high temperature when grain growth occurs. The grain size may also appear different when the grains are actually of different sizes or if the size of the grains is not the same in different directions/orientations. The latter becomes true when the grains are not equiaxed. Therefore, it is common practice to designate an *average grain size* of a material.

The average grain size of a material can be determined in many different ways, but the *linear intercept method* is perhaps the most common and easiest way of measuring the grain size. In this method, one draws straight lines, measuring say about 10 cm, in different random directions on the micrograph. One can then count the number of grain boundaries that these straight lines intersect. If a grain boundary is tangent to the line, then it is counted as 1/2 of an intersection. If a triple point (the point where three grains meet) is intersected, it is counted as 2. By knowing the magnification at which the micrograph was recorded, one can then determine the grain intercept (or grain size). Either this process can be used on the micrographs or one can directly use the measuring eyepiece in the microscope to get the count. However, it is more convenient to use a photomicrograph.

By repeating this measurement a few times and in different directions, one will be able to determine the average (or mean) grain intercept. This intercept is a measure of the grain size. Strictly speaking, there is no actual geometric relationship between the grain intercept and the average grain diameter (or size). However, the intercept measured as above is widely used to represent the grain size of a material.

Example Problem 3.2

Assume that 10-cm-long straight lines have been drawn on a micrograph, recorded at a magnification of 250×. If the average number of grain boundary intersections was counted as 8, determine the mean grain intercept.

Solution 3.2

Since the number of grain boundary intersections in a 10-cm-long straight line was counted as 8, the mean grain intercept is 10 cm/8 = 1.25 cm. But remember that the micrograph was recorded at a magnification of 250×. Therefore, the actual intercept is 1.25 cm/250 = 0.005 cm or 0.005 cm × 10^4 micrometers = 50 micrometers.

Note that micrometers (also referred to as microns in some cases) is abbreviated as μm. Therefore, the mean grain intercept in this case is 50 μm.

Another method commonly used to determine the grain size of materials is the Jeffries' *planimetric method*. In this method, a circle of 79.8 mm diameter (equivalent to a circular area of 5000 mm^2) is drawn on a photomicrograph or used as a template on a projection screen. The number of grains within that area is counted. The number of grains that lie completely within the circle, n_i, and the number of grains intersecting the circle, n_c, are counted. Then the equivalent number of whole grains is obtained from the relationship:

$$n_{eq} = n_i + n_c/2 \tag{3.6}$$

The number of grains per unit area (N_A) is then determined by dividing n_{eq} by the actual area of the specimen investigated, A, that is,

$$N_A = n_{eq}/A \tag{3.7}$$

The value of A is calculated by knowing the magnification M at which the photomicrograph is recorded, using the relationship

$$A = \text{observed area}/M^2 \tag{3.8}$$

The average grain area, $\bar{A}$ in mm^2, is then determined from the relationship

$$\bar{A} = \frac{1}{N_A} \tag{3.9}$$

and the average grain diameter, $\bar{d}$, as

$$\bar{d} = (\bar{A})^{1/2} = \frac{1}{N_A^{1/2}} \tag{3.10}$$

The average grain size of a material may also be expressed in the American Society for Testing and Materials (ASTM) system as

$$n = 2^{N-1} \tag{3.11}$$

where n is the number of grains per square inch at a magnification of 100× and N is the ASTM grain size number. Thus, the number of grains will increase (or the grain size decreases) as the ASTM grain size number increases. In this method, the photomicrograph of a polycrystalline specimen recorded at a magnification of 100× is compared with a set of standard micrographs or graded standard grain-size charts. The charts are indexed for the ASTM grain size numbers ranging from 1 to 8. By trial and error a match is found and the grain size is then appropriately designated

TABLE 3.2
Relationship between ASTM Grain Size Number, Mean Number of Grains per Square Inch, and Actual Grain Size

ASTM Grain Size Number	Number of Grains per Square Inch at 100×		Number of Grains per mm^2 at 1×	Calculated Diameter of Equivalent Spherical Grain	
	Mean	Range		Inches	μm
1	1	0.75–1.5	15.5	0.01	254
2	2	1.5–3.0	31	0.00707	180
3	4	3–6	62	0.005	127
4	8	6–12	124	0.00354	89.8
5	16	12–24	248	0.0025	64
6	32	24–48	496	0.00177	45
7	64	48–96	992	0.00125	32
8	128	96–192	1984	0.000884	22.4
9	256	192–384	3970	0.000625	15.9
10	512	384–768	7940	0.000442	11.2
11	1024	768–1536	15,880	0.000313	7.94
12	2048	1536–3072	31,760	0.000221	5.61

by a number corresponding to the index number of the chart. Table 3.2 gives the relationship between ASTM grain size number, N, mean number of grains per square inch, n, and the actual existing grain size. Sometimes, the actual grains in real crystals are either coarser than ASTM grain size no. 1 or finer than ASTM grain size no. 8. One could then determine the ASTM grain size number, N, using Equation 3.11.

It is not possible that we always observe or record the microstructures at a magnification of 100×; we may be observing/recording at other magnifications also. If the microstructure is at a magnification other than 100×, then the number of grains at 100× can be calculated using the simple relationship

$$\text{Number of grains at } 100\times = \left(\frac{M}{100}\right)^2 \times \text{number of grains at magnification } M \quad (3.12)$$

Note that since we are calculating the number of grains per square inch, we need to square the ratio of the magnifications.

The ASTM grain size scale was established using the English system units. Multiplication of n by 15.5 gives the number of grains per square millimeter at 1×.

Example Problem 3.3

A photomicrograph recorded from a metal specimen shows that there are 32 grains per square inch at a magnification of 100×. What is the ASTM grain size number, N?

Solution 3.3

The equation to be used is 3.11, namely, $n = 2^{N-1}$.

The data given are $n = 32$, and we are asked to find out N. Thus,

$$32 \text{ grains/in.}^2 = 2^{N-1}$$

Taking logarithms on both sides, we get

$$\log 32 = (N - 1) \times \log 2$$

or

$$1.505 = (N - 1) \times 0.3010$$
$$\text{or } N = 6.$$

Example Problem 3.4

A photomicrograph recorded at a magnification of 100× shows that there are 25 grains in a circular area of 2 in. diameter. Calculate the ASTM grain size number.

Solution 3.4

Since the diameter of the circle is 2 in., its area is $\pi \times (2 \text{ in.}/2)^2 = 3.14 \text{ in.}^2$. Since 25 grains were counted in this area, the density of the grains is

$$25 \text{ grains}/3.14 \text{ in.}^2 = 7.96 \text{ grains/in.}^2, \text{ say } 8 \text{ grains/in.}^2$$

Using Equation 3.11,

$$n = 2^{N-1}$$
$$8 \text{ grains/in.}^2 = 2^{N-1}$$

Taking logarithms on both sides,

$$\log 8 = (N - 1) \times \log 2$$

or

$$0.9030 = (N - 1) \times 0.3010$$
$$\text{or } N = 4$$

Example Problem 3.5

A photomicrograph recorded from a metal specimen shows that there are 32 grains per square inch at a magnification of 200×. What is the ASTM grain size number, N?

Solution 3.5

Since the micrograph is now recorded at a magnification of 200×, we need to calculate the number of grains per square inch at a magnification of 100× using Equation 3.12.

Thus, n, number of grains at 100x = $(200/100)^2 \times 32\,\text{grains/in.}^2 = 128\,\text{grains/in.}^2$ Using Equation 3.11, $n = 2^{N-1}$, we have

$$128\ \text{grains/in.}^2 = 2^{N-1}$$

Taking logarithms on both sides, $\log 128 = (N - 1) \times \log 2$

$$2.107 = (N - 1) \times 0.3010$$

or

$$N = 8$$

3.5.4 Volume Fraction of Phases

The properties of materials are determined to a large extent by the grain size in the case of single-phase materials and by the type, size, and volume fraction of phases in the case of two-phase or multiphase materials. Therefore, it becomes important to determine the volume fraction of phases in a multiphase material. The determination of grain sizes, areas or volumes occupied by the different phases, and so on comes under the branch of "quantitative metallography."

The volume fraction of phases can be determined using *point counting*, *lineal analysis*, or *areal analysis* methods. All these are related to each other and one is expected to obtain the same result using any of these three methods. Among these, the point counting method is perhaps the most efficient; that is, it gives the best accuracy with minimum effort. Let us now look at the point counting method.

In the *point counting method*, a clear plastic grid with a large number of systematically spaced points is placed on the photomicrograph, or on the projection screen, or inserted as an eyepiece reticule. In real practice, however, one uses a grid and the points of intersection of the arms are taken as points. The number of points (of intersection) lying on the phase or microconstituent of interest is counted and divided by the total number of grid points to obtain the point fraction. Points lying on the boundary are counted as half-points. This measurement is repeated on a number of regions in the microstructure. The point fraction, P_P, is then given by

$$P_P = \frac{P_\alpha}{P_T} \tag{3.13}$$

where P_α is the number of grid points lying inside the feature of interest, α plus one-half the number of grid points lying on particle boundaries, and P_T is the total number of grid points. Studies have shown that the point fraction, P_P, the area fraction, A_A, and the volume fraction, V_V, are all equal to each other. That is,

$$P_P = A_A = V_V \tag{3.14}$$

There are a few important points worth remembering when one is doing such an analysis. First, point counting is always performed on the minor phase. The fraction of the major phase is then obtained as the difference between 1 and the point fraction of the minor phase. Second, the number of points on the grid should be increased as the volume fraction of the feature of interest decreases. It is suggested that the optimum number of points is $3/V_V$. Therefore, the optimum number of grid points will be 6 if the volume fraction of the phase of interest is 0.5 (i.e., 50%). On the other hand, if V_V decreases to 0.2 (20%), the optimum number of grid points is 15 and this number increases to 30 for 0.1 (10%) and 300 for 0.01 (1%). Third, the fraction obtained needs to be corrected to obtain the weight fraction. The weight fraction will be obtained by multiplying the volume fraction by the density of the phase.

We have described in the above paragraphs the basic principles involved in determining the different microstructural parameters. However, computer programs are available nowadays to calculate the grain size, amounts of different phases present in a multiphase material, and even the orientation of the different grains. These programs reduce the tedium in obtaining the information. However, it is desirable that the beginner does these operations manually to understand the principles and appreciate the physical significance. After gaining enough experience, one could then use the automated programs.

3.5.5 Chemical Composition

The chemical composition of phases cannot be normally obtained from optical microscopy studies. However, we will see later (in Chapter 5) that by observing the microstructural features in plain carbon steels, it is possible to estimate the overall chemical composition of the steel sample. It will be shown that, in plain carbon steels (and this is not true in other types of steels), every 0.1 wt% C (up to the eutectoid composition) will contribute to about 12.5 wt% pearlite in steels. By quantitative microscopy studies, we can determine the volume fraction of the phases (or microconstituents) in the steel sample. That is, one can estimate the proportion of the ferrite and pearlite microconstituents from optical micrographs of hypoeutectoid steels. Therefore, by knowing the volume fraction of pearlite present in the steel sample and realizing that 12.5 wt% pearlite amounts to 0.1 wt% C, one should be able to estimate the overall composition of the steel.

While trying to estimate the overall composition of the plain carbon steel samples from quantitative microscopy studies, it is important to remember that what we obtain from the lever rule is the weight fraction (or percentage) of phases once we know the overall composition and the chemical compositions of the individual phases. However, in the case of quantitative microscopy studies, we obtain the volume fraction of the phases. The weight fraction of the phase is obtained from the volume fraction by multiplying the volume fraction by the density of the phase of interest.

Example Problem 3.6

By performing quantitative microscopy studies on a plain carbon steel sample, it was determined that the steel contains about 40% pearlite and 60% ferrite. Estimate the overall composition of the steel sample.

Solution 3.6

It was mentioned that the steel contained 40% pearlite by volume. Recall that quantitative microscopy studies yield the volume percentage of the phases. The density of ferrite can be taken as that of pure iron, namely, 7.874 g cm^{-3}. The density of pearlite microconstituent is taken as 7.837 g cm^{-3}. Therefore, the total weight of the alloy per unit volume is

$$0.4 \times 7.837\,g\,cm^{-3}\text{ of pearlite} + 0.6 \times 7.874\,g\,cm^{-3}\text{ of ferrite}$$

$$= 3.1348 + 4.7244 = 7.8592\,g$$

Hence, the weight fraction of pearlite in the alloy is 3.1348/7.8592 = 0.3988. That is, the sample contains nearly 40 wt% of pearlite and 60 wt% of ferrite. Note that in this case the volume and weight percentages are not different because the densities of both the microconstituents are almost the same. This may not be true in all cases.

Since it is known that the steel sample will contain about 12.5 wt% pearlite for every 0.1 wt% C, the overall composition of the steel can be estimated from the amount of pearlite present in the microstructure. Therefore, the amount of carbon present in the sample is

$$\frac{40\%\text{ pearlite}}{12.5\%\text{ pearlite}} \times 0.1 = 0.33\%$$

Since it was mentioned that it was a plain carbon steel (only iron and carbon are present in a plain carbon steel), the overall composition of the steel sample is 0.33 wt% carbon and the rest (99.67 wt%) iron.

3.6 SPECIMEN PREPARATION FOR MICROSCOPIC EXAMINATION

It was mentioned earlier that the specimen needs to be properly prepared so that it will display the true microstructural features. For achieving this purpose, the prepared specimen should have the following characteristics:

1. It must be flat, free from scratches, stains, and other imperfections that tend to alter or distort the specimen surface.
2. It should contain all the nonmetallic inclusions intact.
3. It should not show any chipping or galling of hard and brittle intermetallic compounds.
4. It should be free from all traces of disturbed metal.

Specimens with such surfaces are prepared in different stages of grinding, polishing, and etching. Improper preparation of samples is likely to remove all-important inclusions, erode grain boundaries, temper hardened steel specimens, or remove the case-hardened layers in surface-treated specimens, ultimately producing a structure that, superficially at least, will appear microstructurally different from that which is truly representative and characteristic of the material.

3.6.1 Sectioning and Mounting

3.6.1.1 Sectioning

Generally speaking, a specimen for microscopic examination has the physical dimensions of 3/4 to 1 in. diameter by 1/2 in. height, even though this is not always necessary. However, the more important point is that it should be possible to comfortably hold the specimen between your fingers and prepare it for microscopic examination. It should also be realized that not all materials are provided in the required size and shape. Therefore, a specimen of suitable dimensions has to be sectioned from the bulk of the material. Sectioning is the removal of a representative area from the parent piece. The sectioned specimen should be representative of the parent material and the microstructure of the sectioned specimen should not be altered during the process of sectioning, since it is known that the microstructure can be altered either due to the deformation involved or due to the heat generated. In other words, the microstructure of the sectioned specimen and the original parent material should be the same.

A representative specimen may be obtained by either manual or power hack sawing. If the specimen is brittle, it could also be fractured by a hammer. But sawing, shearing, or hammering is not the preferred operation to obtain a representative section of the specimen since these operations produce deformation in the specimen. Therefore, high-speed abrasive cutting offers the best solution to either completely eliminate or minimize the damage caused by deformation and heat. In this process, a bonded abrasive wheel is used. The abrasive material used is either alumina (Al_2O_3) or silicon carbide (SiC), and finer grits result in a softer action and a smoother surface. Bonding is required in order to hold the abrasive grains together in place. Since heat is generated during cutting, sufficient and proper cooling is necessary. Some of the parameters to be optimized to achieve a good sectioning are the type and size of the grit, type and hardness of the bond, wheel speed, and cooling. Low-speed and low-deformation precision sectioning is also occasionally used to obtain small sections of metallic and nonmetallic materials whether they are soft, hard, brittle, ductile, porous, simple, or complex configurations.

3.6.1.2 Mounting

It is also possible that sometimes we come across samples that are not regular, awkwardly shaped, or too tiny to handle with the fingers. Such specimens include chips, wires, small rods, tubing, sheet metal specimens, thin sections, electronic components, and so on. Therefore, such samples are mounted primarily for ease in manipulation and edge preservation. This is most commonly achieved by compression molding techniques.

Mounting can be done by either compression or cold. The *compression mounting* process is used to produce hard mounts in a minimum amount of time. In these methods, the specimen of interest is placed inside a mold with the surface of interest facing down. It is then covered by the bonding material and compressed using pressure. If necessary the whole assembly is heated so that there is good bonding between the specimen and the bonding material. Once the operation is completed, the specimen is ejected. Now the top surface of the specimen can be subjected to the usual procedures of grinding, polishing, and etching.

The mold material may be thermosetting or thermoplastic. Thermosetting media require heat and pressure during the molding cycle and can be ejected at maximum molding temperature. Thermoplastic materials, on the other hand, remain fluid at maximum molding temperatures and become dense and transparent with a decrease in temperature and an increase in pressure. If a specimen cannot withstand heat and/or pressure in compression molding (because the specimen is delicate), cold mounting techniques are preferred. Some of the common molding materials used are Bakelite, Epoxy, and Lucite. This step is not necessary if the specimen is large enough in size and convenient to handle.

3.6.2 Grinding

Grinding can be subdivided into two substages—rough (or coarse) grinding and fine grinding. The main purpose of *rough grinding* is to remove the deformation produced during sectioning of the specimen and to provide an initially flat surface. If the specimen surface is very rough and not flat, then rough grinding can produce a reasonably flat surface by removing gross amounts of surface material. Further, the as-received or sectioned specimen may have sharp edges and scratches. Any burrs on the edges of the specimen will also be removed during this stage. This can be achieved with the help of a file or a motor-driven rotating wheel covered with abrasive belts or discs. During this operation, water is used as a coolant to keep the specimen cool by preventing overheating and to flush away the surface removal products. This helps in keeping the sharp abrasive particles from directly removing the material from the surface of the specimen. Abrasive belts and discs are available in SiC and Al_2O_3 with resin bond for wet or dry operation. The grit size is 50–180; that is, the abrasive particles are in the range of 350–190 μm.

This rough grinding stage may not be necessary in case the sample surface is reasonably smooth and also when we did not use the sectioning stage to obtain a representative sample.

The representative sample (after rough grinding and/or mounting) is then subjected to *fine grinding*. The main purpose of fine grinding is to remove the zone of deformation caused by sectioning and rough grinding and also to remove the scratches from the surface of the specimen. The fine grinding process will limit the depth of deformation on the surface. The abrasives used for fine grinding are SiC, emery, and Al_2O_3. These abrasive grains are bonded to a paper or cloth backing.

In the process of fine grinding, the specimen containing scratches and the heavily deformed layer introduced during the rough grinding stage (which can be up to 10–50 times the depth of scratch produced) is gently rubbed on emery papers of sequentially fine grit size. Initially, a coarse grade paper is used. Care should be taken to ensure that the specimen is rubbed along the length of the paper and only in one direction. Further, the whole specimen surface should touch the emery paper; otherwise, the specimen surface would develop bevels. A common way of fine grinding is to place the emery paper firmly on a smooth and hard surface (e.g., a glass sheet) and gently rub the surface of the specimen in one direction. It may not be a good idea to rub the specimen back and forth. (An experienced user, who can hold the specimen firmly without tilting to one side, may rub the specimen in both the forward and the backward directions.) Therefore, the specimen is first rubbed from a point closest to the operator and in a direction away from the operator. By the time the specimen reaches the end of the emery paper, the specimen is lifted and brought back to the starting point and this sequence is repeated. The process is carried out normally wet. Among the different abrasive particles, SiC has a greater removal rate and can also be used with a lubricant. The process is continued until the scratches due to that particular emery paper are all along one direction on the specimen surface. Once this is achieved, the specimen surface is washed in running water. The process is then repeated on the next finer emery paper in such a way that the new set of parallel scratches is normal to the scratches introduced by the previous (coarser) emery paper. This helps in removing the scratches introduced by the previous emery paper. This process is repeated till the finest emery paper has been used. Remember to rotate the specimen through 90° in moving from one grade of the emery paper to the other and to wash the specimen surface under running water.

In most laboratories, four stations are provided with emery papers containing abrasive particles of grit size 240, 320, 400, and 600. These correspond to particle sizes of 54, 29, 23, and 17 μm, respectively. Occasionally grit no. 120 is also used. Remember to go from a larger particle size to a smaller particle size, that is, from grit size 240 to 320 to 400 and then to 600. Table 3.3 lists the grit numbers and the corresponding particle sizes. Sometimes, you will use emery abrasive and, in that case, the designation is different. Table 3.4 shows the relationship between the grit number and particle size in that case.

3.6.3 Polishing

The main purpose of *polishing* is to remove the deformation zone produced during the earlier operations and to produce a mirror-like surface finish on the specimen. The polishing stage may be considered the most important in the entire sequence of preparing specimens for metallographic examination. This stage can also be subdivided into *rough polishing* and *final polishing.*

During the rough polishing stage, the deformation zone caused by fine grinding is further minimized. And during the final polishing stage, the deformation zone produced by rough polishing is almost completely removed. Any deformation zone produced during the final polishing stage is minimal and this will be removed during the process of etching. Abrasives like diamond, alumina, or ceria are used to remove/minimize the deformation produced in the earlier operations.

The abrasives used should have high hardness, inertness, and low coefficient of friction. Diamond is considered the most effective since these particles retain their shape and size during abrasion and produce a uniform and high rate of material removal with minimal damage induced on the surface. The abrasive particles, usually

TABLE 3.3
Relationship between Grit Numbers and Particle Sizes for Silicon Carbide, Aluminum Oxide, and Garnet

	Particle Size (μm)			
	The United States	FEPA[a]		
Grit No.	Sieve Series	Commercial	F-Series	P-Series
50	350	324		
60	270	274		
80	190	194		
120	115	115		
150	80	97		
180	70	81		
220	62	58		
240	54	49	49.3	58.5
280	35	41	41.5	52.2
320	29	34	34.4	46.2
400	23	29	23.0	35.0
600	17	20	14.3	22.8

[a] Federation Europeane des Fabricants des Produits Abrasifs.

up to 10 μm in size, are suspended in different types of media, usually water. The suspension medium contributes to both lubrication and removal rate of material.

Polishing is carried out by holding the specimen against a horizontal rotating disc. The disc is covered with a velvet cloth, which is covered by the solution containing the suspension of the abrasive particles. It is a common practice to have three or four different stations of polishing, each with a wheel that is covered with abrasive particles of different sizes. Generally, the sizes of the abrasive particles are 5, 1, and 0.5 μm. During polishing, the velvet cloth is periodically supplied with the abrasive

TABLE 3.4
Relationship between Grit Number and Emery Particle Size

Grit No.	Particle Size (μm)
3	85
2	70
1	50
1/0	33
2/0	30
3/0	28
4/0	25

particles in the form of a suspension, and the specimen is rotated around its own axis. The polishing wheel normally rotates in the counterclockwise direction and the specimen should be rotated in the clockwise direction. Maximum feasible pressure should be applied to produce the maximum rate of material removal. Just like in the grinding stage, it is a good practice to wash the specimen surface under running water, when moving from one polishing station to another. Again, you should move from one station that uses coarse particles to the next that uses finer particles. After polishing, the specimen should be quickly dried to prevent staining or corrosion. After rinsing, the specimen is flooded with high-quality alcohol and dried in a stream of warm dry air.

Automatic polishing machines are available nowadays and these take away the tedium of manually preparing the specimens. However, manual polishing has always a better control and the operator could decide on the extent of polishing required.

The preparation of specimens by the mechanical methods described above is tedious and painstaking, especially in the preparation of soft metals. In these cases, a good specimen may not be obtained owing to the inevitable formation of disturbed metal on the surface of the specimen during polishing. *Electrolytic polishing* is an important alternative to mechanical polishing, since electrolytic polishing precludes the formation of additional disturbed material on the ground surface of a specimen. The main disadvantage associated with electrolytic polishing is the complete or partial removal of nonmetallic inclusions by chemical reaction between such inclusions and the electrolyte.

At the end of the polishing stage, all the scratches on the surface of the sample should disappear and the specimen surface should have a mirror-like surface. You should be able to see your reflection in a well-polished specimen as though you are looking into a mirror!

Figure 3.14 shows a series of photomicrographs depicting the surface condition of a 1040 steel (plain carbon steel containing 0.4 wt% C) specimen. Note that immediately after rough grinding to make the specimen surface flat, one observes a series of parallel scratches introduced on the surface (Figure 3.14a). On polishing with successively finer diamond grit, both the width and the depth of scratches are reduced. At the end of the final polishing stage, the specimen surface is completely devoid of all scratches (Figure 3.14d).

3.6.4 Etching

The purpose of *etching* is to make visible the many structural features of the material. However, the presence of scratches and other surface defects, nonmetallic inclusions, graphite flakes in cast irons, and so on in materials can be seen without etching on a polished surface. The *etching reagents* (also known as *etchants*), suitable for etching metallographic specimens, are composed of organic or inorganic acids, alkalies of various kinds, and other complex substances in solution with an appropriate solvent such as water, alcohol, glycerin, glycol, or mixtures of these solvents.

It is the acid present in the etching solution that reacts with the microconstituents of the structure and produces different contrasts. Contrast is the difference in visual

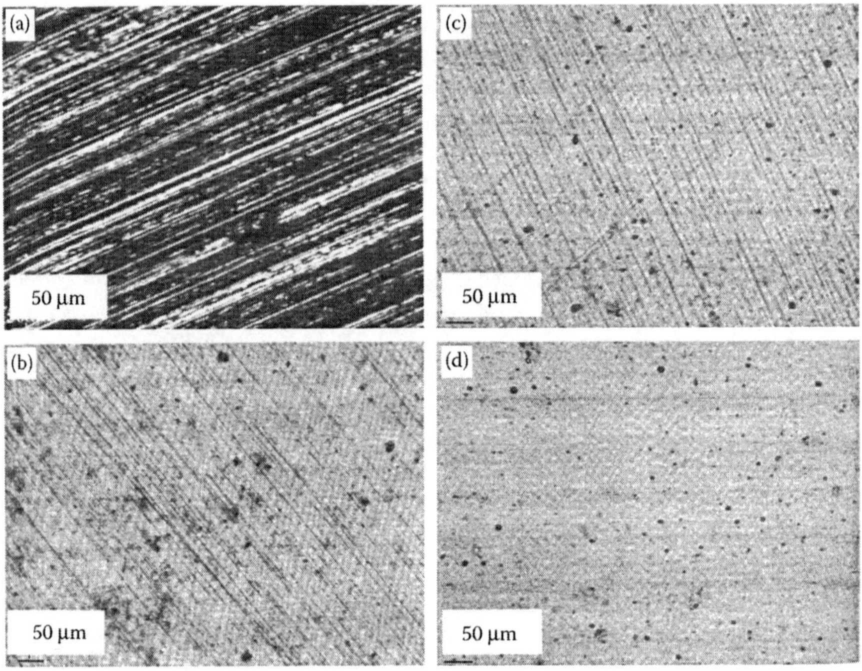

FIGURE 3.14 Optical micrographs of 1040 steel after grinding and polishing: (a) after rough grinding to achieve a planar surface, (b) after polishing with a 6 μm diamond grit, (c) after polishing with a 1 μm diamond grit, and (d) after polishing with a 0.25 μm diamond grit. (After Brandon, D. and W. D. Kaplan, 1999. *Microstructural Characterization of Materials*, Chichester, UK: John Wiley & Sons, Inc. With permission.)

properties that makes a part of the microstructure distinguishable from other parts and the background. However, for easy handling and also to control the reaction rate, the acid is diluted by adding water, alcohol, or any other solvent. Remember that you need to pour acid into water and not vice versa! The nature and extent of attack will be different depending on the reactivity of the phase or microconstituent. We had seen earlier that the grain boundaries get attacked to a greater extent than the grains. Similarly, different phases present in the samples get attacked to different levels depending on their reactivity.

The process of etching is carried out as follows. The etching reagent is prepared and kept in a small porcelain or petri dish. The specimen should be immersed in the etchant solution with the polished surface touching the etchant solution. Care should be taken in handling the etchant solution since this is corrosive. A good practice is to hold the specimen with a pair of tongs and dip the specimen in the etching solution. The specimen surface should be carefully examined from time to time and the specimen is removed from the etchant when the specimen surface (which had a mirror-like finish before etching) becomes slightly dull. At this stage, one will be able to see the grain structure of the specimen with the unaided eye, although this requires some practice. You should also make sure that the specimen surface does not become colored! After etching, the specimen should be thoroughly washed under running water and dried with alcohol. Alternately, a cotton swab could be taken, dipped in the etchant solution and applied to the polished specimen surface.

Both the time for which the specimen is etched and the temperature at which the etching operation is carried out are important. The etching action becomes stronger if etching is carried out at higher temperatures and/or for longer times. Etching also becomes stronger if the concentration of the acid in the etchant is higher. A majority of the etchants are used at room temperature and, therefore, the *time of etching* needs to be controlled. Depending on the metal to be etched and the etching reagent selected, the time of etching may vary from a few seconds to 30 min or longer. Many etchants have compositions such that the time required for proper etching is long; this enables a better control of the etching process. If the etching time is too short, then enough reaction may not take place and therefore the microstructural details are not seen clearly. If the specimen is etched for too long, or with very highly reactive etchants, then the microstructural features may be obliterated and the specimen surface is colored. Additionally, the specimen may develop *etch pits* (deep holes in the specimen surface that mar the microstructural appearance).

The etching time is also dependent on the magnification at which the microstructure is to be observed. Etching for a time to reveal maximum details at high magnification is much too short a time to reveal fully the microstructural details at low magnifications. If the microstructure is to be photographed, the etching time should be slightly less than that required to produce optimum contrast for visual examination. It is a good practice to etch for a slightly shorter time, observe the microstructure, and then increase the etching time, if necessary. On the other hand, if the specimen is overetched, one may have to repolish the specimen and etch the sample again for a shorter time.

Electrolytic etching is particularly useful to reveal the microstructural features of materials such as thermocouple materials, severely cold-worked metals, heat- and corrosion-resistant alloys, and alloys that exhibit surface passivity during conventional etching.

The *mechanism of etching* is different depending on whether we are considering a single-phase or a multiphase material. In *multiphase alloys*, the different phases in the material react differently to the etching reagent. The phase with the higher potential is anodic, or electropositive, to the other and consequently dissolves readily during normal etching. Because of preferential dissolution of the anodic phase during etching, this phase is somewhat roughened and depressed from the plane of the polished surface. Consequently, this phase appears dark. In the case of *single-phase alloys* (and pure metals), structural contrast is established and grain boundaries are revealed mainly because of differences in the rate at which the different grains are attacked by the reagent. The different rate of attack is due to orientation differences of the grains with respect to the plane of the polished surface.

We can understand the above effect with reference to Figure 3.15. Note that we are considering contrast arising from three different grains. Because these grains are oriented crystallographically differently, the extent of etching of these grains is also different. Assuming that the middle grain (B) is most affected, the least amount of light will get fully reflected back to the objective lens and so this grain will appear

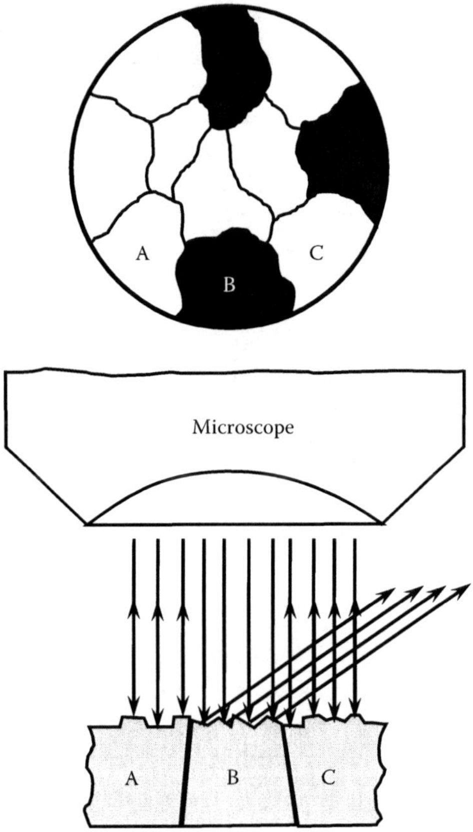

FIGURE 3.15 Mechanism of showing contrast in a single-phase material that has been etched after polishing. The etching characteristics and the contrast differ from grain to grain depending on their crystallographic orientation and the amount of light reflected back into the objective lens from the surfaces of the grains.

dark. On the other hand, the grain on the left (A) is less affected than the middle grain, and the one on the right (C) is least affected. As a result, the amount of light reflected back into the objective lens is maximum. Consequently, the grains A and C will appear brighter than the middle grain.

It is important to remember that the disturbed metal (formed at different stages of specimen preparation) is first removed to reveal the true structural characteristics of the material. This may require alternate etching and polishing several times—generally three cycles. Polishing should be carried out with care, not to introduce additional disturbed metal.

A number of different reagents are available to etch different types of samples. One etchant cannot be used to etch all types of samples. For example, an etchant ideally suited for one alloy may be completely unsuitable for another alloy. On the other hand, different etchants may be used for the same sample. The nature of details and the clarity (resolution) with which they could be seen will be different depending on the type of etchant used. Nital (a mixture of *nit*ric acid and ethyl *al*cohol) and picral (a mixture of *picr*ic acid and ethyl *al*cohol) are the most commonly used etchants to reveal the microstructure of steel samples. The most common concentrations used are 2% nital and 4% picral. While nital is ideally suited to reveal the ferrite grain boundaries, picral seems to be better suited to fully reveal the microstructural features of pearlite. A mixture of ammonium hydroxide and hydrogen peroxide is suitable to etch copper and α-brass samples. Table 3.5 lists some of the more common etching solutions used. The reader should consult standard reference books (*ASM Handbooks* or vendors' manuals) for an exhaustive list of etchants used to reveal the microstructural features of different types of materials.

Figure 3.16 shows a series of three optical micrographs showing the effects of etching time on the microstructure of a 1040 steel sample etched with nital. Note that after etching the sample lightly (etched for a short time, or underetched), the microstructural features are not clear at all (Figure 3.16a). On the other hand, if the specimen is overetched (etched for a long time), all the microstructural features are obliterated and etch pits are also developed (Figure 3.16c). The true microstructural features are revealed only on etching for the right time (Figure 3.16b).

3.7 SOME TYPICAL MICROSTRUCTURES

We will now describe the microstructural features of some common materials. All the microstructures are pretty and esthetically very pleasing to the eye. It should be realized that the variety of microstructures one could get will depend on the material, its composition, and the mechanical and/or thermal treatment the sample has experienced. Here, however, we will describe only a few examples. Even though the microstructure developed will not depend on whether the sample is ferrous or nonferrous, we will classify this section into two categories.

3.7.1 Nonferrous Alloy Samples

Figure 3.17a shows the microstructure of a cast aluminum bronze (Cu–Al alloy). Since the material is in the cast condition, one can clearly see the dendritic (tree-like structure) pattern. Figure 3.17b shows the microstructure of a cast Cu–4.5 wt% P alloy observed in an OM. Since this is a hypoeutectic alloy composition, one is expected to observe the primary phase and the eutectic. Further, since this is a cast alloy, it will show variation in composition (or segregation) from point to point in the specimen. The primary α-phase (solid solution of phosphorus in copper) is present in the form of dendrites in which the copper concentration decreases from the center (where it is maximum) toward the rim of the dendrite. The eutectic microconstituent in between the dendrites contains the α-phase and the Cu_3P intermetallic. The α-phase in the eutectic is segregation free due to the nature of the eutectic transformation. The microstructural features in a specimen of this type can be better viewed using cross-polarized light plus a sensitive tint plate, when the different phases appear in color. Figure 3.17c shows the microstructure of a galvalume coating where again one can see the microstructural features characteristic of a cast structure.

TABLE 3.5
Chemical Etching Reagents for Microscopic Examination of Different Materials

Material	Etchant	Composition	Procedure for Use/Comments
Al and Al alloys	HF etch	1 mL HF (48%), 200 mL H_2O	Swab for 15 s or immerse for 30–45 s
	Keller's reagent	2 mL HF (48%), 3 mL HCl (conc.), 5 mL HNO_3 (conc.), 190 mL H_2O	Immerse for 8–15 s, wash in a stream of warm water, and blow dry. Do not remove etching products from the surface
Carbon steels	Nital	2 mL HNO_3 and 98 mL ethanol or methanol (95% or absolute)	Gives maximum contrast between pearlite and ferrite or cementite network; differentiates ferrite from martensite
	Picral	4 g picric acid, 100 mL ethanol or methanol (95% or absolute), and 4 or 5 drops 17% zephiran chloride (wetting agent)	For all grades of carbon steels (annealed, normalized, quenched, quenched and tempered, spheroidized)
	Vilella's reagent	5 mL HCl, 1 g picric acid, and 100 mL ethanol or methanol (95% or absolute)	For revealing austenitic grain size in quenched and quenched and tempered steels
`Cu and Cu alloys		20 mL NH_4OH, 0–20 mL H_2O, 8–20 mL 3% H_2O_2	Immersion or swabbing for 1 min; preferred for brasses
Mg and Mg alloys	Glycol	1 mL HNO_3 (conc.), 24 mL water, 75 mL ethylene glycol	Swab or immerse the specimen for a few seconds to 1 min; shows general structure
	Acetic–picral	5 mL acetic acid, 6 g picric acid, 10 mL H_2O, 100 mL ethanol (95%)	Immerse the specimen face up with gentle agitation until the face turns brown. Wash in a stream of alcohol, dry with a blast of air. A universal etchant. Defines grain boundaries, reveals cold work and twinning readily
Ni and Ni alloys		1 part HNO_3 (conc.), 1 part acetic acid (glacial). Use fresh solution	For revealing grain boundaries. Immerse or swab the specimen for 5–20 s
	Glyceregia	10 mL HNO_3 (conc.), 20 mL HCl (conc.), 30–40 mL glycerol	Etch by immersing or swabbing the specimen for 30 s to 5 min
Stainless steels	Glyceregia	3 parts glycerol, 2–5 parts HCl, 1 part HNO_3	Popular etch for all stainless grades. Use fresh, never store. Use with care under hood
	Dilute aqua regia	15 mL HCl, 5 mL HNO_3, 100 mL H_2O	Uniform etching of austenite, outlines carbides, σ-phase, and ferrite

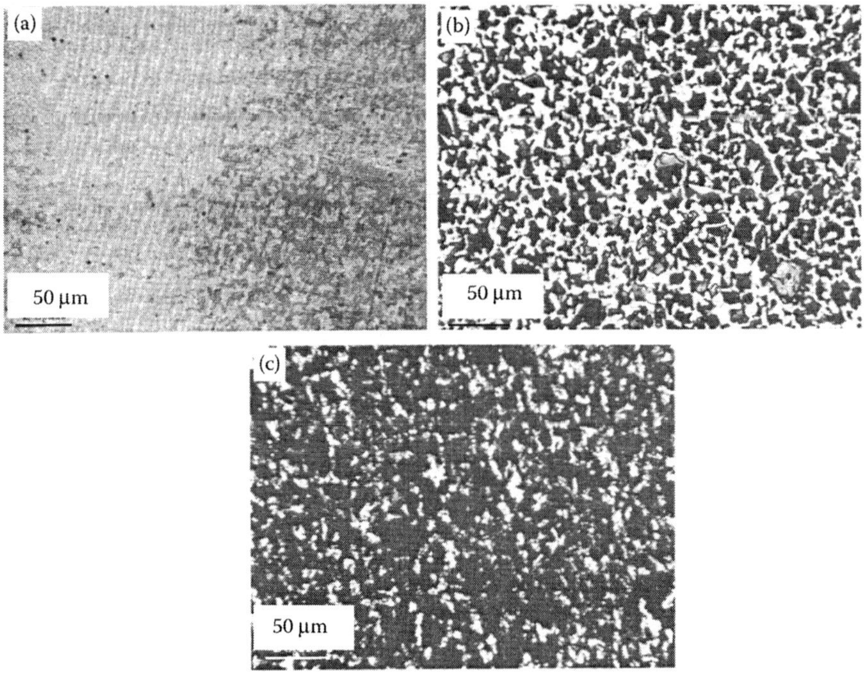

FIGURE 3.16 Effects of etching time on the microstructure of a 1040 steel specimen. The optical micrographs show (a) underetched, (b) properly etched, and (c) overetched specimens. While the microstructural features are not well resolved in the underetched sample, they are completely destroyed in the overetched sample. (After Brandon, D. and W. D. Kaplan, 1999. *Microstructural Characterization of Materials*, Chichester, UK: John Wiley & Sons, Inc. With permission.)

Figure 3.18 shows the microstructure of a wrought (worked at a reasonably high temperature) copper sample. The microstructure of this sample is quite different from that of a cast structure. Instead of the dendritic structure one now sees an equiaxed structure of the grains. During the working of the sample, the cast structure is broken down and because of the relatively high temperature of working the material will go through the stages of recovery, recrystallization (formation of new strain-free grains), and grain growth. That is why reasonably large grains are formed that are equiaxed in shape. Further, since copper has a low stacking fault energy (about 40 mJ m^{-2}), annealing twins (parallel bands running across the full width of the grains) are present inside the grains.

Figure 3.19 presents two micrographs of annealed Cu–30 wt% Zn (α-brass) samples at two different magnifications. The specimens were annealed at a high temperature of 704°C. In both the micrographs, one can clearly see equiaxed grains containing annealing twins. The presence of annealing twins is more clearly brought out in the specimen observed at the higher magnification. It may also be noted that, in comparison with pure copper, the density of annealing twins in α-brass is higher due to the decreased stacking fault energy of α-brass.

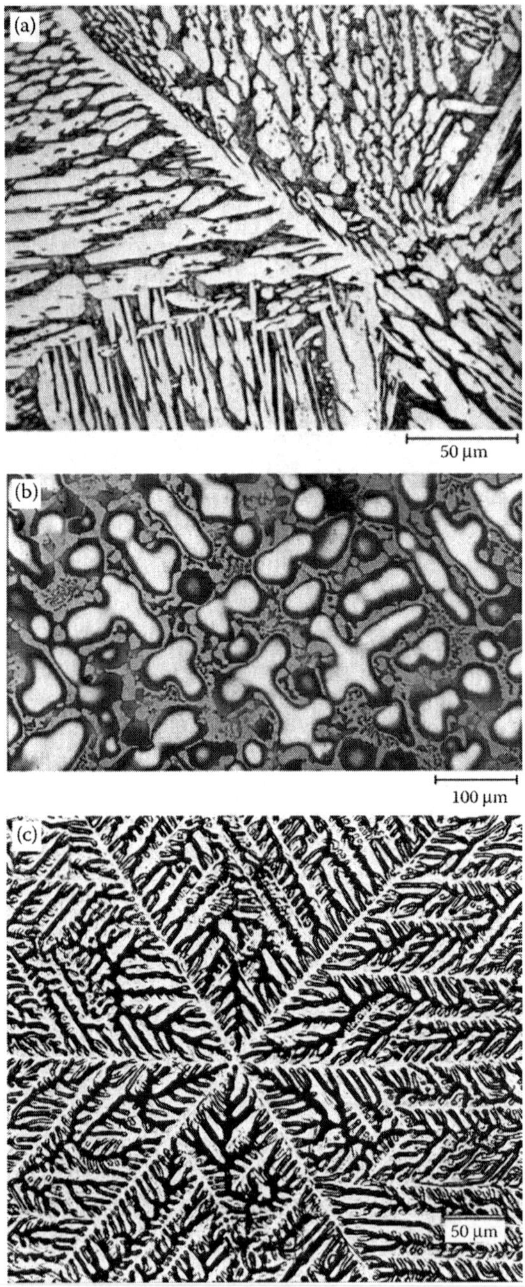

FIGURE 3.17 (a) Photomicrograph of a cast aluminum bronze. (b) Photomicrograph of a cast Cu–4.5 wt% P alloy showing dendrites of the α-phase and the eutectic microconstituent. (c) Photograph of a galvalume coating (coating of zinc on aluminum) clearly showing the dendritic pattern on the surface. (Photos courtesy of ASM International, Materials Park, OH. With permission.)

FIGURE 3.18 Photomicrograph of a wrought copper sample. Note that this microstructure is very different from the cast structures shown in Figure 3.17. The annealed copper sample shows equiaxed grains. Further, since copper has a low stacking-fault energy, annealing twins are present in the grains. (Photo courtesy of ASM International, Materials Park, OH. With permission.)

Figure 3.20 shows the microstructure of pure zinc. Since this is a pure metal (and therefore a single-phase material) one expects to see only grain boundaries in this polycrystalline material. Note that zinc is a soft metal and it has an HCP structure. Since deformation in HCP metals occurs through a twinning process, mechanical twins have been introduced into this zinc sample during sample preparation.

From the above description it is clear that the microstructural features are different as a function of composition, the mechanical or thermal treatment the sample has experienced, and whether it is in a cast or annealed condition.

3.7.2 Ferrous Alloy Samples

We will now describe the microstructural features of some ferrous alloy samples. Figure 3.21 shows the microstructures of steel samples with different carbon contents. The phases present in steels and their microstructures in different equilibrium and nonequilibrium conditions will be explained in much greater detail later in Chapters 5 and 6. When the carbon content in the steel sample is very low (and other interstitial elements are not present) the steel will only contain a single-phase α-ferrite (interstitial solid solution of carbon in α-iron) and the microstructure will appear like that shown in Figure 3.21a. Here one sees only grains and grain boundaries. However, when the carbon content is increased to higher levels, additional phases start to appear. For example, Figure 3.21b shows the microstructure of a plain carbon steel containing 0.4 wt% carbon. At this composition, the steel contains two phases, namely, α-ferrite and cementite (Fe_3C compound). However, part of the α-ferrite and cementite coexists as a single microconstituent referred to as pearlite. This appears as a dark area (if it is not resolved) or as a lamellar structure (alternate layers of ferrite and cementite) if it is resolved. Thus, the microstructure in Figure 3.21b can be

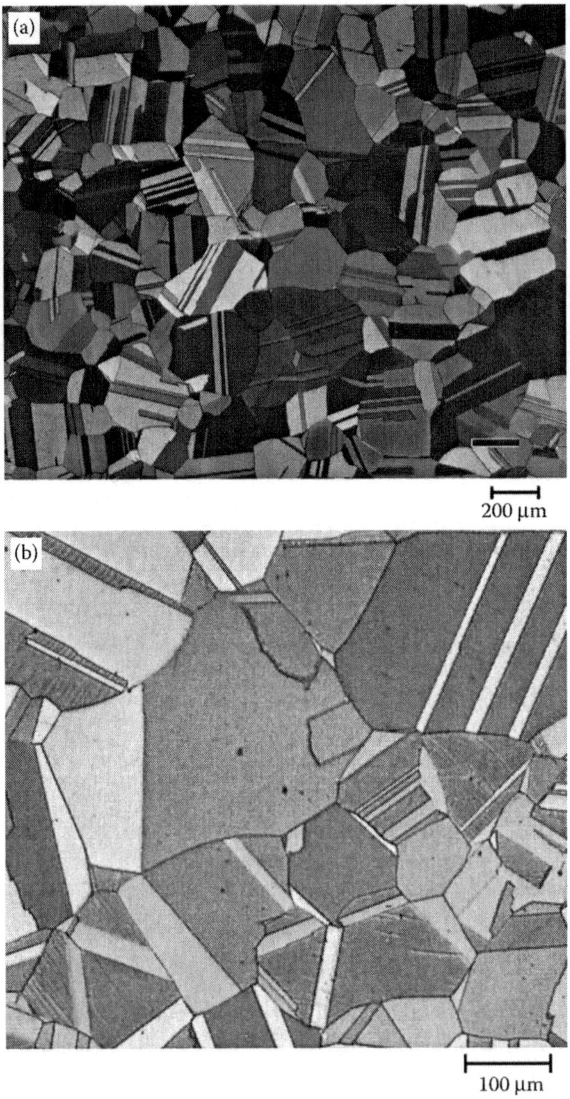

FIGURE 3.19 Photomicrographs of annealed Cu–30 wt% Zn alloy (α-brass) at two different magnifications. (a) Low magnification and (b) high magnification. One can clearly see annealing twins in all the grains. Since the stacking-fault energy of α-brass is lower than that of pure copper, the density of annealing twins is higher in the α-brass samples. (Photos courtesy of ASM International, Materials Park, OH. With permission.)

described as consisting of α-ferrite and pearlite. With further increase in carbon content, the amount of pearlite continues to increase up to the eutectoid composition. Thus, at the eutectoid composition of 0.76 wt% carbon, the microstructure will consist of only pearlite (Figure 3.21c). However, when the carbon content is higher than the eutectoid composition (hypereutectoid steels), the microstructure will consist of

FIGURE 3.20 Photomicrograph of pure metal zinc. The grains are elongated in this specimen. The presence of mechanical twins inside the grains may additionally be noted. (Photo courtesy of ASM International, Materials Park, OH. With permission.)

cementite and pearlite. Figure 3.21d shows the microstructure of a 1.3 wt% C steel showing well-resolved pearlite and the white-etching cementite phase along the boundaries of the pearlite nodules.

When a steel sample is quenched rapidly from the austenitic region, the austenite phase will undergo a diffusionless transformation. The product of this transformation is referred to as martensite. We will see more about this transformation and the properties of martensite in Chapter 6. Figure 3.22 shows the microstructure of martensite obtained by quenching a 1.86 wt% C steel from the austenitic region. The martensite (appearing in the lenticular shape) as seen in this micrograph will be referred to as plate martensite. In addition to the martensite, the microstructure is also showing the presence of ledeburite (a eutectic mixture of austenite and cementite) at the grain junctions (where three grain boundaries meet) and austenite (the white-etching matrix region). The ledeburite microconstituent can be easily identified because of its characteristic appearance of the distribution of fine particles of cementite in the austenite matrix.

Figures 3.23a and b shows the microstructures of two austenitic stainless-steel samples—304 and 310 stainless steels. Austenitic stainless steels have an FCC structure and low stacking-fault energy. Consequently, these steels contain annealing twins. Thus, it may be noted that FCC materials with low stacking-fault energies show the presence of annealing twins in their microstructures.

The second phase in any material need not always be along the grain boundaries. Thus, for example, by suitably heat treating the steel samples at reasonably high temperatures, the cementite phase may become spheroidal in shape (to reduce the total surface area and hence the energy). Figure 3.24 shows the microstructure of an alloy steel (52100 Cr alloy steel) that was heated at 770°C in 10 h, held at that

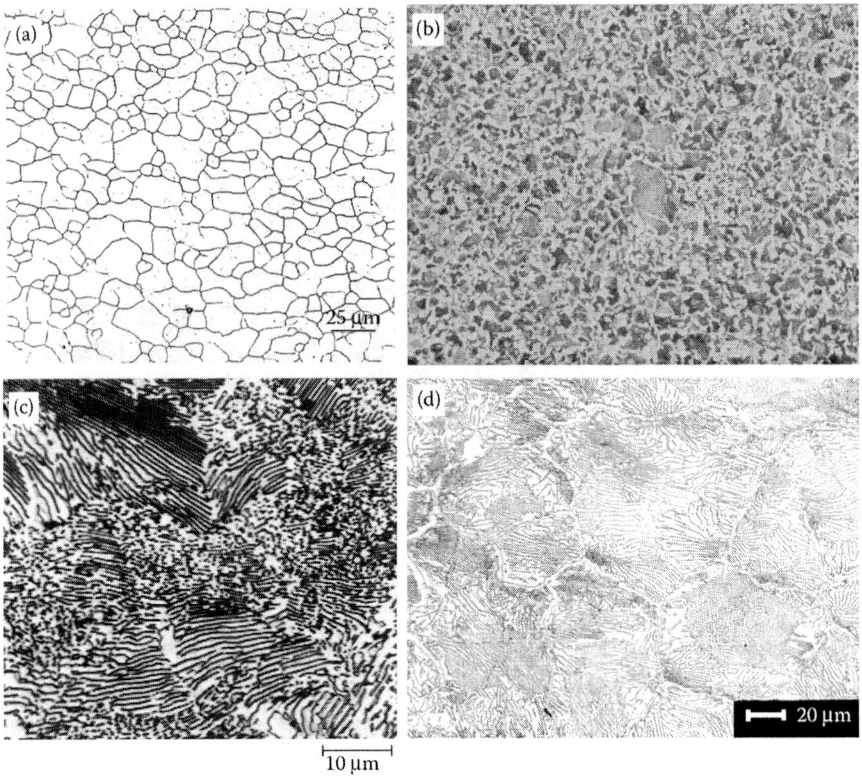

FIGURE 3.21 Photomicrographs of steels containing different carbon contents. (a) Very low carbon interstitial-free steel showing only α-ferrite grains. (b) Steel containing 0.4 wt% C showing the presence of α-ferrite (white etching) and pearlite (dark-etching region). (c) Microstructure of a eutectoid steel (0.76 wt% C) showing the presence of only pearlite. (d) Microstructure of a hypereutectoid (1.3 wt% C) steel showing the presence of well-resolved pearlite nodules surrounded by cementite. (Photos (a) and (c) courtesy of ASM International, Materials Park, OH. With permission. Photos (b) and (d) courtesy of Vander Voort, G. F. Consultant—Struers, Inc. With permission.)

temperature for 5 h, and slowly cooled at a rate of 10°C h^{-1} to 650°C and then furnace cooled to room temperature. Note that the cementite phase is now in the form of spheroids dispersed in a matrix of ferrite.

Figure 3.25 shows the microstructure of a gray cast iron that was just polished. One can see the coarse graphite flakes dispersed in a ferrite matrix. The graphite flakes in cast irons can be seen without the necessity of etching the samples.

3.8 PRECAUTIONS

It is most important to remember that extreme cleanliness has to be maintained during every stage of the preparation of specimens for metallographic examination.

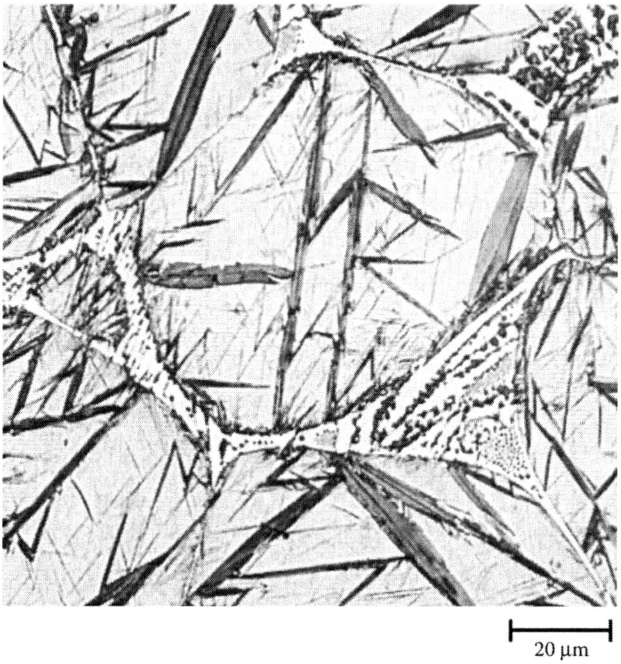

FIGURE 3.22 Microstructure of a 1.86 wt% C steel showing plate martensite within austenite grains and ledebuite in the grain junctions. (Photo courtesy of ASM International, Materials Park, OH. With permission.)

Also remember to wear safety glasses during the specimen preparation process. The abrasive particles should not be transferred from one stage of grinding (or polishing) to another. Therefore, the specimen needs to be thoroughly washed when moving from one station to another. Above all, it should be realized that specimen preparation is an art! However, following the steps described above will greatly aid in obtaining a specimen that is truly representative of the bulk material and one that is most suitable for microscopic examination.

3.9 EXPERIMENTAL PROCEDURE

Take different types of specimen materials—aluminum, copper, α-brass, and steel. Let all of them be in the annealed condition. Prepare these specimens for microscopic examination using the procedure described above. Etch these specimens using the appropriate etching reagents and observe their microstructures under a microscope.

You will notice that aluminum, copper, and α-brass are single-phase materials, while steel is a multiphase material. The single-phase materials will show equiaxed grain structure. Again, you will notice a difference between the two pure metals aluminum and copper. Even though both of them have the FCC crystal structure, copper has a very low stacking-fault energy and, therefore, it will show the presence of annealing twins inside the grains. Aluminum, on the other hand, has a high

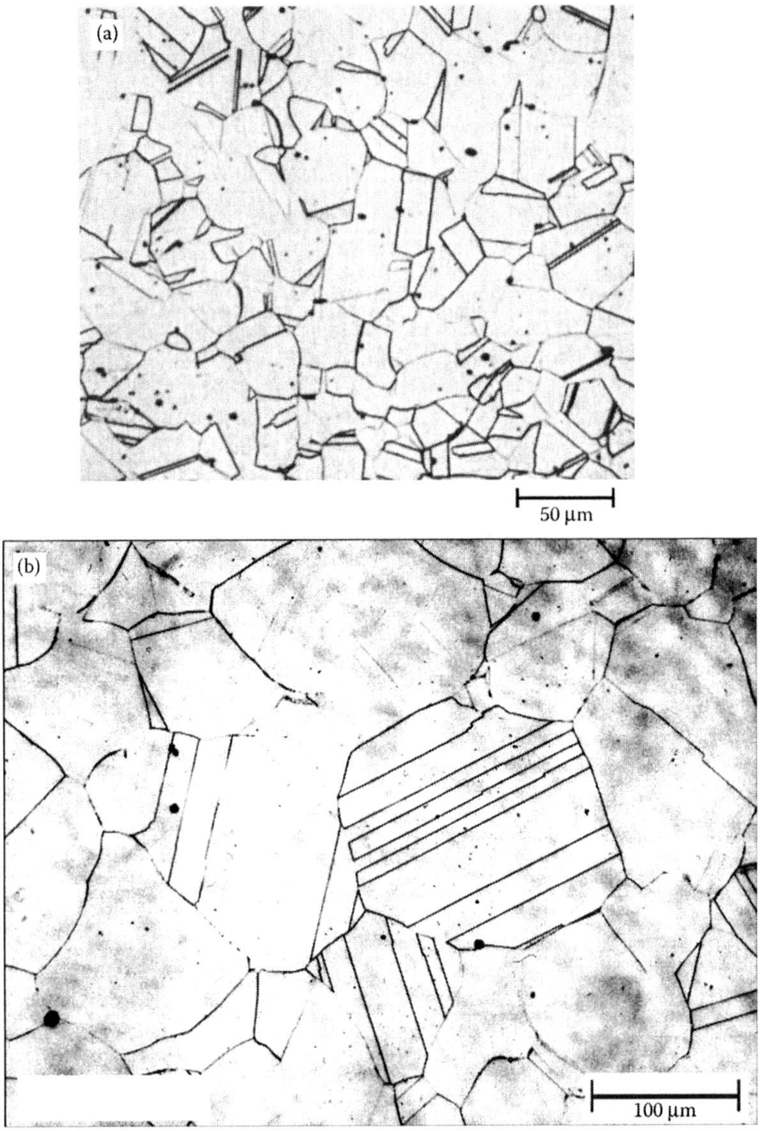

FIGURE 3.23 Photomicrographs of austenitic stainless steels: (a) 304 stainless steel and (b) 310 stainless steel. Both the steels contain annealing twins in their microstructures. (Photos courtesy of ASM International, Materials Park, OH. With permission.)

stacking-fault energy and, therefore, we will not see annealing twins inside the equiaxed grains of aluminum. α-brass is a solid solution alloy of zinc in copper, which also has the FCC structure. Since the stacking-fault energy of α-brass is lower than that of copper, the density of annealing twins in α-brass is higher than that in pure copper.

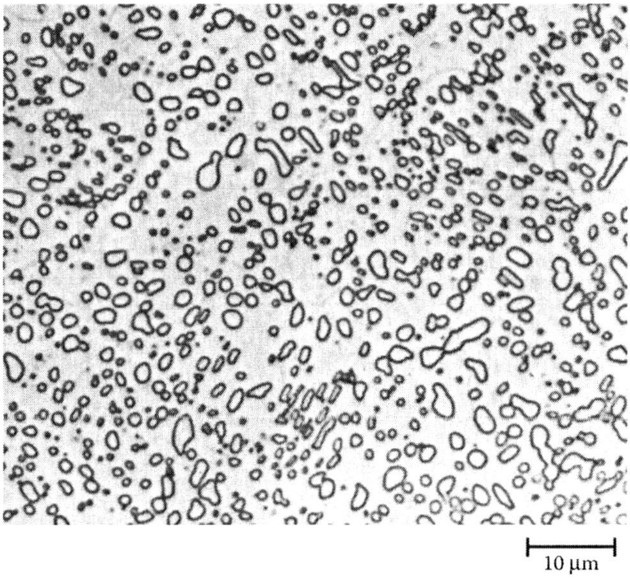

FIGURE 3.24 Photomicrograph of 52100 chromium alloy steel suitably heat treated to produce the cementite phase as spheroids dispersed in a ferrite matrix. (Photo courtesy of ASM International, Materials Park, OH. With permission.)

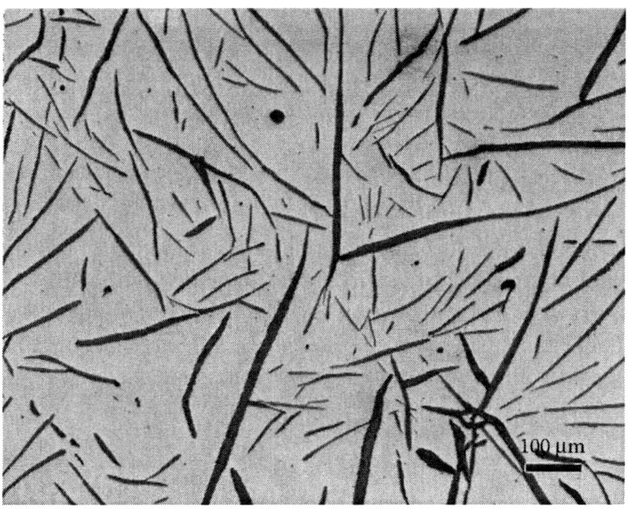

FIGURE 3.25 Photomicrograph of as-polished gray cast iron specimen showing coarse graphite flakes in a ferrite matrix. The graphite flakes in cast iron can be observed without the necessity for etching the sample. (Photo courtesy of ASM International, Materials Park, OH. With permission.)

The steel specimen will show the presence of two phases, namely, α-ferrite (an interstitial solid solution of carbon in α-iron) and cementite (Fe_3C compound). However, depending on the composition of the steel sample, whether it is hypoeutectoid or hypereutectoid, the actual microconstituents in the microstructure may be ferrite and pearlite or cementite and pearlite, respectively. In a eutectoid composition steel, only pearlite is present.

It may also be a good idea to determine the grain size of the single-phase materials using the linear intercept method described earlier and to calculate the proportion of phases (or microconstituents) in the steel sample.

This experiment will provide you full experience in preparing specimens for microscopic examination (grinding, polishing, and etching), the use of an OM, observing microstructures at different magnifications, and interpretation of microstructural features in different types of specimens.

If the appropriate equipment is available in the laboratory, it may be a good idea to take a specimen of odd shape (or very small in size) and mount it. One could then go through the process of grinding, polishing, etching, and microscopic observation as above.

EXERCISES

3.1 What is the basic difference between a biological and a metallurgical microscope?

3.2 How does one obtain a magnified image in the microscope?

3.3 Name the three important components of a metallurgical microscope.

3.4 What are the different light sources available for microscopic examination of materials samples?

3.5 What factors determine the choice of a suitable light source?

3.6 Name some common lens defects.

3.7 What is chromatic aberration?

3.8 How can the chromatic aberration be corrected?

3.9 What is spherical aberration?

3.10 Describe how the spherical aberration can be corrected or minimized.

3.11 How will you obtain the magnification in a microscope?

3.12 What is a measuring-type eyepiece?

3.13 What is numerical aperture?

3.14 On what factors does the numerical aperture depend?

3.15 How can one increase the numerical aperture of a lens without changing the lens system?

3.16 What is resolving power or limit of resolution?

3.17 How can the limit of resolution be improved?

3.18 Define empty magnification.

3.19 What type of information can one get from the microstructure of a sample?

3.20 What is grain size?

3.21 How can the grain size of a material be determined from optical micrographs?

3.22 What is ASTM grain size number?

3.23 If the ASTM grain size number, N, is expressed by the equation: $n = 2^{N-1}$, where n is the number of grains per square inch at 100×, calculate the actual grain size for $N = 3$.
3.24 If 40 grains were observed in a square inch area in a micrograph when viewed at 200×, calculate the ASTM grain size number.
3.25 Calculate the ASTM grain size number if 40 grains were observed in a square inch area in a micrograph when viewed at 50×.
3.26 What is the effect of grain size on the strength of a material?
3.27 Briefly explain how you can determine the number of phases from the microstructure of a material.
3.28 Draw the typical microstructures of annealed and heavily cold-worked single-phase materials.
3.29 If the microstructure of a plain carbon steel shows that it contains 60 vol% pearlite and 40 vol% ferrite, estimate the carbon content of the steel.
3.30 Why is it necessary to prepare a specimen for metallographic examination?
3.31 What are the requirements of a good specimen?
3.32 How can a hard specimen be removed from a large piece of metal without damaging it?
3.33 What is the sequence of steps in preparing a good specimen for optical metallography?
3.34 Why is polishing required after fine grinding the specimen?
3.35 What is the purpose of etching?
3.36 Describe how contrast develops between grains.
3.37 What is the difference in the etching mechanism between a single-phase and a two-phase alloy?
3.38 What are the effects of time and temperature on etching a metallographic specimen?
3.39 Name an etchant for steel samples. What is its composition?
3.40 Describe how the microstructure of a single-phase material differs from that of a two-phase material.

FURTHER READING

Bramfitt, B. L. and A. O. Benscoter, 2002. *Metallographer's Guide; Practices and Procedures for Irons and Steels*. Materials Park, OH: ASM International.

Brandon, D. and W. D. Kaplan, 1999. *Microstructural Characterization of Materials*. Chichester, UK: John Wiley & Sons.

Callister, Jr., W. D. and D. G. Rethwisch. 2010. *Materials Science and Engineering, An Introduction*, 8th edition. New York, NY: John Wiley & Sons, Inc.

Modin, H. and S. Modin, 1973. *Metallurgical Microscopy*. London: Butterworths.

Petzow, G. 1999. *Metallographic Etching*. Materials Park, OH: ASM International.

Richardson, J. H., 1971. *Optical Microscopy for the Materials Science*. New York, NY: Marcel Dekker.

Rostoker, W. and J. R. Dvorak, 1990. *Interpretation of Metallographic Structures*, 3rd edition. San Diego, CA: Academic Press.

Russ, J. C., 1995. *The Image Processing Handbook*. Boca Raton, FL: CRC Press.

Samuels, L. E., 1982. *Metallographic Polishing by Mechanical Methods*. Materials Park, OH: ASM International.

Samuels, L. E., 1999. *Light Microscopy of Carbon Steels*. Materials Park, OH: ASM International.

Scott, D. A., 1991. *Metallography and Microstructure of Ancient and Historic Metals*. Marin del Rey, CA: The Getty Conservation Institute.

Vander Voort, G. F., 1984. *Metallography Principles and Practice*. New York, NY: McGraw-Hill, Inc.

4 Scanning Electron Microscopy

4.1 INTRODUCTION

We had seen in the previous chapter that optical microscopy is a very powerful and useful technique to observe microstructural details of different types of specimens. But, the technique of optical microscopy suffers from a few limitations. One of the most serious of them is the limit of resolution that could be achieved. It was mentioned that the limit of resolution is given by the expression

$$\text{Limit of Resolution} = \frac{\lambda}{2NA} \tag{4.1}$$

where λ is the wavelength of the light beam used for illuminating the object and NA is the numerical aperture of the objective lens, which in turn is given by the expression, $NA = \mu \sin \alpha$, where μ is the refractive index of the medium between the objective lens and the object (usually air) and α is one-half the maximum angle of the light beam that could be collected by the objective lens. Therefore, the limit of resolution can be rewritten as

$$\text{Limit of resolution} = \frac{\lambda}{2\mu \sin \alpha} \tag{4.2}$$

But, a more accurate analysis suggests that the above equation should be

$$\text{Limit of resolution} = \frac{0.61\lambda}{\mu \sin \alpha} \tag{4.3}$$

Consequently, the smallest limit of resolution that can be achieved in the optical microscope using visible light is only about 200 nm, assuming that we are using violet light ($\lambda = 380$–430 nm) and a value of $\alpha = 30°$. This value of 200 nm is too large to observe many of the important microstructural features that are important in materials. For example, precipitates and second-phase particles in precipitation-hardened or dispersion-strengthened alloys, and so on, are much finer and are on the order of a few tens of nanometers. For observing these features and also those which are still finer (Guinier–Preston zones in precipitation-hardening systems (a few nanometers in size), dislocations, dispersoids, etc.) we need to use more powerful microscopes which could provide higher magnifications and lower limits of resolution.

The *scanning electron microscope* (most commonly abbreviated as SEM) is one of the more versatile instruments for microstructural investigations. Compared to the optical microscope, the SEM is capable of achieving magnifications of 10,000× to 150,000× and the limits of resolution can be brought down to about 10 nm in routine instruments and as low as 1 nm in the present-day high-resolution instruments. The TEM, on the other hand, can achieve much higher magnifications and much lower limits of resolution than the SEM. But, we may not require the use of TEM for every type of investigation. Further, preparation of specimens for TEM is involved, time-consuming and interpretation of images is not straightforward. We need to have knowledge of the contrast theory to properly interpret the microstructural features observed in the TEM. Hence, it can be considered that the SEM ably bridges the gap between the optical microscope and the TEM, to provide intermediate magnifications and resolving power.

The most important advantage of an SEM, in comparison with the optical microscope is its depth of focus. The depth of focus depends upon the magnification at which the image is observed. It is about 1 μm at a magnification of 10,000× and reaches up to 2 mm at a low magnification of 10×. Whereas the optical microscope generates a two-dimensional image of the surface of a bulk specimen and the TEM also generates a two-dimensional image from a thin planar slice taken from a bulk material, the SEM provides information on the three-dimensional topography of the surface. That is, we will be able to see in focus points of different height at the same time. Consequently, the images produced in an SEM have a three-dimensional appearance. Interpretation of SEM images is relatively easy, there is not much of specimen preparation, and the instrument is relatively easy to operate. In addition to the above, it is also possible to achieve semiquantitative or even quantitative analysis of composition from a small volume of material with good lateral resolution. In some of the more recent SEMs it is also possible to obtain crystallographic information of the material which was earlier possible only with the TEM. Because of all these advantages, SEM has become one of the most popular techniques for investigating microstructural features of materials—both biological and inorganic. It has also been found to be of immense use in failure analysis of materials.

4.2 BASIC DESIGN OF THE SEM

Unlike the optical microscope which we had discussed earlier, the source of "illumination" in the SEM is an electron beam. Since an electron beam cannot be focused by glass lenses, we use electromagnetic lenses to focus the electron beam onto the specimen. The signals from the specimen are then collected by an electron detector, amplified, and the image is displayed on a high-quality television screen. Provision is also made to record the image either digitally or on a photographic film. Since the electron beam has a limited path length in air, the whole setup is enclosed in a vacuum chamber. The vacuum level in the chamber should be better than 10^{-6}–10^{-10} Torr, depending on the type of electron source used. A diffusion pump or a turbomolecular pump is used to generate the required vacuum in the SEM chamber. Hydrocarbon

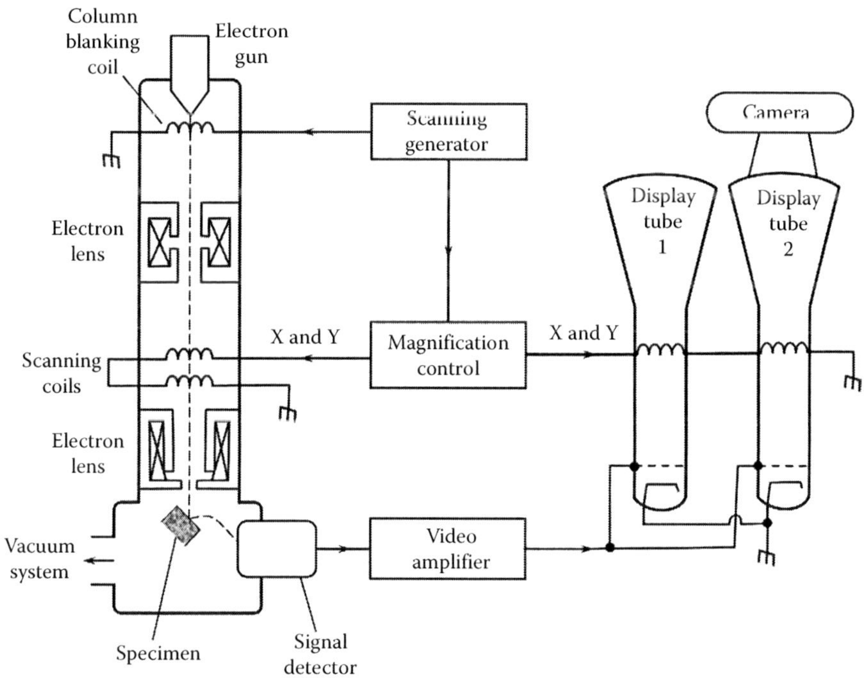

FIGURE 4.1 Schematic of the typical design of an SEM.

contamination from the pump oil can be eliminated, if a turbomolecular pump is used. Figure 4.1 shows the principal features of an SEM. The important components of an SEM are the electron source and the electron detector. We will also have the lenses and other accessories. After discussing the basic components of the SEM, we will discuss how image formation occurs in the SEM and the ways of interpreting the images.

4.3 ELECTRON SOURCE

Electrons are generated when a negatively charged tungsten filament (called the cathode) is heated to high temperatures—*thermionic emission*. Temperatures as high as 3000°C are required to produce a sufficiently bright source. The electron beam current densities in this case are typically of the order of 5×10^4 A m^{-2}. It is easy to work with these filaments, but they need to be frequently replaced because of evaporation. Even though the electron beam is sufficiently bright for most common applications, a higher current density can be obtained by using a *lanthanum hexaboride* (*LaB*$_6$) *crystal*. The *current density* in this case can be as high as 1×10^6 A m^{-2}. Since the work function (energy to be supplied to an electron to escape from the surface) for LaB$_6$ is lower than for tungsten, these filaments can be operated at lower temperatures. However, LaB$_6$ filaments require a much better vacuum than when using tungsten filaments to achieve a longer lifetime. Still higher current densities of the order of 1×10^{10} A m^{-2} can be obtained when a *field-emission source* is used. Field-emission sources use very sharp tips (with a diameter of 50–100 nm). The strong electric field created at the tip extracts electrons from the source even at low temperatures through the phenomenon of tunneling. Consequently, the field-emission source is also called a cold cathode. The field-emission guns are the choice of an electron source in the present-day instruments even though they are difficult to work with, they require very high vacuum, and occasional cleaning and sharpening through thermal flashing. This is because of the vastly increased resolution and brightness of the source even at low voltages.

The electrons produced by any one of the above methods need to be accelerated through a potential difference. Most SEMs use an accelerating voltage of 1–50 kV (1 kV = 1000 V). Through this process the electrons acquire energy, and the energy of the electrons can be calculated using the simple equation

$$\text{Energy} = e \times V \tag{4.4}$$

where e is the charge on the electron (1.602×10^{-19} C) and V is the applied voltage expressed in volts. Thus, the energy of the electron beam increases with increasing applied voltage and these electrons will then be referred to as high-energy electrons. The higher the energy of the electrons, the larger will be the volume of the material from which we can gather information.

The electron beam from the source passes through a series of electromagnetic lenses, and each lens is associated with an aperture that defines the divergence of the electron beam. The top lenses are called the *condenser lenses*. The divergence of the beam is a function of the focal length of the lens, which in turn is a function of the current passing through the lens coils. The higher the current passing through the lens, the shorter is the focal length and the higher the divergence. By increasing the current passing through the first lens, the size of the image produced is reduced (that is why we use the term "spot size" for this control). Because of the increased divergence of the beam, it is also possible to control the beam current.

The defocused electron beam from the condenser lens then reaches the *probe lens*, which functions in a manner somewhat similar to the objective lens in the optical microscope (or the transmission electron microscope). This lens focuses the electron beam on to the surface of the specimen. But, unlike in the other microscopes, the probe lens in the SEM does not collect the signal from the specimen. Information about the specimen is collected from the inelastic scattering processes that occur when the electron beam interacts with the specimen. These signals are then collected by an electron detector.

The electron beam scans the specimen in a manner much similar to that in a cathode ray tube used for image formation on a television (TV) screen. A scan generator controlling the current of the scanning coils deflects the beam along closely spaced lines. The magnification is changed by adjusting the current in the deflection coils, and can be in the range of 10 to 100,000×. The electrons excited by the electron beam and emitted from the specimen are collected by an electron detector. The current of the electrons hitting the detector is small and needs to be amplified to obtain reasonable brightness in the image. The image is then recorded for subsequent analysis and interpretation.

FIGURE 4.2 Photograph of a commercial JSM-7001F SEM. (Photo courtesy of JEOL. With permission.)

The sample in the SEM is attached to the specimen stage. It is possible to translate the specimen in the *X*- and *Y*-directions, as well as a tilt with respect to the beam axis. It is also possible to rotate the specimen about an axis normal to the specimen's surface. The specimen can also be moved in the *Z*-direction for adjusting the distance (called the *working distance*) between the specimen surface and the probe lens. Thus, a number of positions in the specimen at different angles of tilt/orientation can be observed in the SEM. Figure 4.2 shows a photograph of a commercial SEM.

4.4 ELECTRON BEAM–SPECIMEN INTERACTIONS

The electron beam that is generated by the electron source and passing through the lenses before it hits the specimen is referred to as the *primary electron beam.* When the primary electron beam interacts with the specimen, several different types of signals are produced. Figure 4.3 shows the type of signals that are generated and the depth in the specimen from which they are generated. Note that we are not considering the elastically scattered electrons which get diffracted from the crystal planes. Instead, we are only considering the inelastic scattering processes. These signals can be used to form images and also to analyze the specimen for its chemical composition.

The signals that are generated due to interaction between the primary electron beam and the specimen include:

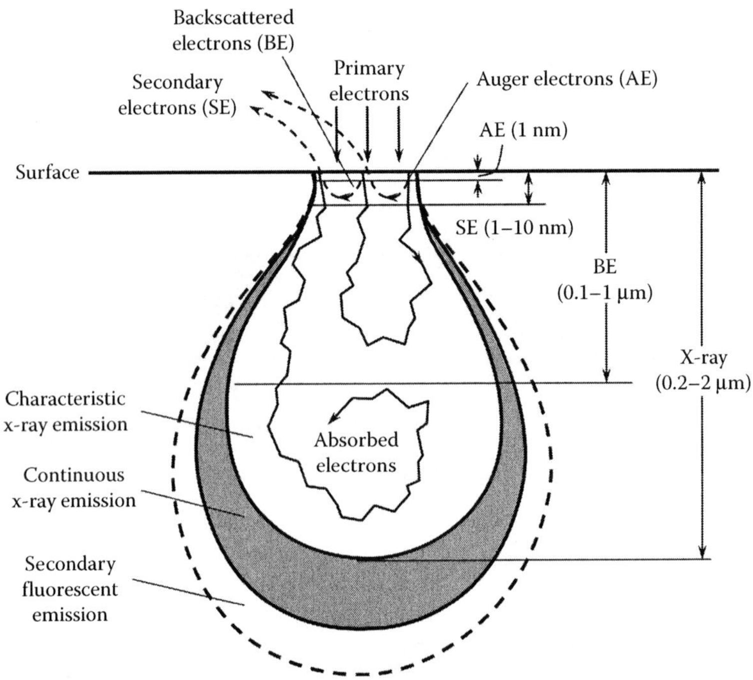

FIGURE 4.3 Types of electron beam–specimen interactions. Although a large number of signals are generated, only some of them are commonly used. The most commonly used signals are secondary electrons, backscattered electrons, and x-rays.

- Secondary electrons
- Backscattered electrons
- X-rays
- Auger electrons
- Cathodoluminescence
- Electron beam image current

Of these, the first three are more commonly used to image the specimen surface and also for chemical analysis. So, let us look at these in some more detail.

4.4.1 Secondary Electrons

Secondary electrons are formed due to interaction between the primary electrons and the loosely bound electrons in the atoms of the specimen. The number of secondary electrons generated per primary electron is termed the *secondary electron emission coefficient*, and it is usually greater than one. In fact, it may reach values of several hundred. The energy of the secondary electrons is independent of the energy of the incident electrons and of the specimen material. Although the secondary electrons have a range of energies ranging from as low as about 3 eV and sometimes reaching up to about 100–200 eV, the typical range is 10–50 eV. These secondary electrons that could be detected are generated from near the surface of the material, approximately up to 10 nm, but generally 0.5–1 nm. It is true that the secondary electrons are generated at greater depths also. But, due to their low energy, these get absorbed by the specimen before they escape from the surface and reach the detector.

The yield of secondary electrons is not strongly dependent on the atomic number of the material on which the primary electrons are incident. Instead, it is the angle between the incident beam and the specimen surface, that is, *surface topography* of the specimen, which is most important in determining the secondary electron yield. Therefore, the secondary electrons are used to determine the surface topography of specimens.

Since the envelope of the excited volume moves closer to the surface when the beam hits the surface at a small angle, the secondary electron yield increases. Consequently, the yield increases as the angle between the incident beam and the specimen surface decreases. Any region protruding from the specimen surface (positive radius of curvature) increases the chances of the secondary electrons escaping and consequently being detected. Because of the low energy of the secondary electrons, they are easily deflected by a low-bias voltage and are collected with a very high efficiency. Figure 4.4 shows a typical SEM image recorded using the secondary electron signal. Notice the large depth of focus and the three-dimensional appearance of the surface.

4.4.2 Backscattered Electrons

Backscattered electrons are produced by single large-angle or multiple small-angle scattering events. The proportion of the backscattered electrons depends mainly on the atomic number of the material, or the average atomic number in the case of phases containing more than one element (that is why the contrast produced this way is referred to as *atomic number contrast*), but is also dependent on the surface

FIGURE 4.4 Typical secondary electron image. Note that the surface topography of the specimen is clearly visible.

inclination, primary energy of incident electrons, and the number of outer-shell electrons. The backscattered electrons originate in a surface layer, usually 0.1–1 μm; the depth increasing with increasing energy of the incident electron beam and decreasing atomic number. When the incident electron beam hits a material with a high atomic number, many electrons are backscattered by atoms close to the surface, with little change in energy. On the other hand, when materials with low atomic numbers are hit by the primary electrons, fewer electrons are backscattered and more energy is lost. Consequently, phases with a higher average atomic number appear "bright" and those with lower average atomic number appear "dark" in the images. The average energy of the backscattered electrons is less than that of the primary electrons, but of the same order of magnitude. These are usually detected in an annular region close to the probe lens pole-pieces. Figure 4.5 shows a typical backscattered electron image. Note that the phases with higher average atomic number appear "bright" in the image.

4.4.3 X-Rays

As seen in Chapter 2 (X-ray Diffraction), characteristic x-rays are generated when the primary electron beam has sufficient energy to knock out an electron from the atom. Depending on the shell, which the knocked-out electron was occupying and the shell from which the electron fills that hole, we will have different types of x-rays, for example, K_α, K_β, L_α, L_β, and so on. We had also seen in Chapter 2 that we use x-rays of well-defined wavelength to obtain diffraction patterns. But, in the SEM we will utilize all the characteristic x-ray beams to obtain chemical information about the material.

The x-rays generated by the specimen can be utilized in three different ways. Firstly, the spectrum can be analyzed by *wavelength-dispersive spectroscopy* (WDS), which is used exclusively in electron probe microanalyzers (EPMA). On the other hand, the *energy-dispersive spectroscopy* (EDS), is more easily implemented in

FIGURE 4.5 Typical backscattered electron image. The micrograph is from a WC-Co alloy. Note that WC being heavy (higher atomic number) appears bright in the micrograph whereas Co being light (lower atomic number) appears dark.

commercial scanning or transmission electron microscopes. In this method, characteristic x-rays are collected with the beam stationary at a specific location (point or area) on the sample surface. The x-ray collection times are typically 60–100 s. The third way of utilizing the x-rays is as an x-ray dot image or *elemental map*, in which each characteristic photon detected is recorded as a white dot (Figure 4.6). The position of the dot on the screen corresponds to the coordinates of the beam as it is scanned over a selected area of the specimen surface. Several elements may be detected simultaneously. Each of these is displayed in different windows, which are usually color coded for the different elements.

A lithium-drifted silicon solid-state detector transfers the x-ray impulses to a multichannel analyzer. The processed information is displayed on the screen which enables rapid qualitative and quantitative analysis of the composition of the specimen. A typical EDS spectrum is displayed in Figure 4.7. In this spectrum, the number of x-ray photons detected (intensity of the x-ray beam) is plotted on the *Y*-axis, while the energy of the x-ray beam is plotted on the *X*-axis. (Remember from Chapter 2 that characteristic x-radiations have specific wavelengths and therefore specific energies.) Standard energy values of the energies of the different characteristic x-ray

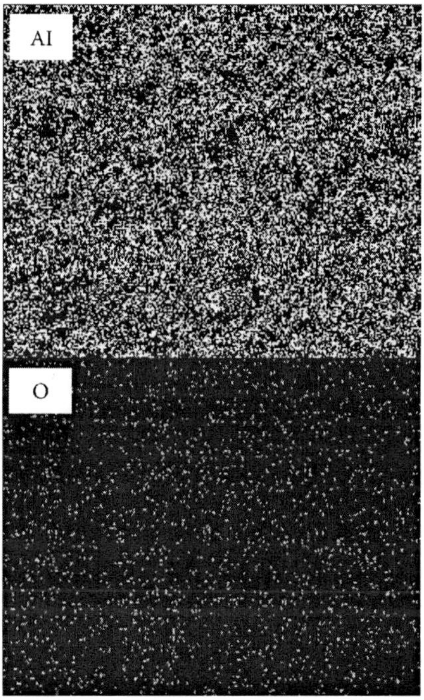

FIGURE 4.6 X-ray elemental map from an Al–Al_2O_3 nanocomposite specimen showing the aluminum and oxygen maps. The white dots in the aluminum map show the positions of aluminum atoms and the white dots in the oxygen map show the positions of the oxygen atoms. These micrographs show that both the atoms are uniformly distributed in the specimen. If there is segregation of one component, there will be higher concentration of the white dots corresponding to that element.

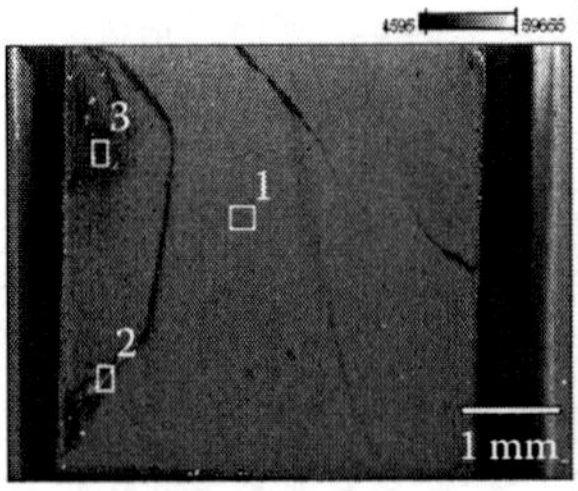

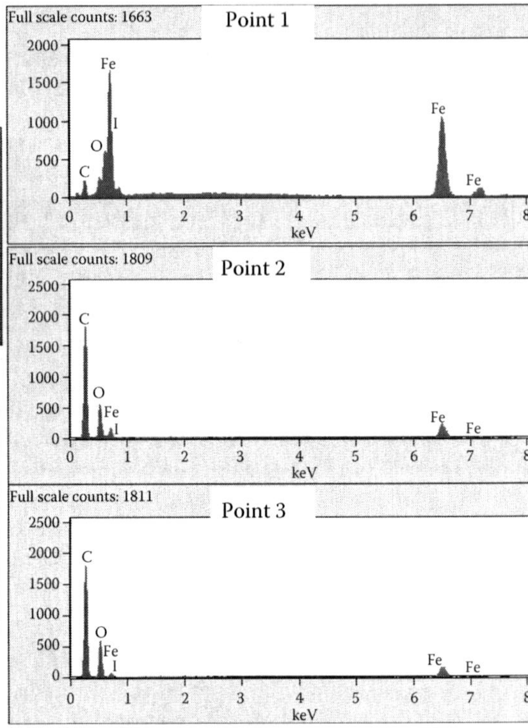

Point #	C–K	O–K	Fe–K
1	10.79 ± 0.95	3.05 ± 0.59	85.60 ± 1.78
2	61.53 ± 0.57	21.29 ± 0.94	17.17 ± 0.86
3	61.71 ± 0.56	24.65 ± 0.86	13.64 ± 0.82

FIGURE 4.7 X-ray spectra obtained in the SEM. The intensity of the x-ray beam is plotted on the *Y*-axis, while the energy of the x-ray beam is indicated on the *X*-axis. Thus, by knowing the energy of the x-ray beam corresponding to any of the peaks in the spectrum, the chemical identity of the element can be determined. The intensity of the x-ray peak is related to the amount of the element present. Thus, we can determine the chemical composition of the different phases both qualitatively and quantitatively. The figure shows the results obtained from three different points in an Fe-based alloy. The SEM micrograph, EDS spectra, and the tabulated compositions are presented.

wavelengths are known and are available in standard reference books. Therefore, by knowing the energy corresponding to each peak, the chemical identity of the element present in the material can be determined. The relative intensities of the peaks are proportional to the amounts of the elements present in the sample, even though some corrections are required to be made before one can get quantitative analysis of the elements present in the sample. Thus, both qualitative and quantitative chemical analysis of the elements present in the sample can be carried out.

It is relatively easy to detect the presence of elements having a high atomic number. But, it is difficult to detect elements with low atomic numbers, especially when they are present in small quantities. Due to the interference of the beryllium or plastic window in front of the detector, elements having atomic numbers (Z) less than 11 (i.e., sodium) cannot be analyzed in a routine SEM. But, SEMs equipped with ultrathin windows or windowless detectors can detect elements down to $Z = 5$

(boron). Chemical analysis is obtained from a depth of about 1–10 μm, with approximately the same lateral resolution. This value can be increased with increasing energy of the incident electron beam and decreasing specific density (atomic weight) of the elements present in the specimen. The relative errors in quantitative analysis are approximately 10%.

4.5 SPECIMEN PREPARATION

Specimens for observation in the SEM need not be extensively prepared. For example, they need not have a mirror-like surface finish as required for optical microscopy examination. Nor, is it necessary to have extremely thin (about 100 nm) specimens as in transmission electron microscopy. However, to obtain good images, it is necessary that the specimen (1) has a contamination-free surface (free from any organic residues such as oil and grease), (2) is stable under vacuum system and under the electron beam, (3) gives sufficiently high yield of electrons, and (4) does not build up an electrostatic charge on the surface. The last requirement is most crucial since the associated charge instability will lead to unstable emission of secondary electrons. This, in turn, will destroy the stability and resolution of the image. The problem of "charging" gets aggravated in ceramic (or any other nonconducting) materials because if the samples do not conduct they soon charge up under the beam. One way of preventing (or at least minimizing) this effect is to operate the SEM at low voltages. Alternately, the nonconducting specimen surface could be coated with a thin layer of an electrically conducting film. Usually a thin layer of carbon, gold, platinum, or gold–palladium alloy (a few nanometers in thickness, generally between 10 and 100 nm) is deposited for this purpose. In general, it is a good practice, to have a conducting tape attaching the sample to the specimen holder. Apart from the above requirements, it is also necessary that the specimen is able to fit into the sample chamber and that there is enough space for it to be tilted and/or rotated freely.

The specimen surface is also occasionally sputter-coated with an electrically conducting film to improve contrast in the image. A gold–palladium alloy is generally used for this purpose. One difficulty in using specimens with sputter-coated films on their surface is that these films interfere with chemical analysis. That is, the elements present in the sputter coating will also be included in the chemical analysis. Consequently, the chemical composition of the specimen reported will appear to be different from the original values. Therefore, if chemical analysis is required, then it is better to do the analysis first and then perform microstructural observations after coating the surface with a conducting film. If both microstructural observation and chemical analysis are important for a nonconducting alloy, then a compromise is to perform the SEM work at low voltages.

The size of the specimen is immaterial as long as it can be accommodated inside the specimen chamber. To enable the specimen to be rotated and/or tilted, the specimen is usually fixed onto a specimen holder using adhesive conducting tapes. To observe powder specimens, the powder is sprinkled onto an adhesive tape and this is secured onto the specimen holder. Remember that whenever porous specimens, or specimens that have been mounted using bakelite or other mount materials are used in the SEM, it is possible that they outgas. Therefore, it is a good practice to place such samples in an external vacuum chamber, degas the samples, and then transfer them to the SEM chamber.

4.6 APPLICATIONS

The SEM has found extensive applications in the fields of materials science and engineering, geology, physics, chemistry, life sciences, and in fact almost every branch of science and engineering. This is because it is relatively easy to operate the SEM and obtain useful information about microstructure and chemical composition of the samples. In fact, sometimes people use the SEM as a substitute for an optical microscope since no extensive specimen preparation is required for the SEM. But, the most important use of SEM is, based on its large depth of focus, in studies of fractography and failure analysis of a wide variety of materials including metals, ceramics, semiconductors, polymers, and composites. In the semiconductor industry, SEM is an indispensable tool for studying the integrity of devices. Let us now look at some important applications of SEM in materials science and engineering.

4.6.1 Fractography

The SEM is most commonly used in investigating the failure of materials through study of fracture surfaces. A descriptive treatment of observation of fracture surfaces is known as *fractography*. The SEM has become an indispensable tool in fractographic studies because the SEM has

- A large depth of focus
- Three-dimensional appearance of the fractured surfaces
- Very little specimen preparation required for observation
- Ability to achieve a wide range of magnifications
- Possibility of obtaining chemical composition and consequently identifying the phase/feature responsible for failure/fracture

Further, it has been possible to predict the reason for failure of materials by observing the microstructure of fracture surfaces.

The type of fracture is determined by the appearance of the fracture surface as viewed in the SEM. For example, when cracks propagate along grain boundaries of the material, then the fracture is known as *intergranular fracture*. On the other hand, when cracks propagate through the grains, the fracture is known as *intragranular fracture* (*transgranular* or *transcrystalline* fracture are the other terms used). It is possible to determine whether a material has failed by intergranular or intragranular mode by observing the fracture surfaces. In the case of intergranular fracture, it is easy to see cracks along the grain boundaries, that is, the grains are separated along the boundaries.

As we will see later, materials can fail either in a ductile manner (when the specimen experiences significant plastic deformation) or in a brittle manner (when the specimen does not experience any plastic deformation). When the material experiences plastic deformation, there is a change in the shape of the material. For example,

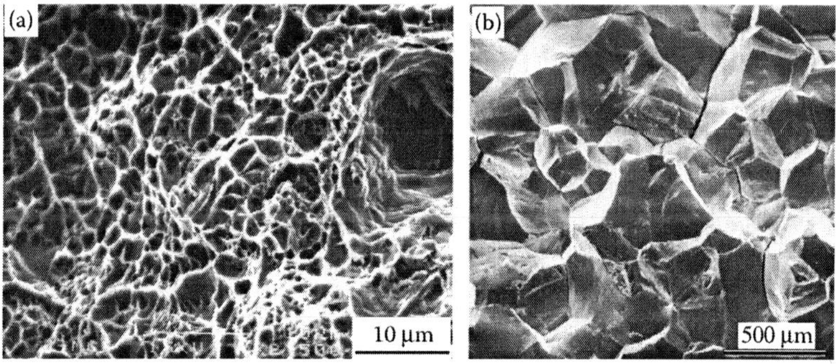

FIGURE 4.8 Typical fracture appearances of materials. (a) Ductile fracture showing the presence of dimples. (b) Brittle fracture showing the appearance of planar cleavage facets which give sharp contrast.

during tension, the specimen gets elongated, and eventually fails through coalescence of microvoids. The size of these microvoids is related to the grain size of the material. That is, the voids get pulled apart in the direction of loading. As a result, one would observe dimples on the surface of a specimen that has fractured in a ductile manner. On the other hand, during brittle fracture, no plastic deformation occurs in the specimen. The crack propagation takes place very rapidly and in a direction almost perpendicular to the direction of the applied load and therefore, the fractured surface is relatively flat. These features can be clearly seen in the two fractographs showing the ductile and brittle fracture of materials (Figure 4.8). Note the presence of dimples in Figure 4.8a showing ductile fracture and intergranular fracture in Figure 4.8b.

Apart from these two types of fracture (ductile and brittle) that occur in materials during a tensile test, materials can also fail during fatigue, impact, creep, corrosion, wear, and many other situations including combinations of these effects. The technique of fractography has been found to be very useful in determining the reason(s) for the material failure. We will see in Chapter 11 (Fatigue Testing) that fatigue cracks start from a point and then they propagate during subsequent cycles. A clear manifestation of this effect is the presence of ridges and fatigue striations. Figure 4.9 shows an SEM fractograph of the surface of a specimen that has failed by fatigue. It may be noted from this micrograph that a crack had formed at the top edge. Further, the smooth region near the top of the micrograph corresponds to the area over which the crack had propagated slowly. Subsequently, rapid propagation of the cracks had occurred over the area having a dull and fibrous texture. This constitutes the major portion of the micrograph. Thus, by observing the microstructural details of the fractured surface, it is possible to determine the reason for failure and also the point of origin of failure. Similar conclusions can be drawn from other types of failure. As another example, Figure 4.10 shows fatigue striations on a fractured aluminum sample. The SEM is also very valuable in investigating failure of integrated circuits in the semiconductor industry (Figure 4.11).

Another important advantage of using SEM in failure investigations is that, in addition to identifying the microstructural features, it is also possible to obtain the

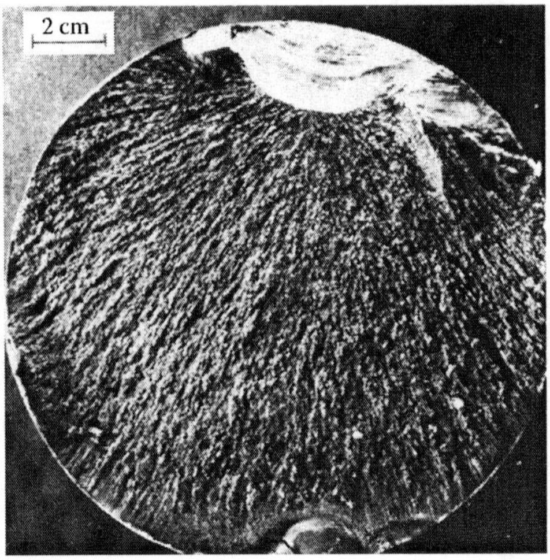

FIGURE 4.9 Fractograph of a specimen that has failed by fatigue. (Photo courtesy of ASM International, Materials Park, OH. With permission.)

chemical composition of the phases present using the EDS technique. That is, one can identify the phases responsible for the failure of materials.

4.6.2 Microstructural Features

The SEM can also be used to study the microstructural features of materials—grain size, shape, distribution of phases, and so on. The SEM is capable of producing much

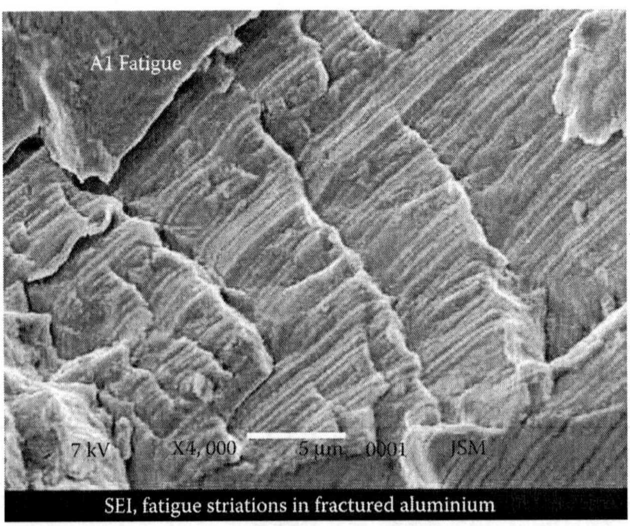

FIGURE 4.10 Secondary electron image showing fatigue striations in a fractured aluminum specimen. (Photo courtesy of JEOL. With permission.)

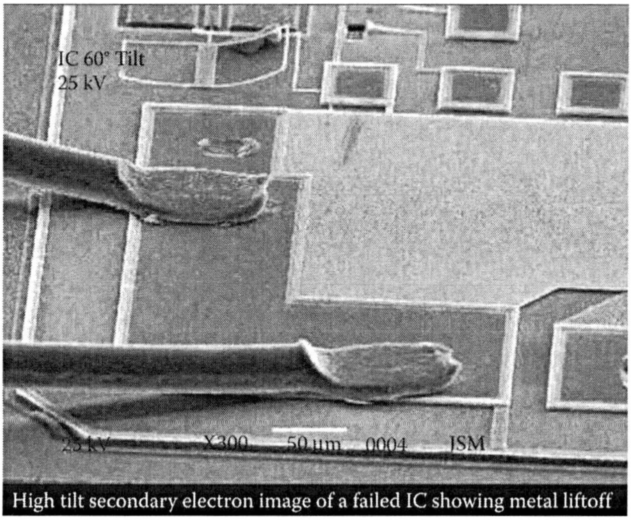

FIGURE 4.11 Secondary electron image of a failed integrated circuit showing metal lift-off. (Photo courtesy of JEOL. With permission.)

higher magnifications and lower limits of resolution than is possible in an optical microscope. Further, we have seen above that atomic number contrast can easily identify the presence of heavy elements in any area of the sample. One can also conduct image analysis and chemical analysis of the different phases and this enables one to more completely describe the microstructure of the material.

4.7 EXPERIMENTAL PROCEDURE

It is possible that your laboratory has more than one SEM. Get familiar with the instrument and identify the different controls. It is important that you try to learn how to operate the vacuum system, method of loading of the samples, and also observation and recording of microstructures. Note that you can change the brightness and contrast of the micrographs as well as determine the chemical composition of the different phases present in the sample.

A good exercise will be to look at the fracture surfaces of samples that have failed in tension, impact, and fatigue testing experiments. Observe and record these microstructures and determine the type of fracture—whether it is ductile or brittle. Further, you can also observe some of the standard samples that have failed in different modes.

EXERCISES

4.1 If we have an optical microscope, why do we need a scanning electron microscope (SEM)?
4.2 Name the biggest advantage of an SEM.
4.3 What type of radiation is used in an SEM?
4.4 How can you generate a beam of electrons?

4.5 What type of signals are generated when a beam of electrons impinges on a metal specimen?
4.6 How are secondary electrons produced?
4.7 How are backscattered electrons produced?
4.8 Differentiate between secondary and backscattered electrons.
4.9 What type of information can be obtained from the secondary electrons?
4.10 What type of information can be obtained from the backscattered electrons?
4.11 Which signal will you use to obtain topographic information?
4.12 Which signal will you use to obtain mass information?
4.13 What signal would you use to obtain information from a fractured surface?
4.14 Why do we get a better resolution in the SEM than in an optical microscope?
4.15 Can we obtain chemical composition using an SEM?
4.16 Can we use ceramic samples as they are in an SEM? Explain your answer.
4.17 Why are some specimens coated with a thin layer of gold before inserting them in the SEM?
4.18 How can we examine nonconducting ceramic samples in an SEM?
4.19 How can we obtain chemical analysis using an SEM?
4.20 What are the requirements of a good SEM specimen?
4.21 What are some typical applications of an SEM?
4.22 What is the characteristic feature in an SEM from a fatigue-failed specimen?
4.23 An engineering component failed during service. List some possible reasons as to why it could have failed and suggest what technique you would use to determine whether the component failed in a ductile or a brittle manner. How would you decide the nature of failure?

FURTHER READING

Brandon, D. and W. D. Kaplan. 1999. *Microstructural Characterization of Materials*. Chichester, UK: John Wiley & Sons.

Goldstein, J. I., D. E. Newbury, P. Echlin, D. C. Joy, C. E. Lyman, E. Lifshin, L. Sawyer, and J. R. Michael. 2003. *Scanning Electron Microscopy and X-Ray Microanalysis*. New York, NY: Springer.

5 The Iron–Carbon Phase Diagram and Microstructures of Steels

5.1 INTRODUCTION

Ferrous-based alloys, especially those based on the *iron–carbon system*, are perhaps the most used engineering materials in the industry with more than a billion tons consumed annually. These include plain carbon steels, alloy steels, stainless steels, tool steels, and cast irons. The iron–carbon phase diagram is very useful for investigating the effect of addition of carbon on the *allotropic changes* in pure iron. These allotropic changes are the basis for altering the microstructures and properties of *steels* (iron–carbon alloys which contain <2.1 wt% carbon) and *cast irons* (iron–carbon alloys containing from 2.1 to 6.67 wt% carbon). If iron were not one of the most abundant elements in the Earth's crust, or if it did not undergo the allotropic transformations on heating or cooling, its alloys would not have achieved their pre-eminent position as structural materials. More than anything else, it is the ability to vary the properties of iron-based alloys by varying the conditions under which the allotropic transformation takes place that makes these alloys so important. The *phase diagram* also helps in predicting the phases and microstructures of Fe–C alloys when they are subjected to heating and/or cooling under equilibrium conditions. Steels offer the most opportunities to modify the microstructure and mechanical properties by different heat treatments because of the phase transformations occurring in them in the solid state. In this chapter, we will discuss the Fe–C (or more appropriately, the Fe–Fe_3C) phase diagram and see what types of microstructures are developed at different carbon contents under equilibrium conditions. In Chapter 6, we will discuss the types of microstructures obtained on cooling the steel samples from elevated temperatures under nonequilibrium conditions. We will also discuss the different types of heat treatments that can be applied to steels and the nature of microstructures produced in steels and the resultant properties.

But, before we go into details of the Fe–Fe_3C phase diagram, it would be useful to understand some basic principles of phase diagrams.

5.2 PHASE DIAGRAMS

Phase diagrams are graphical representations of the different phases present in an alloy system at different compositions, temperatures, and pressures. Stated differently, a phase diagram enables one to identify the temperature–composition (and temperature–pressure or composition–pressure) regions in which different phases are thermodynamically stable. It is possible from such phase diagrams to obtain the number of phases, their compositions, and the relative proportions of the phases as functions of temperature, pressure, and overall composition of the material. Even though a great majority of engineering alloys exist in metastable or nonequilibrium state, the tendency for the materials is always to move toward equilibrium provided that the right conditions are made available. Thus, knowledge of phase diagrams is very useful. A phase diagram is also referred to as a *constitution diagram* or an *equilibrium diagram*. Since the phase diagram usually represents the status or constitution of an alloy under equilibrium conditions, it is also occasionally referred to as an *equilibrium phase diagram*. This distinguishes it from the *nonequilibrium phase diagrams* which have been becoming popular and useful in recent times due to the development of several different nonequilibrium processing techniques. But, note that if nothing is specified, a phase diagram is an equilibrium phase diagram.

5.3 REPRESENTATION OF PHASE DIAGRAMS

Figure 5.1 represents the phase diagram of a pure element. The temperature is plotted on the abscissa (X-axis) and the pressure on the ordinate (Y-axis). The phase diagram indicates the regions in which the solid, liquid, and vapor phases are stable

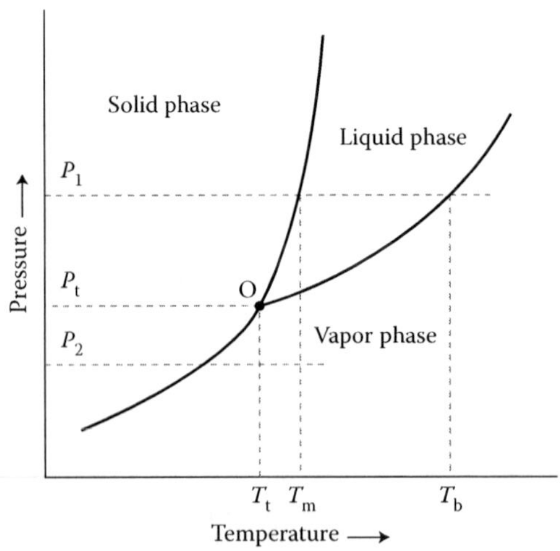

FIGURE 5.1 A typical phase diagram of a pure element. The pressure–temperature regions over which the solid, liquid, and vapor phases are stable are indicated. At point O, all the three phases—solid, liquid, and vapor—coexist. This point is referred to as a triple point.

at different temperatures and pressures. At a particular pressure, say P_1, as the temperature is increased, the solid will transform into the liquid phase at the melting point T_m, and if the temperature continues to be increased, the element will transform into the vapor phase at the boiling point T_b. At a lower pressure, say P_2, the element may directly transform from the solid to the vapor phase as the temperature is increased. But, at some intermediate pressure, P_t, for example, as the temperature is increased, we will reach a temperature T_t, at which all the three phases—solid, liquid, and vapor—coexist. This point, O, with the coordinates (P_t, T_t) is called the *triple point*. A triple point is one at which three different phases coexist—in this case solid, liquid, and vapor. Sometimes this point is also referred to as an *invariant point* since the pressure and temperature at which all the three phases coexist are fixed; nothing can be varied.

In actual practice, however, pure metals are not frequently used as engineering materials; we deal with alloys containing two or more elements present in them. If we now consider an alloy containing two metals, for example, we need to plot the phase diagram showing the variation of pressure, temperature, and composition. Representing this three-dimensional situation on a two-dimensional plane is not easy. But, luckily for us, most materials operations are carried out at atmospheric pressure and therefore pressure is not an important variable. In other words, we could assume that pressure is constant and represent the effects of temperature and composition only. That is, we are considering a cross section of the temperature–pressure–composition diagram at a constant pressure of usually 1 atmosphere.

A typical phase diagram between two metals is presented in Figure 5.2. In this phase diagram, the temperature is represented on the *Y*-axis and the composition on the *X*-axis. Metal A is represented on the left-hand side of the diagram and metal B on the right-hand side. The composition of metal B increases from 0% on the left-hand side to 100% on the right-hand side. Similarly, the composition of metal A decreases from 100% on the left-hand side to 0% on the right-hand side. The temperature increases from bottom to top.

The temperature may be represented either in degrees Celsius (°C), Fahrenheit (°F), or Kelvin (K). The temperatures can be easily converted from one scale into another using the following relationships:

$$K = {}^\circ C + 273 \tag{5.1}$$

$$^\circ C = \frac{F - 32}{1.8} \tag{5.2}$$

For example, 100°C will correspond to 373 K or 212°F, and 700°F will correspond to 371°C or 644 K. (Note that in the SI units, the symbol ° (representing degrees) is not used when the temperature is expressed in Kelvin.) Table 5.1 lists the equivalent temperatures in the Celsius, Fahrenheit, and Kelvin scales.

The composition of the alloy can be expressed in either atomic percentage, weight percentage, or mole percentage. *Atomic percentage* represents the number of atoms

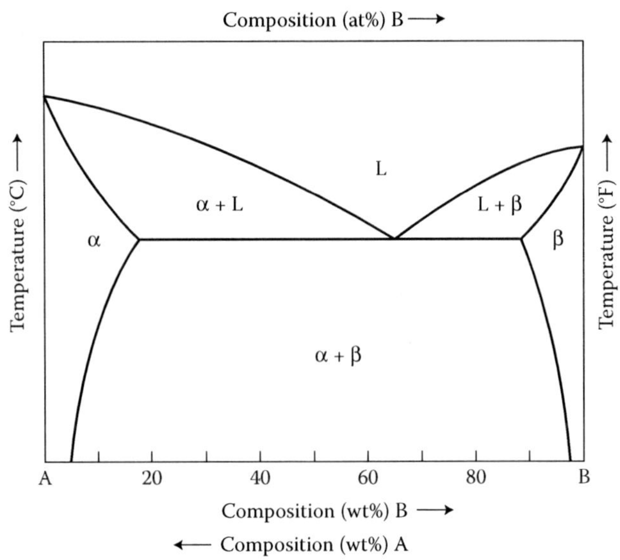

FIGURE 5.2 A typical binary phase diagram between two metals A and B. A eutectic reaction occurs in this alloy system.

of one kind as a fraction of the total number of atoms in the alloy, and is usually designated as at%. As an example, if an alloy contains two components and there are 90 atoms of A type and 10 atoms of B type, then

$$\text{at\% B} = \frac{10 \text{ atoms of B}}{90 \text{ atoms of A} + 10 \text{ atoms of B}} \times 100 = \frac{10}{100} \times 100 = 10\% \tag{5.3}$$

The *weight percentage*, on the other hand, represents the weight of one kind of atoms as a fraction of the total weight of the alloy, and is designated as wt%. Thus, if an alloy contains 70 g of metal A and 30 g of metal B, then the weight percentage of metal B can be calculated as

$$\text{wt\% B} = \frac{30 \text{ g of B}}{70 \text{ g of A} + 30 \text{ g of B}} \times 100 = \frac{30}{100} \times 100 = 30\% \tag{5.4}$$

The mole percentage similarly is the number of moles of one kind of component as a fraction of the total number of moles in the system. This is represented as mol%. The use of at% or wt% is more common for metallic or nonmetallic elements, whereas mol% is more commonly used to express the composition when mixtures of compounds are involved. For example, we use terms such as 30 at% Ni, 0.6 wt% C, or 20 mol% NiO. The values of at% and wt% can be converted from one into the

TABLE 5.1
Equivalent Temperatures in Celsius, Fahrenheit, and Kelvin Scales

Degrees Celsius (°C)	Degrees Fahrenheit (°F)	Degrees Kelvin (K)
−273	−459	0
−200	−328	73
−150	−238	123
−100	−148	173
−50	−58	223
−40	−40	233
−30	−22	243
−20	4	253
−10	14	263
0	32	273
10	50	283
20	68	293
30	86	303
40	104	313
50	122	323
60	140	333
70	158	343
80	176	353
90	194	363
100	212	373
150	302	423
200	392	473
250	482	523
300	572	573
350	662	623
400	752	673
450	842	723
500	932	773
550	1022	823
600	1112	873
650	1202	923
700	1292	973
750	1382	1023
800	1472	1073
850	1562	1123
900	1652	1173

continued

TABLE 5.1 (continued)
Equivalent Temperatures in Celsius, Fahrenheit, and Kelvin Scales

Degrees Celsius (°C)	Degrees Fahrenheit (°F)	Degrees Kelvin (K)
950	1742	1223
1000	1832	1273
1100	2012	1373
1200	2192	1473
1300	2372	1573
1400	2552	1673
1500	2732	1773
1600	2912	1873
1700	3092	1973
1800	3272	2073
1900	3452	2173
2000	3632	2273

other knowing the atomic weights of the individual components and using the following relationships:

$$(\text{at\%})_1 = \frac{(\text{wt\%/atomic weight})_1}{(\text{wt\%/atomic weight})_1 + (\text{wt\%/atomic weight})_2} \times 100 \tag{5.5}$$

$$(\text{wt\%})_1 = \frac{(\text{at\%} \times \text{atomic weight})_1}{(\text{at\%} \times \text{atomic weight})_1 + (\text{at\%} \times \text{atomic weight})_2} \times 100 \tag{5.6}$$

These relationships can be easily extended to alloy systems that contain more than two components as

$$(\text{at\%})_1 = \frac{(\text{wt\%/atomic weight})_1}{(\text{wt\%/atomic weight})_1 + (\text{wt\%/atomic weight})_2 + (\text{wt\%/atomic weight})_3 + \cdots} \times 100 \tag{5.7}$$

Example Problem 5.1

What is the composition, in weight percentage, of an alloy that consists of 90 at% gold and 10 at% copper?

Solution 5.1

We need the atomic weights of gold and copper to solve this problem. From standard reference books, the atomic weights of gold and copper are taken as 196.9 and 63.55 g mol^{-1}, respectively. Using Equation 5.6, we obtain

$$(\text{wt\%})_{\text{Cu}} = \frac{(\text{at\%} \times \text{atomic weight})_{\text{Cu}}}{(\text{at\%} \times \text{atomic weight})_{\text{Cu}} + (\text{at\%} \times \text{atomic weight})_{\text{Au}}} \times 100$$

$$= \frac{10 \times 63.55 \text{ g mol}^{-1}}{(10 \times 63.55 \text{g mol}^{-1}) + (90 \times 196.9 \text{g mol}^{-1})} \times 100$$

$$= 3.46 \text{ wt\% Cu}$$

Similarly, the wt% of Au can also be calculated using Equation 5.6. It will be

$$(\text{wt\%})_{\text{Au}} = \frac{(\text{at\%} \times \text{atomic weight})_{\text{Au}}}{(\text{at\%} \times \text{atomic weight})_{\text{Cu}} + (\text{at\%} \times \text{atomic weight})_{\text{Au}}} \times 100$$

$$= \frac{90 \times 196.9 \text{g mol}^{-1}}{(10 \times 63.55 \text{g mol}^{-1}) + (90 \times 196.9 \text{g mol}^{-1})} \times 100$$

$$= 96.54 \text{ wt\% Cu}$$

Note: Since we know that there are only two components present in the alloy, the wt% of the second component can be calculated as the difference between 100 and the wt% of the first component. Accordingly, the wt% of gold in the alloy can be calculated as the difference between 100 and the wt% of copper. Therefore,

$$\text{wt\% Au} = 100\text{–}3.46 \text{ (wt\% Cu)} = 96.54 \text{ wt\%}$$

Note that the result obtained, as is to be expected, is the same as above.

From the above example, it may be seen that even though the amount of copper is 10 at%, its wt% is much less because the atomic weight of copper is much less than that of gold. That is, the weight of copper atoms is much less than that of gold atoms. In other words, a lighter element will weigh much less than a heavier element! The weight of an atom can be calculated by dividing the atomic weight with Avogadro's number. Thus, the weight of a Cu atom is

$$\frac{63.55 \text{ g mol}^{-1}}{6.023 \times 10^{23} \text{ atoms mol}^{-1}} = 1.055 \times 10^{-22} \text{ g}$$

Similarly, the weight of an Au atom is

$$\frac{196.9 \text{ g mol}^{-1}}{6.023 \times 10^{23} \text{ atoms mol}^{-1}} = 3.269 \times 10^{-22} \text{ g.}$$

Example Problem 5.2

What is the composition, in atomic percentage, of an alloy that consists of 80 wt% Al and 20 wt% Si?

Solution 5.2

To solve this problem also, we need the atomic weights of Al and Si, which are 26.98 and 28.09 g mol^{-1}, respectively. Using Equation 5.5, we obtain

$$\begin{aligned}(\text{at\%})_{\text{Si}} &= \frac{(\text{wt\%/atomic weight})_{\text{Si}}}{(\text{wt\%/atomic weight})_{\text{Si}} + (\text{wt\%/atomic weight})_{\text{Al}}} \times 100 \\ &= \frac{(20 \text{ wt\%}/28.09 \text{ g mol}^{-1})_{\text{Si}}}{(20 \text{ wt\%}/28.09 \text{ g mol}^{-1})_{\text{Si}} + (80 \text{ wt\%}/26.98 \text{ g mol}^{-1})_{\text{Al}}} \times 100 \\ &= 19.36 \text{ at\% Si}\end{aligned}$$

Since in this case also, we have only two components, namely, Al and Si, the at% of Al can be obtained as the difference between 100 and the at% of Si. Therefore,

$$\text{at\% Al} = 100 - 19.36 \text{ (at\% Si)} = 80.64 \text{ at\%}$$

The above calculation brings out another important result, namely, that if the atomic weights of the two elements involved are close to each other, then their at% and wt% are very close to each other, that is, there exists only a small difference in the at% and wt% of an element.

Phase diagrams are classified as unary, binary, ternary, quaternary, quinary, and so on, depending on whether the number of components involved is one, two, three, four, five, and so forth. Although most engineering alloys contain a large number of components (the quantity of some of the components may be very small, less than 1%), representation of phase diagrams with more than two components on a two-dimensional plane is not easy. However, a convenient way of representing higher-order (more than two components) phase diagrams in two dimensions, for example, on a paper, is by considering only a section of the phase diagram. Consequently, a *ternary phase diagram* (containing three components) is represented by plotting the phase fields as a function of composition at a constant temperature (referred to as isothermal sections, and also assuming that the pressure is constant). That is, the complete constitution of the alloys is obtained by stacking one isothermal section over the other in the order of increasing temperature. Thus, the ternary phase diagram, at a constant temperature, is represented by an equilateral triangle, the vertices of which correspond to the pure components. Figure 5.3 represents a schematic of a ternary phase diagram.

Binary and ternary phase diagrams of common and commercially important alloys may be found in standard reference books. See, for example, *Binary Alloy Phase*

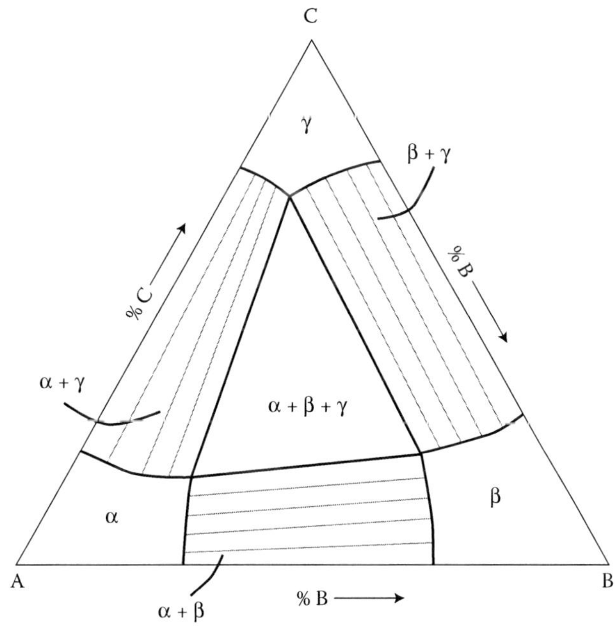

FIGURE 5.3 Schematic of a ternary phase diagram among metals A, B, and C. The horizontal lines in the two-phase fields represent the tie lines, which help in determining the compositions and amounts of the different phases present.

Diagrams edited by T. B. Massalski and published by ASM International in 1990 in three different volumes. Individual volumes containing binary phase diagrams based on a specific metal, for example, Al, Cu, Fe, Mg, Ni, and so on, have also been published by ASM International. Compendiums of ternary phase diagrams have also been published by ASM International and VCH Publishers.

5.4 THE PHASE RULE

We have so far been using terms like "phase," "component," and so forth. Let us now look into these terms a little more critically and see what we really mean by them. A *phase* may be defined as a homogeneous portion of a system that is physically distinct, chemically homogeneous, and mechanically separable from other portions of the system. A phase can exist in the solid, liquid, or vapor states. Two different phases will have a boundary between them and the physical, chemical, and mechanical characteristics will vary in a discontinuous and abrupt manner across the boundary. The liquid phase of any material is usually represented by the letter "L," whereas the solid phases are represented by the lower-case Greek letters—α, β, γ, δ, and so on.

A *component* is a distinct chemical substance from which the phase is formed. It may be understood as a chemically distinct and essentially indivisible substance. Components are normally pure elements or stable compounds. The combination of several components may constitute a phase. For example, when Ni dissolves in Cu to form a solid solution, it is one phase even though it contains two components. Thus, a phase may have more than one component but the reverse is not true. That is, a component cannot have more than one phase. But, it is possible that the same component may exist as another phase at a different temperature and/or pressure. For example, water may exist as ice (below 0°C), water (between 0°C and 100°C) or as steam (water vapor, at temperatures >100°C), even though it is one component. Thus, Cu, Au, Al, Si, and so on, which are pure metals, are components. Stable compounds such as H_2O, NaCl, NiO are also components even though they contain more than one element.

The relationship between the number of components and phases under the given conditions of temperature and pressure is determined by the *Gibbs phase rule*. The *phase rule** determines the number of phases (P) that can coexist at equilibrium with the minimum number of components (C) that can be used to form the system and the *degrees of freedom* (F). The phase rule is expressed as

$$F = C + 2 - P \tag{5.8}$$

The degrees of freedom (F) represent the number of independent variables—temperature, composition, or pressure—that can be changed without changing the number of phases for the given number of components.

The phase rule mentioned above can also be rewritten as $P + F = C + 2$ to easily remember it by the mnemonic: **P**olice **F**orce is equal to **C**ops + **2**.

Most engineering materials are processed at or near atmospheric pressure and so we normally deal with the condensed phases only (liquid and solid). Consequently, pressure can be considered as fixed at 1 atm, that is, pressure is no longer a variable.

* Josiah Willard Gibbs (1839–1903) was perhaps the most distinguished American physicist, chemist, and mathematician of the nineteenth century. He was born in New Haven, Connecticut on February 11, 1839 and spent all his life there except for three years when he went to study in Europe. Gibbs studied in Yale College and was awarded the first PhD degree in engineering in the United States in 1863 from the Sheffield Scientific School at Yale. After teaching for three years at Yale, he went to study in Paris, Berlin, and Heidelberg and returned to Yale in 1869 to be appointed a Professor of Mathematical Physics in 1871. The appointment was unpaid at first, since he did not publish anything yet. Majority of his work was published in the *Transactions of the Connecticut Academy of Sciences*, a rather obscure journal edited by his brother-in-law. Between 1876 and 1878 Gibbs wrote a series of papers on the graphical analysis of multiphase chemical systems, which were eventually published together in a monograph titled *On the Equilibrium of Heterogeneous Substances*. This is considered his most important contribution to physical sciences, and which is unquestionably among the greatest and most enduring monuments of the wonderful scientific activity of the nineteenth century. In this work, he defined what is now called the Gibbs free energy as the maximum amount of useful work that can be extracted from a system in the absence of any volume change or heat interaction, a concept which is of fundamental importance in chemical reactions. This work did not get much recognition until it was translated into German in 1892 by Wilhelm Ostwald and into French in 1899 by Henri Louis Le Chatelier. Gibbs' contributions, however, were not fully recognized until sometime after the 1923 publication of Gilbert N. Lewis and Merle Randall's *Thermodynamics and the Free Energy of Chemical Substances*, which introduced Gibbs' methods to chemists worldwide. These methods also became much of the foundation for chemical engineering. Several Nobelists were influenced by the work of Gibbs. In 1901, Gibbs was awarded the highest possible honor granted by the international scientific community of his day, granted to only one scientist each year: the Copley Medal of the Royal Society of London. The American Chemical Society established the Willard Gibbs Medal in his memory in 1910. Gibbs died on April 28, 1903 in New Haven, Connecticut.

Therefore, the phase rule can be written in a modified form (for the condensed systems) as

$$F = C + 1 - P \tag{5.9}$$

It is this form of the phase rule that we will use to determine the degrees of freedom in any given system.

It is important to realize that the phase rule needs to be obeyed in every region of every phase diagram. (The English language has a proverb: every rule has its exceptions. But, there are no exceptions to the phase rule! It has been stated that Gibbs phase rule, a general theorem of thermodynamics, is so fundamental that it is unlikely ever to be overthrown.)

5.5 APPLICATION OF THE PHASE RULE

Let us now demonstrate how the phase rule can be used to determine the accuracy of the phase diagram. As mentioned earlier, there are no exceptions to the phase rule and every region of the phase diagram has to obey the phase rule.

Consider the hypothetical binary phase diagram as shown in Figure 5.4a. Note that this diagram is very similar to that presented in Figure 5.2. Also, recall that a binary phase diagram involves two components and let these two components be A and B.

Let us now apply the phase rule (Equation 5.9) to different regions in the phase diagram (Figure 5.4a). Consider point "1" corresponding to the melting point of pure component A. From thermodynamic principles, we know that both the solid and liquid phases are in equilibrium at the melting point. Therefore, at the melting point of A, liquid A and solid A are in equilibrium and therefore the number of phases, $P = 2$. Further, since we are considering only the pure component, $C = 1$. Substitution of these values of C and P in Equation 5.9 gives

$$F = 1 + 1 - 2 = 0 \tag{5.10}$$

Therefore, the degrees of freedom are zero. That is, we cannot change the temperature (and we cannot anyway change the composition since we are dealing with a single component) and hope to have the two phases (liquid A and solid A) in equilibrium. A similar argument applies to the melting point of pure component B.

Let us now consider a situation when the binary alloy exists as a single phase. Consider point "2" corresponding to the liquid phase, L. This is a single-phase region and therefore $P = 1$. Further, the number of components is 2, that is, $C = 2$. Substitution of these values into Equation 5.9 gives the result

$$F = 2 + 1 - 1 = 2 \tag{5.11}$$

Thus, there are two degrees of freedom. That is, to fully describe the alloy, we can specify up to two variables and these are composition and temperature. In other words, we can maintain the single-phase condition of the alloy while varying either the temperature or composition or both. A similar situation exists in the other single-phase regions, namely, the α or β phase regions.

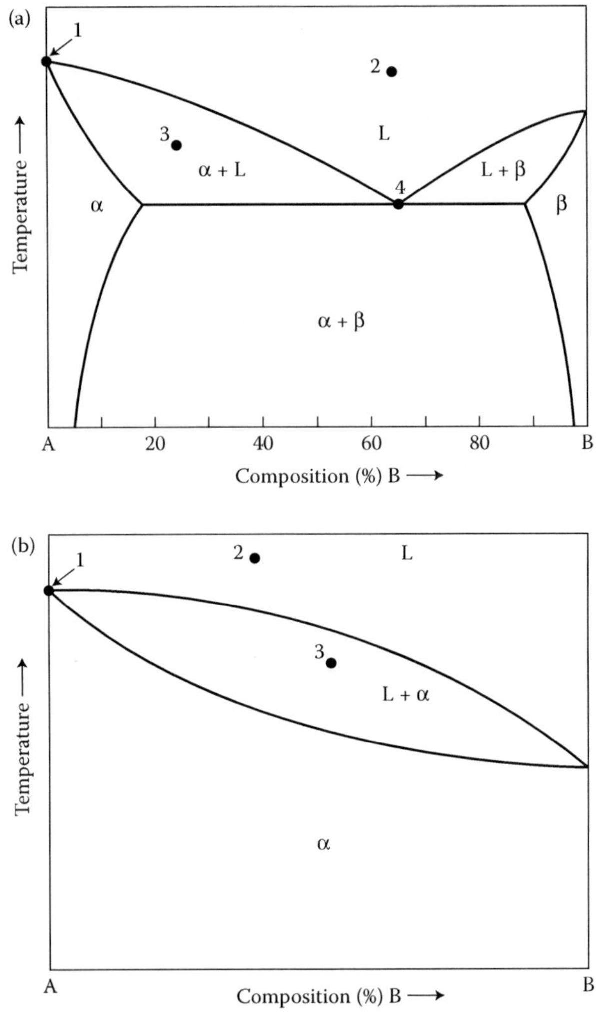

FIGURE 5.4 Schematics of two different binary phase diagrams to determine whether the phase rule is followed in every portion of the phase diagram. (a) Eutectic system and (b) isomorphous system.

Let us now apply the phase rule to point "3," where the liquid phase, L and the solid phase, α coexist. At this point, we have two components (A and B), that is, $C = 2$, and there are two phases (α and L), that is, $P = 2$. Substitution of these values into Equation 5.9 yields

$$F = 2 + 1 - 2 = 1 \tag{5.12}$$

Therefore, we can specify either the temperature or the composition, but not both, to completely define the system. As an example, let us assume that we have specified the temperature in the two-phase (α + L) region. At that temperature, the compositions of the α and L phases are fixed, that is, the compositions cannot be changed. Instead of defining the temperature, as described above, we could also define the composition of any one of the phases in this two-phase region. Let us assume that the composition of the α phase in equilibrium with the L phase is fixed. Then, the composition of the L phase also is fixed. That is, both the temperature and composition of the phases are fixed.

Let us now concentrate on point "4." At this point, the alloy under equilibrium contains three phases, namely, L, α, and β, that is, $P = 3$. Since this is a binary alloy system, $C = 2$. Substituting these values into Equation 5.9, we get

$$F = 2 + 1 - 3 = 0 \tag{5.13}$$

When $F = 0$, it means that there are no degrees of freedom. That is, we cannot change either the composition or the temperature. Points at which $F = 0$ are known as *invariant points*. The melting points of the two components, for example, are invariant points. In more complex phase diagrams than that shown in Figure 5.4a, we will come across many more invariant points.

A similar situation exists in other types of phase diagrams also. For example, consider the binary isomorphous phase diagram represented in Figure 5.4b. An isomorphous system is one in which the two components are completely soluble in each other in the solid state. (It is assumed, and it is true, that most metals are completely soluble in each other in the liquid state.) Here also, points "1," "2," and "3" have similar situations as described above for Figure 5.4a. But, we cannot have a situation like point "4" in Figure 5.4a in Figure 5.4b, because the maximum number of phases present in this phase diagram is only 2. Therefore, irrespective of the complexity of the phase diagram, it can always be subdivided into smaller portions, where the number of phases is 1, 2, or 3.

According to Equation 5.9, in a binary alloy, $F = 0$, that is, there are no degrees of freedom, only when the number of phases in equilibrium is 3. Note that for the invariant points, $F = 0$ and so neither the composition nor the temperature can be changed, while maintaining equilibrium. The invariant points in any alloy system help us in identifying the maximum number of phases that could coexist under equilibrium conditions at a given composition and temperature. Recalling Equation 5.9,

$$F = C + 1 - P$$

note that for the invariant points, that is, when $F = 0$,

$$0 = C + 1 - P, \quad \text{or}$$

$$P = C + 1 \tag{5.14}$$

That means, the maximum number of phases that could exist in an alloy system (at a given temperature and alloy composition) is the number of components plus 1. That is, the maximum number of phases that could exist in a binary system ($C = 2$),

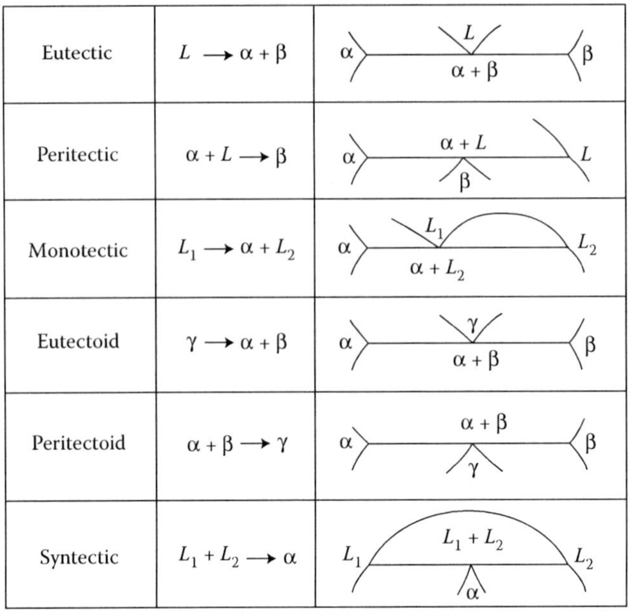

FIGURE 5.5 Schematic representation of several different types of invariant reactions that occur in binary alloy systems.

at a given temperature and composition, is (2 + 1 =) 3, that in a ternary system ($C = 3$) it is (3 + 1 =) 4, in a quaternary system ($C = 4$) it is 5, and so on.

Reactions that take place at a constant temperature between phases (three in the case of binary systems, four in the case of a ternary alloy systems, etc.) are called *invariant reactions.* Many different types of invariant reactions occur in alloy systems. The common invariant reactions involving three different phases and occurring in binary alloy systems are represented in Figure 5.5. We will see some of the invariant reactions (peritectic, eutectic, and eutectoid) occurring in the Fe–C phase diagram, to be discussed later.

5.5.1 Determination of Compositions of Phases

The compositions of phases can be easily determined from the phase diagram. Let us consider the phase diagram shown in Figure 5.6, with the liquidus and solidus lines labeled. Note that this is similar to the phase diagram shown in Figure 5.4b. If we are in a single-phase region, then the composition of the phase is the same as the overall composition of the alloy. The determination is, however, nontrivial when two phases coexist. For example, let us assume that we are interested in determining the compositions of the two phases at a temperature T when the overall composition of the alloy is C_o. The alloy now contains two phases, namely, liquid (L) and solid solution (α). We can use the following procedure to determine the compositions of the two phases.

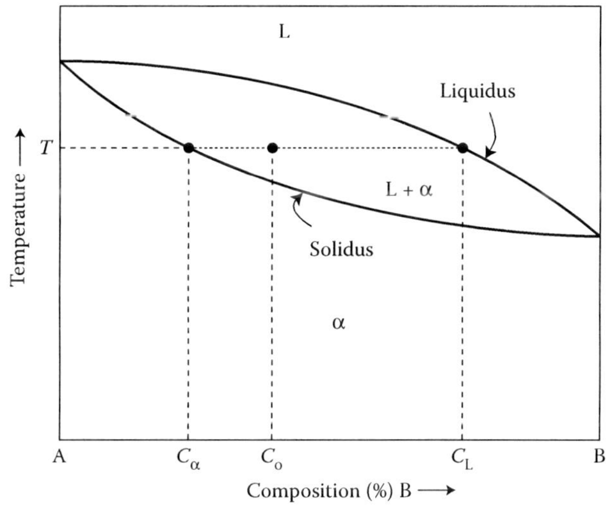

FIGURE 5.6 Determination of compositions and amounts (proportions) of phases in a binary alloy system. The composition of the phase is the same as the overall composition of the alloy, if it exists as a single phase. In a two-phase region, the compositions of the two phases are determined by drawing a tie line (a line representing a constant temperature) and finding the points of intersection of the tie line with the phase boundaries. The amounts of phases are determined using the lever rule.

1. Draw a *tie line* (a line representing a constant temperature, also referred to as an *isotherm*, and extending the full-width of the two-phase region).
2. Determine the points of intersection of the *tie line* with the phase boundaries.
3. The point of intersection of the tie line with the *liquidus* line (the locus of temperatures corresponding to different compositions above which liquid is the only stable phase, that is, the L/(α + L) boundary) gives the composition of the liquid phase, C_L. The composition of the phase can be determined by dropping a vertical line from that point to the composition axis and reading off the composition.
4. Similarly, the point of intersection of the tie line with the *solidus* line (the locus of temperatures corresponding to different compositions below which solid is the only stable phase, that is, the (α + L)/α boundary) gives the composition of the solid phase, C_α.

Thus, for an alloy of overall composition C_o, the compositions of the solid and liquid phases at temperature T are C_α and C_L, respectively. A corollary of the above procedure is that, irrespective of the overall composition of the alloy, the compositions of the two phases are fixed at any given temperature. But, they are different at different temperatures.

A procedure similar to the above can be used in any two-phase region of a phase diagram, irrespective of the nature of the phase diagram.

5.5.2 Determination of Amounts of Phases

In addition to determining the compositions of the coexisting phases as described above, it is also possible to calculate the amounts of the phases, either in terms of fractions or percentages. The determination is again trivial in the case of single-phase regions. Since, only one phase is present, its fraction is 1.0 or the percentage amount is 100%. The solution is a little more difficult in the case of two-phase regions and it is here that it is also more useful. This can be done using the concept of *Lever Rule*.

The procedure is described again with reference to Figure 5.6, wherein we have a phase diagram representing complete solid solubility of one component in the other in both the solid and liquid states, that is, an isomorphous system. But, the calculations are equally applicable to any system which contains a mixture of two phases—whether two liquids, two solids, or one solid and one liquid phase. Let us choose the overall composition of the alloy again as C_o and assume that we are interested in determining the fractions (or percentages) of the two phases at a temperature *T*. The following procedure can be adopted to obtain the result:

1. Draw the tie line (a horizontal line representing a constant temperature) at the desired temperature *T* and extend it to the entire width of the two-phase region.
2. Determine the compositions of the solid and liquid phases as points of intersections of the tie line with the solidus and liquidus lines, respectively. These are C_α and C_L, respectively, for the α and *L* phases. This step is the same as described above in Section 5.5.1.
3. The fraction of the liquid phase is determined by taking the length of the tie line away from the liquid phase (liquidus point) and dividing it by the total length of the tie line. The length of the tie line can be measured either in terms of the actual lengths using a ruler (but, this will not provide accurate values) or by subtracting the compositions. Thus, in the present case, the length of the tie line away from the liquid phase is $C_o – C_\alpha$ and the total length of the tie line is $C_L – C_\alpha$. Thus, the fraction of the liquid phase is

$$\text{Fraction of the liquid phase, } f_L = \frac{C_o - C_\alpha}{C_L - C_\alpha} \quad (5.15)$$

4. The percentage of the L phase is obtained by multiplying the fraction of the phase with 100. That is,

$$\text{Percentage of the liquid phase} = f_L \times 100 = \frac{C_o - C_\alpha}{C_L - C_\alpha} \times 100 \quad (5.16)$$

5. The fraction of the other phase (α, in this case) is also determined in a similar way. The fraction of the α phase is obtained by dividing the length of the lever away from the solidus point (solid phase α) with the total length of the lever. Therefore,

$$\text{Fraction of the } \alpha \text{ phase, } f_\alpha = \frac{C_L - C_o}{C_L - C_\alpha} \tag{5.17}$$

$$\text{Percentage of the } \alpha \text{ phase} = f_\alpha \times 100 = \frac{C_L - C_o}{C_L - C_\alpha} \times 100 \tag{5.18}$$

6. Alternately, the fraction of the α phase can also be determined as the difference between 1 and the fraction of the liquid phase. And, the percentage of the α phase is obtained as the difference between 100 and the percentage of the liquid phase.

Note that the percentage of the phase will be the weight percentage of the phase when the compositions of the two components are expressed in wt% of the solute (or solvent) element.

5.6 DERIVATION OF LEVER RULE

The lever rule equations mentioned above (Equations 5.15 through 5.18) can be derived by considering the mass balance between the constituents. Let us consider the phase diagram represented in Figure 5.6, and choose the overall alloy composition as C_o. Let us also assume that we are interested in determining the proportion of the phases at a temperature *T*.

Since the alloy contains only two phases, namely, L and α, the sum of the fractions of the two phases must be equal to 1, that is,

$$f_\alpha + f_L = 1 \tag{5.19}$$

Note that C_o of the solute B is distributed between both the liquid (L) and solid (α) phases. If the wt% of B in the α phase C_α and that in the liquid phase is C_L, then we can write

$$C_o = f_\alpha C_\alpha + f_L C_L \tag{5.20}$$

Since we know that $f_\alpha + f_L = 1$, we can write that $f_L = 1 - f_\alpha$. Substituting this value of f_L into Equation 5.20, we get

$$C_o = f_\alpha C_\alpha + (1 - f_\alpha) C_L \tag{5.21}$$

Expanding the terms,

$$C_o = f_\alpha C_\alpha + C_L - f_\alpha C_L$$

or

$$f_\alpha C_L - f_\alpha C_\alpha = C_L - C_o$$

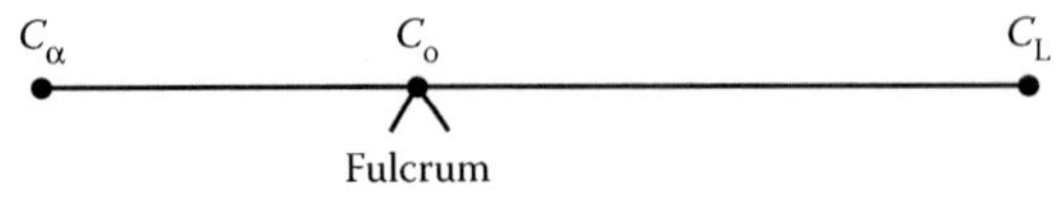

FIGURE 5.7 Concept of a lever to calculate phase proportions in a two-phase mixture. The overall composition of the alloy is represented as the fulcrum of the lever and the compositions of the two phases as the two ends of the lever.

that is,

$$f_\alpha(C_L - C_\alpha) = C_L - C_o$$

Rearranging the above terms, we obtain

$$f_\alpha = \frac{C_L - C_o}{C_L - C_\alpha}$$

and this equation is identical to Equation 5.17 above.

The above expressions for determining the fractions of the phases can also be appreciated by considering the three compositions—C_o, C_α, and C_L—to represent the fulcrum and the two ends of a lever as shown in Figure 5.7. Imagine that two weights, each representing the weight fraction (or percentage) of the phases (*not* the compositions of the phases), are suspended from the two ends of the lever. To maintain the balance for the lever, that is, to maintain the lever in a horizontal position, the fulcrum will be shifted toward the heavier side of the weights. That is why, to calculate the weight fractions of the phases, one measures the length of the segment of the lever away from the actual phase of interest. Obviously, if the two weights are equal, the fulcrum will be exactly in the middle of the lever.

Example Problem 5.3

If a binary Pb–Sn alloy containing 40 wt% Sn is maintained at a temperature of 200°C, it contains two phases. The phases are the α solid solution of Sn in Pb containing 17 wt% Sn and the Pb–Sn liquid phase containing 55 wt% Sn. Calculate the proportions of the two phases in the alloy.

Solution 5.3

The solution to this problem requires using the Lever Rule. Using the designation employed in deriving the lever rule, for the alloy under consideration,

$$C_o = 40 \text{ wt\%Sn}, \quad C_\alpha = 17 \text{ wt\% Sn}, \quad \text{and} \quad C_L = 55 \text{ wt\% Sn}$$

Using Equation 5.15, we can calculate the fraction of the liquid phase as

$$\text{Fraction of the liquid phase, } f_L = \frac{C_o - C_\alpha}{C_L - C_\alpha}$$

$$= \frac{40 - 17}{55 - 17} = 0.605 \text{ or } 60.5\%$$

Since the alloy contains only two phases, the fraction of the α phase is $1-0.605 = 0.395$ or 39.5%.

It is important to realize that the fraction or *percentage* of the phase is different from the *composition* of the phase. The chemical composition of the phase is given as the wt% or at% of the solute (or solvent) element. Note that the total solute (or solvent) content of the alloy is distributed between the two phases. On the other hand, the percentage of the phase is the fraction of the phase as a ratio of the total weight of the alloy (containing both the phases). That is why these two have different values.

When two metals (or for that matter any two elements or compounds) are mixed together, several different types of new phases may form depending on the composition and temperature at which this mixing occurs and the thermodynamic behavior of the system. The types of phases formed include solid solutions, intermetallic phases, compounds, or mixtures of these. The complexity of the phase diagrams depends on the interaction behavior of the components. Thus, a phase diagram may be as simple as that shown in Figure 5.6 or it may contain two or more solid solutions (Figure 5.2), and even some intermetallic or compound phases.

Recall that when we mentioned that the maximum number of phases that could be present in equilibrium is $C + 1$, where C represents the number of components, it applies only to a given composition and temperature. But, other different phases could exist at other compositions and temperatures. Thus, in the full phase diagram, a binary alloy system may contain many different phases. For example, the binary Cu–Al phase diagram is very complex and contains 16 different phases and the binary Au–Mn system is even more complex and contains 42 different phases!

Figure 5.8 shows some typical phase diagrams. An isomorphous phase diagram was represented in Figure 5.6. It may be noted from Figure 5.8 that a eutectic reaction is represented in Figure 5.8a in the Pb–Sn system, a peritectic reaction in Figure 5.8b between Ni and Ru, and a series of peritectic reactions and a eutectoid reaction in Figure 5.8c in the Cu–Zn system. Note that in these phase diagrams, the different reactions are taking place at a constant temperature (even though the temperatures are different in different alloy systems), known as the invariant temperature. Also recall that at the invariant points, three phases are in equilibrium in a binary alloy system and therefore the number of degrees of freedom is zero. In addition to the invariant reactions presented in these phase diagrams, other reactions are also possible. Figure 5.5 has presented several examples of the invariant reactions involving three phases that occur in binary alloy systems.

With this background information on phase diagrams, let us now look at the iron–carbon phase diagram, its special features, and the types of microstructures obtained at different compositions.

5.7 THE IRON–CARBON PHASE DIAGRAM

Pure iron exists in three different *allotropic forms*—α, γ, and δ. *Allotropy* (or *polymorphism*) is the phenomenon wherein elements or compounds may exist in more than one crystalline form under different conditions of temperature and/or pressure.

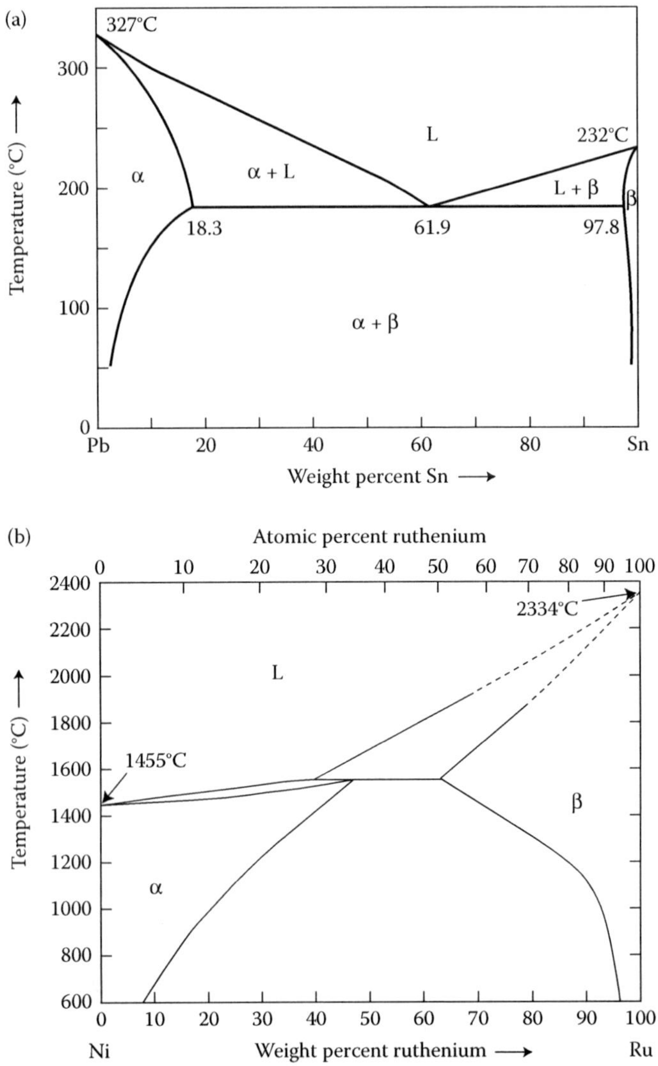

FIGURE 5.8 Different types of phase diagrams showing eutectic, peritectic, and a series of peritectic reactions. (a) Pb–Sn phase diagram features a eutectic reaction, (b) Ni–Ru phase diagram features a peritectic reaction, and (c) Cu–Zn phase diagram shows a series of peritectic reactions and a eutectoid reaction. (After ASM International, Materials Park, OH. With permission.)

α-Fe has a BCC structure and is stable up to 912°C, γ-Fe has an FCC structure and is stable between 912°C and 1394°C, and δ-Fe, which also has a BCC structure, is stable from 1394°C up to the melting point of iron, namely, 1538°C. The main difference between α-Fe and δ-Fe is that since δ-Fe exists at a higher temperature, it has a slightly larger lattice parameter than α-Fe. Iron also exhibits a magnetic transition from the ferromagnetic to the paramagnetic state at 770°C. Assuming that this

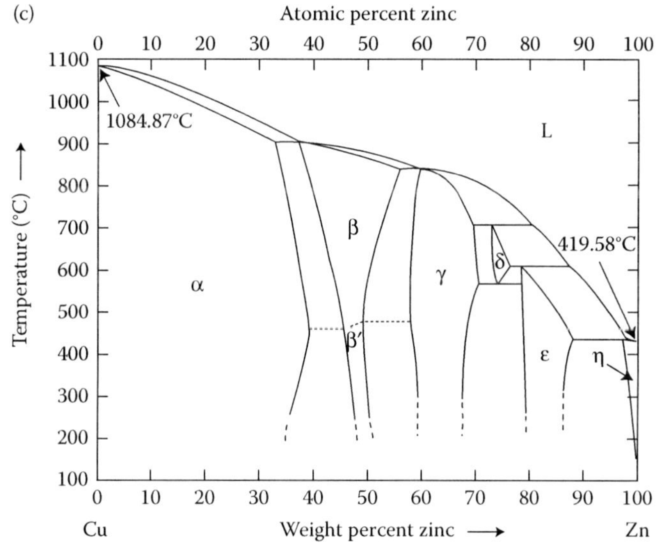

FIGURE 5.8 *Continued*

magnetic change represented a true structural change, the paramagnetic phase was earlier labeled (wrongly) as the β-phase. But, since this is not a true allotropic transformation, this designation of the paramagnetic phase as the β-phase of iron has now been dropped. That is why, in the current literature, you will only find the α, γ, and δ phases and not the β phase.

5.7.1 Solid Solubility Limits

Figure 5.9 shows the Fe–C phase diagram. The phase diagram features four solid phases and one liquid phase. The four solid phases are α-Fe (α-ferrite), γ-Fe (austenite),* δ-Fe (δ-ferrite), and Fe_3C (cementite). α-Fe (α-ferrite) is an interstitial solid solution of carbon in α-iron with a BCC structure and has a maximum solid solubility of 0.021 wt% C in Fe at the eutectoid temperature of 740°C. γ-Fe (austenite) is an interstitial solid solution of carbon in γ-Fe with an FCC structure and has a maximum solid solubility of 2.1 wt% C at the eutectic temperature of 1153°C. δ-Fe (δ-ferrite) is an interstitial solid solution of carbon in δ-Fe with a BCC structure and

* William Chandler Roberts-Austen (1843–1902), in whose honor the γ-Fe solid solution was named Austenite, was born on March 3, 1843 in Kennington, UK and studied at the Royal School of Mines (RSM) in London. He was a metallurgist noted for his research on the physical properties of metals and their alloys. In 1870 he was appointed to the Chair of Metallurgy at the RSM and started his research into the structure and properties of steel. He developed procedures for the analysis of alloy constituents and an automatic recording pyrometer used to record temperature changes in furnaces and molten materials. He became a world authority on the technical aspects of minting coins. His work had many practical and industrial applications. He was professor of metallurgy at the Royal School of Mines in London from 1882 to 1902, made a Fellow of the Royal Society in 1875, and was knighted in 1899. He passed away on November 22, 1902 in London.

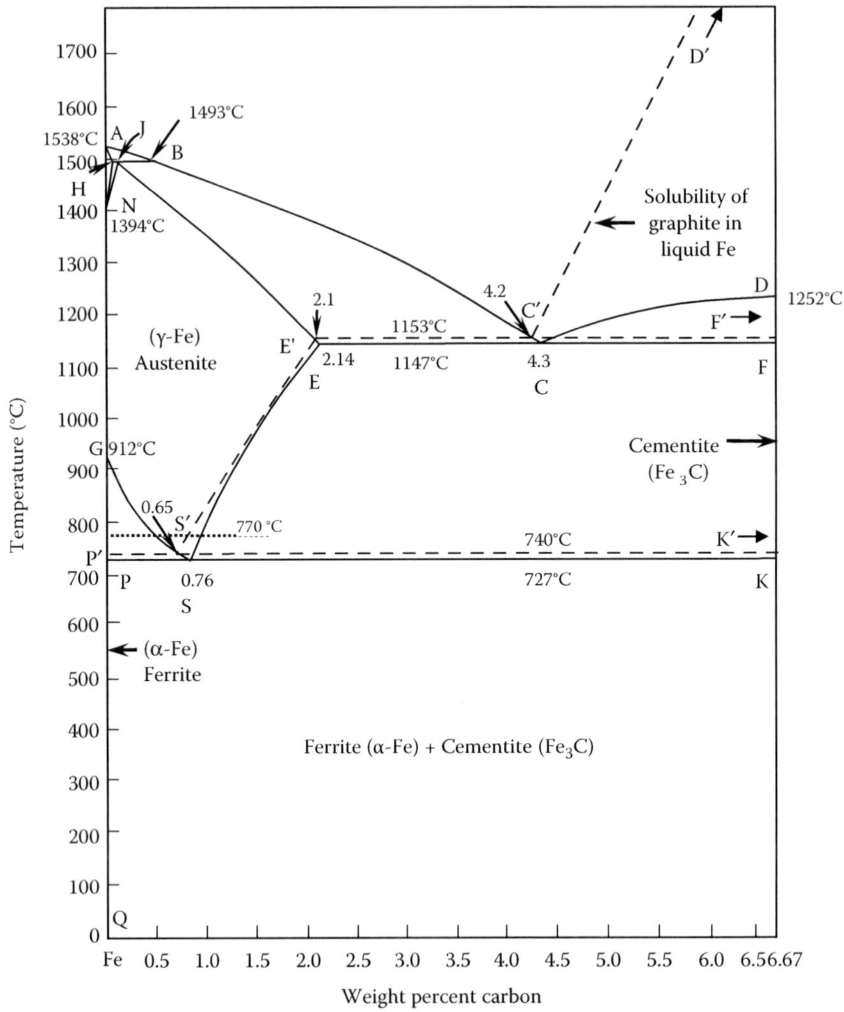

FIGURE 5.9 The Fe–C phase diagram. Note that the continuous lines represent the Fe–Fe_3C phase diagram and that the dotted lines the Fe–C phase diagram. (After ASM International, Materials Park, OH. With permission.)

has a maximum solid solubility of 0.09 wt% C at the peritectic temperature of 1493°C. Cementite (Fe_3C) is an iron–carbon compound with an orthorhombic crystal structure and contains 6.67 wt% or 25 at% carbon.

From now onward, we will be using the symbols α, γ, and δ only (and drop–Fe) to represent the α-ferrite (α-Fe), austenite (γ-Fe), and δ-ferrite (δ-Fe) phases, respectively. Further, since δ-ferrite is not of much practical importance, if the term ferrite is mentioned, it will automatically refer to α-ferrite. Crystal structure details of the iron–carbon alloy phases are listed in Table 5.2. Recall that Fe–C alloys containing

TABLE 5.2
Crystal Structure Details of the Iron–Carbon Alloy Phases

Phase	Composition wt% (at%) C	Crystal Structure	Lattice Parameters (nm)		
			a	b	c
Pure iron	0 (0)	BCC	0.28665	—	—
α-Ferrite	0–0.021 (0.096)	BCC	0.28665[a]	—	—
γ-Austenite	0–2.1 (9.06)	FCC	0.36544[a]	—	—
δ-Ferrite	0–0.09 (0.40)	BCC	0.29315[a]	—	—
Fe_3C (cementite)	6.67 (25)	Orthorhombic	0.5090	0.6748	0.4523
Carbon[b]	100	Hexagonal	0.247	—	0.693
Carbon	100	Diamond cubic	0.3570	—	—

Note: The structure of pure Fe is BCC at room temperature. But, as mentioned in the text, Fe also exists in the γ-form, with an FCC structure, between 912°C and 1394°C, and in the δ-form, with a BCC structure, from 1394°C up to the melting point of 1538°C.

[a] The lattice parameters are quoted for pure Fe and not for the solid solutions. The lattice parameter of the solid solution will be larger than that of the pure metal.

[b] Other forms of carbon also exist.

up to 2.1 wt% C are referred to as steels and those containing from 2.1 to 6.67 wt% C as cast irons.

Even though it is not common nowadays, all the important (or critical) points in the Fe–C phase diagram were used to be labeled with letters. For example, point A represents the melting point of pure iron, point G represents the $\alpha \rightarrow \gamma$ transformation temperature in pure iron, point S represents the eutectoid composition, and so on. This labeling is also included in Figure 5.9. The compositions and temperatures corresponding to these critical points are listed in Table 5.3.

5.7.2 Special Features of the Fe–C Phase Diagram

The Fe–C phase diagram is special in more than one way. First, the diagram is drawn only up to 6.67 wt% C, which corresponds to the composition of cementite, Fe_3C. Normally, a phase diagram is represented from one pure component to the other. That is, in this case, it should have been represented from pure iron up to pure carbon. But, Fe–C alloys beyond 6.67 wt% C are commercially not useful. Therefore, the phase diagram is represented between pure Fe and Fe_3C (cementite) as the two components. Consequently, this diagram is frequently referred to as the Fe–Fe_3C phase diagram (or the pseudobinary Fe–C phase diagram). Note that the invariant reaction temperatures and compositions are slightly different between the Fe–C and Fe–Fe_3C phase diagrams.

Second, Fe_3C (cementite) is a metastable phase. That is, given sufficient time, Fe_3C decomposes into its more stable components of Fe and C. However, for all practical purposes, Fe_3C may be considered a "stable" phase since it takes an impractically

TABLE 5.3
Important Compositions and Temperatures in the Fe–Fe_3C and Fe–C Phase Diagrams

	Fe–Fe_3C Phase Diagram			Fe–C Phase Diagram	
Point	Temperature (°C)	Composition, wt% (at%)	Point	Temperature (°C)	Composition, wt% (at%)
A	1538	0.0 (0.0)	A	1538	0.0 (0.0)
B	1493	0.53 (2.43)	B	1493	0.53 (2.43)
C	1147	4.3 (17.3)	C′	1153	4.2 (17.1)
D	1250	6.67 (25)	D′	[a]	100 (100)
E	1147	2.14 (9.23)	E′	1153	2.1 (9.06)
F	1147	6.67 (25)	F′	1153	100 (100)
G	912	0.0 (0.0)	G	912	0.0 (0.0)
H	1493	0.09 (0.4)	H	1493	0.09 (0.4)
J	1493	0.16 (0.74)	J	1493	0.16 (0.74)
K	727	6.67 (25)	K′	740	100 (100)
L	20	6.67 (25)	L′	20	100 (100)
M	768	0.0 (0.0)	M	768	0.0 (0.0)
N	1394	0.0 (0.0)	N	1394	0.0 (0.0)
P	727	0.022 (0.104)	P′	740	0.021 (0.096)
Q	20	0.006 (0.028)	Q	20	0.006 (0.028)
S	727	0.76 (3.46)	S′	740	0.65 (2.97)

[a] The exact value is not known.

long time for Fe_3C to transform to the equilibrium phases at room temperature. For this reason, this phase diagram is also referred to as the metastable Fe–C equilibrium diagram.

Because of the above two reasons, it is the Fe–Fe_3C phase diagram that is commonly used to determine the phase boundaries and the temperature regimes in which the different phases are stable. Note again that these compositions and temperatures are used to plan heat treatment operations. Consequently, from now onward we will use the compositions and temperatures pertaining to the Fe–Fe_3C phase diagram and *not* the Fe–C phase diagram. The important compositions and temperatures pertaining to the Fe–Fe_3C system are also summarized in Table 5.3.

5.7.3 Invariant Reactions

The Fe–Fe_3C phase diagram features three invariant reactions—peritectic, eutectic, and eutectoid. These reactions are reversible in the sense that they occur both on heating and cooling. The only difference is that if the heating or cooling is not conducted very slowly (resembling equilibrium conditions), the actual temperature at which the invariant reaction occurs may be higher during heating and lower during cooling. The magnitude of increase (or decrease) in temperature depends on the rate of heating (or cooling). (This change in reaction temperature is only due to kinetic effects, but thermodynamically the reaction should still occur at the temperature indicated in the phase diagram.) The higher/lower the temperature at which the reaction occurs, the higher is the heating/cooling rate.

In the *peritectic reaction*, Fe–C liquid with 0.53 wt% C and δ-ferrite with 0.09 wt% C combine to produce the γ (austenite) phase containing 0.16 wt% C on cooling at the peritectic temperature of 1493°C. This reaction can be represented as

$$\underset{0.53\text{ wt\% C}}{L} + \underset{0.09\text{ wt\% C}}{\delta} \xleftrightarrow{1493^\circ\text{C}} \underset{0.16\text{ wt\% C}}{\gamma} \tag{5.22}$$

In the *eutectic reaction* a liquid with 4.3 wt% C transforms to two solid phases, namely, γ (austenite) with 2.14 wt% C and Fe_3C (cementite) containing 6.67 wt% C, on cooling at the eutectic temperature of 1147°C. This reaction is represented as

$$\underset{4.30\text{ wt\% C}}{L} \xleftrightarrow{1147^\circ\text{C}} \underset{2.14\text{ wt\% C}}{\gamma} + \underset{6.67\text{ wt\% C}}{Fe_3C} \tag{5.23}$$

The third invariant reaction in the Fe–C phase diagram, and undoubtedly the most important for heat treatment purposes, is the *eutectoid reaction*. In this reaction, γ (austenite) with 0.76 wt% C transforms to two solid phases—α (ferrite) with 0.022 wt% C and Fe_3C (cementite) with 6.67 wt% C—on cooling at the eutectoid temperature of 727°C. This reaction is represented as

$$\underset{0.76\text{ wt\% C}}{\gamma} \xleftrightarrow{727^\circ\text{C}} \underset{0.022\text{ wt\% C}}{\alpha} + \underset{6.67\text{ wt\% C}}{Fe_3C} \tag{5.24}$$

It may be noted that in both the eutectic and eutectoid reactions two new solid phases are formed on cooling from a high-temperature phase. The only difference is that in the eutectic reaction, the high-temperature phase is a liquid while in the eutectoid reaction, the high-temperature phase is a solid. That is, the eutectoid reaction is an all solid-state reaction. In fact, all solid-state invariant reactions are designated with the suffix-*oid*. For example, we can also have a peritectoid reaction or a monotectoid reaction.

Of the three invariant reactions in the Fe–C phase diagram, the peritectic reaction is the least important for commercial purposes, since steels are very rarely heated to the high temperatures at which the peritectic reaction occurs. The eutectoid reaction is the most important since steels are heated to above the eutectoid temperature so that the steel exists in the austenitic condition, from where it is cooled at a controlled rate to obtain different microstructures and consequently achieve different mechanical properties. The eutectic reaction is important in cast irons.

Steels containing less than the eutectoid carbon content (<0.76 wt% C) are referred to as *hypoeutectoid steels* and those that contain more than the eutectoid carbon content (>0.76 wt% C) are referred to as *hypereutectoid steels*. Steels that contain the eutectoid carbon content (0.76 wt% C) are referred to as *eutectoid steels*. The transformation temperatures (or phase boundaries) are also commonly referred to by letter symbols. For example, the line PSK (the eutectoid temperature) is referred to as A_1, and the line GS as A_3, the temperature above which austenite is the only stable phase in hypoeutectoid steels. Similarly, the line SE is referred to as A_{cm}, the temperature above which austenite is the only stable phase in hypereutectoid steels (i.e., cementite dissolves in austenite and is no longer present). Depending on whether the steels are being heated or cooled, these designations are labeled with the subscripts c and r. That is, the A_1 temperature is labeled A_{c1} when the steel is heated or A_{r1} when the steel is cooled. The subscripts c and r are abbreviations for the French words *chauffage* (for heating) and *refroidissement* (for cooling).

5.8 COOLING BEHAVIOR AND MICROSTRUCTURAL DEVELOPMENT

Let us now look at the types of microstructure that develop when steels of different carbon content are cooled from the austenitic region. Recall that austenite (γ-phase) is the high-temperature phase existing above 912°C in pure iron and is the interstitial solid solution of carbon in γ-iron. It may also be noted from the Fe–Fe_3C (or Fe–C) phase diagram that the austenite phase becomes stable at lower temperatures as the carbon content is increased up to the eutectoid composition of 0.76 wt% C. For discussing the cooling behavior of steels, we will choose three steel compositions—a hypoeutectoid steel containing 0.2 wt% C, the eutectoid steel containing 0.76 wt% C, and a hypereutectoid steel containing 1.1 wt% C. To facilitate the discussions, an enlarged version of the eutectoid portion of the Fe–Fe_3C phase diagram is shown in Figure 5.10.

To produce the austenite phase (a process referred to as *austenitization*), a steel specimen is heated to a temperature about 30–50°C above the A_3 temperature for hypoeutectoid steels and about 50°C above the A_{cm} temperature for hypereutectoid steels. Even though the specimens could be cooled from the austenitization temperature at different rates in practice, we will now only consider the microstructures developed during equilibrium cooling (very slow cooling, i.e., at any given instance, the temperature of the specimen is assumed to be the same throughout the sample and that no temperature gradient exists). A situation closest to this ideal behavior is obtained in practice when the power to the furnace is turned off after the austenitization treatment and the specimen is allowed to cool inside the furnace. At each temperature considered, we would like to evaluate the (1) number and type of phases, (2) compositions of the individual phases, and (3) amounts (or proportions) of the phases. Additionally, we would look at the microstructure of the specimen at each of these temperatures.

5.8.1 HYPOEUTECTOID (0.2 WT% C) STEEL

The changes in the number and amounts of phases that occur on cooling this steel sample will be described with reference to Figure 5.11. Let us assume that the steel specimen is austenitized at a temperature of 880°C and cooled extremely slowly (approaching equilibrium cooling conditions) down to room temperature. At a temperature of 880°C, indicated by point "1" in the phase diagram, the steel sample exists in the austenitic condition. Thus, the number of phases is only one and that is

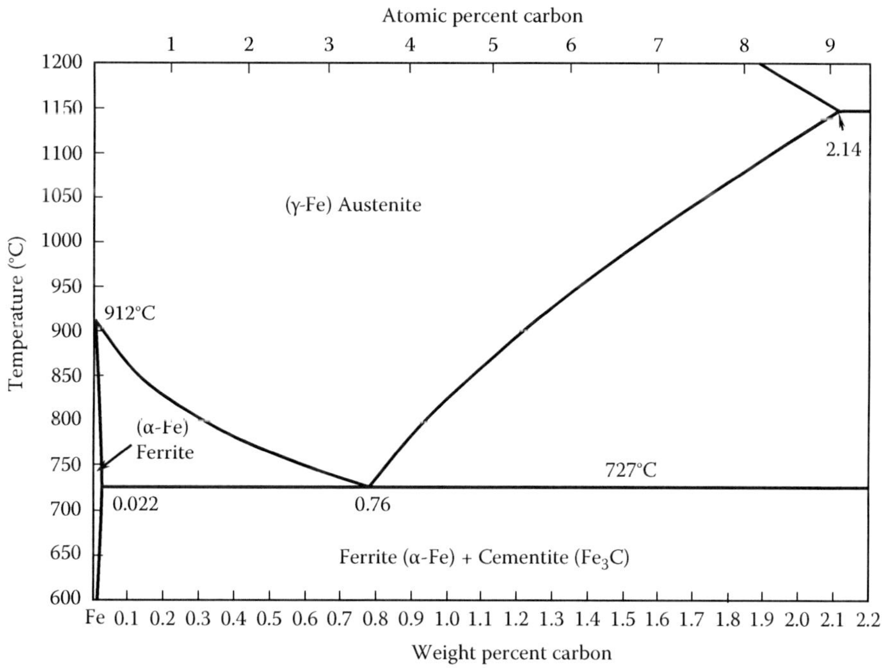

FIGURE 5.10 An enlarged portion in the neighborhood of the eutectoid reaction of the Fe–Fe_3C phase diagram.

austenite. The composition of the austenite phase is 0.2 wt% C (and the balance 99.8 wt% is Fe). Since austenite is the only phase present at this temperature, its amount is 100%. A representative schematic microstructure of this sample at this temperature is shown in Figure 5.12a. It represents an equiaxed microstructure of austenite grains with the presence of annealing twins (parallel bands running across the full width of the grains) in some of them. Even though both ferrite and austenite (or for that matter any other single-phase material) contains equiaxed grain structure in the slowly cooled condition, the austenite phase is characterized by the presence of annealing twins due to its low stacking fault energy.

On cooling this steel sample to lower temperatures, ferrite (recall that it is α-ferrite) starts forming at a temperature of about 830°C, the temperature at which the dotted line representing the composition of this steel intersects the A_{r3} line. That is, the alloy will now contain two phases, namely, austenite and ferrite. This two-phase (ferrite + austenite) constitution of the alloy will continue till the eutectoid temperature (A_{r1}) is reached. Thus, at any temperature between 830°C and 727°C, the steel sample contains two phases, namely, ferrite and austenite.

As described earlier, at any temperature in the two-phase region, the compositions of the two phases can be obtained by drawing a tie line (a line representing a constant temperature and going across the full width of the two-phase field) and finding the points of intersection of this tie line with the phase boundaries. For example, at a temperature of 800°C, represented by point "2," the ferrite phase has a composition

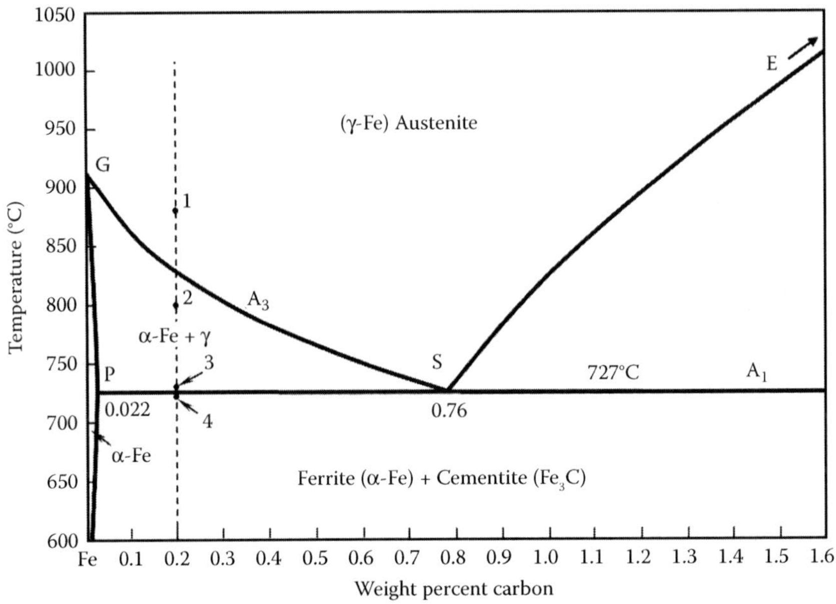

FIGURE 5.11 Fe-rich portion of the Fe–Fe_3C phase diagram featuring the eutectoid reaction to describe the cooling behavior of a 0.2 wt% C steel sample. The composition of the chosen steel is indicated by a vertical dotted line. The points at which the number and amounts of phases are determined are also indicated.

of about 0.02 wt% C (and the balance Fe) and the austenite phase about 0.32 wt% C. It is to be noted that with decreasing temperature, the carbon content of the ferrite phase increases very slightly (for all practical purposes, it remains constant) and follows the line GP and that of austenite increases (much more rapidly) and follows the line GS. Consequently, by the time the steel sample cools down to the A_{r1} (eutectoid) temperature, the ferrite phase will have 0.022 wt% C and the austenite phase 0.76 wt% C. This is true whatever the initial overall composition of the steel specimen is as long as it lies in between the composition range of 0.022 and 0.76 wt% C.

The proportions of the two phases can be found by using the lever rule. Thus, at 800°C, the 0.2 wt% C steel contains

$$\% \text{ Ferrite } (\alpha) = \frac{0.32 \text{ wt\%C} - 0.2 \text{ wt\%C}}{0.32 \text{ wt\%C} - 0.02 \text{ wt\%C}} \times 100 = 40\% \tag{5.25}$$

and

$$\% \text{ Austenite } (\gamma) = \frac{0.2 \text{ wt\%C} - 0.02 \text{ wt\%C}}{0.32 \text{ wt\%C} - 0.02 \text{ wt\%C}} \times 100 = 60\% \tag{5.26}$$

A schematic of the microstructure of the 0.2 wt% C steel specimen at this temperature is shown in Figure 5.12b. It consists of a mixture of both ferrite and

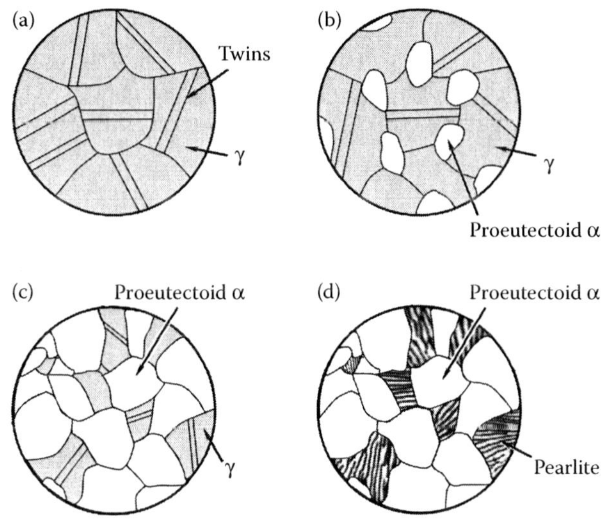

FIGURE 5.12 Schematics of microstructures developed on slowly cooling a 0.2 wt% C steel sample from the austenitic region to room temperature. The microstructures shown in (a), (b), (c), and (d) correspond to points "1," "2," "3," and "4," respectively in Figure 5.11. (a) At 880°C, the specimen shows the presence of a single phase, namely, austenite. (b) At 800°C, the specimen contains two phases, namely, austenite and proeutectoid ferrite. Even though both the phases appear as equiaxed grains, these two can be distinguished from each other by the presence of annealing twins in the austenite phase and their absence in the ferrite phase. (c) Microstructure of the specimen just above the eutectoid temperature, say 728°C. Even though the phases present are still austenite and ferrite, the relative proportion of the ferrite phase is much more than what it was at 800°C. (d) Microstructure of the specimen just below the eutectoid temperature, say 726°C. The specimen now contains ferrite and cementite phases. However part of ferrite and all the cementite coexist as pearlite. Consequently, it is appropriate to state that the specimen contains (proeutectoid) ferrite and pearlite (a lamellar mixture of eutectoid ferrite and eutectoid cementite).

austenite phases. The traditional way of describing this microstructure is that the ferrite phase has precipitated in a matrix of austenite. Both the phases are present as equiaxed grains, but the austenite phase can be distinguished from the ferrite phase by the presence of annealing twins in the former. The ferrite that has formed at this temperature (or at any temperature above the eutectoid temperature, i.e., between 912°C and 727°C) needs to be differentiated from the ferrite that forms as a result of the eutectoid reaction. Thus, the ferrite that has formed above the eutectoid reaction temperature, that is, before the eutectoid reaction has occurred, is referred to as *proeutectoid ferrite* (or *primary ferrite*) and the one that has formed as a result of the eutectoid reaction as *eutectoid ferrite*.

At a temperature just above the eutectoid temperature, say 728°C, represented by point "3," the ferrite and austenite phases have the compositions of 0.022 and 0.76 wt% C, respectively. The proportion of the two phases can be determined by using the lever rule as 75.9% ferrite and 24.1% austenite. A schematic of the microstructure of the specimen at this temperature is shown in Figure 5.12c. Note that this microstructure is not significantly different from that shown in Figure 5.12b. Comparing the schematics of the microstructures depicted in Figure 5.12b and c, it may be noted that both the microstructures contain the same two phases, namely, austenite and ferrite. But, the proportion of the two phases is different in the two microstructures. The amount of ferrite has increased (and the amount of austenite had decreased) on cooling to lower temperatures in the two-phase field region. Consequently, the amount of ferrite is much higher at 728°C than what it was at 800°C.

The nature of the phases, their compositions, and their proportion change dramatically on cooling the steel specimen to temperatures below the eutectoid temperature. This is because the austenite phase has now undergone a eutectoid transformation and has formed the ferrite and cementite phases. These will be referred to as eutectoid ferrite and eutectoid cementite. Let us now consider the situation at a temperature of 726°C, indicated by point "4," just below the eutectoid temperature. From the phase diagram, it is clear that the steel alloy now contains two phases, and these are ferrite and cementite. By drawing the tie line we can determine the composition of the ferrite phase as 0.022 wt% C and that of cementite as 6.67 wt% C. The proportions of the two phases can be calculated, using the lever rule, as

$$\% \text{ Ferrite } (\alpha) = \frac{6.67 \text{ wt\% C} - 0.2 \text{ wt\% C}}{6.67 \text{ wt\% C} - 0.022 \text{ wt\% C}} \times 100 = 97.3\% \tag{5.27}$$

and

$$\% \text{ Cementite } (\text{Fe}_3\text{C}) = \frac{0.2 \text{ wt\% C} - 0.022 \text{ wt\% C}}{6.67 \text{ wt\% C} - 0.022 \text{ wt\% C}} \times 100 = 2.7\% \tag{5.28}$$

A schematic of the microstructure of the steel specimen at this point is shown in Figure 5.12d. Note that the microstructure at this stage is very similar to that shown in Figure 5.12c except that the austenite phase has now transformed to pearlite. As mentioned earlier, the eutectoid reaction at 727°C and 0.76 wt% C involves the simultaneous formation of ferrite and cementite phases from the high-temperature austenite phase. This mixture of ferrite and cementite at the eutectoid composition is called *pearlite*. The two phases have fixed composition (0.022 wt% C for ferrite and 6.67 wt% C for cementite) and also a fixed proportion. The proportion of the ferrite and cementite phases in a steel of eutectoid composition can be calculated using the lever rule as

$$\% \text{ Ferrite } (\alpha) = \frac{6.67 \text{ wt\% C} - 0.76 \text{ wt\% C}}{6.67 \text{ wt\% C} - 0.022 \text{ wt\% C}} \times 100 = 88.9\% \tag{5.29}$$

and

$$\% \text{ Cementite } (\text{Fe}_3\text{C}) = \frac{0.76 \text{ wt\% C} - 0.022 \text{ wt\% C}}{6.67 \text{ wt\% C} - 0.022 \text{ wt\% C}} \times 100 = 11.1\% \tag{5.30}$$

In the microstructure of steel samples, pearlite appears as a lamellar structure of alternate layers (lamellae) of ferrite and cementite. Pearlite is considered as a microconstituent and not as a phase (this is a mixture of two phases, and has a distinctive appearance in the microstructure). Note that, a microconstituent may contain more than one phase, but a phase will not contain more than one microconstituent. Stated differently, a phase is one microconstituent, but a microconstituent need not be one phase. This microconstituent in steels, obtained as a result of the eutectoid reaction, is termed pearlite, since the polished and etched microconstituent shines like the mother-of-pearl. As pearlite contains a large amount of ferrite (the ratio of the ferrite to cementite phases in pearlite is 88.9:11.1, or approximately 8:1), the lamellar thickness of ferrite is much larger than that of cementite.

Even though it is mentioned that pearlite is a lamellar structure of alternate layers of ferrite and cementite and that the thickness of the ferrite phase is much larger than that of cementite, pearlite may not always appear resolved, especially when viewed at low magnifications. At low magnifications, it may simply appear as a dark mass. Only at relatively high magnifications, pearlite will appear resolved, that is, one can clearly see the ferrite and cementite phases separately. The minimum magnification at which the lamellae will appear resolved depends on the *interlamellar spacing*, that is, the distance between two cementite lamellae (or between two ferrite lamellae).

Example Problem 5.4

Calculate the minimum magnification required to resolve pearlite that has an interlamellar spacing of 0.2 μm.

Solution 5.4

It is known that the resolving power of the naked eye is about 0.07 mm. That is, the eye can clearly see two objects individually if they are separated by a distance of at least 0.07 mm (70 μm). But, the observation will be much more comfortable if the separation is larger. Therefore, to be able to see the lamellae resolved, we need a minimum magnification of

$$\frac{70\ \mu m}{0.2\ \mu m} = 350\times$$

Even though this is the minimum magnification required, it is always desirable to observe the microstructure at a higher magnification for comfortable vision. Also note that the pearlite structure can be seen resolved at a lower magnification only when the interlamellar spacing is large, and that higher magnifications are required when the interlamellar spacing is small.

Recall that the mixture of ferrite and cementite formed as a result of the eutectoid reaction constitutes pearlite. Therefore, even though we had calculated that the 0.2 wt% C steel contained 97.3% ferrite and 2.7% cementite, the microstructure of the specimen will only show two distinct parts—a plain area and a lamellar area. These distinctive parts are referred to as microconstituents. That is, the ferrite and cementite phases are redistributed into these two microconstituents. Therefore, for all practical purposes, it can be considered that the steel now contains ferrite and pearlite. Since it was calculated that the amount of (proeutectoid or primary) ferrite in a 0.2 wt% C steel was 75.9% just above the eutectoid temperature, it will continue to be present below the eutectoid temperature also. The 24.1% austenite that was present above the eutectoid temperature will transform to the pearlite microconstituent below the eutectoid temperature. A schematic of the microstructure of this steel sample just below the eutectoid temperature (at 726°C), containing ferrite and pearlite, is shown in Figure 5.12d.

Example Problem 5.5

a. Calculate the amount of ferrite and cementite present in a 0.3 wt% C steel specimen slowly cooled to a temperature of 726°C.
b. Calculate the amount of proeutectoid ferrite and eutectoid ferrite in this sample.

Solution 5.5

a. From the Fe–Fe_3C phase diagram in Figure 5.11, the composition of the ferrite and cementite phases at 726°C are 0.022 and 6.67 wt% C, respectively. Since the composition of the steel sample has been mentioned as 0.3 wt% C, the proportions of the ferrite and cementite phases can be calculated using the lever rule as

$$\%\ \text{Ferrite}\ (\alpha) = \frac{6.67\,\text{wt\%C} - 0.3\,\text{wt\%C}}{6.67\,\text{wt\%C} - 0.022\,\text{wt\%C}} \times 100 = 95.8\%$$

and

$$\%\ \text{Cementite}\ (\text{Fe}_3\text{C}) = \frac{0.3\,\text{wt\%C} - 0.022\,\text{wt\%C}}{6.67\,\text{wt\%C} - 0.022\,\text{wt\%C}} \times 100 = 4.2\%$$

b. A steel specimen with the eutectoid carbon content (0.76 wt%) will have 100% pearlite, and pearlite contains 88.9% eutectoid ferrite in it. Therefore, we can first calculate the amount of pearlite in the 0.3 wt% C steel sample and from that the amount of eutectoid ferrite in it can be calculated. Note that 100% pearlite is formed only at the eutectoid composition of 0.76 wt% C when austenite transforms to pearlite. The amount of pearlite in the slowly-cooled 0.3 wt% C sample is:

$$\%\ \text{Pearlite} = \frac{0.3\,\text{wt\%C} - 0.022\,\text{wt\%C}}{0.76\,\text{wt\%C} - 0.022\,\text{wt\%C}} \times 100 = 37.7\%$$

Since pearlite contains 88.9% eutectoid ferrite in it, the amount of eutectoid ferrite in the pearlite microconstituent is $37.7 \times 0.889 = 33.5\%$. Therefore, the amount of proeutectoid ferrite in the sample is

$$95.8 - 33.5 = 62.3\%.$$

The same result can also be obtained by simply subtracting the amount of pearlite from 100%, that is, 100 − 37.7 = 62.3%, since the sample contains only two microconstituents, namely, proeutectoid ferrite and pearlite.

This situation will continue until the specimen reaches room temperature. That is, the steel continues to have only two phases (ferrite and cementite), but the microstructure will show the presence of proeutectoid ferrite and pearlite as the microconstituents. There will not be any significant difference in the composition or proportion of the phases (and microconstituents) as one cools the sample from the eutectoid temperature to room temperature since the phase boundaries representing the ferrite (line PQ) and cementite are almost vertical, that is, compositions of the ferrite and cementite phases do not change with temperature. (The proportions could be different in other alloy systems if the phase boundaries (solvus lines) have a slope, that is, their compositions change with temperature.)

Figure 5.13 presents a photomicrograph of the actual microstructure of the 0.2 wt% C steel specimen slowly cooled from the austenitic region. The lightly etched equiaxed grains represent the ferrite phase and the dark-etched part the pearlite. Note that pearlite is not resolved in this micrograph.

A summary of the number and types of phases, their compositions, and phase proportions for the 0.2 wt% C steel sample as it is slowly cooled to room temperature is presented in Table 5.4. Situations very similar to those described above will be obtained in steels containing lower (down to 0.022 wt% C) or higher concentrations of carbon, up to the eutectoid composition (0.76 wt% C). The same two microconstituents—proeutectoid ferrite and pearlite—will be present in all the samples. The composition of the ferrite and cementite phases will also be the same. But, the only difference is that the proportions of the ferrite and pearlite microconstituents will change with the composition of the steel specimen selected. The amount of

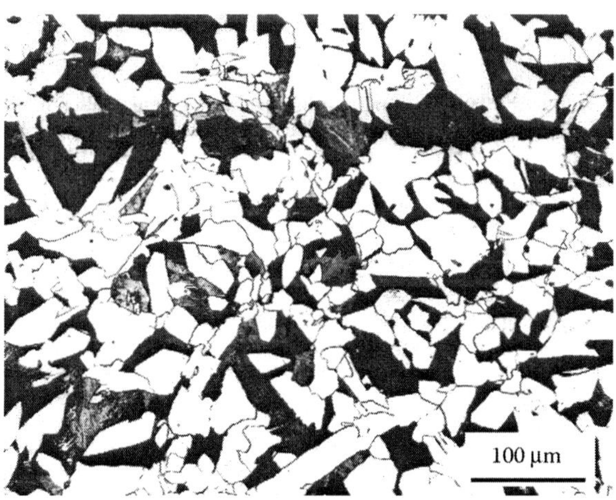

FIGURE 5.13 Optical micrograph representing the room temperature microstructure of a slowly cooled 0.2 wt% C steel specimen etched with 4% picral + 2% nital.

TABLE 5.4
Number, Nature, Compositions, and Proportions of the Phases Present at Different Temperatures as a Hypoeutectoid Steel Containing 0.2 wt% C is Slowly Cooled from the Austenitic State to Room Temperature

Temperature (°C)	Number of Phases	Nature of the Phases	Compositions of the Phase(s) (wt% C)	Percentage of the Phase(s)
880 (Point "1")[a]	1	Austenite	0.2	100
800 (Point "2")	2	Austenite	0.32	60
		Ferrite	0.02	40
728 (Point "3")	2	Austenite	0.76	24.1
		Ferrite	0.022	75.9
726[b] (Point "4")	2	Ferrite	0.022	97.3
		Cementite	6.67	2.7

[a] In Figure 5.11.
[b] At this temperature, the specimen contains two microconstituents, namely, ferrite and pearlite. The microstructure will now consist of 75.9% proeutectoid ferrite and 24.1% pearlite.

pearlite increases with increasing carbon content and reaches 100% at the eutectoid composition. Figure 5.14 presents the variation of pearlite content with carbon content in hypoeutectoid steels calculated for a temperature just below the eutectoid temperature. As a first approximation, it may be considered that the amount of pearlite increases by about 12.5% for each increase of 0.1 wt% carbon content. Thus, 0.2 wt% C steel contains ~25% pearlite, 0.4 wt% C steel contains ~50% pearlite, 0.6 wt% C

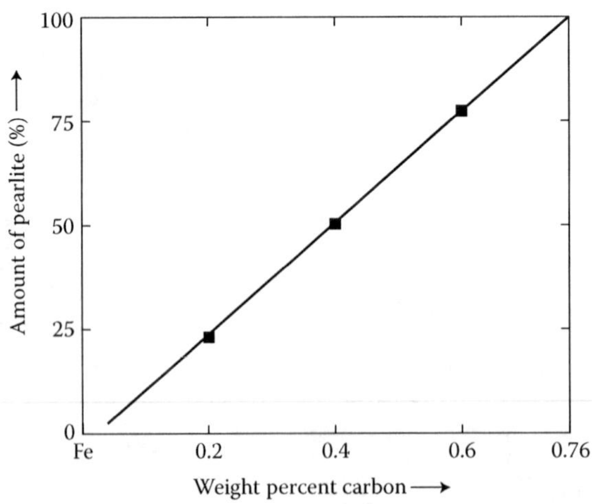

FIGURE 5.14 Variation of the amount of pearlite with carbon content in hypoeutectoid steels.

steel contains ~75% pearlite, and the eutectoid steel (0.76 wt% or ~0.8 wt% C) contains 100% pearlite. Even though the proportions of the proeutectoid ferrite and pearlite change very slightly with temperature, the amounts are not any significantly different from the values mentioned above. For example, the room temperature solid solubility of C in Fe is virtually zero. Therefore, the amount of pearlite in the 0.2 wt% C steel sample at room temperature is 26.3%, which is not significantly different from the value of 24.1% calculated at a temperature just below the eutectoid temperature.

5.8.2 Eutectoid (0.76 wt% C) Steel

Let us now discuss the cooling behavior of a steel sample with the eutectoid composition (0.76 wt% C) (Figure 5.15). Since the austenite transformation temperature for this steel is 727°C, complete austenitization can be achieved at a temperature of 760–780°C. At this austenitization temperature, represented by point "1," the steel sample contains only one phase, namely, austenite, with the eutectoid composition (0.76 wt% C). Since this is the only phase present in the sample, its amount is 100%. The microstructure of the sample will consist of equiaxed grains of austenite showing the presence of annealing twins in some of them. It will look very similar to the microstructure shown in Figure 5.12a. This austenite phase will continue to be stable

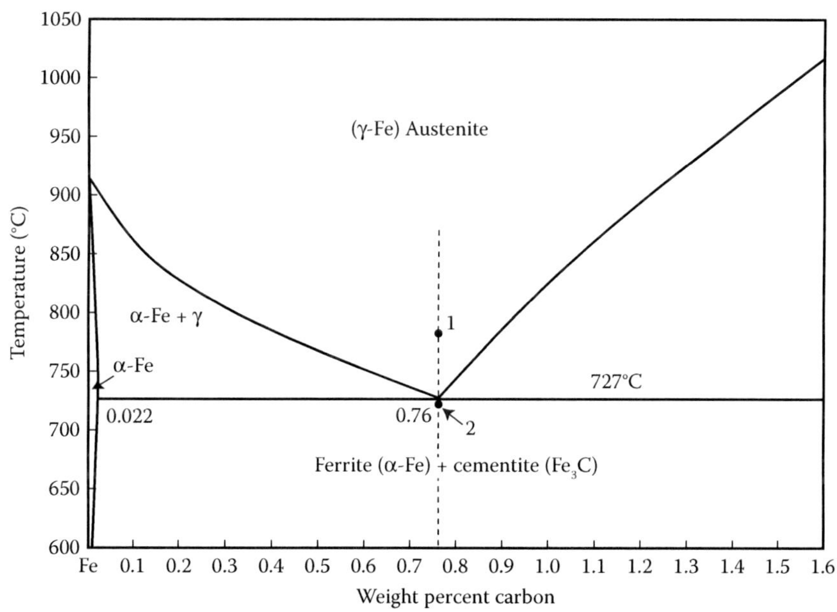

FIGURE 5.15 Fe-rich portion of the Fe–Fe_3C phase diagram featuring the eutectoid reaction to describe the cooling behavior of a eutectoid steel sample. The eutectoid composition (0.76 wt% C) is indicated by a vertical dotted line. The points at which the number and amounts of phases are determined are also indicated.

until we reach the eutectoid transformation temperature of 727°C. But, on cooling to a temperature just below the eutectoid temperature, represented by point "2," it will completely transform to pearlite. Therefore, below the eutectoid transformation temperature, the steel contains two phases, namely, ferrite and cementite, but only one microconstituent (pearlite) and its amount is obviously 100%. This same situation will continue until the specimen reaches down to room temperature.

Recall that pearlite is made up of two phases—ferrite and cementite in the approximate ratio of 8:1, but it is only one microconstituent. The pearlite microconstituent will consist of 88.9% ferrite and 11.1% cementite, as calculated in Equations 5.29 and 5.30. We should also realize that this ratio of these two constituent phases in pearlite is the same irrespective of the overall composition of the steel samples. (Note, however, that even if the steel contains other phases such as proeutectoid ferrite or proeutectoid cementite if the steel under consideration is hypoeutectoid or hypereutectoid, pearlite will always contain the same proportion of ferrite and cementite phases. This is because a completely pearlitic structure will form *only* when a steel of 0.76 wt% C (eutectoid composition) is cooled to temperatures below the eutectoid temperature.) The microstructure of the eutectoid steel is shown in Figure 5.16. Note that the lamellar structure of pearlite (made up of ferrite and cementite) is seen very clearly resolved.

Recall that the pearlitic structure containing ferrite (with 0.022 wt% C) and cementite (with 6.67 wt% C) is obtained from a steel sample containing the eutectoid

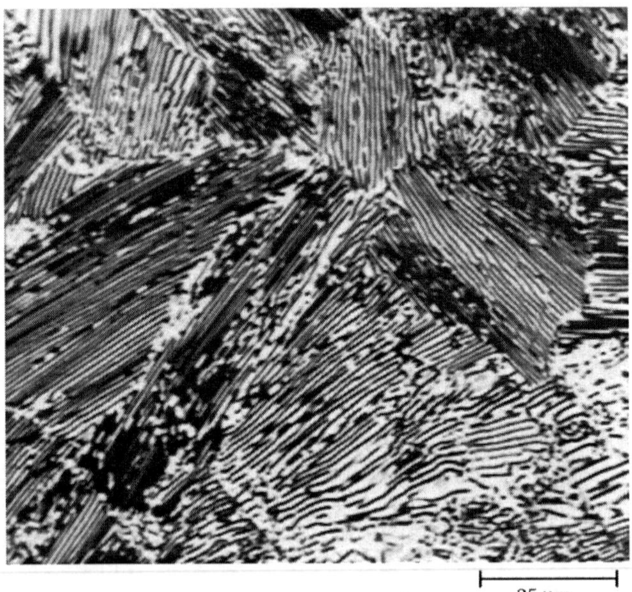

FIGURE 5.16 Microstructure of pearlite in a eutectoid (0.76 wt% C) steel showing a well-resolved pearlitic structure, that is, the individual lamellae of ferrite and cementite are clearly resolved. Picral etch. (Photo courtesy of ASM International, Materials Park, OH. With permission.)

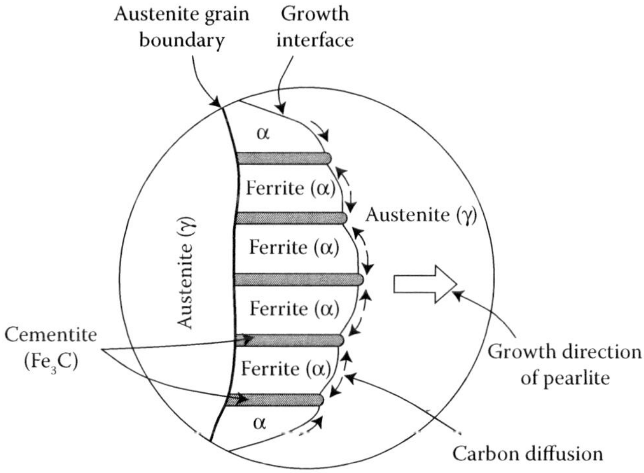

FIGURE 5.17 Schematic representation of the formation of alternate layers of ferrite and cementite in a pearlite colony from the austenite phase. The direction of the diffusion of C atoms is indicated by arrows.

C content. In other words, the carbon content present in the high-temperature austenite phase needs to be redistributed to ferrite with a much lower C content and cementite with a much higher C content. Such redistribution of the solute is possible only through diffusional processes. Figure 5.17 shows schematically how this is possible. Let us assume that the austenite phase contains an average of 0.76 wt% C. But, it is possible that some local fluctuations in composition are present such that the C content is momentarily very high at some point. Under these conditions, it is possible that the cementite phase nucleates. (Alternately, if there is a large depletion of the C content, the ferrite phase could nucleate.) Assuming that the cementite phase has nucleated, there will be depletion of C content in the neighborhood and therefore the situation is conducive to the formation of the ferrite phase. The C atoms will now diffuse away from the ferrite phase enriching the neighborhood with C atoms and therefore the cementite phase will again form. Thus, the ferrite and cementite phases nucleate alternately resulting in the formation of the lamellar structure of pearlite. The transformation from austenite to pearlite is complete when the pearlite colonies completely exhaust the austenite phase.

5.8.3 Hypereutectoid (1.1 wt% C) Steel

Let us now discuss the cooling behavior of a steel sample with a carbon content of 1.1 wt% C, which is higher than that of the eutectoid composition (Figure 5.18). For this steel sample, a completely austenitic phase can be obtained only when the specimen is heated to a temperature higher than 870°C (above the A_{cm} temperature). Therefore, let us assume that the specimen was austenitized at a temperature of 900°C and that the specimen is cooled slowly to room temperature from this temperature. At 900°C, indicated by point “1,” the steel exists as a single-phase austenite

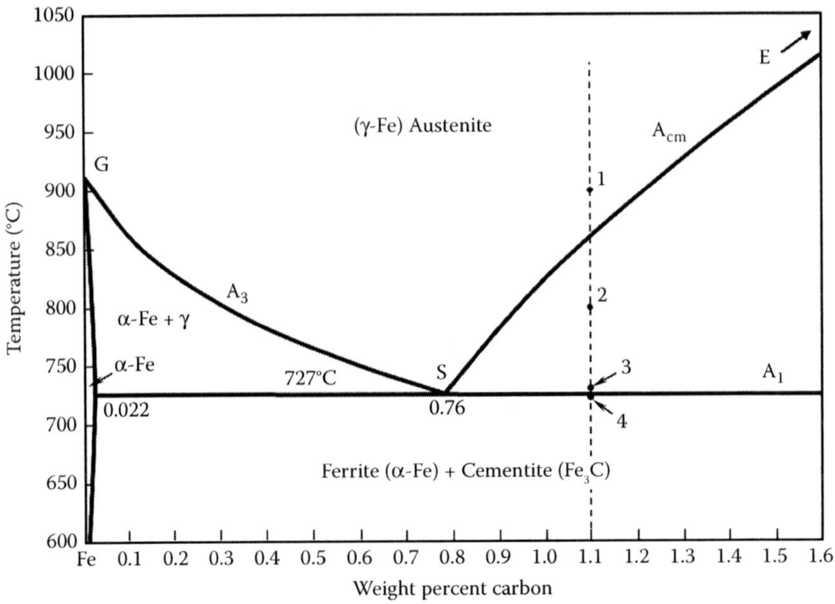

FIGURE 5.18 Fe-rich portion of the Fe–Fe_3C phase diagram featuring the eutectoid reaction to describe the cooling behavior of a 1.1 wt% C steel sample. The selected composition is indicated by a vertical dotted line. The points at which the number and amounts of phases are determined are also indicated.

with the composition of 1.1 wt% C. Since there is only one phase, its amount is 100%. The microstructure of this steel at 900°C consists of equiaxed austenite grains with the presence of annealing twins in some of them. A typical schematic microstructure is shown in Figure 5.19a. Note that this microstructure is very similar to that shown in Figure 5.12a.

When this steel specimen is cooled to lower temperatures, proeutectoid (or primary) cementite starts forming at about 870°C. At this temperature, the amount of proeutectoid cementite is almost zero, but its amount increases as the temperature is further lowered. On cooling to temperatures lower than 870°C, the steel exists as a mixture of two phases. In this two-phase region, the two phases are austenite and (proeutectoid) cementite and their compositions and proportions can be calculated using the tie line and lever rule, respectively. Also note that, on cooling, the composition of the austenite phase changes along the line ES, whereas the composition of the cementite phase is constant, since its composition is represented by a vertical line in the phase diagram.

As an example, let us consider the situation at 800°C, indicated by point "2." The composition of austenite can be determined by drawing a tie line at this temperature and the point of intersection of this tie line with the line ES will give the composition of austenite as 0.9 wt% C. The composition of the cementite (Fe_3C) phase is fixed at 6.67 wt% C.

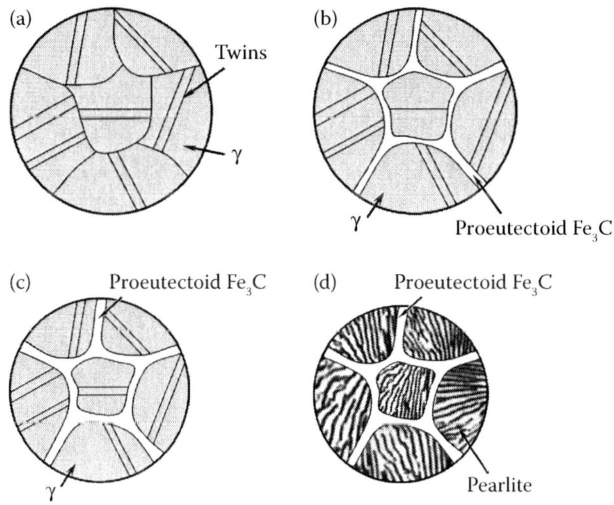

FIGURE 5.19 Schematic of microstructures developed in a hypereutectoid (1.1 wt% C) steel slowly cooled from the austenitic region to room temperature. (a) At 900°C, the steel contains a single phase (austenite). Note that this microstructure is very similar to that shown in Figure 5.12a. (b) Microstructure at 800°C, when the steel contains proeutectoid cementite and austenite. (c) Microstructure at a temperature just above the eutectoid temperature, say at 728°C. (d) Microstructure at a temperature just below the eutectoid temperature, say at 726°C, when the steel specimen contains proeutectoid cementite and pearlite.

The proportion of the two phases can now be calculated using the lever rule as

$$\%\ \text{Austenite}\ (\gamma) = \frac{6.67\ \text{wt\% C} - 1.1\ \text{wt\% C}}{6.67\ \text{wt\% C} - 0.9\ \text{wt\% C}} \times 100 = 96.5\% \tag{5.31}$$

and

$$\%\ \text{Cementite}\ (\text{Fe}_3\text{C}) = \frac{1.1\ \text{wt\% C} - 0.9\ \text{wt\% C}}{6.67\ \text{wt\% C} - 0.9\ \text{wt\% C}} \times 100 = 3.5\% \tag{5.32}$$

The microstructure at this temperature will be as shown schematically in Figure 5.19b. The white-etching phase present along the grain boundaries is proeutectoid cementite, while the equiaxed grains with annealing twins in some of them represent the austenite phase. A similar situation will exist as we continue to cool the steel sample to lower temperatures. The composition of the austenite phase follows the line ES and consequently the amount of the cementite phase increases as the eutectoid temperature is approached. Just above the eutectoid temperature, say at a temperature of 728°C, indicated by point "3," the austenite phase will have 0.76 wt% C and the cementite 6.67 wt% C. Accordingly, the amounts of the austenite and cementite phases at this temperature are 94.2% and 5.8%, respectively. The microstructure of the steel sample at this temperature is shown schematically in Figure 5.19c.

It consists of equiaxed grains of austenite, with annealing twins in some of them, and the cementite phase is present as a white-etching phase along the grain boundaries of austenite. This microstructure is very similar to that represented in Figure 5.19b, except that in Figure 5.19c, the relative amount of the cementite phase is slightly higher.

As we cool the specimen to below the eutectoid temperature, represented by point "4," the proeutectoid cementite phase will continue as-is, but the austenite phase will transform to pearlite. The microstructure will now contain a large fraction of pearlite (94.2%) and a small amount of cementite along the boundaries of the pearlite nodules. Again, pearlite appears as a dark unresolved microconstituent at low magnifications, and as alternate lamellae of ferrite and cementite in a well-resolved condition at higher magnifications. A schematic microstructure corresponding to this condition is shown in Figure 5.19d.

Figure 5.20 shows an optical micrograph of a slowly cooled 1.1 wt% C steel specimen and clearly shows all the microstructural features described above—proeutectoid cementite along the grain boundaries of well-resolved pearlite nodules.

Figure 5.21 shows the microstructures of steels with four other compositions—three hypoeutectoid steels with 0.1 wt% C, 0.4 wt% C, and 0.6 wt% C and one hypereutectoid steel with 1.4 wt% C. It may be noted that the pearlite content increases with increasing carbon content in the hypoeutectoid steels and the white-etching cementite layer in the hypereutectoid steel is thicker than what it was in the 1.1 wt% C steel (Figure 5.20).

Figure 5.22 shows schematically the type and amount of microconstituents present in different regions of the steel portion of the Fe–Fe_3C phase diagram. While the hypoeutectoid steels contain proeutectoid ferrite and pearlite, the eutectoid steel contains only pearlite, and the hypereutectoid steel contains proeutectoid cementite and pearlite.

FIGURE 5.20 Microstructure of a slowly cooled hypereutectoid steel specimen containing 1.1 wt% C. (Photo courtesy of ASM International, Materials Park, OH. With permission.)

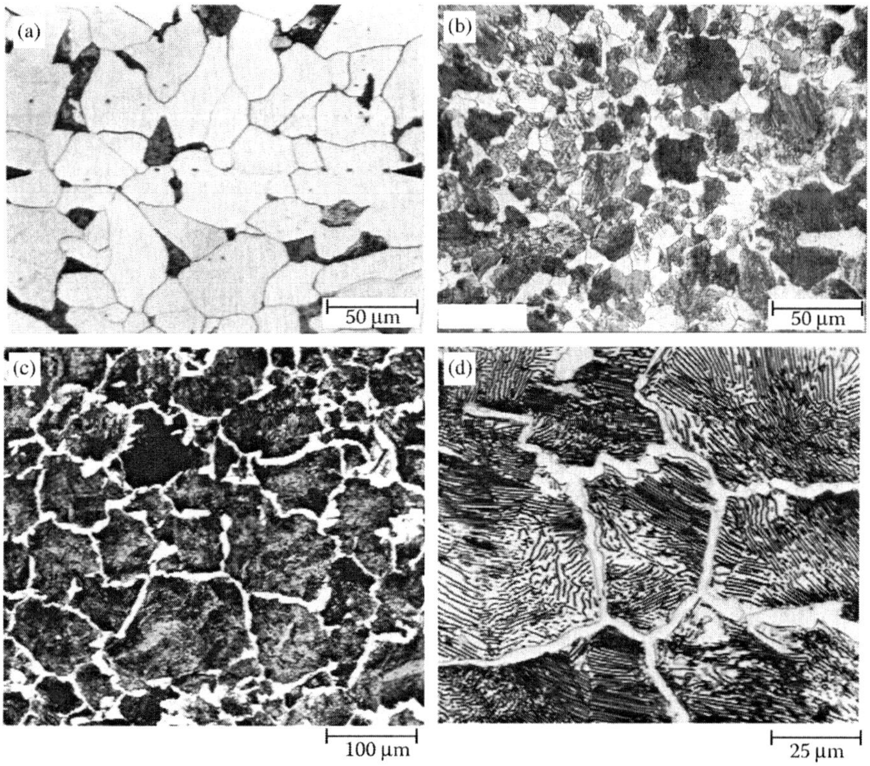

FIGURE 5.21 Microstructures of slowly cooled (a) 0.1 wt% C, (b) 0.4 wt% C, (c) 0.6 wt% C, and (d) 1.4 wt% C steel specimens. While specimens (a) and (b) were etched with 2% nital, specimens (c) and (d) were etched with 4% picral. (Photos courtesy of ASM International, Materials Park, OH. With permission.)

5.9 DIFFERENTIATION BETWEEN PROEUTECTOID FERRITE AND PROEUTECTOID CEMENTITE

Note that in Figure 5.13, proeutectoid ferrite etches light and that the proeutectoid cementite also etches light as shown in Figure 5.20. That is, both ferrite and cementite appear white under the microscope. Further, both proeutectoid ferrite and proeutectoid cementite could be present surrounding the pearlite nodules in the hypoeutectoid and hypereutectoid steels, respectively. Therefore, if only a micrograph (or the microstructure) of the steel specimen or a steel specimen is provided and the composition is not mentioned, it will be difficult to decide whether the white-etching phase in the microstructure is proeutectoid ferrite or proeutectoid cementite and to determine whether the given steel is hypoeutectoid or hypereutectoid in nature. Even though there are other methods to decide this (e.g., the composition of the phase can be determined by microprobe analysis or by energy-dispersive spectrometry in an SEM), the microstructural appearance can also be used to our advantage.

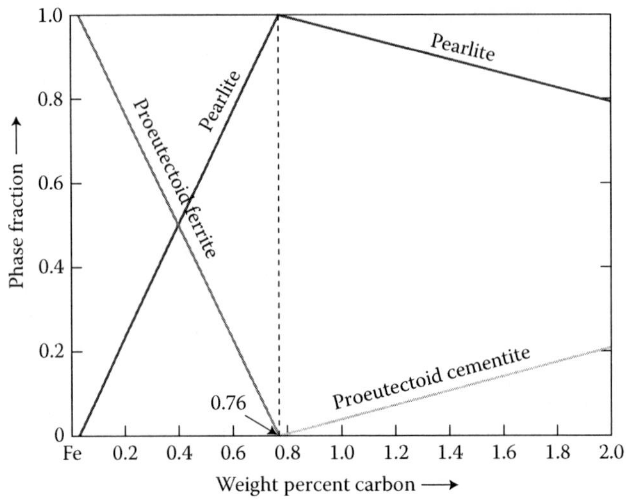

FIGURE 5.22 Schematic representation of the microconstituents present in steel specimens of different carbon contents.

The way to differentiate between proeutectoid ferrite and proeutectoid cementite can be explained by considering steels with two different compositions—one a hypoeutectoid steel containing 0.6 wt% C and the other a hypereutectoid steel containing 1.0 wt% C. That is, we choose two steels both approximately 0.2 wt% C away from the eutectoid composition on either side of the eutectoid point. Let us now calculate the amount of the proeutectoid (or primary) phase (ferrite in the hypoeutectoid and cementite in the hypereutectoid steel) using the lever rule. The result is

$$\begin{aligned}&\%\ \text{Proeutectoid ferrite in } 0.6\ \text{wt\% C steel}\\&=\frac{0.76\ \text{wt\% C}-0.6\ \text{wt\% C}}{0.76\ \text{wt\% C}-0.022\ \text{wt\% C}}\times 100=21.7\%\end{aligned} \tag{5.33}$$

$$\begin{aligned}&\%\ \text{Proeutectoid cementite in } 1.0\ \text{wt\% C steel}\\&=\frac{1.0\ \text{wt\% C}-0.76\ \text{wt\% C}}{6.67\ \text{wt\% C}-0.76\ \text{wt\% C}}\times 100=4.1\%\end{aligned} \tag{5.34}$$

Thus, you will notice that the amount of proeutectoid ferrite is much higher than the amount of the proeutectoid cementite. The amount of proeutectoid cementite will not increase significantly even if we have 1.5 wt% C (still it is only 12.5%) (see, e.g., Figure 5.21d). But, practical engineering steels rarely contain that much of carbon in them. Therefore, the proeutectoid ferrite phase will appear thicker along the boundaries of pearlite colonies in hypoeutectoid steels, while the proeutectoid cementite phase will appear much thinner. This is a common observation, even though one could theoretically have a situation where the proeutectoid ferrite and proeutectoid cementite contents are the same. But, if one obtains a thin layer of a white-etching phase in a plain carbon steel sample, it is more likely that it is a hypereutectoid steel and if the white-etching phase is thicker, then it is more likely to be a hypoeutectoid steel.

Another important point to remember is that in reality, the ferrite phase in high-carbon (but still less than the eutectoid composition) steels, particularly beyond about 0.65 wt% C, is not seen clearly. In other words, it appears as though the whole structure is made up of only pearlite. Thus, if one observes a thin white-etching phase around the pearlite nodules, the steel is most likely hypereutectoid in nature.

5.10 MICROSTRUCTURAL OBSERVATION

Steel specimens (or for that matter, any metallic specimen) need to be properly prepared to observe their microstructures under a microscope. The nature of preparation and the type of specimen required will be different for different types of microscopes. For example, a properly prepared specimen of any thickness is suitable for optical or SEM studies. But, an extremely thin specimen (about 100 nm in thickness) is required for TEM investigations.

The usual steps in the preparation of good specimens for optical metallographic observation were described earlier in Chapter 3. These include grinding of the surface to minimize the surface roughness, polishing it to produce a mirror-like surface finish, and etching it to reveal the microstructural details. As mentioned before (Chapter 3), different specimens have to be etched by different chemical reagents (etchants) to reveal their microstructural features. Plain carbon steel specimens are most commonly etched with *nital* (a mixture of *ni*tric acid and ethyl *al*cohol). Generally, 2% nital (2% concentrated nitric acid and the rest ethyl alcohol by volume) is used, but occasionally, 5% nital is also used. It is an excellent reagent for revealing the microstructure of martensite (a nonequilibrium hard phase in steels and we will discuss this later in Chapter 6), and is also very good for revealing ferrite in a martensitic matrix and to bring out ferrite grain boundaries in low-carbon steels. But, nital is not always the best reagent to use to properly reveal *all* the microstructural features. *Picral* (a mixture of *picr*ic acid and ethyl *al*cohol) is an excellent reagent to reveal the microstructure of ferrite–cementite constituents—pearlite and bainite (another nonequilibrium phase)—in steels. The use of picral is prohibited by some organizations because picric acid can be made to explode under certain conditions. However, picral-related accidents are less common than for nital. Vilella's reagent, which also contains picric acid, is exceptionally valuable for certain compositions and microstructures. Sometimes, austenitic stainless steels are etched to bring out the grain boundaries very well, without revealing the twins (Figure 5.23). Such a microstructure makes grain size measurements much easier.

As noted earlier, both the ferrite and cementite phases appear light under a microscope. Cementite actually appears as a light-yellow compound when polished topographically flat. But, pearlite (which is a lamellar mixture of ferrite and cementite) appears dark if it is not resolved. Even if pearlite is resolved, it still appears as though the cementite phase is etching dark. This is due to the fact that the incident light is not reflected by the rounded layer edges of cementite. Ferrite appears white because the light is reflected back into the microscope by its flat surfaces.

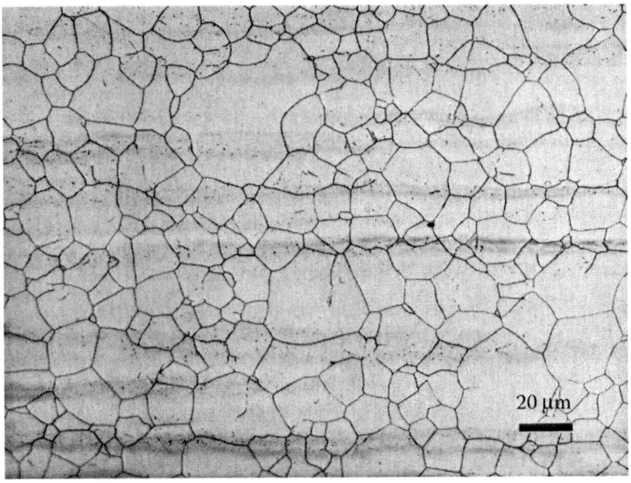

FIGURE 5.23 Microstructure of a 316L austenitic stainless-steel sample solution treated at 950°C and etched with 60% HNO_3 at 1.4 V DC for 2 min. The grain boundaries are clearly seen, but not the annealing twins. (Photo courtesy of Vander Voort, G. F. Consultant—Struers, Inc. With permission.)

5.11 EXPERIMENTAL PROCEDURE

1. Take a series of steel samples of different carbon content. It may be a good idea to take samples of the eutectoid composition and also hypoeutectoid and hypereutectoid composition.
2. Prepare all the samples for metallographic examination and observe their microstructures.
3. Observe the microstructures at different magnifications and record them.
4. Identify the phases (or microconstituents) present in the samples and label them on the micrographs.
5. Using quantitative microscopy methods, determine the grain sizes and amounts of the different microconstituents present in the sample.
6. Try to estimate the carbon content of the samples, if they are plain carbon steel samples.

EXERCISES

5.1 What is a phase diagram?

5.2 What is an invariant point?

5.3 What is a triple point?

5.4 If an alloy of copper and zinc contains 40 wt% zinc, express the composition of the alloy in atomic percentage.

5.5 If an alloy of copper and gold contains 10 wt% gold, express the composition of the alloy in atomic percentage.

5.6 If an alloy of copper and aluminum contains 20 at% aluminum, express the composition of the alloy in weight percentage.

5.7 If an alloy of copper and tin contains 20 at% tin, express the composition of the alloy in weight percentage.
5.8 Define the terms phase and component.
5.9 What is a binary alloy system?
5.10 Define solidus temperature.
5.11 Define liquidus temperature.
5.12 State Gibbs phase rule as applied to condensed phase systems.
5.13 Apply the Gibbs phase rule to different points in the Ni–Ru phase diagram (Figure 5.8b) and verify whether the phase rule is obeyed in this system.
5.14 Identify the invariant points in the Ni–Ru phase diagram (Figure 5.8b).
5.15 Determine the compositions of the liquid and α phases in the Pb–Sn phase diagram (Figure 5.8a) at a temperature of 250°C.
5.16 Determine the compositions of the α and β phases in the Pb–Sn phase diagram (Figure 5.8a) at a temperature of 100°C.
5.17 What is a tie line?
5.18 Determine the amounts of the liquid and α phases for an alloy of Pb–40 wt% Sn in the Pb–Sn phase diagram (Figure 5.8a) at a temperature of 250°C.
5.19 Determine the amounts of the α and β phases for an alloy of Pb–50 wt% Sn in the Pb–Sn phase diagram (Figure 5.8a) at a temperature of 100°C.
5.20 What is lever rule?
5.21 What information can one derive from the lever rule?
5.22 What is allotropy?
5.23 How many allotropic forms of solid iron exist?
5.24 What is a solid solution?
5.25 Differentiate between substitutional and interstitial solid solutions.
5.26 What is the maximum carbon content in steels?
5.27 What is ferrite?
5.28 What is cementite?
5.29 What is austenite?
5.30 What is a hypoeutectoid steel?
5.31 What is a hypereutectoid steel?
5.32 What is a peritectic reaction?
5.33 What is a eutectoid reaction?
5.34 How will you differentiate microstructurally between ferrite and austenite?
5.35 What is proeutectoid ferrite?
5.36 Differentiate between eutectoid ferrite and proeutectoid ferrite.
5.37 What is pearlite?
5.38 Define interlamellar spacing.
5.39 What is proeutectoid cementite?
5.40 Based on microstructural features, how will you differentiate between proeutectoid ferrite and proeutectoid cementite.
5.41 Name two common etchants for steel samples.
5.42 Assuming that the solid solubility of carbon in ferrite is zero and that cementite contains 6.67 wt% carbon, calculate the relative amounts of ferrite and cementite in a eutectoid composition (0.76 wt% carbon) steel sample at room temperature.

5.43 What is the name of the microconstituent obtained on slow cooling of eutectoid steel from the austenite region to room temperature?

5.44 What phases are present at room temperature and under equilibrium conditions in a hypoeutectoid steel?

5.45 What phases are present at room temperature and under equilibrium conditions in a hypereutectoid steel?

5.46 If the microstructure a steel sample cooled slowly from the austenite region contains 75 wt% ferrite and 25 wt% pearlite, calculate the approximate carbon content of the steel.

5.47 A plain carbon steel contains 90 wt% ferrite and 10 wt% cementite. What is its average carbon content in weight percent?

5.48 A plain carbon steel contains 48.2 wt% proeutectoid ferrite. What is its average carbon content in weight percent?

5.49 If the microstructure of a steel sample cooled slowly from the austenitic region contains 25 wt% ferrite and 75 wt% pearlite, calculate the approximate carbon content of the steel.

5.50 How does pearlite appear under a microscope?

FURTHER READING

Bhadeshia, H. K. D. H. and R. W. K. Honeycombe, 2006. *Steels: Microstructure and Properties*, 3rd edition. London, UK: Butterworth-Heinemann.

Brooks, C. R. 1996. *Principles of the Heat Treatment of Plain Carbon and Low Alloy Steels*. Materials Park, OH: ASM International.

Callister, Jr., W. D. and D. G. Rethwisch, 2010. *Materials Science and Engineering, An Introduction*, 8th edition. New York, NY: John Wiley & Sons, Inc.

Krauss, G. 2005. *Steels: Heat Treatment and Processing Principles*. Materials Park, OH: ASM International.

Leslie, W. C. 1991. *The Physical Metallurgy of Steels*. Herndon, VA: Tech Books.

Massalski, T. B. (ed.) 1990. *Binary Alloy Phase Diagrams*. Materials Park, OH: ASM International.

Samuels, L. E. 1999. *Light Microscopy of Carbon Steels*. Materials Park, OH: ASM International.

Sinha, A. K. 1989. *Ferrous Physical Metallurgy*. Boston, MA: Butterworths.

Vander Voort, G. F. 1984. *Metallography Principles and Practice*. New York, NY: McGraw-Hill, Inc.

6 Heat Treatment of Steels

6.1 INTRODUCTION

The details of the Fe–C phase diagram, the possible equilibrium phase transformations, and the different types of microstructures developed during equilibrium cooling from the austenitic region were described in Chapter 5. In this chapter we will discuss the different types of heat treatments that can be given to these Fe–C alloys to produce microstructures that are substantially different from those achieved during equilibrium cooling, and the properties that can be obtained. We will focus our attention only on steels, that is, Fe–C alloys containing up to a maximum of about 2 wt% C, but generally up to about 1.2–1.3 wt% C.

Heat treatment is defined as the heating *and* cooling of a solid metal or alloy in such a way as to obtain the desired microstructure and/or properties. The heat treatment procedure usually involves heating the material to a high-enough temperature to cause a phase transformation, holding the specimen at temperature for a predetermined length of time, and then cooling it back to room temperature at a *controlled rate* to achieve the desired microstructure. Thus, heating of the metal to an elevated temperature (without any phase transformation occurring) to increase the flow characteristics (or processability) of the material is not considered a heat treatment. But, sometimes people call this treatment as annealing!

Almost all engineering components (except for some cast irons) undergo at least one type of heat treatment before they are put into use. This is true for both cast and wrought (worked/fabricated) components. Although heat treatment operations could be carried out both on pure metals and alloys, the variety of heat treatments that could be conducted on pure metals is rather limited. But, heat treatment of alloys provides us with an opportunity to get a large variety of microstructures and mechanical properties. Majority of the heat treatments depend on phase transformations that take place in an alloy system in the solid state. Consequently, a metal or an alloy that does not undergo any phase transformation in the solid state cannot be heat treated effectively to obtain the desired properties.

Both ferrous and nonferrous alloys are subjected to heat treatment procedures. But, in this chapter we will only deal with the types of heat treatment operations as applied to ferrous alloys (and more specifically steels), which will produce a variety of microstructures and hence different properties. In fact, the steel technology depends heavily on the possibility to modify their microstructures and properties. In this connection, the eutectoid reaction in steels (the austenite phase producing a mixture of ferrite and cementite phases, namely, pearlite on cooling) plays a major role.

It is very important to choose the proper heat treatment for the given component. A heat treatment to achieve certain properties in a particular component made of an alloy may not be suitable for the same component made of another alloy. The properties obtained could be totally different.

6.2 REACTION RATES

We had seen earlier in Chapter 5 that different types of microstructures could be obtained when steel samples of different compositions are cooled to different temperatures and under equilibrium conditions. The nature and amounts of the phases were found to be different depending on the carbon content of the steel sample. But, the nature and amounts of phases in a steel sample of a given composition can be drastically altered by allowing the austenite phase to transform at different cooling rates, that is, under *nonequilibrium cooling* conditions.

For the austenite phase to transform under nonequilibrium conditions, it is first necessary to retain the austenite phase at temperatures lower than the eutectoid transformation temperature, without any transformation occurring. This process is referred to as *undercooling* of the austenite phase. Note that it is, in principle, easy to undercool a phase to a lower temperature in the solid state than from the liquid state, since the reaction rates during solid-state transformations are slow, that is, they take a long time for the reaction to be completed. On the other hand, the liquid-to-solid transformations are faster, and therefore it is more difficult to undercool the liquids to very low temperatures. Thus, by cooling the steel sample rapidly (*quenching*) from the austenitic condition (solid state), to a temperature above that when the martensitic transformation starts occurring, the austenite phase will be retained as-is without any transformation taking place. If this undercooled austenite phase is then allowed to transform at any temperature, the amount (or fraction) of the transformation product increases with time. Further, the nature of the phase will also be different depending on the temperature at which the transformation occurs.

The time required for the undercooled austenite phase to transform to other phases will be a function of the temperature at which the transformation is allowed to take place. In general, the rates of reaction during solid-state transformations can be represented by the *Avrami equation*

$$y = 1 - \exp(-kt^n) \tag{6.1}$$

where y is the fraction of the phase transformed, t is the time, and k and n are time-independent constants for the reaction. Since most transformations are of the nucleation and growth type, at any given temperature, the amount of the phase transformed increases with time. A typical transformation curve showing variation of the amount of the product phase formed as a function of time is shown in Figure 6.1.

Majority of the transformations from one phase to another (or to a mixture of phases) involve two stages—nucleation and growth. That is, the new product phase does not form until the nucleus of the new phase that has formed had time to grow to a measurable level. That is why you will notice in Figure 6.1 that the reaction does not occur for some time. This minimum time required for the transformation to start is referred to as the *incubation time*. This is the time required for the process of nucleation to occur. Since the growth of nuclei occurs after the incubation time, the

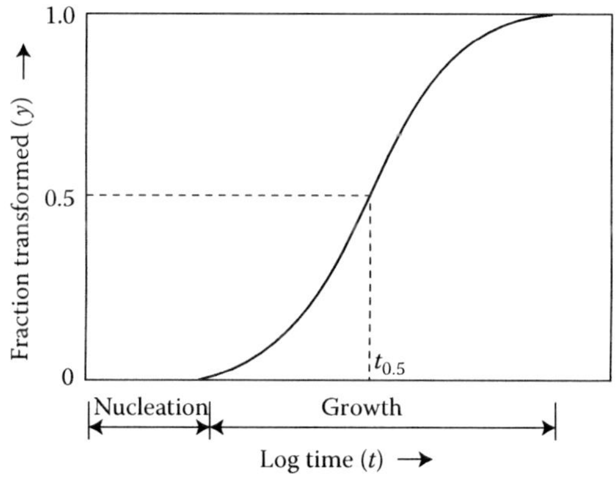

FIGURE 6.1 Fraction of phase transformed as a function of time at a constant temperature. The plot shows an S-shaped curve. Note that there is an incubation time, before which no transformation takes place.

reaction starts slowly (but the rate of reaction increases), then picks up speed, and eventually the reaction slows down (the reaction rate decreases). Consequently, the plot appears to resemble the letter "S" and that is why these plots are referred to as *S-curves*. One can measure the time required for a particular fraction of the phase to be transformed (say, 0.5, but you can choose any value you like) to determine the reaction rate. The relationship between the time required for a given fraction of the phase to be transformed and the temperature at which the transformation occurs can be combined together into one equation

$$\frac{1}{t_{0.5}} = Ae^{-Q/RT} \tag{6.2}$$

where $t_{0.5}$ is the time required for 50% of the transformation to be completed at temperature T (expressed in degrees Kelvin), R is the universal gas constant (8.314 J mol^{-1} K^{-1} or 1.987 cal mol^{-1} K^{-1}), and Q is the *activation energy* for the transformation to occur. The reciprocal of the time ($1/t$) required for the transformation to take place is also known as the rate (r). Thus, the above equation is also known as the *rate equation*, or more commonly as the Arrhenius-type equation. Equation 6.2 can now be rewritten as

$$r = Ae^{-Q/RT} \tag{6.3}$$

When natural logarithms are taken on both the sides of Equation 6.3, we get

$$\ln r = \ln A - \left(\frac{Q}{RT}\right) \tag{6.4}$$

The above expression represents an equation for a straight line. Therefore, when we plot ln r versus $1/T$, we get a straight line (Figure 6.2) with a slope equal to $-Q/R$ and an intercept equal to ln A. From the slope, the activation energy, Q for the transformation can be calculated.*

It is important to know the concept of activation energy. Activation energy is the energy that needs to be supplied to the system for the reaction to proceed forward. In the absence of this energy, usually thermal in nature, that is, obtained by raising the temperature, the reaction (or transformation) will not take place. The value of the activation energy provides a clue to the mechanism by which the transformation is taking place. For example, if the activation energy for the transformation is the same as the energy required for the diffusion of solute atoms in the lattice, then one can conclude that the phase transformation is taking place through diffusion of solute atoms in the lattice. The higher the activation energy for the process, more energy needs to be supplied for the transformation to take place. That is, comparing two processes, the one that requires the lower activation energy occurs more easily. If more energy is supplied than required, then the reaction will go faster. In such a situation, it is stated that the *kinetics* of the reaction are faster. That is why the kinetics of the reaction are faster at higher temperatures.

---

* Svante August Arrhenius (1859–1927) was a Swedish scientist and one of the founders of the science of Physical Chemistry. He was born on February 19, 1859 in Vik, Sweden. He started his doctoral work at the University of Uppsala, but found the chemistry professors there rigid and uninspiring. So, he moved to the Royal Swedish Academy of Sciences in Stockholm to work with Erik Edlund and graduated in 1884. His doctoral thesis did not impress his professors in the University, but others such as Rudolf Clausius, Wilhelm Ostwald, and Jacobus Henricus van't Hoff, who were also to become founding fathers of physical chemistry, were very impressed. The most important idea in the dissertation was his explanation of the fact that a salt dissolved in water dissociates into charged particles (ions) even in the absence of an electric current, although neither the pure salt nor pure water is a conductor. This was at variance with the theory of Faraday who believed that ions were produced only in the process of electrolysis. Later extensions of this very work earned Arrhenius the Nobel Prize in Chemistry in 1903. Except for a short period of three years when he worked with Ostwald in Riga, Kohlrausch in Würzburg, Boltzmann in Graz and van't Hoff in Amsterdam, he spent all his life in Sweden. He became a Professor at Stockholm University in 1895 and rector in 1896. Arrhenius became involved in setting up the Nobel Institutes and Nobel Prizes and was very influential in arranging prizes for his friends and denying them to his enemies. He became a Fellow of the Royal Society of London in 1910, was awarded the Royal Society's Davy Medal (1911), the Faraday Medal of the Chemical Society (1914), and the Franklin Medal in 1920. He received many honorary degrees from many Universities. In 1905, he was made Director of the newly created Nobel Institute for Chemistry where he remained until his retirement in 1927. He died at Stockholm on October 2, 1927.

The Arrhenius equation gives the dependence of the rate constant k of chemical reactions on the temperature and activation energy. The equation was first proposed by J. H. van't Hoff in 1884; five years later, Arrhenius provided a physical justification and interpretation for it. In 1889, Arrhenius explained the fact that most reactions require added heat energy to proceed by formulating the concept of activation energy, an energy barrier that must be overcome before two molecules will react. The Arrhenius equation gives the quantitative basis of the relationship between the activation energy and the rate at which a reaction proceeds. It can be used to model the temperature dependence of diffusion coefficients, population of crystal vacancies, creep rates, and many other thermally induced processes/reactions. Arrhenius had also developed the green house law that states: *if the quantity of carbonic acid increases in geometric progression, the augmentation of the temperature will increase nearly in arithmetic progression.* He estimated that halving of CO_2 would decrease the temperatures by 4–5°C and a doubling of CO_2 would cause a temperature rise of 5–6°C. This question is still of a main concern of Earth scientists today.

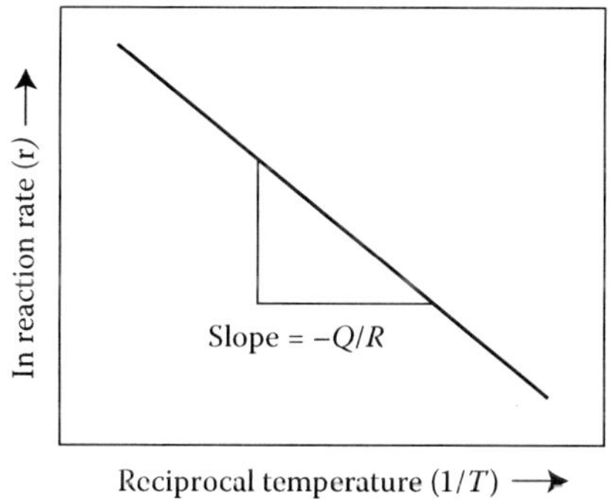

FIGURE 6.2 Schematic plot of reaction rate versus reciprocal of temperature at which the transformation is occurring. Note that a straight line is obtained when ln r is plotted against $(1/T)$. The slope of the straight line is $(-Q/R)$, from which Q, the activation energy for the transformation can be calculated.

6.2.1 Rate of Nucleation

We have mentioned above that a phase transformation involves both nucleation and growth. But, the rates at which nucleation and growth occur as a function of temperature are different. For easy understanding, let us consider the liquid-to-solid transformation, in which we have a homogeneous liquid phase, and that by decreasing the temperature to below the transformation temperature (melting point or solidus temperature), the liquid will transform into the solid (α phase). This transformation will take place by first forming a nucleus of the product α phase from the parent liquid phase and then its growth (Figure 6.3). (Remember, however, that we will be discussing the heat treatment of steels, a process that takes place completely in the solid state.)

Once a nucleus of reasonable size forms, the free energy of the system decreases since the free energy of the product phase is always lower than that of the parent phase. Assuming that the nucleus has a spherical shape with a radius r, the volume of the nucleus is $(4/3)\pi r^3$. If the free energy change per unit volume is represented by

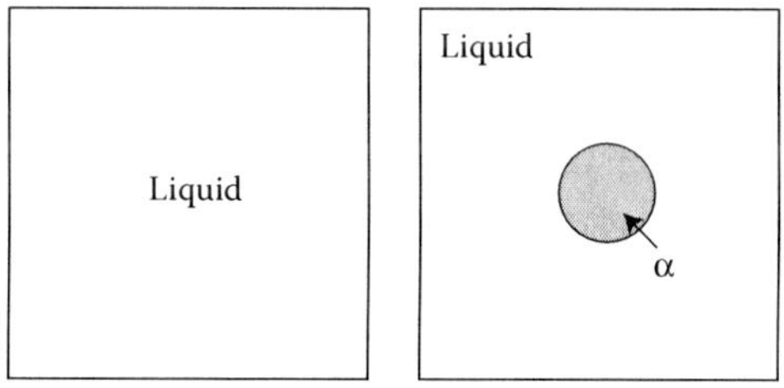

FIGURE 6.3 Nucleation of the solid phase α from the liquid phase. While formation of the new phase decreases the free energy of the system, creation of an interface between the liquid and α phases raises the free energy of the system. Therefore, there is a critical size of the nucleus above which only it can grow and that the phase transformation can occur.

ΔG_v, then the decrease in the volume free energy of the system, due to the formation of the nucleus, is $(4/3)\pi r^3 \times \Delta G_v$.

Associated with the formation of a nucleus of the product phase, however, is the creation of an interface between the parent and product phases. The free energy of the system increases due to the formation of the interface. This interfacial energy per unit area is designated as γ. Thus, the increase in free energy of the system due to formation of the interface is $4\pi r^2 \times \gamma$. Consequently, the balance between the volume free energy, which decreases the free energy of the system and the interfacial energy, which increases the free energy of the system, determines whether there is a net decrease in the free energy or not. The transformation will take place only when there is a net decrease in the free energy. The total change in the free energy of the system is the sum of these two terms and is given by

$$\Delta G = -\frac{4}{3}\pi r^3 \times \Delta G_v + 4\pi r^2 \times \gamma \tag{6.5}$$

Figure 6.4 shows a plot of these two contributions to the overall change in the free energy of the system. The process of nucleation will continue if the nucleus has a size larger than r^* and a minimum energy of ΔG^* is provided. Below the critical value of r^*, continued increase in its size results in an increase in the free energy and therefore, the nucleus will not grow. These subcritical nuclei are referred to as

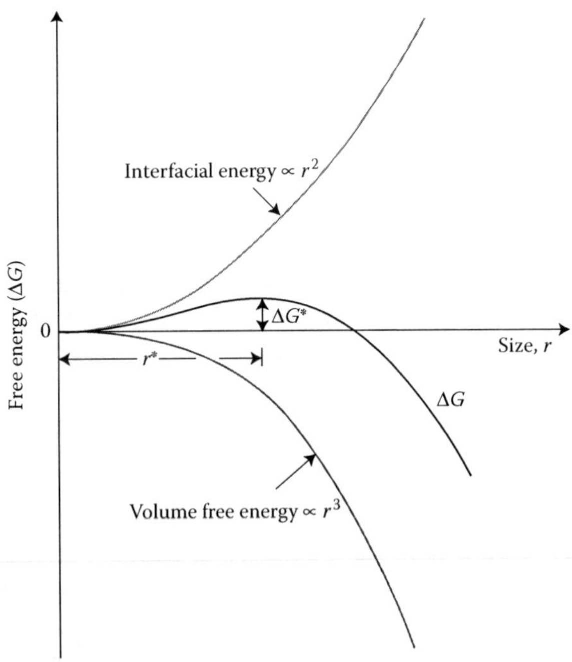

FIGURE 6.4 The change in free energy associated with the formation of a nucleus of a sphere of radius r. The values of r^* and ΔG^* are also indicated.

embryos. The free energy of the system decreases with an increase in the size of the nucleus, only when the nucleus has a size larger than r^*.

The critical size of the nucleus that is required for the nucleation to proceed forward, r^* is obtained by differentiating the above Equation 6.5 with respect to r, and then equating the derivative to zero. Accordingly,

$$\frac{d\Delta G}{dr} = -\frac{4}{3} \times 3\pi r^2 \times \Delta G_v + 4 \times 2\pi r \times \gamma \tag{6.6}$$

Equating the above equation to zero, we obtain

$$-4\pi r^{*2} \times \Delta G_v + 8\pi r^* \times \gamma = 0 \tag{6.7}$$

or,

$$r^* = \frac{2\gamma}{\Delta G_v} \tag{6.8}$$

It is also possible to calculate the critical energy, ΔG^* that needs to be supplied for the nucleus to grow. This can be obtained by substituting the value of r^* in Equation 6.5, when we get

$$\Delta G^* = -\frac{4}{3}\pi\left(\frac{2\gamma}{\Delta G_v}\right)^3 \times \Delta G_v + 4\pi\left(\frac{2\gamma}{\Delta G_v}\right)^2 \times \gamma \tag{6.9}$$

$$\Delta G^* = -\frac{32}{3}\pi \times \frac{\gamma^3}{(\Delta G_v)^2} + 16\pi \times \frac{\gamma^3}{(\Delta G_v)^2} \tag{6.10}$$

or,

$$\Delta G^* = \frac{16\pi\gamma^3}{3(\Delta G_v)^2} \tag{6.11}$$

The values of ΔG^* and r^* are also indicated in Figure 6.4.

Without mentioning it specifically, we have assumed in the above discussion that the probability of a nucleus to form is the same everywhere in the liquid phase. This situation is referred to as *homogeneous nucleation*. But, in reality homogeneous nucleation is difficult. Since solidification usually occurs when the liquid metal is poured into a solid mold, the mold walls act as preferential sites for nucleation to occur. Impurities and other extraneous particles present in the liquid can also act as sites for nucleation. If nucleation occurs in this way on mold walls or at impurities (or on defects in crystals during solid-state transformations), then this type of nucleation is referred to as *heterogeneous nucleation*. Without going into any details, however, let us realize that the activation energy for heterogeneous nucleation is much smaller than that for homogeneous nucleation. Therefore, heterogeneous nucleation takes place more easily than homogeneous nucleation. But, there is no difference in the critical nucleus size between these two types of nucleation processes.

Also note that the free energy change per unit volume, ΔG_v, which is the driving force for nucleation to take place, increases as the degree of undercooling, ΔT, increases. ΔT is the difference in the temperature at which the transformation is taking place and the equilibrium transformation temperature. Thus, if the melting point of aluminum is 660°C and solidification is taking place at 655°C, then the undercooling is 660–655 = 5°C. Therefore, Equations 6.8 and 6.11 can be rewritten in terms of ΔT as

$$r^* \propto \frac{1}{\Delta T} \quad \text{and} \quad \Delta G^* \propto \frac{1}{(\Delta T)^2} \tag{6.12}$$

From the above expressions in Equation 6.12 it is clear that the critical size of the nucleus to form is smaller and the activation energy for its formation is also lower, the larger the amount of undercooling, ΔT. Consequently, nucleation can take place even when the radius of the nucleus is smaller and the activation energy available is lower provided we have achieved a large amount of undercooling. In other words, the rate of nucleation increases with an increase in the degree of undercooling, that is, at lower temperatures, as indicated in Figure 6.5.

Opposing this trend is the fact that diffusivity is also a function of temperature. The formation and growth of the nucleus requires diffusion and the diffusion of atoms is faster at higher temperatures and slower at lower temperatures. The

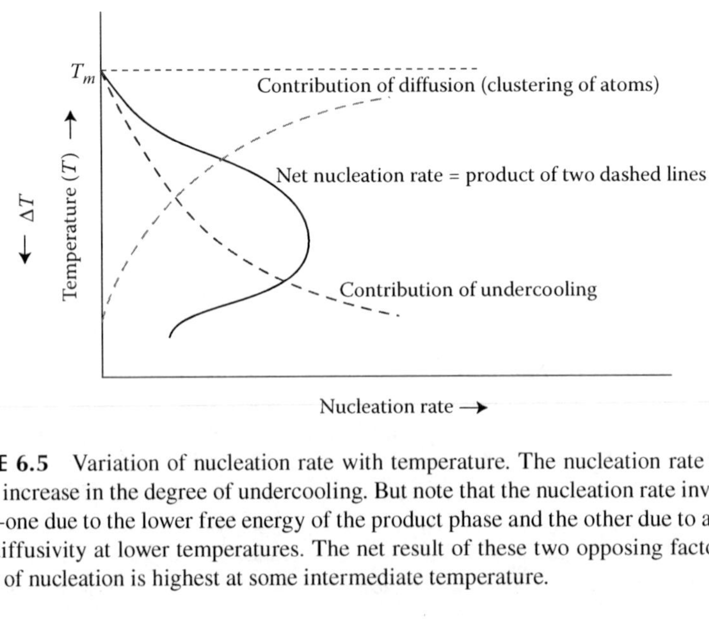

FIGURE 6.5 Variation of nucleation rate with temperature. The nucleation rate increases with an increase in the degree of undercooling. But note that the nucleation rate involves two terms—one due to the lower free energy of the product phase and the other due to a decrease in the diffusivity at lower temperatures. The net result of these two opposing factors is that the rate of nucleation is highest at some intermediate temperature.

diffusion coefficient or diffusivity, D varies with temperature, T according to the equation

$$D = D_o e^{-Q/RT} \quad (6.13)$$

where D_o is a preexponential constant and Q is the activation energy for diffusion to take place. Note that this is also an Arrhenius-type of equation. The contribution of diffusion to the process of nucleation is also represented in Figure 6.5. Thus, the net nucleation rate during transformation is a product of these two opposing trends. Thus, at lower temperatures the nucleation rate is determined by the balance between the increased rate due to increased undercooling and the decreased rate due to lower diffusivity. On the other hand, at higher temperatures, the nucleation rate is determined by the balance between the decreased rate due to the smaller amount of undercooling and the higher value due to increased diffusivity. Consequently, the nucleation rate increases with decreasing temperature up to a certain value and then starts to decrease. The nucleation rate is the highest at some intermediate temperature below the transformation temperature. The variation of nucleation rate as a function of ΔT is shown in Figure 6.5.

We have so far discussed the fact that nucleation can occur either in a homogeneous or heterogeneous manner, but focused our attention only on the liquid-to-solid transformation. The situation will be quite different when one solid phase transforms into another solid phase. This is because the strain effects due to the volume differences between the parent and product phases also contribute to the free energy changes in the transformation. That is, the change in free energy during the solid–solid transformation may be written as

$$\Delta G = -\frac{4}{3}\pi r^3 \times \Delta G_v + 4\pi r^2 \times \gamma + \Delta G_s \quad (6.14)$$

where ΔG_s represents the strain energy contribution to the free energy change during the transformation. Because of this additional contribution, the activation energy for the formation of the critical nucleus will be different between the liquid-to-solid and solid-to-solid phase transformations. Except that the details for the solid-to-solid transformations are different from the liquid-to-solid transformations, the general trends are not different among these different types of nucleation processes.

6.2.2 Rate of Growth and Rate of Transformation

Majority of phase transformations involve both nucleation and growth of the product phase. As mentioned earlier, growth requires diffusion of atoms and diffusion is faster at higher temperatures. Therefore, the rate of growth of the product phase is higher at higher temperatures. The variation of both the rate of nucleation and the rate of growth are plotted in Figure 6.6. Notice that the rate of growth decreases continuously with temperature, while the rate of nucleation is highest at some intermediate temperature. By taking the product of the rate of nucleation and rate of

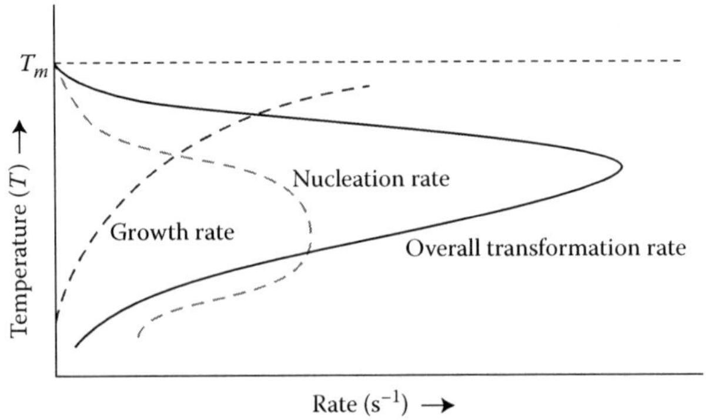

FIGURE 6.6 Plot showing the rate of nucleation, rate of growth, and rate of transformation as a function of temperature.

growth, it is possible to obtain the rate of transformation. The variation of the rate of transformation as a function of temperature also reaches a maximum at some intermediate temperature below the transformation temperature as shown in Figure 6.6. Stated differently, it means that the times required for the transformation (to start and to finish) are long at both high and low temperatures and are the least at some intermediate temperature.

6.3 ISOTHERMAL TRANSFORMATION DIAGRAMS

In the previous sections we have seen how the rate of transformation changes with temperature. We had also considered the variation of the fraction of the phase transformed as a function of time at any given temperature. Therefore, if we now combine the effects of both time and temperature on the fraction of the phase transformed, it would provide useful information about the phase transformation. The result is a *transformation diagram*, which provides information about the times required for a transformation to start and finish at any given temperature. This diagram provides information about the kinetics of the transformation as well as the nature and amounts of transformation products at any given temperature. Even though these transformation diagrams can be constructed for any type of material, these are frequently determined for steels because of their technological importance.

The combination of time, temperature, and amount of transformation occurring in a steel of given composition can be most conveniently represented in the form of a ***Time–Temperature–Transformation diagram***. In short, these diagrams are referred to as *T–T–T diagrams*. Since this diagram is obtained by allowing the transformation to take place at a constant temperature (isothermal, iso = same and thermal = temperature), this diagram is also referred to as an ***Isothermal–Transformation diagram***. We will also see later that the shape of this diagram appears like the letter "C" and so these are also called *C-curves*. In the *T–T–T* diagrams, the ordinate (*Y*-axis) represents the temperature and the abscissa (*X*-axis) represents time, usually in seconds. Time is plotted on a log scale because of the long times required for the transformation to be completed at both high temperatures and low temperatures. Occasionally, times corresponding to 1 minute, 1 hour, 1 day, and 1 week are also indicated on the *T–T–T* diagram, for easy identification of the time period.

The *T–T–T* diagram has to be determined for each composition of the steel separately. Let us now see how the *T–T–T* diagram is experimentally determined. Let us also assume that we wish to determine the *T–T–T* diagram for eutectoid steel. First, a large number of steel samples of the eutectoid composition are heated into the austenitic region. These samples are then held there for a time long enough for complete homogenization to be achieved. (Since steels at room temperature are usually mixtures of ferrite and cementite, it is required for these phases to completely transform into the homogeneous austenite phase, the starting phase for the isothermal transformation process.) A rule of thumb is that to bring about complete homogenization the steel sample needs to be held for 1 h at the austenitization temperature for every 1-in. thickness of the sample. After homogenization, the samples are quenched to a temperature below the eutectoid transformation temperature, at which the transformation has to be studied. The homogenized and quenched samples are then taken out from the quenching medium at regular intervals to study the progress of transformation. For example, one specimen is taken out after some time, say 2 s and then quenched in ice brine. Another specimen is taken out after some more time, say 5 s, another after 10 s, 20 s, 30 s, and so on and quenched into ice brine solution. All these ice-brine-quenched specimens are then prepared for metallographic examination and the microstructures observed. The amount of the transformation that has occurred (from austenite to pearlite in this case) is estimated by quantitative microscopy methods (see Chapter 3: Optical Microscopy for details). From these results, one could plot the percentage of the phase transformed as a function of time at the transformation temperature studied. Figure 6.7 shows the different steps involved in determining this plot of percentage of phase transformed at a constant temperature as a function of time. Note that this plot of fraction of pearlite formed versus time is similar to that shown in Figure 6.1.

This procedure is repeated at other temperatures and the times required for the start of transformation and that for the finish of transformation are noted at each temperature. The locus of these times forms the *T–T–T* diagram for this steel (Figure 6.8). Note that determination of the *T–T–T* diagram requires lot of samples, extensive metallographic specimen preparation, and pains-taking microstructural observations. Consequently, it is an expensive and time-consuming process. More than a hundred specimens often are required to determine a complete *T–T–T* diagram for a steel of given composition. Such diagrams need to be constructed for each steel composition.

The *T–T–T* diagram reflects the kinetics of the eutectoid transformation. Since the transformation of austenite to pearlite involves both nucleation of the ferrite and cementite phases and their growth, the kinetics of the phase transformation will be similar to that shown in Figure 6.6. Therefore, the transformation is slow both at high and low temperatures (i.e., it takes a long time both for the start and finish of transformation) and is the fastest at some intermediate temperature. This is the temperature indicated by the nose or knee of the C-curve and is at about 550°C for eutectoid steel.

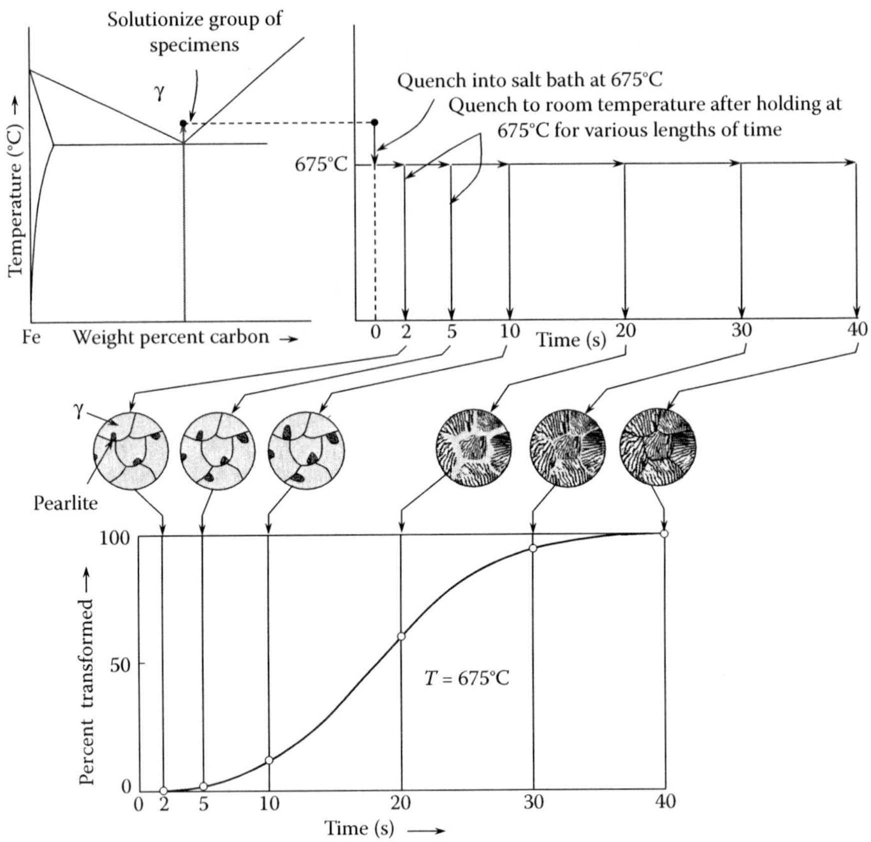

FIGURE 6.7 Method of determining the amount of phase transformed at a constant temperature as a function of time for a eutectoid steel. A large number of steel samples of the eutectoid composition are first austenitized in the austenitic region. These samples are then quenched into a salt bath maintained at a temperature below the eutectoid temperature, at which the transformation behavior is to be studied. Samples are then taken out at regular intervals and quenched to room temperature and the amount of pearlite formed is determined for each time. The amount of pearlite formed is plotted as a function of time.

The *T–T–T* diagrams make it possible to select steels and design heat treatments that will either produce desirable microstructures of ferrite and cementite, or avoid diffusion-controlled transformations, and thereby produce martensitic microstructures of maximum hardness. The greatest use of *T–T–T* diagrams is in predicting the nature and amount of phases present at any given temperature and time combination. Figure 6.9a shows a complete *T–T–T* diagram of eutectoid steel determined in a manner described above. Note, however, that when we deal with hypoeutectoid or hypereutectoid steels, there is a proeutectoid phase—ferrite in the hypoeutectoid steels and cementite in hypereutectoid steels—and this starts forming above the eutectoid transformation temperature. Therefore, the *T–T–T* diagrams for noneutectoid steels

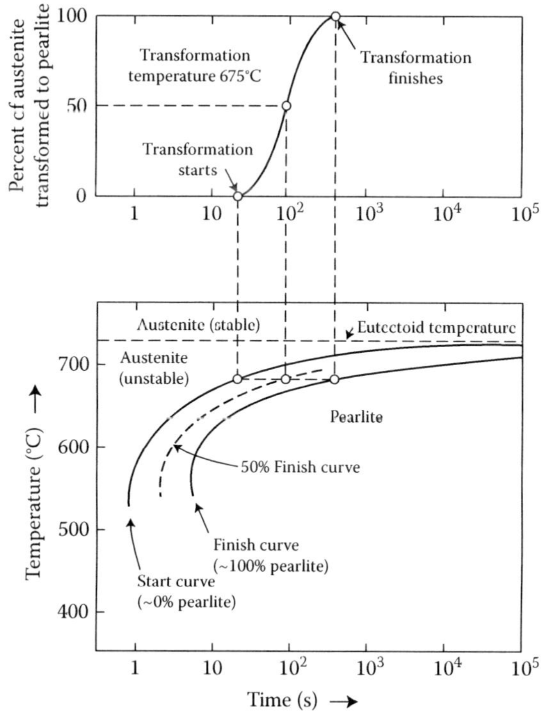

FIGURE 6.8 From a plot similar to that shown in Figure 6.7, the time to start and the time to finish the transformation are determined at the chosen temperature. By repeating these experiments at different temperatures, the times for the start and finish of transformation are noted at each temperature. The isothermal transformation diagram or the *T–T–T* diagram is then plotted from the time, temperature, and amount of transformation data obtained above.

will appear slightly different from that shown in Figure 6.9a. A *T–T–T* diagram for a hypereutectoid steel is shown in Figure 6.9b.

6.4 TRANSFORMATION PRODUCTS

Let us now discuss about the nature and amount of the transformation product depending on the temperature at which the transformation is allowed to take place and the times for which the transformation occurs.

6.4.1 Pearlite

Depending on the transformation temperature, the nature of the transformation product is different. Further, the amount of the transformation product depends on the time spent at that temperature. Above the nose of the *T–T–T* curve, the product is pearlite consisting of alternate lamellae of ferrite and cementite. The lamellar spacing of pearlite is determined by the transformation temperature, and the higher the

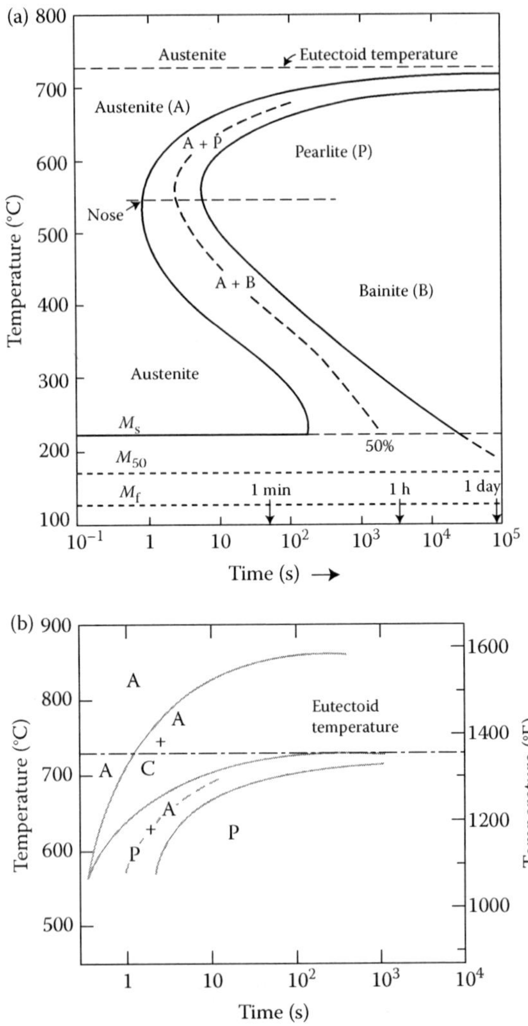

FIGURE 6.9 *T–T–T* diagrams for (a) eutectoid and (b) hypereutectoid steels. Whereas no transformation takes place in the eutectoid steel until it is cooled to below the eutectoid temperature, proeutectoid cementite starts forming above the eutectoid temperature in the case of hypereutectoid steels.

transformation temperature the larger is the interlamellar spacing due to the increased diffusivity of atoms. Consequently, the pearlite formed at higher temperatures will be referred to as *coarse pearlite*. Pearlite formed at lower temperatures will have smaller interlamellar spacing and so the pearlite will be referred to as *fine pearlite*. Figure 6.10 shows the typical microstructures of coarse pearlite and fine pearlite. Note that the difference between these two types of pearlite is only in the interlamellar spacing, that is, the spacing between two cementite lamellae or between two ferrite lamellae. The nature and proportion of the phases (ferrite and cementite) in both coarse and fine pearlite, however, are the same. Optical microscopic examination suggests that the alternate lamellae of ferrite and cementite will be seen clearly resolved in the case of coarse pearlite. These lamellae may not be easily resolved in the case of fine pearlite. Consequently, when we have fine pearlite, the pearlite microconstituent will appear like a black mass, if the individual lamellae are not resolved. Even fine pearlite can be resolved at higher magnifications.

6.4.2 Bainite

When the transformation takes place at temperatures below the nose of the T-T-T curve the transformation product is quite different from pearlite and is no longer lamellar in nature. Even though the transformation product still contains ferrite and cementite (or carbide) phases, it is now called *bainite.* Since the pearlite formed at low temperatures (but still above the nose of the *T–T–T* curve) will have a fine lamellar spacing, the total interfacial area between the ferrite and cementite phases is so large that the steel will be in a high-energy state. The total internal energy of the steel could be reduced by adopting a different morphology for the constituent phases. Consequently, the microstructure of bainite is significantly different from that of the lamellar structure of pearlite.*

Bainite can be described as a nonlamellar structure of ferrite and cementite. Bainite formation at relatively higher temperatures takes place according to the following mechanism. Ferrite having a very low solubility of carbon forms first as a fine plate at the austenite grain boundaries. Clusters of these fine plates are referred to as sheaves. The carbon rejected from the austenite due to formation of ferrite enriches the neighboring austenite phase which eventually precipitates as cementite in between these ferrite plates. Depending on the carbon content of the steels, the cementite between the ferrite plates may exist either as continuous layers or as discrete particles.

* Edgar Collins Bain (1891–1971) was a pioneering American metallurgist who was born near La Rue, Marion county, OH on September 14, 1891. On receiving the B.S. degree in chemical engineering from the Ohio State University in 1912, he entered government service first with the U.S. Geological Survey and then with the National Bureau of Standards (now called the National Institute for Standards and Technology). After working at several different establishments, Bain was asked to head the Physical Metallurgy Division of the United States Steel Corporation in 1928. About this time, he wrote a classic paper, along with E. S. Davenport, on "Transformation of Austenite at Constant Subcritical Temperatures," presenting for the first time the now famous *T–T–T* diagram. One of the results of this investigation was the discovery of hitherto unknown microstructure. During late summer of 1934, a group of Bain's close associates in the laboratory sauntered into his office and presented to him a superb, large, neatly mounted micrograph at 1000× of a new, isothermally formed microstructure of specimens of isothermally transformed steel. It was designated in large capital letters BAINITE and below, "First noted by, and so named, in 1934 in honor of E. C. Bain by the research staff of the U.S. Steel Corporation."

In January 1935, he was appointed vice president of Research and Technology for the U.S. Steel Corporation. In 1954, he was elected to the National Academy of Sciences. Two years later, he was named Chairman-designate of the Division of Engineering and Industrial Research of the National Research Council and then served as its Chairman during 1957 and 1958. The Edgar C. Bain Laboratory for Fundamental Research in Monroeville, PA is a true testament of his contributions to the science of steel. He died on November 27, 1971 in Edgeworth, PA.

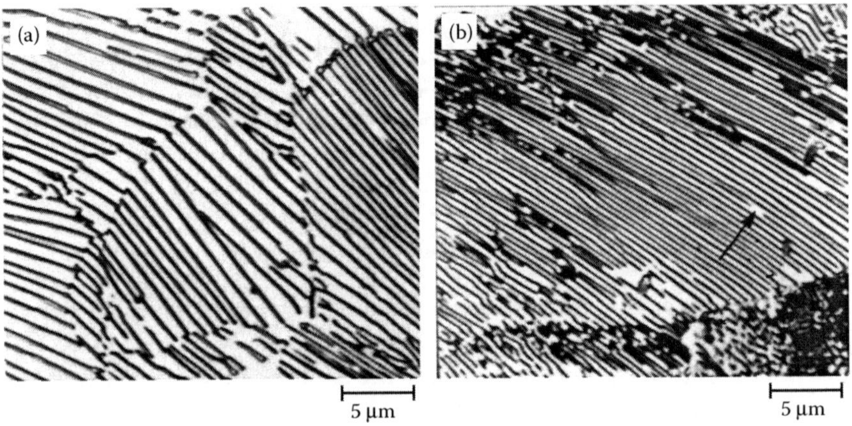

FIGURE 6.10 Microstructures of (a) coarse pearlite and (b) fine pearlite in eutectoid composition steel. Both the specimens were etched with 4% picral. (Photos courtesy of ASM International, Materials Park, OH. With permission.)

Bainite formation also takes place by nucleation and growth processes. But, the carbon diffusion distance during the bainitic transformation is less than that during the pearlitic transformation. Like in the case of pearlitic transformation where we have coarse and fine pearlite depending on the transformation temperature, the bainite formed at relatively higher temperatures (but still below the nose of the T-T-T curve) will be referred to as *upper bainite*. Because of its microstructural appearance, it is also called *feathery bainite*. We have described above the mechanism for the formation of upper bainite. The bainite formed at lower temperatures will be referred to as *lower bainite* or *acicular bainite*. The microstructure of lower bainite is not significantly different from that of upper bainite. In the lower bainite also we have ferrite and cementite. But, the cementite particles will be present not only in between the ferrite plates, but they also precipitate inside the ferrite plates. While the cementite particles in between the ferrite plates form due to the enrichment of austenite by carbon, precipitation of cementite particles inside the ferrite plates takes place from the supersaturated ferrite phase. Further, the carbide in lower bainite need not be cementite; some transition carbide phases may also form. Figure 6.11 shows the typical microstructures of upper bainite and lower bainite in a 4360 steel isothermally transformed at 495°C (upper bainite) and 300°C (lower bainite).

6.4.3 Martensite

At very low temperatures, of the order of 200°C or lower, the above-described diffusional transformations may not take place at all. This is because the diffusion of carbon to bring about the compositional changes to form ferrite–cementite mixtures becomes very slow. Instead, a diffusionless or *martensitic transformation* takes place when the austenite phase is allowed to transform at very low temperatures. The product phase, called *martensite*, forms without involving any carbon

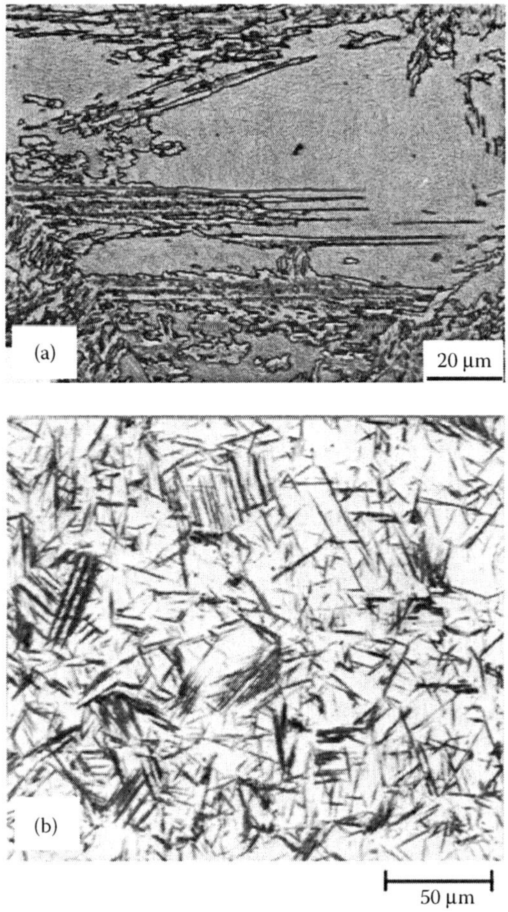

FIGURE 6.11 Microstructures of (a) upper bainite and (b) lower bainite obtained by isothermally transforming a 4360 steel at 495°C to produce upper bainite and at 300°C to produce lower bainite. (Photos courtesy of ASM International, Materials Park, OH. With permission.)

diffusion. Consequently, the composition of martensite is the same as the composition of austenite from which it had formed. Further, this transformation does not follow the Avrami transformation kinetics, that is, the transformation rate is not described by a sigmoidal curve. The growth rate in this type of transformation is so high (about 10^7 m s^{-1}, or approaching the speed of sound) that nucleation becomes the controlling step. The martensitic transformation is also known as *athermal* or *displacive* transformation. It is known as athermal transformation because diffusion is not involved and the amount of martensite phase formed depends only on temperature and not on time. As a result, the amount of martensite formed will not increase with time at any temperature. (Note that this is quite different from the pearlitic or bainitic transformations we have seen above, where the amount of the phase formed increases with time at any given temperature.) It is also known as displacive transformation because the martensite phase is formed only by a slight displacement of atoms in the lattice, usually much less than one interatomic distance.*

6.4.4 Morphology of Martensite

The martensite phase is easily identified by its characteristic lenticular shape (in the shape of a lens) in the optical microscope. Figure 6.12 shows a typical photomicrograph of a steel sample showing the martensitic structure. The lenticular (or needle)-shaped phase is martensite. The white portions in the micrograph represent the austenite phase that did not transform to martensite. The martensite plate (or lath) spans the whole width of the austenite grain and a large number of such laths form in one grain. (These details cannot, however, be discerned in the above photomicrograph.) Even though it may appear that the martensite laths are forming at random, a well-defined crystallographic relationship exists between the parent austenite and product martensite phases. Such relationships can be established with the help of TEM investigations. The plane in the austenite phase on which the martensite lath forms is called the *habit plane*, and the Miller indices of the habit plane change with the carbon content of the steel. The indices of the habit plane are indicated in terms of the austenite (γ) phase. For example, the habit plane is $\{111\}_\gamma$ in low-carbon steels (0–0.4 wt% C), $\{225\}_\gamma$ in medium-carbon steels (0.5–1.4 wt% C), and $\{259\}_\gamma$ in high-carbon steels (1.5–1.8 wt% C).

The morphology of martensite is different in steels of different carbon content. Even though the presence of martensite can be identified using optical microscopy methods, detailed microstructural features of martensite can only be obtained

* Adolf Martens (1850–1914) was a German scientist in whose honor the hard phase obtained on quenching steels was named martensite. Martens was born on March 6, 1850 in Beckendorf in the district of Mecklenburg, Germany. He received his engineering education at the Royal "Gewerbe-akademie" (Royal Industrial Academy), Berlin and joined the engineering staff of the Royal Prussian Railways in 1871. Along with personalities like Osmond, Roberts-Austen, Sorby, Tschernoff, he is considered one of the founders of metallography in the nineteenth century. While working at the Royal Prussian Railways, he could perform his metallographic work only in his spare time and at his own expense. For his studies, he developed by himself the experimental tools he needed, including an optical microscope with a new illumination technique. Since it was very difficult at that time to obtain micrographs of good quality, Martens made very accurate drawings of the microstructures with the help of a special optical prism attached to the microscope. Later, in cooperation with the Carl–Zeiss optical factory he designed "what was to be the best equipment for metallographic photomicrography for several decades." Although Adolf Martens did not publish any studies connected directly with the microstructural constituent of steel which bears his name, in view of his immense contributions to microstructural studies, Osmond termed the "martensitic" structure of steel after him. Martens was appointed Professor in 1889, and also became a Member of the Royal Academy of Sciences in Berlin in 1904. By combining the Royal Mechanical-Technical Testing Institute, the Chemical Testing Institute, and the Test Office for Civil Engineering Materials, he founded the Royal Material Testing Office (now known as Federal Institute for Materials Research and Testing—BAM) under his directorship in 1904. BAM has now instituted the Adlof Martens Fellowship program in his honor. He died on July 24, 1914.

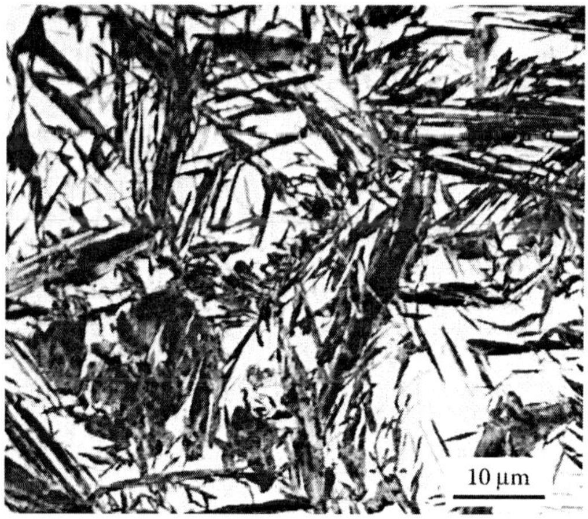

FIGURE 6.12 Microstructure of martensite. The needle-shaped grains are the martensite phase and the white regions are austenite that did not transform to martensite. The austenite phase retained at room temperature without transformation to martensite is called retained austenite.

through TEM methods. From such electron microscopy methods, it has been determined that martensite in low-carbon steels is characterized by plates or laths, about 0.5 μm wide containing a high dislocation density. Martensite in medium-carbon steels is characterized by lenticular plates which contain microtwinning and dislocations. On the other hand, martensite in high-carbon steels is still lenticular in shape, but it is heavily twinned.

6.4.5 Mechanism of Martensite Formation

Martensite starts forming from untransformed austenite at a temperature referred to as the martensite start or M_s temperature. The amount of martensite formed increases as the temperature is lowered and is completed at a temperature referred to as the martensite finish or M_f temperature. That is, the amount of martensite increases with a decrease in temperature, the amount being almost negligible or zero at M_s and 100% at M_f. At any intermediate temperature, between M_s and M_f, the amount of martensite is between 0% and 100%, and is determined only by the temperature at which the transformation is taking place. The time at temperature does not play any role. That is, if the steel is maintained at a temperature between M_s and M_f, the amount of martensite does not increase with time. Both the M_s and M_f temperatures are indicated by horizontal lines in the *T–T–T* diagram (Figure 6.9a). It is also important to remember that martensite can form only from the austenite phase and not from any other transformed product such as pearlite or bainite. Since the austenite to pearlite or bainite transformation has to be avoided to achieve martensite formation, the austenite phase has to be cooled very rapidly without any transformation from occurring. In other words, the austenite phase has to be cooled above a critical cooling rate. This *critical cooling rate*, which needs to be exceeded for the martensite phase to form, can be determined by drawing a tangent at the nose to the transformation start curve in the *T–T–T* diagram.

The martensite phase has a body-centered tetragonal crystal structure. Its *c* parameter increases and *a* parameter decreases with increasing carbon content; but, the axial ratio (*c*/*a* value) increases with increasing carbon content (Figure 6.13). It is very difficult to detect the tetragonality in low-carbon martensites (<0.2 wt% C) and therefore the martensite in very low-carbon steels can be effectively considered as a supersaturated solid solution of carbon in α-Fe. The tetragonality, however, becomes significant at higher carbon contents; the *c*/*a* ratio is still only about 1.05 for martensite containing 1.0 wt% C.

As shown in Figure 6.14, a body-centered tetragonal unit cell can be outlined from two adjacent FCC unit cells. But, the axial ratio (*c*/*a*) of the tetragonal unit cell is now $(a/a\sqrt{2}/2) = \sqrt{2}$ (= 1.414). *X*, *Y*, and *Z* represent the axes of the FCC unit cell whereas *X′*, *Y′*, and *Z′* are the axes of the BCT unit cell obtained from the FCC unit cell. By a homogeneous contraction of the *Z′*-axis by about 20% and expansion of the *X′*- and *Y′*-axes by about 12% (through twinning or slip processes), the correct axial ratio observed for the martensitic phase can be achieved.

Carbon atoms in the austenite phase (FCC crystal structure) occupy the 0,0,1/2-type octahedral interstitial sites (Figure 6.15). Even though there is one octahedral interstitial site for each iron atom in the FCC structure, the amount of available carbon in the steel is much less and therefore not all the interstitial sites in the FCC lattice are occupied; only a few of them are occupied by carbon. Once the steel is quenched from the austenitic region, these carbon atoms get trapped in the FCC lattice along the *Z*-direction. Since the preferential sites occupied by carbon are along

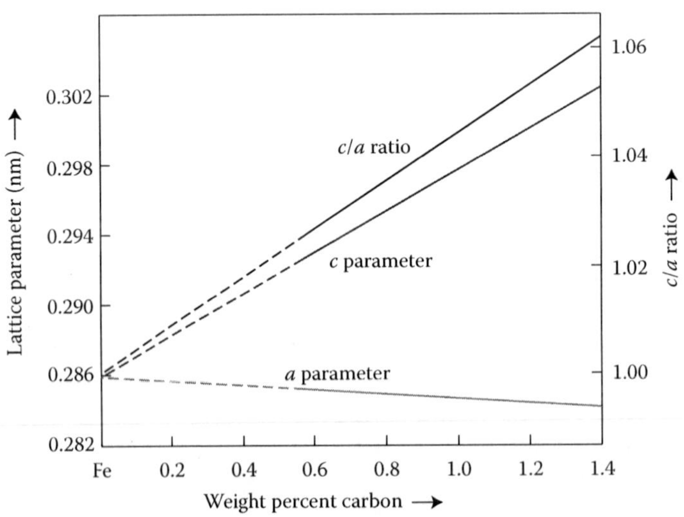

FIGURE 6.13 Variation of the lattice parameters of martensite as a function of carbon content in steels.

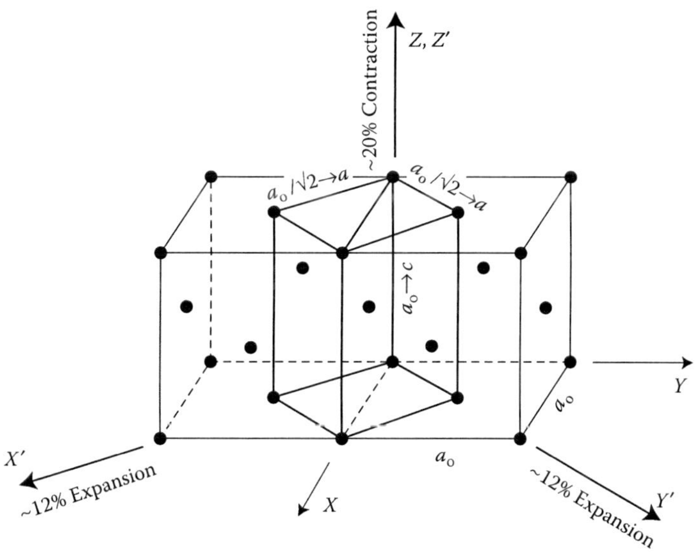

FIGURE 6.14 Lattice correspondence for the formation of BCT martensite from the FCC austenite phase. A body-centered tetragonal unit cell (outlined in heavy lines) can be obtained from the FCC unit cell of austenite. The BCT unit cell of martensite with the correct axial ratio (c/a) can be obtained by a homogeneous expansion of about 12% along the X'- and Y'-axes and about 20% compression along the Z'-axis of the BCT unit cell outlined from the FCC unit cell.

the Z-axis of the FCC structure, the lattice parameter along the Z-direction increases, resulting in the formation of a tetragonal structure. With increasing carbon content in the austenite phase, a larger number of carbon atoms are trapped in these interstitial sites and therefore the c parameter increases even more causing the c/a ratio to be larger.

6.5 RETAINED AUSTENITE

Even though we had defined the M_f temperature as the temperature at which the martensitic transformation is complete (i.e., 100% martensite phase has formed and that no more martensite will form on cooling the steel to lower temperatures), in practice it is impossible to obtain a situation when *all* the austenite will completely transform to martensite even below the M_f temperature. In other words, some austenite remains untransformed and this is referred to as *retained austenite*. The mechanism of the formation of retained austenite is related to the volume change on forming martensite from austenite. Since the FCC structure of austenite is close-packed and the BCT structure of martensite is more open, there is a volume expansion on formation of martensite from austenite. The martensite plates formed during the transformation surround the austenite phase and exert an enormous amount of

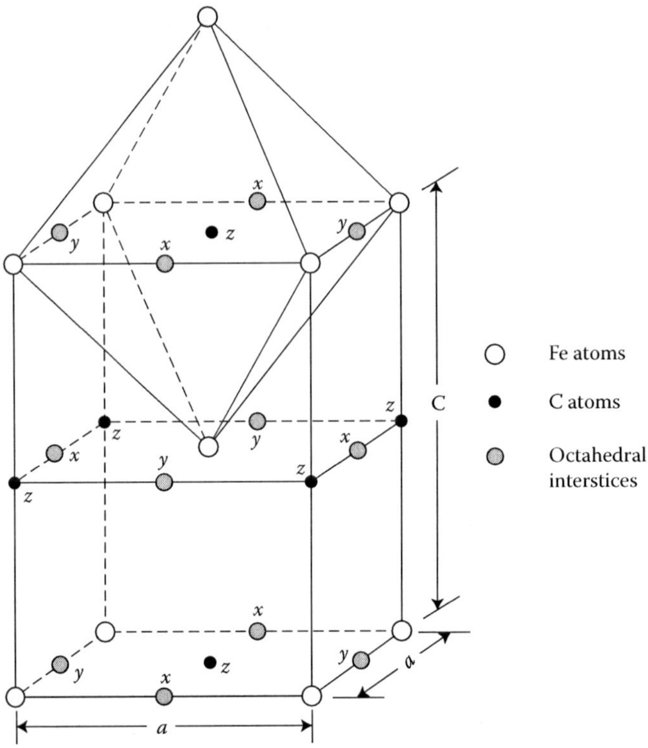

FIGURE 6.15 The unit cell of BCT martensite showing the location of the Fe and C atoms. C atoms occupy the octahedral interstitial sites. Even though the lattice contains one octahedral interstitial site per Fe atom, due to the limited amount of C atoms available in the alloy, only a few of these sites are occupied. The preferred sites of occupation of the C atoms are along the *Z*-axis of the unit cell. The other octahedral positions along the *X*- and *Y*-axes in the lattice, indicated by *x* and *y*, are not occupied.

pressure. Under these high-pressure conditions, the austenite phase gets stabilized and will not be able to transform to martensite. Further transformation can occur only when this pressure is released by the cracking of the specimen, provided the temperature is below M_f. In fact, one can frequently see cracks in specimens containing martensite that have been quenched severely from the austenite phase; these cracks are referred to as *quench cracks*.

Both the M_s and M_f temperatures decrease with an increase in carbon content. The alloying elements present in steel also have an effect on the M_s temperature. All alloying elements, except for cobalt, lower the M_s temperature. Even though a number of equations have been developed to relate the M_s temperature to the composition of steel, the following equation gives a reasonable estimate of the M_s temperature:

$$M_s(°C) = 539 - 423(\%C) - 30.4(\%Mn) - 12.1(\%Cr) - 17.7(\%Ni) - 7.5(\%Mo) \quad (6.15)$$

where the composition is expressed in weight percentage. The rate of decrease of M_f temperature is much faster than that of M_s temperature. Accordingly, it is possible that the M_f temperature of some high-carbon steels is below room temperature. Therefore, on quenching to room temperature, a high-carbon steel may not completely transform to martensite. Such steels should be given a cryogenic (or at least a subzero) treatment to produce a fully martensitic structure.

The presence of retained austenite in quenched steel samples has both beneficial and deleterious effects. The presence of retained austenite in steel improves the fatigue properties, increases toughness, and also the plane strain fracture toughness. But, its presence reduces the mechanical strength and hardness and also the dimensional stability. In the case of tool steels, the retained austenite may be fully transformed to martensite by subzero treatments (cooling of the steel to very low temperatures). Sometimes steels are subjected to cryogenic treatment (cooling to liquid nitrogen temperatures) to achieve a similar objective.

Retained austenite decreases the full potential hardness of the steel. A serious problem will occur when the quenched steel containing martensite is tempered (reheated to obtain some ductility and toughness in the steel). After tempering, since martensite decomposes, the pressure exerted on the austenite is less and therefore, it can transform to martensite if the steel is cooled to below the M_s temperature. (If the temperature is above M_s, the austenite could transform to other phase mixtures such as bainite or pearlite.) This martensitic transformation will increase the hardness of the steel instead of achieving the intended decrease in hardness and strength.

Martensitic transformations can also take place in nonferrous alloys, for example, in some copper-based alloys and nickel–titanium alloys or the so-called shape memory alloys. But, the nonferrous martensites are not hard and strong since interstitial elements such as carbon are not present, which get trapped in the lattice to increase the hardness. The martensite transformation terminology has, however, become so generic now that the high-temperature phase is referred to as austenite and the low-temperature phase as martensite in all martensitic transformations irrespective of whether they occur in ferrous or nonferrous alloys.

6.6 ISOTHERMAL TREATMENTS

Figure 6.16 shows the *T–T–T* diagram for a steel of eutectoid composition. Using this diagram, it is possible for us to predict what types of phases or phase combinations are produced as a function of time at different temperatures. We can also estimate the proportion of the phases produced on transformation. Let us assume that the steel sample is austenitized at 760°C for 1 h and that we will consider transformations taking place in this sample on allowing the transformation to take place isothermally (at a constant temperature).

Treatment 1: The austenitized steel is cooled rapidly to 680°C from the austenitization temperature and held there. Since this temperature is below the eutectoid transformation temperature and above the M_s temperature, the specimen will be in the metastable austenitic condition up to about 30 s, when transformation to coarse pearlite will start. The amount of coarse pearlite will increase with time (according to the Avrami kinetics) and pearlite formation is complete at about 630 s. There

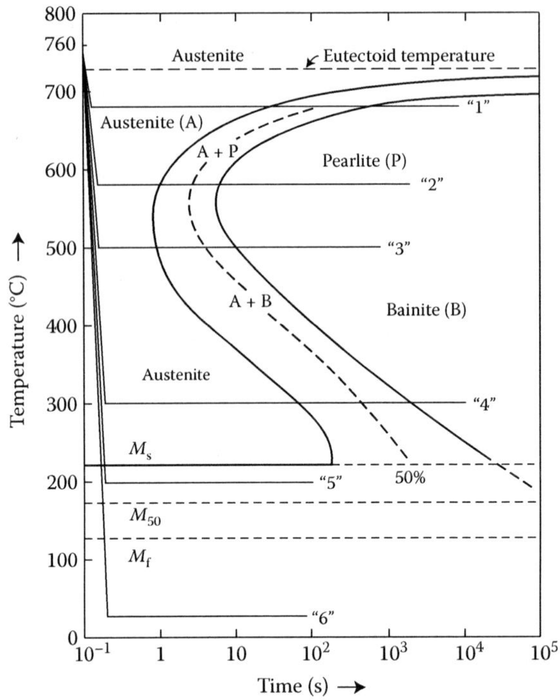

FIGURE 6.16 *T–T–T* diagram for eutectoid steel. The nature and amount of phases for different isothermal treatments can be determined using this diagram. The numbers "1," "2," "3," "4," "5," and "6" refer to the six isothermal treatments described in the text.

is no change either in the nature or amount of pearlite even if the specimen is held at this temperature for a longer time. We will have 100% coarse pearlite. This microstructure will continue to be present on cooling the steel sample to room temperature.

Treatment 2: The austenitized steel is cooled rapidly to 580°C and held there. The specimen will continue to be in the metastable austenitic condition for about 1 s when fine pearlite will start forming. The amount of fine pearlite will increase with time until the transformation is complete at about 7 s. Longer holding at this temperature will not change the constitution of the steel specimen. The specimen will now contain 100% fine pearlite.

Treatment 3: The austenitized steel is cooled rapidly to 500°C and held there. Note that this temperature is below the nose of the *T–T–T* curve. The specimen will be in the metastable austenitic condition for up to 1 s when upper (or feathery) bainite will start forming. The amount of the upper bainite will continue to increase with time up to 10 s, when the full amount of the upper bainite has formed. No more changes will occur even on continued holding at this temperature. The specimen will now contain 100% upper (or feathery) bainite.

Treatment 4: The austenitized steel is cooled rapidly to 300°C and held there. The specimen will be in the metastable austenitic condition for a little over 70 s, when the lower (acicular) bainite starts forming. With continued holding, the amount of lower bainite will increase with time and all the lower bainite would have formed by about 2000 s (about 30 min). Beyond this time, no more changes will occur. The specimen will now contain 100% lower (or acicular) bainite.

Treatment 5: The austenitized steel is cooled rapidly to 200°C and held there. Since the specimen is now at a temperature lower than M_s (the martensite start temperature), the austenite phase will transform to martensite. However, since this temperature is close to the M_s temperature (220°C), and approximately 1/4 the distance from M_s between the M_s and M_f temperatures, the amount of martensite will be approximately 25%. Longer holding at this temperature will not increase the amount of martensite (since the martensitic transformation is athermal in nature). So, the specimen will now contain about 25% martensite and 75% retained austenite.

Treatment 6: The austenitized steel is cooled rapidly to room temperature and held there. Since this temperature is well below the M_f temperature (the martensite finish temperature), all the austenite would have transformed into martensite (keeping in mind that a small amount of retained austenite would always be present). Keeping the sample for a long time at room temperature will not change the microstructure or the nature of the phase(s) present in the sample. The specimen will contain mostly martensite (plus a small amount of retained austenite). Even though martensite is metastable in nature, no significant transformation to the equilibrium phases takes place at room temperature.

It is also possible to have other types of treatment such as combinations of some of the above treatments. For example, let us consider the eutectoid steel specimen austenitized at 760°C for 1 h and quenched rapidly to 580°C. We had seen in Treatment # 2 that fine pearlite will start forming at 1 s. Let us see what would happen if we had held this sample at 600°C for 3 s and then quenched rapidly to room temperature. Holding the sample for 3 s would produce approximately 50% of fine pearlite and the remaining 50% would be the untransformed austenite. Since the sample is now quenched to room temperature, the austenite will transform into martensite. The fine pearlite which had formed at the higher temperature would not change; it will continue to be present as-is. Therefore, the final result at room temperature is that the sample will contain 50% fine pearlite and 50% martensite.

The constitution of samples heat treated in different ways can be similarly found by using the *T–T–T* diagram. It is a good practice to work out different possible combinations of heat treatment involving isothermal treatment and quenching operations to obtain different types of microconstituents (and microstructures) in the sample.

The *T–T–T* diagrams for off-eutectoid composition steels (i.e., hypoeutectoid or hypereutectoid steels) are more complicated than those for the eutectoid steel. This is because at temperatures higher than the eutectoid, the primary phase (either proeutectoid ferrite or proeutectoid cementite) starts forming before the pearlite formation occurs. Consequently, one notices an additional curved line extending from the temperature at which the proeutectoid phase forms up to the knee of the *T–T–T* curve. Another less obvious observation is that, in comparison with the M_s and M_f temperatures in the eutectoid steels, the M_s and M_f temperatures are higher in the hypoeutectoid and lower in the hypereutectoid steels. (We had noted above that the M_s and M_f temperatures decrease with increasing carbon content.) Remember that

T–T–T diagrams are different for steels of different composition. Therefore, they have to be established for *each* composition. *T–T–T* diagrams for all important steels have been determined and may be found in standard reference books such as *Atlas of Isothermal Transformation and Cooling Transformation* edited by Boyer (1997). Standard Heat Treatment books also contain the *T–T–T* diagrams for some steels.

6.7 EFFECT OF ALLOYING ELEMENTS ON THE *T–T–T* DIAGRAM

Addition of alloying elements to steel generally delays the austenite-to-pearlite or austenite-to-bainite transformation. Consequently, the start and finish curves of the *T–T–T* diagram are shifted to longer times. High carbon content, increased amount of alloying elements, and large austenite grain size all delay the pearlite transformation and therefore, the *T–T–T* diagrams are shifted to longer times. The extent of delay of the pearlite transformation is different for different alloying elements. Austenite stabilizers such as Cu and Ni have only a marginal effect. It has been reported that Al, Co, and Si shift the start of the proeutectoid ferrite reaction in the *T–T–T* diagram to shorter times, that is, the reaction takes place faster.

Not only the position, but the shape of the *T–T–T* curve is also affected by the alloying elements, especially those which are carbide formers. The pearlite transformation is delayed much more than the bainite transformation. Consequently, the pearlite nose is shifted much farther to the right than the bainite nose (Figure 6.17). A consequence of the different shifts of the pearlite and bainite transformation curves is that it is possible to produce bainite in an alloy steel on continuous cooling, while that is not possible in a plain carbon steel (see below for more details).

6.8 CONTINUOUS COOLING TRANSFORMATION DIAGRAMS

We have seen above how the *T–T–T* or the isothermal transformation diagrams are established and how they are used to interpret the formation of different phases on subjecting the steels to isothermal transformation processes. But, in practice, however, it is rare that we hold the specimens at a constant temperature and follow the progress of transformation as a function of time. That is, we do not subject the specimens to isothermal transformation behavior. Instead, we normally allow the specimens to cool continuously from a high temperature to a low temperature. Note that such a cooling process involves a range of cooling rates since the cooling rate will be different in different temperature regimes. Therefore, we need to develop more realistic diagrams, which will provide information on continuous cooling of samples. Such diagrams are referred to as *Cooling Transformation* (*CT*) or *Continuous Cooling Transformation* (*CCT*) *diagrams*. These diagrams determine the nature and amounts of phases in steels on cooling them continuously from a high temperate to room temperature, as practiced in an industrial or normal laboratory situation.

The *CCT* diagrams can be developed by continuously cooling the samples from the austenitization treatment temperature at a predetermined cooling rate. As a guideline, if we cool the sample very slowly (say at a rate of 5°C s^{-1}), then we will obtain coarse pearlite in the microstructure. Instead, if the sample is cooled at a

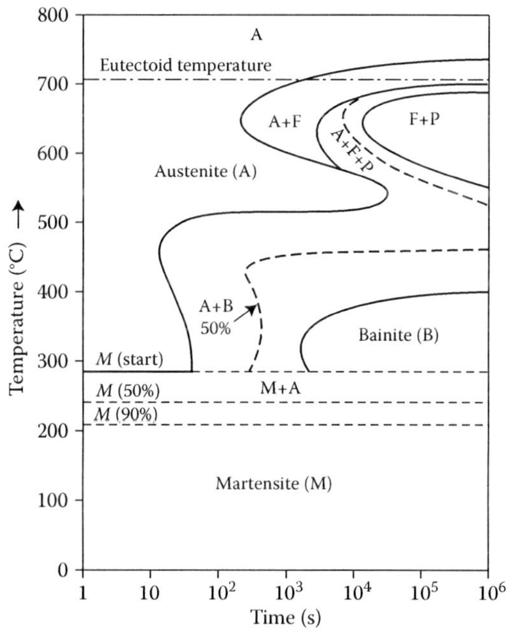

FIGURE 6.17 *T–T–T* diagram for an alloy steel (type 4340 containing 1.85 wt% Ni, 0.8 wt% Cr, and 0.25 wt% Mo in addition to about 0.4 wt% C, the rest being iron). Addition of alloying elements shift the transformation curves to longer times. But, the shift is more for the pearlite transformation curves in comparison to those of bainite transformation.

higher rate, for example, 30°C s^{-1}, then we will get fine pearlite. On the other hand, if we cool the sample rapidly, say at 200°C s^{-1}, then we will obtain martensite.

The method of developing the *CCT* diagrams is also involved. Cylindrical rod specimens are subjected to different cooling rates and the times required for the start and finish of transformation are determined. The onset of transformation is detected using techniques such as microscopy, dilatometry, magnetic permeability, or others. The nature of phases formed can be determined using metallographic techniques.

Luckily, the *CCT* diagrams and *T–T–T* diagrams are related to each other. Figure 6.18 shows a comparison of the *CCT* and *T–T–T* diagrams for a steel of eutectoid composition. Note that the temperatures for the start and finish of transformations for the *CCT* diagrams are lower and the times are longer in comparison to the *T–T–T* diagrams. That is, the continuous cooling transformation curves are shifted to longer times and lower temperatures.

There are also other important differences between the *T–T–T* and *CCT* diagrams. For example, it is not possible to obtain the bainite structure fully in the eutectoid steel, or for that matter in any plain carbon steel (steel containing only Fe and C), on continuously cooling it to room temperature. This is because by the time the steel reaches the bainitic region, the steel would have (at least partially) transformed to pearlite. And we had seen earlier that we could obtain bainite only from the austenite and not from pearlite. In other words, it is not possible for a plain

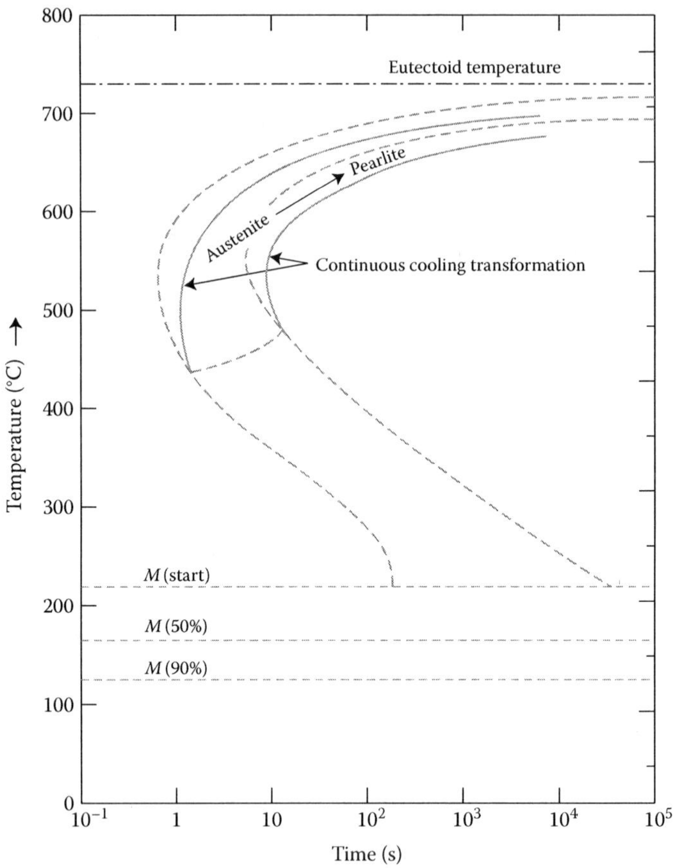

FIGURE 6.18 Superimposition of the *T–T–T* and *CCT* diagrams for a eutectoid steel. Note that, in comparison to the *T–T–T* diagram, the temperatures are lower and times are longer for the *CCT* diagrams.

carbon steel to reach the bainite formation region during continuous cooling without allowing the austenite to first transform to pearlite. Consequently, one cannot obtain bainite during continuous cooling of Fe–C alloys. On the other hand, one can obtain bainite in Fe–C alloys during isothermal transformation. The situation will, however, be different in the case of alloy steels where the alloying elements shift the pearlite and bainite transformation regions differently to different times. The M_s and M_f temperatures are, however, the same in both isothermal and continuous cooling diagrams.

6.9 TYPES OF HEAT TREATMENT

Development of the desired properties in steels depends heavily on the microstructure of the steel, which in turn depends on the type of heat treatment the steel has experienced. There are a number of different types of heat treatment processes and each of them has a specific purpose. Consequently, specific terms have been assigned to these heat treatment procedures on the basis of what they are expected to achieve. The important and commonly employed heat treatment procedures, as applied to steels, are:

- Annealing
- Normalizing
- Quenching
- Tempering

Other less frequently applied heat treatment procedures applied to steels are:

- Marquenching
- Austempering
- Martempering, and so on

But, in this chapter, we will concentrate on the more commonly used heat treatment procedures.

Depending on the requirements for obtaining the appropriate microstructure and mechanical properties, a combination of the above-listed heat treatment procedures also is employed. For example, the steel could be quenched from the austenitic region to produce the martensite phase. But, martensite is hard and brittle. Therefore, to increase the toughness of steel containing martensite, it could be tempered after quenching. A variety of other heat treatment operations are also conducted on nonferrous alloys. One such important heat treatment, to improve the strength of a two-phase material is known as *precipitation hardening.* In this heat treatment, a two-phase alloy is first heated to an elevated temperature where the alloy will exist in a single-phase region. It is then quenched to room temperature to retain the single phase in a supersaturated condition, and then aged (annealed) to allow the second phase to precipitate out as fine particles. This treatment increases the hardness and strength of the alloy significantly.

Similarly, the phenomenon of *polymorphism* (existence of a single-phase material in different crystal structures at different temperatures or pressures) in ZrO_2 and other materials helps in designing a suitable heat treatment procedure to improve the toughness of the inherently brittle ZrO_2. This is referred to as *transformation toughening* and depends on the significant volume increase (~9%) when the high-temperature tetragonal phase transforms to the low-temperature monoclinic form. But, in this chapter we will concentrate only on the common heat treatment operations carried out on steels.

Heat treatments can be carried out basically in two different ways. One way is to heat the material to a predetermined temperature and study the changes in the microstructure and properties of the alloy as a function of time at that temperature. This will be referred to as *isothermal heat treatment* since the material is maintained at the same temperature during transformation. In this case, the changes in material are studied as a function of time at a constant temperature. The other way is to heat the material continuously to an elevated temperature and maintain the material at that temperature for a fixed length of time. Such heat treatments are carried out at different temperatures but for the same amount of time at every temperature. This type of treatment is referred to as *isochronal* (iso = same and chronal = time) *heat treatment*, since the material is maintained for the same length of time at each temperature. In this method, the changes in the material are studied as a function of temperature for a constant time. Sometimes, this type of heat treatment is also referred to as nonisothermal heat treatment. Isothermal heat treatments are not practical and common. Therefore, in reality, heat treatments are carried out by continuous heating and/or cooling of specimens. That is why we looked at the continuous cooling transformation diagrams also. Let us now discuss the different types of heat treatment that are carried out on steels and the resultant microstructures and mechanical properties.

6.9.1 Annealing

Annealing is one of the most common heat treatments applied to materials (steels and nonferrous alloys alike). The most important result of *annealing* is to induce softening and to allow further working of the material. Thus, annealing is a very general heat treatment. But, there are different types of annealing treatments—*full annealing*, *stress-relief annealing*, *recrystallization annealing*, *homogenization annealing*, and so on depending on the end purpose for which annealing is carried out. If nothing is specified, annealing means full annealing. Annealing has been defined as any heating operation above a certain temperature, holding the sample at this temperature for a predetermined amount of time, and subsequently cooling the sample very slowly to room temperature.

As applied to steels, annealing is a heat treatment that involves heating the steel sample to the austenitic region, holding there until the specimen is homogeneous, and then slowly cooling it to room temperature. The cooling from the austenitic region is done very slowly, usually at the rate of about 10^{-2}°C s^{-1}. Such slow cooling is achieved when the hot sample is allowed to cool in the furnace by turning off power to the furnace. This is commonly referred to as *furnace cooling*. Figure 6.19 superimposes the cooling curve corresponding to annealing on the *T–T–T* diagram.

The temperatures employed for austenitization (to obtain a homogeneous austenite phase) are different for different steels and depend on the carbon content of the steel. They are usually about 40°C above the Ac_3 temperature (the GS line in the Fe–Fe_3C diagram in Figure 5.9) for hypoeutectoid steels or above the Ac_1 temperature (eutectoid temperature) for hypereutectoid steels. By heating the hypoeutectoid steels to 40°C above the Ac_3 temperature the steel becomes completely austenitic. But, in the case of hypereutectoid steels, the steel exists only as a mixture of austenite and cementite on heating it to about 40°C above the eutectoid temperature. This is because heating the hypereutectoid steels into the austenitic region (above the A_{cm} temperature) is not desirable. On subsequent slow cooling from the austenitic region, the proeutectoid cementite phase starts precipitating on the grain

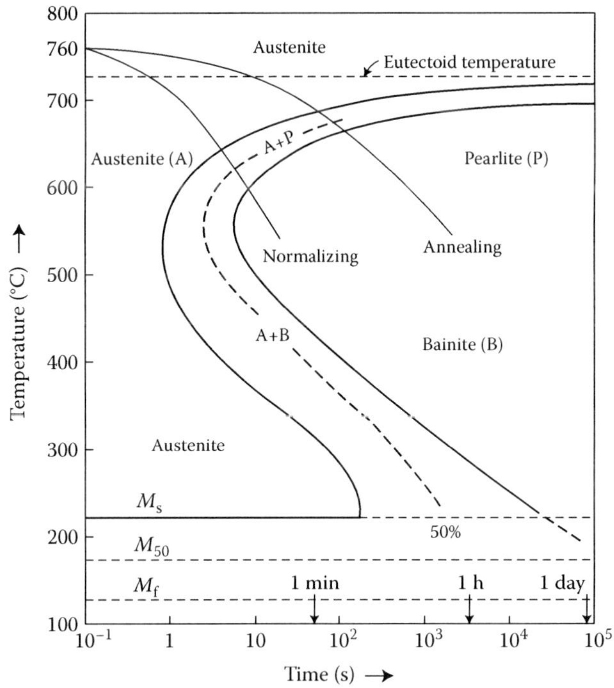

FIGURE 6.19 Superimposition of cooling curves corresponding to annealing and normalizing heat treatments on the *T–T–T* diagram.

boundaries of prior austenite grains. Being hard and brittle, the cementite phase provides an easy fracture path along the grain boundaries. Consequently, the steel will have poor mechanical properties. Figure 6.20 shows the temperature ranges used for full annealing treatments.

The time of holding the sample for homogenization is usually 1 h for every 1-in. thickness of the sample. This will allow complete dissolution of the carbide phase present in the steel to form the homogeneous austenite phase.

Depending on the composition of the steel, the microconstituents present in the sample will be different on annealing. Fully annealed eutectoid steel will have only coarse pearlite. On the other hand, fully annealed hypoeutectoid steel will contain coarse-grained equiaxed proeutectoid ferrite and coarse pearlite, while fully annealed hypereutectoid steel will have proeutectoid cementite and coarse pearlite. Typical microstructures of fully annealed 0.45 wt% C (hypoeutectoid), 0.76 wt% C (eutectoid), and 1.2 wt% C (hypereutectoid) steel samples are shown in Figure 6.21. The large interlamellar spacing in pearlite should be particularly noted. Note also the presence of proeutectoid cementite along the grain boundaries of the pearlite colonies in the hypereutectoid steel.

Fully annealed steels are soft, easily machinable, possess good hot workability characteristics, and also respond uniformly to subsequent hardening operations.

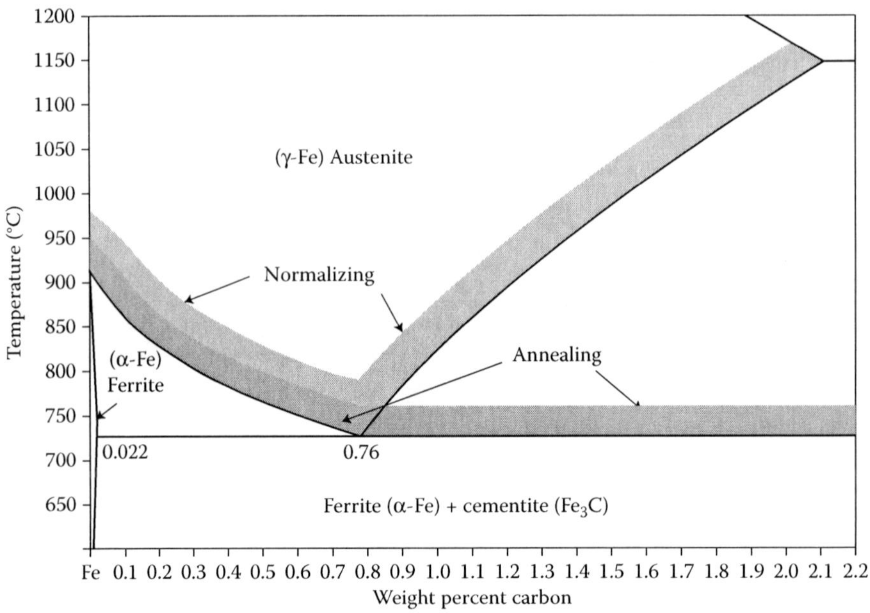

FIGURE 6.20 The Fe–Fe_3C phase diagram showing the temperature regions to which steels are heated for annealing and normalizing purposes.

6.9.2 Normalizing

Normalizing is another common heat treatment applied to steels to produce a uniform microstructure. In the process of *normalizing* also, the steel is heated into the austenitic region for homogenization and then cooled faster than during the annealing process, usually in air. However, there are differences between annealing and normalizing heat treatments.

The temperatures at which homogenization is carried out in the austenite phase are somewhat higher for normalizing than for annealing (Figure 6.20). The temperatures are about 50–70°C above the Ac_3 temperature for hypoeutectoid steels. In the case of hypereutectoid steels, the steel is heated above the Ac_{cm} temperature to ensure dissolution of the carbides and carbide networks that might have formed during prior processing. (Recall that for annealing, the hypereutectoid steel is heated only into the austenite + cementite region and not into the homogeneous austenitic region.) The use of higher austenitizing temperatures in the normalizing operation results in grater uniformity of composition and structure of austenite.

The austenitized steel is cooled in still air, at rates of the order of 0.1–1°C s^{-1}. The best practice of cooling in air is to suspend the parts in air or to place them on special beds so that the cooling air surrounds all the faces of the parts. It is a bad practice to dump all the parts into a pile. In such a case, only some of the parts are cooled by air. The cooling rate obtained during normalizing is higher than that during annealing and the cooling curve corresponding to normalizing is also

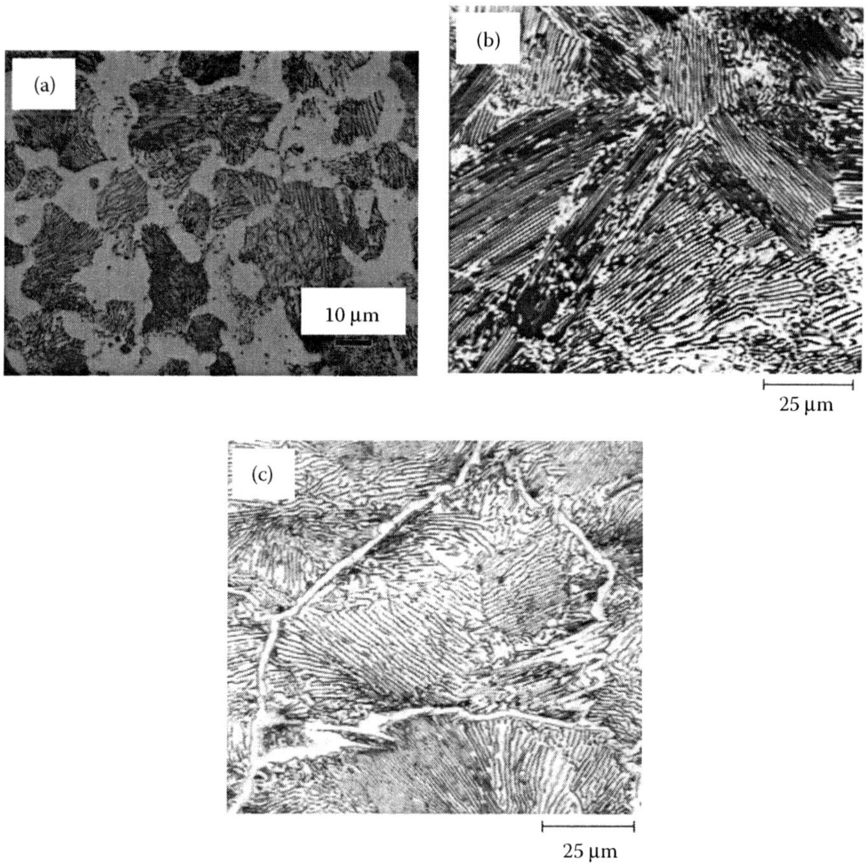

FIGURE 6.21 Typical microstructures of fully annealed (a) 0.45 wt% C, (b) 0.76 wt% C, and (c) 1.2 wt% C steel samples. Note the large interlamellar spacing of pearlite in all the samples. Additionally, note that the 0.45 wt% C steel sample shows the presence of a large amount of proeutectoid ferrite. The 0.76 wt% C steel contains only pearlite. On the other hand, the 1.2 wt% C steel sample contains a small amount of proeutectoid cementite along the grain boundaries of pearlite nodules. All the specimens were etched with 4% picral. (Photo (a) courtesy of Vander Voort, G. F., Consultant—Struers, Inc. With permission; Photos (b) and (c) courtesy of ASM International, Materials Park, OH. With permission.)

represented in Figure 6.19. Because of the faster cooling rate experienced by the steel, it has a refined microstructure. As a result, the steel is stronger and still possesses good machinability.

The typical microstructure of a plain carbon steel sample in the normalized condition is pearlite (for eutectoid steel), proeutectoid ferrite and pearlite (for hypoeutectoid steels), and proeutectoid cementite and pearlite (for hypereutectoid steels). Typical microstructures of normalized 0.45 wt% C (hypoeutectoid), 0.76 wt% C (eutectoid), and 0.95 wt% C (hypereutectoid) steel samples are shown in Figure 6.22.

Note that the interlamellar spacing in pearlite is now smaller than that in the annealed samples.

Majority of the common engineering steels show the above microstructural features of fine pearlite in steels subjected to the normalizing heat treatment. However, we had seen earlier that the *T–T–T* curve shifts to longer times with the addition of alloying elements. Therefore, the critical cooling rate required for the formation of martensite in alloy steels may be much lower than that in plain carbon steels. Hence, there is a possibility for the martensite phase to form on normalizing highly alloyed steels. Such steels in which the martensite phase forms on normalizing (air cooling) are referred to as *air-hardening steels*.

The normalizing heat treatment is more economical than annealing since furnaces are not required to control the cooling rate (instead, the steel samples are left outside the furnace to cool in still air). Further, the productivity is higher. But, realize that the objectives of these two heat treatments—annealing and normalizing—are

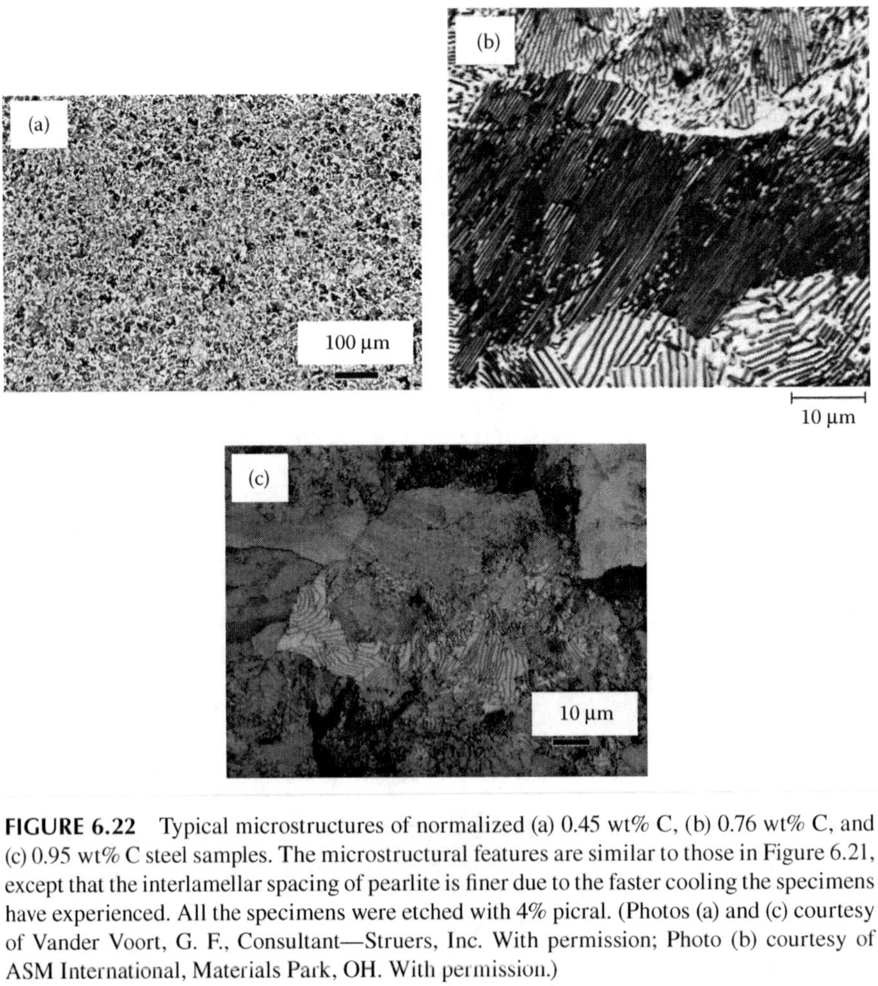

FIGURE 6.22 Typical microstructures of normalized (a) 0.45 wt% C, (b) 0.76 wt% C, and (c) 0.95 wt% C steel samples. The microstructural features are similar to those in Figure 6.21, except that the interlamellar spacing of pearlite is finer due to the faster cooling the specimens have experienced. All the specimens were etched with 4% picral. (Photos (a) and (c) courtesy of Vander Voort, G. F., Consultant—Struers, Inc. With permission; Photo (b) courtesy of ASM International, Materials Park, OH. With permission.)

different. Whereas normalizing is used to refine the grain sizes of specimens that were subjected to a high-temperature treatment, homogenize the sample, and allow smoother machining, annealing is done to induce softness, facilitate machinability, and eliminate internal or residual stresses. Normalizing produces harder and stronger steel than annealing.

6.9.3 Quenching

Quenching is another important heat treatment given to steels. According to history, quenching of steels was first mentioned by Homer around 800 B.C.—"As when a man who works as a blacksmith plunges a screaming great axe blade or adze into cold water, treating it for temper, since this is the way steel is made strong."

Quenching is accomplished by quickly transferring the hot sample into a coolant, usually referred to as a quenchant. The quenchant may be maintained at, above, or below room temperature depending on the necessity and the quenching rate desired. The common quenchants employed in practice are water, brine (water in which salt is dissolved), and oils, even though polymer solutions, molten salts, and molten lead (or other metals with low melting points) are used for specialized purposes. The presence of salt in water is expected to increase the cooling rate achieved.

A large variety of quenchants have been used over the years all over the world. The origin of some of these is shrouded in mystery. According to Chinese texts, "The Han River is sluggish and weak and is not suitable for quench-hardening. The Shu River is bold and vigorous … ." The unique and ingenious method of quenching the Japanese swords produces the characteristic curvature of the blade. The first mention of quenching in medieval Europe was by the twelfth century German Monk Theophilus. His recommendations were very specific: "Tools are given a harder tempering in the urine of a small red-headed boy than in ordinary water. Other recommendations included urine of goats fed ferns for three days, distilled vinegar, distilled urine, oil, etc. Quenching in oil is expected to provide a softer quench."

The usual product of quenching of steels is martensite. But, remember that formation of martensite requires that the cooling rate is above a critical value. Therefore, if the cooling rate achieved during quenching is above the critical value, then martensite formation will occur; otherwise nonmartensitic structures such as pearlite or bainite will form. Figure 6.23 shows the microstructures of martensite in steel samples of different carbon contents.

The process of quenching is done by putting the hot sample in a bucket of the cooling medium—water, brine, or oil, or any other quenching medium. When the hot sample is placed in the quenching medium, the heat from the sample is extracted by the quenching medium in three different stages. The first stage involves the formation of a thin, continuous, and insulating blanket of water vapor or steam around the sample. Since heat transfer to the surrounding parts can only take place by radiation or conduction through the vapor blanket, heat extraction is slow and consequently the cooling rate is low. During the second stage, the vapor blanket surrounding the sample boils and allows the water to come into direct contact with the steel sample. Thus, the cooling rate in this stage is the highest. During the third stage,

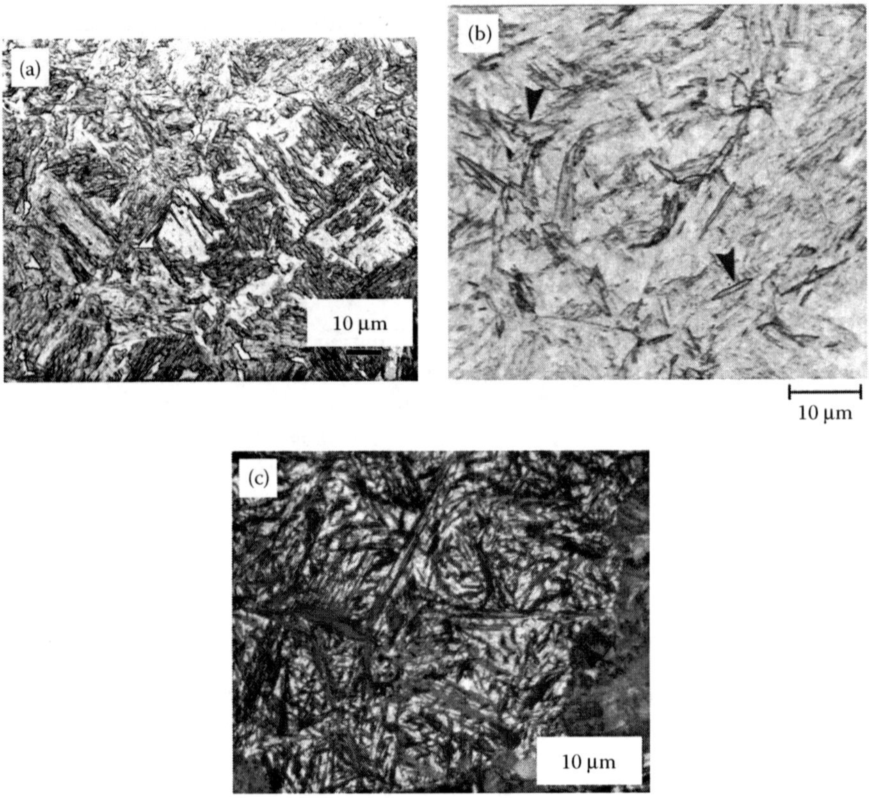

FIGURE 6.23 Microstructure water-quenched steels showing martensite with different morphologies. (a) lath martensite in 8115 alloy steel (0.15 wt% C), (b) 0.6 wt% C steel sample showing plate martensite (arrow shows a typical example of plate martensite) among lath martensite, and (c) plate martensite in a high-carbon W1 tool steel (1.05 wt% C) sample. The etchants were 2% nital for (a), sodium bisulfite for (b), and Klemm's I reagent for (c). (Photos (a) and (c) courtesy of Vander Voort, G. F., Consultant Stuers, Inc. With permission; Photo (b) courtesy of ASM International, Materials Park, OH. With permission.)

when the surface temperature of the hot sample falls below the boiling point of the quenching medium, vaporization stops and heat extraction is controlled by conduction and convection processes. At this stage again the cooling rate is low.

Amongst all the common quenchants used, the highest cooling rate is achieved when the samples are quenched in brine solution. (This has also been described by Giambattista della Porta (ca.1535–1615) when he wrote "If you quench red-hot iron in distilled vinegar, it will grow hard. The same will happen if you do it into distilled urine, by reason of the salt. If you temper it with dew that in the month of May is found on vetches leaves, it will grow most hard. For what is collected above them is salt.") In the decreasing order of efficiency, the other quenchants are water, synthetic polymer solutions, oils, molten salt baths, and air. (The quenchants are arranged in increasing order of their efficiency—severity of quench—in Chapter 7.) The efficiency of quenching may be increased by agitation of the quenching medium since the vapor blanket is continuously broken down. Other factors of the quenchant that influence the attained cooling rate are: type, temperature, density, viscosity, and wetting characteristics.

Apart from the quenching medium, other factors that determine the cooling rate are mass or section thickness of the sample, thermal diffusivity, thermal conductivity, and volume specific heat of the sample.

While water and oil are the most common quenching media used in practice, molten salts or molten metals are used for specialized purposes. The greatest advantages of the molten salts or metals are that the specimens reach the temperature of the quenchant quickly and there is very little chance of surface oxidation occurring. But, proper precautions need to be taken while using salt baths or molten metal baths. While molten salts can spurt when in contact with water, and salts containing cyanide or barium are poisonous, molten lead can also be toxic.

Quenching involves rapid heat removal from the sample to the neighboring quenching medium. Depending on the section thickness of the sample, the rate of heat removal could be significantly different at the surface and in the interior. Consequently, the temperatures at the surface and the interior could be different and this leads to development of *thermal stresses*. The magnitude of these thermal stresses could be in excess of 1000 MPa. The higher the quenching rate, the higher is the magnitude of these stresses.

As mentioned earlier, transformation of austenite to martensite involves volume expansion. The volume expansion is about 4% in the case of 1 wt% C steel. This volume change, due to the transformation, also develops stresses in the material and these are referred to as *transformation stresses*. The combination of the thermal and transformation stresses leads to cracking of the samples and this is known as *quench cracking*. Quench cracking can be avoided by slow cooling, but in this case the sample may not completely transform to martensite. Addition of alloying elements normally increases the hardenability of steels (see Chapter 7 for details) and in that case slower cooling rates may also produce complete hardening through the cross section of the sample. Thus, quench cracking may be avoided by selecting steel with a high hardenability.

6.9.4 Tempering

The as-quenched steel contains martensite which is very hard and brittle. However, for many applications, we require steel that possesses a good combination of strength, ductility, and toughness. Ductility and toughness can be induced into the steel by reheating the as-quenched steel to an elevated temperature, but below the A_{c1} (eutectoid) temperature. Such a process of reheating the quenched steel containing martensite to induce some toughness in the material is referred to as *tempering*.

Depending on the carbon content and quenching conditions, the quenched steel contains martensite and some retained austenite. Both these are metastable phases at room temperature and therefore they undergo some changes on reheating (tempering). During tempering, the steel undergoes several structural changes, and associated with these structural changes, the mechanical properties also change. The types of changes that occur during tempering differ depending on the carbon content of the steel. Traditionally, the tempering process has been subdivided into four stages.

The temperature ranges for these stages of tempering have been noted to overlap depending on the tempering times employed. Occasionally, a pre-tempering stage is also reported. In this stage, carbon atoms segregate to dislocations and various boundaries during quenching and/or while holding the sample at or near room temperature. This clustering will precede carbide formation during tempering and occurs between 0°C and 90°C.

In the first stage of tempering (100–250°C), martensite starts decomposing to *tempered martensite*. In high-carbon steels, the tempered material contains a low-carbon (usually about 0.25 wt% C) martensite and a transition carbide, termed as *ε-carbide*. The tempered martensite has lower carbon content than that of the as-quenched martensite and the ε-carbide has a formula corresponding to $Fe_{2.4}C$, which contains about 8 wt% C. The amount of carbon present in the low-carbon martensite is about 0.25 wt% and seems to be independent of the initial carbon content of the as-quenched martensite. On increasing the tempering temperature to about 200°C, the ε-carbide particles grow into well-developed arrays of needle-like morphology with well-established orientation relationships between ε-carbide and tempered martensite. The tempered martensite is as hard and strong as the as-quenched martensite, but is more ductile and tough.

In the second stage of tempering (200–300°C), the retained austenite transforms into bainite. In the low-temperature region, the (lower) bainite consists of ferrite and ε-carbide, while at higher temperatures, the (upper) bainite will consist of ferrite and cementite.

In the third stage of tempering (250–400°C), the tetragonality of martensite disappears completely and we will have only ferrite, that is, there is no supersaturation of carbon. The carbide present is also cementite, but some authors claim that a transition carbide (Hägg carbide or Chi, χ carbide) is present in this condition. The cementite phase nucleates at (a) the interfaces between the matrix and the ε-carbide, (b) along the twins within the martensite plates, and (c) grain boundary regions.

In the fourth stage of tempering (and some people do not consider this as one of the stages of tempering but as a separate heat treatment process), coarsening of the cementite particles takes place and they become spheroidized. This process is referred to as *spheroidization*. This process starts at a temperature of 300–400°C and continues up to about 700°C. At high temperatures, the spheroidal cementite particles are uniformly dispersed in a ferrite matrix, obtained through a process of recrystallization. The cementite particles transform into spheroids to minimize the surface energy. However, the kinetics of spheroidization are slow in high-carbon steels, due to a higher number density of cementite particles. Figure 6.24 shows the microstructure of steel that has been quenched and subsequently tempered to produce a spheroidized structure. The cementite particles, in this condition, are present as spheroids in a ferrite matrix. The effect of tempering temperature on the mechanical properties (yield strength, tensile strength, and reduction in area of cross-section) is shown in Figure 6.25 for an oil-quenched 4340 alloy steel.

The progress of tempering can be followed by examining the microstructural changes in the samples. Martensite and pearlitic structures can be easily identified

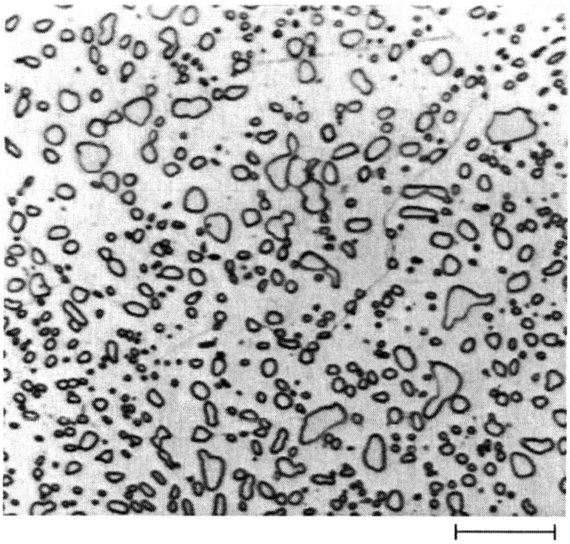

FIGURE 6.24 Microstructure showing spheroidization of cementite particles in a 1.0 wt% carbon steel. The small "rounded" particles are cementite, which are dispersed in a ferrite matrix. Picral etch. (Photo courtesy of ASM International, Materials Park, OH. With permission.)

from optical micrographs. But, it is not easy to follow the microstructural changes that occur during the different stages of tempering using optical microscopy methods. Further, the internal structure of martensite or the microstructural features of bainite cannot also be easily identified. We need a TEM for this purpose. In addition to the microscopic techniques described above, one can also use electrical resistivity measurements or mechanical property measurements to follow the progress of tempering. One can also use the dilatometry technique for this purpose by following the dimensional changes in the specimen. The specific volume of the specimen decreases during the first stage, increases significantly during the second stage, and shrinks again during the third stage of tempering. These volume changes are related to the presence of different phases in the microstructure at different stages of tempering.

6.10 TEMPER EMBRITTLEMENT

It was mentioned above that steels are tempered to induce some ductility and toughness into them by transforming the as-quenched martensite which is hard and brittle into less brittle transformation products of ferrite-cementite (or carbide) aggregates. But, depending on the steel composition and tempering temperature, the process of tempering may lead to a reduction in the toughness of steel. For example, if an alloy steel is quenched and tempered and then slowly cooled through,

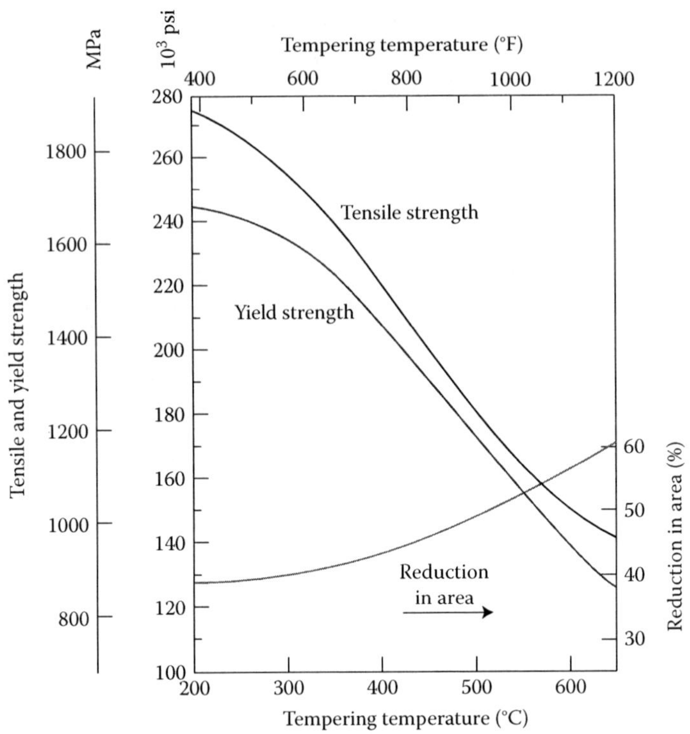

FIGURE 6.25 Variation of yield strength, tensile strength, and reduction in area as a function of the tempering temperature for an oil-quenched alloy steel 4340. (After Callister, Jr., W. D. and D. G. Rethwisch, 2010. *Materials Science and Engineering, An Introduction*, 8th edition. New York, NY: John Wiley & Sons, Inc.)

or held, in the temperature range of 375–575°C, then the fracture toughness of this steel is very low. Such a metallurgical phenomenon, which has been a mystery for a long time, is termed *temper embrittlement.* This phenomenon of temper embrittlement is also known as the *500°C embrittlement*, since it occurs at a temperature around 500°C.

Even though temper embrittlement is a long-standing metallurgical problem, and the conditions under which this occurs are known, it is only recently that the reasons for the occurrence of this phenomenon are becoming clear. This is essentially due to the new theoretical approaches and the availability of equipment capable of detecting grain boundary segregation, for example, Auger electron spectroscopy. It is now established that the phenomenon of temper embrittlement is due to the segregation of certain elements present in steel to the prior austenite grain boundaries, thus lowering the cohesion of these boundaries. The effect of temper embrittlement is revealed through a measure of the notched-bar impact energy of the steel that was tempered in this critical temperature region. The ductile–brittle transition temperature is raised and the shelf energy is lowered (see Chapter 10: Impact Testing). Observations of the fracture surfaces of the steel reveal that the transgranular fracture is replaced by the intergranular mode below the transition temperature.

It is the segregation of metalloid elements such as P, Sn, As, and Sb to the prior austenite grain boundaries that have a marked effect on the phenomenon of temper embrittlement. The propensity of segregation is in the order of Sb > Sn > P. It is also known that temper embrittlement is most prone in steels that contain an appreciable amount of substitutional alloying elements such as Mn, Ni, or Cr. As a consequence, plain carbon steels are not considered to be susceptible to temper embrittlement, especially when the Mn content is low, <0.5%. Small amounts of Mo and W prevent embrittlement, but larger amounts enhance it.

Since the ductile–brittle transition temperature of materials is affected by the austenitic grain size (see Chapter 10: Impact Testing), the tendency toward temper embrittlement is also changed by the grain size. For a given concentration of impurity elements, the amount of embrittlement is reduced with a decrease in the grain size.

A simple way to avoid temper embrittlement is to avoid slow cooling of the steel through the critical temperature range after it was tempered at a higher temperature. This can be done by quenching the steel after it was tempered. This fast-cooling prevents the segregation of the interstitial elements to the grain boundaries. Further, if the steel is already embrittled, its toughness can be restored by reheating the steel to a temperature above about 600°C and then quenching it to a low temperature. Yet another way, which may not always be possible, is to choose the steel composition properly avoiding the presence of the elements responsible for the temper embrittlement.

6.11 PROPERTIES OF HEAT-TREATED STEELS

The mechanical properties of materials are primarily controlled by the microstructural parameters. We have seen above that the microstructures of steels can be significantly altered by heat treatments. Therefore, the mechanical properties also will be significantly altered, based on the type of microstructure.

Amongst the different phases present in steel under "equilibrium" conditions, cementite is much harder and more brittle than ferrite. Consequently, pearlite, which is a mixture of both ferrite and cementite (in approximately 8:1 ratio) is harder than ferrite but softer than cementite. Further, the hardness and strength of steels increase, while the impact energy, percentage elongation, and reduction in area decrease, with increasing carbon content (Figure 6.26). This is due to the presence of increased amount of cementite with increasing amount of carbon in steels. In addition to the carbon content, the interlamellar spacing in pearlite also determines the strength properties of steel. Fine pearlite (with a smaller interlamellar spacing) is much harder and stronger than coarse pearlite (with a larger interlamellar spacing). This is because the hard cementite lamellae hinder deformation. The ferrite–cementite mixture of spheroidized steels (spheroidal cementite particles dispersed in a ferrite matrix) is much softer than pearlite. In fact, this is the softest and most ductile condition achievable in steel containing both ferrite and cementite. This softness is most important for cold deforming of low- and medium-carbon steels and machining of high-carbon steels. Figure 6.27 shows the hardness variation of martensite, fine

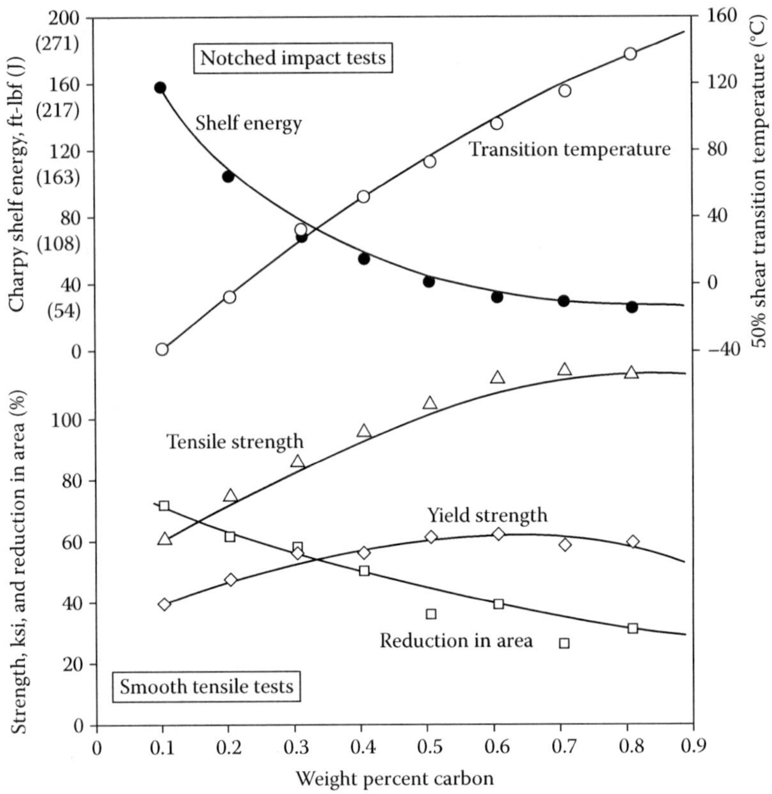

FIGURE 6.26 Yield strength, tensile strength, percentage reduction in area, Charpy impact energy, and ductile–brittle transition temperature as a function of carbon content in plain carbon steels.

pearlite obtained by air cooling, and spheroidite with the carbon content. Note that spheroidite is much softer and more ductile, and consequently tougher, than pearlite.

In the heat-treated condition, as-quenched steels are very hard (with a hardness of about 600 VHN or 65 R_C). The normalized steels which contain finer pearlite colonies are much harder than the fully annealed steels which contain coarse pearlite. Note that steels containing martensite are the hardest and those containing spheroidite are the softest.

6.12 EXPERIMENTAL PROCEDURE

1. Choose steels of different carbon content and alloying elements. As an example, one could choose steels containing 0.4 wt% C (hypoeutectoid composition) and 0.76 wt% C (eutectoid composition), and 1.0 wt% C (hypereutectoid composition). Note that for each composition, we would

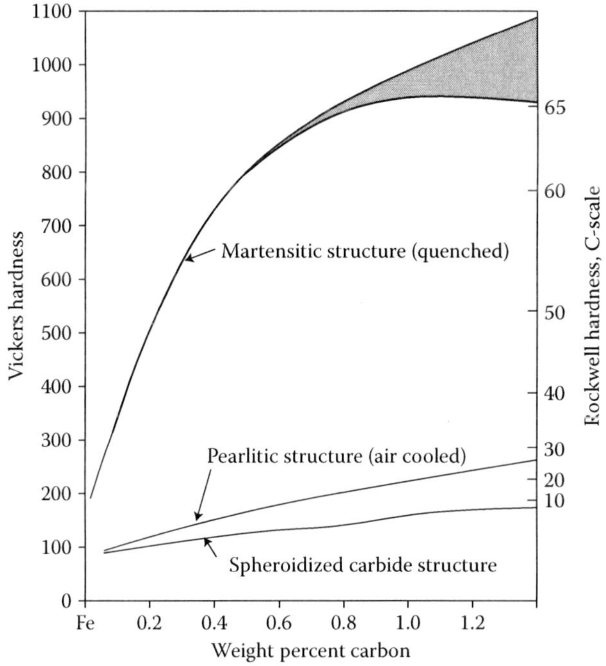

FIGURE 6.27 Vickers and Rockwell hardness as a function of carbon content for plain carbon steels containing martensite, fine pearlite, or spheroidite microstructures.

like to do different heat treatments and therefore, we need a large number of samples.

2. Take nine plain carbon steel samples each of different composition and austenitize them at the appropriate temperature for 1 h each.
3 Quench five samples in water, one in brine solution, and one in oil. Normalize the eighth sample (by allowing to cool in air) and anneal the ninth sample (by allowing the sample to cool in the furnace itself).
4. Choose one of the five samples in the water-quenched condition, and all the others (quenched in brine or oil and normalized or annealed) for metallographic examination. Prepare all the specimens for metallographic examination using the procedures described earlier.
5. Record all the microstructures at appropriate magnifications. Make sure that the water-quenched steel sample contains martensite.
6. Choose the other four water-quenched steel samples and temper (reheat) them at 300°C, 400°C, 500°C, and 700°C for 1 h each.
7. Quench the tempered steel samples (in step # 6) in cold water. Quenching is permitted here because no further phase transformations occur in these samples.
8. Prepare the above tempered and quenched samples for metallographic examination and record their microstructures. Label the different microconstituents present in the different specimens.

9. Measure the hardness values of all the specimens (quenched, normalized, annealed, and quenched and tempered) using a Vickers or Rockwell hardness tester.
10. Tabulate the hardness values of all the specimens, as shown below. Remember that you need to measure the hardness of each sample at least 4–5 times to obtain a meaningful and reliable value.
11. Plot the variation of hardness with tempering temperature. Also include the hardness of the as-quenched specimen in the plot.
12. Interpret the microstructures and the hardness versus tempering temperature plot in terms of the changes that take place during tempering of the specimens.

Note: The above procedure can be repeated on steels of other composition and on alloy steel samples also.

It may be a good idea to distribute the samples of different composition to different groups and/or batches of students, since it will be very difficult for one batch to complete heat treatment, metallographic specimen preparation, microstructural recording, and hardness measurement for all the specimens.

Results of Hardness Measurements

Composition of the Steel Sample	Heat Treatment	Hardness	Microstructural Details
0.4 wt% C (e.g.)	Annealed		Coarse pearlitic structure
	Normalized		
	Oil quenched		
	Water quenched		
	Water quenched and tempered at 200°C for 1 h		
	Water quenched and tempered at 300°C for 1 h		
	Water quenched and tempered at 400°C for 1 h		
	Water quenched and tempered at 500°C for 1 h		
	Spheroidized at 700°C for 1 h		

Results for the other compositions can also be tabulated in a similar manner.

EXERCISES

6.1 Why are steels heat treated?
6.2 What is activation energy?
6.3 Differentiate between homogeneous and heterogeneous nucleation.
6.4 Why is a critical size required for the nucleus to grow?
6.5 What is the effect of undercooling on the critical nucleus size and the activation energy for its formation?
6.6 Why does a plot of nucleation rate versus temperature show a peak?
6.7 Differentiate between isothermal and isochronal treatments.
6.8 What is an isothermal transformation diagram?
6.9 What do the three *T*s represent in the *T–T–T* diagram?

6.10 What is pearlite?
6.11 Differentiate between coarse pearlite and fine pearlite.
6.12 What is bainite?
6.13 Differentiate between upper bainite and lower bainite.
6.14 How does upper bainite form in steels?
6.15 What is martensite?
6.16 How can you obtain martensite in a steel sample?
6.17 What is habit plane?
6.18 What do M_s and M_f temperatures represent?
6.19 What is the critical cooling rate for formation of martensite with regard to the *T–T–T* diagrams?
6.20 Will the amount of martensite increase with time at a constant temperature?
6.21 Will you get more martensite if the steel is quenched at a higher cooling rate than the critical cooling rate?
6.22 Is there an advantage of cooling the steel at very high cooling rates?
6.23 Why is the crystal structure of martensite tetragonal?
6.24 Why is retained austenite present even in water-quenched steel above the critical cooling rate?
6.25 What is the effect of retained austenite on the mechanical properties of steel?
6.26 Can martensitic transformation take place in nonferrous alloys?
6.27 Why are nonferrous martensites soft in comparison with ferrous martensites?
6.28 Draw the *T–T–T* diagram for eutectoid composition steel and label the different phase fields and the axes.
6.29 Draw the *T–T–T* diagram for a eutectoid composition steel sample and indicate the cooling paths to obtain coarse pearlite and martensite.
6.30 What is the effect of alloying additions on the positions of the *T–T–T* curves?
6.31 Differentiate between *T–T–T* and continuous cooling transformation diagrams.
6.32 List some common heat treatments as applied to steels.
6.33 What is annealing of steel?
6.34 What is the purpose of annealing of steel?
6.35 What type of microstructure is developed in eutectoid steel when it is annealed?
6.36 What is normalizing of steel?
6.37 What type of microstructure is developed in eutectoid steel when it is normalized?
6.38 Distinguish between fine pearlite and coarse pearlite.
6.39 Is the steel harder in the annealed or normalized condition?
6.40 How can you obtain martensite in a steel sample by normalizing?
6.41 Differentiate between normalizing and annealing.
6.42 What type of microstructure is developed in eutectoid steel when it is water quenched?
6.43 Why is steel martensite hard?
6.44 Grade the cooling rates from the highest to the lowest in the following treatments: (a) water quenching, (b) annealing, (c) oil quenching, and (d) normalizing.
6.45 Why is steel tempered after it is hardened?

6.46 What are the different stages of tempering?
6.47 Briefly describe the microstructural changes that take place during tempering.
6.48 What is the effect of tempering on the mechanical properties of steel?
6.49 What is temper embrittlement?
6.50 Why does temper embrittlement occur in steels?
6.51 How can temper embrittlement be avoided?
6.52 What is spheroidization?

FURTHER READING

Bhadeshia, H. K. D. H. and R. W. K. Honeycombe, 2006. *Steels: Microstructure and Properties*, 3rd edition. London, UK: Butterworths-Heinemann.

Boyer, H. (ed.) 1997. *Atlas of Isothermal Transformation and Cooling Transformation Diagrams*. Materials Park, OH: ASM International.

Brooks, C. R. 1996. *Principles of the Heat Treatment of Plain Carbon and Low Alloy Steels*. Materials Park, OH: ASM International.

Callister, Jr., W. D. and D. G. Rethwisch, 2010. *Materials Science and Engineering, An Introduction*, 8th edition. New York, NY: John Wiley & Sons, Inc.

Krauss, G. 2005. *Steels: Heat Treatment and Processing Principles*. Materials Park, OH: ASM International.

Leslie, W. C. 1991. *The Physical Metallurgy of Steels*. Herndon, VA: Tech Books.

Samuels, L. E. 1999. *Light Microscopy of Carbon Steels*. Materials Park, OH: ASM International.

Sinha, A. K. 1989. *Ferrous Physical Metallurgy*. Boston, MA: Butterworths.

Vander Voort, G. F. 1984. *Metallography Principles and Practice*. New York, NY: McGraw-Hill, Inc.

7 Hardenability of Steels

7.1 INTRODUCTION

We had seen earlier that when steel samples are quenched from the austenitic region into water at room temperature (i.e., to a temperature below the martensite start (M_s) temperature), they undergo a martensitic transformation. We had also noted that a *critical cooling rate* was required for this transformation to occur. That is, for the martensite phase to form, the specimen must experience a cooling rate above a critical (or minimum) value. Therefore, if a large steel specimen is quenched from the austenitic phase region, it is possible that the cooling rate at the surface is above the critical value required for the formation of martensite and so martensite would form. However, it is possible that the cooling rate in the interior of the sample is less than the critical value and therefore, martensite would not form. Consequently, there is a limited depth in the sample, up to which only the critical cooling rate experienced by the sample is above the critical value and therefore, the martensite phase would form only up to that depth in the sample. Since martensite is a hard phase in steels, its presence increases the hardness of the sample and therefore, the formation of martensite and hardening are considered synonymous. If the pearlitic transformation (forming a mixture of ferrite and cementite) takes place, the hardness obtained is lower than when martensite is formed. In other words, the hardness of the sample is determined by the amount of martensite present in the sample. Traditionally, a steel is considered hardened, if it contains at least 50% martensite; if the steel contains less than 50% of martensite, it is considered unhardened.

The cooling rate experienced by a sample is a function of the size of the sample, its ability to transfer heat from the interior to the surface (thermal diffusivity) and the ability of the quenching medium to remove heat from the surface of the sample. But, all these factors can be combined together and expressed in terms of *hardenability* of the sample. Traditionally, the concept of hardenability has been associated only with steels.

7.2 DEFINITION OF HARDENABILITY

Hardenability has been defined in different ways by different people. A simple definition of hardenability is that it is the ability of a steel sample to be hardened by rapid cooling (quenching). Another way of defining it is "the property, in ferrous alloys, that determines the depth and distribution of hardness produced by quenching." Since the hardness in a steel sample is obtained due to formation of martensite, a more comprehensive definition of hardenability is "the capacity of a steel specimen to transform partially or completely from austenite to some percentage of martensite at a given depth when cooled under some given conditions." Traditionally, a measure of the hardenability of steel is the diameter of a cylinder which will have just 50% martensite at the center when quenched into an ideal quench. We will see later what we mean by an ideal quench.

The last definition accurately describes the physical process resulting in hardening of the sample. Also note that the definition deals with (1) some percentage of martensite, (2) depth of the sample, and (3) cooling under some given conditions. Therefore, for a full description of hardenability, all these three factors need to be considered.

It is also important to remember that hardenability and hardness are two entirely different things. We will see later (in Chapter 8: Hardness Testing) that hardness is the ability of a material to resist an applied force and involves the use of an indenter under static load. The ability of the material to resist plastic deformation depends on the carbon content and the microstructure of the steel. Thus, the same steel can have different hardness values depending on its microstructure, which is affected by the cooling rate. On the other hand, hardenability is defined above as the ability of a steel sample to be hardened by the formation of martensite through a given heat treatment. As we will see later, hardenability is used to describe the heat treatment response of steels using either hardness measurements or microstructure, both of which are interrelated. Hardenability is concerned with the depth of hardening or the hardness profile obtained, and not the ability to achieve a particular hardness.

Hardenability is an intrinsic material property of the steel and depends on the chemical composition, austenite grain size, and chemical homogeneity. On the other hand, the hardness of the sample is determined mainly by the carbon content and microstructure of the steel. That is, the presence of other alloying additions does not contribute to the hardness of the steel. But, the alloying elements increase the hardenability of the steel samples.

7.3 DISTRIBUTION OF HARDNESS

When a steel sample is quenched from the austenitic region in a quenching medium such as water or oil, it is easy to visualize that the quenching medium extracts heat first from the surface of the sample. The heat from within the sample needs to be transferred to the surface through the metal piece and then the heat is removed from the surface by the quenching medium. The thermal diffusivity of the steel determines its ability to transfer heat from the interior to the surface. Thus, for a given quenching medium, the thermal diffusivity of the sample determines the temperature distribution at different points in the specimen. Consequently, regions away from the surface of the specimen, for example, in the center of the specimen, cool very slowly and this allows diffusion-controlled transformations (resulting in the formation of ferrite and cementite mixtures, i.e., pearlite, bainite, etc.) to take place. That is, the martensitic transformation will not take place and therefore, the hardness of the sample is low. This situation becomes more important and serious in large-sized samples. Unfortunately, there is not much that can be done to change the thermal diffusivity of steels. Hence, we could only control the cooling rate achieved (and formation of martensite) by a proper choice of the quenching medium.

Figure 7.1a shows how the hardness varies as a function of distance from the surface of the sample in a water-quenched plain carbon steel SAE 1045. (SAE stands for the Society of Automotive Engineers who classify steels according to their chemical composition). The typical chemical composition of the 1045 steel is 0.48 C, 0.60 Mn, 0.022 P, 0.016 S, and 0.17 Si (all the values are in weight percentages). These plots are obtained by taking cylindrical steel rods of different diameters, quenching them in water, and measuring their hardness across their cross sections. Note that the highest hardness is achieved only on the surface of the steel and that too only in small cross sections. The hardness dropped off very rapidly as one goes from the surface into the interior of the sample and also as the size of the steel sample increased. The hardness distribution as a function of distance from the surface in water-quenched alloy steel SAE 6140, again as a function of the sample size, is shown in Figure 7.1b. The alloy steel SAE 6140 has the chemical composition: 0.42 C, 0.73 Mn, 0.027 P, 0.023 S, 0.25 Si, 0.94 Cr, and 0.17 V (all by weight percentages). Although the trend is the same in both the plain carbon and alloy steels, the actual hardness value obtained is higher in the alloy steel at larger sizes and also the rate of decrease is slower from the

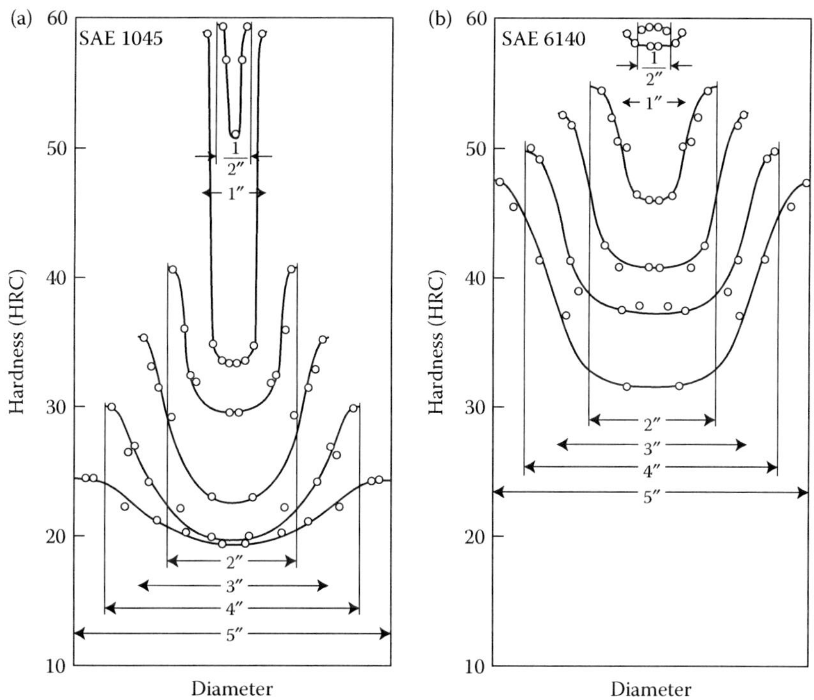

FIGURE 7.1 Hardness distribution in water-quenched samples of (a) SAE 1045 plain carbon steel and (b) SAE 6140 alloy steel. Note that hardness is highest at the surface of small samples and that the hardness drops significantly with increase in diameter and also as one moves into the interior of the sample. Note also that the high hardness is retained to a greater depth in the alloy steel samples. (After Grossmann, M. A. and E. C. Bain, 1964. *Principles of Heat Treatment*, 5th edition. Materials Park, OH: ASM International.)

surface into the interior of the sample. That is, in comparison with the plain carbon steel, the hardness is retained to a greater depth in the alloy steel samples.

A very similar trend is also obtained when these samples are quenched in a different quenching medium, for example, oil. An important difference, however, is that the actual value of the hardness obtained at any given position, is lower in the oil-quenched samples than in the water-quenched samples.

Heat transfer at the interface of the quenching medium and the steel sample is a complex process and is determined by the emissivity of the sample (i.e., its ability to radiate heat from the surface) and the convection currents within the quenching medium.

7.4 SEVERITY OF QUENCH

The efficiency or effectiveness of the different quenching media is ranked by a parameter known as *severity of quench*, and is designated by the letter "*H*." The higher the cooling rate achieved, the higher is the severity of quench. The value of *H* for a given quenching medium can be determined experimentally as follows:

Take a series of steel rods of given composition and of different diameters and quench them in the chosen quench medium, say water. Measure the hardness of the samples across the diameter. As mentioned earlier, the steel will be considered hardened if it contains 50% martensite. On the other hand, if the martensite content is less than 50%, the steel will be considered unhardened.

The criterion why a steel is considered hardened if it contains at least 50% martensite is based on two facts. First, if the microstructure of steel containing varying amounts of martensite and pearlite is observed, it is noticed that the etching response of the steel is significantly different in the martensitic and pearlitic regions. More specifically, the difference is sharp at the 50% martensite–50% pearlite region. Second, if the fracture surface of the steel containing varying amounts of martensite and pearlite is observed, it is noticed that the region containing 50% martensite correlated well with the transition from very smooth or faceted intergranular fracture associated with a predominantly martensitic structure to a rough, transgranular surface associated with ductile fracture of the softer nonmartensitic transformation products. In addition to the above two factors, it has also been noted that the hardness of the steel sample changes rapidly as the 50% martensite region is approached. Since it is possible to clearly identify the region corresponding to 50% martensite formation based on hardness and both microstructure and fracture appearance, it has been conventionally accepted that the steel is hardened if it contained at least 50% martensite.

With increasing diameter of the sample, it will be noted that a larger portion of the interior will be unhardened. Schematic representations of the hardened portions are shown in Figure 7.2 by the shaded areas and the unshaded areas represent the unhardened parts of the sample. The diameters of the unhardened cores, D_u are measured and the ratio of the unhardened core diameter to the diameter of the rod, that is, D_u/D is plotted as a function of the bar diameter, D (Figure 7.3). A similar experiment can be conducted when the steel samples of the same composition are quenched in other quenching media such as oil, brine, and so on. Since lower quenching rates will not

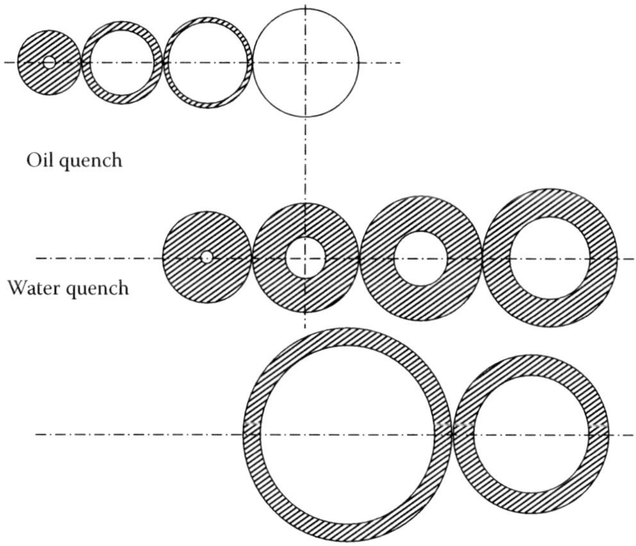

FIGURE 7.2 Schematic representation of the extent of hardening in water-quenched or oil-quenched steel samples. The shaded portions represent the hardened case of the sample, that is, they contain more than 50% martensite. Note the depth of hardening decreases with increasing diameter of the rod.

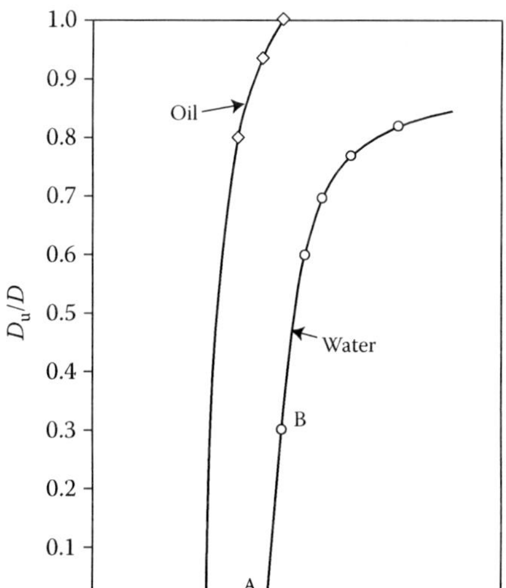

FIGURE 7.3 Plot showing the ratio of unhardened to hardened diameters (D_u/D) as a function of bar diameter (D) for different quenching media. The slope is steeper the slower the quenching rate. (After Grossmann, M. A. and E. C. Bain, 1964. *Principles of Heat Treatment*, 5th edition. Materials Park, OH: ASM International.)

be able to produce hardening in the interior of the thicker sections, the slopes of the curves for the slower quenching rates will be steeper. These curves can now be superimposed on experimentally determined plots of D_u/D versus HD (available in standard Heat Treatment reference books, Grossmann and Bain, 1964; ASM Handbook, 1991), where H represents the severity of quench. By matching the slopes of these curves with those of the experimentally determined curves of D_u/D versus HD, the value of HD can be found. By dividing the HD value with the corresponding D value, the value of H can be obtained. You will notice that the H values are different for different quenching media.

There are a large number of quenching media available in the market. These include water, different types of oils, soap water, liquid nitrogen, vacuum, aqueous solutions, mercury, brine, and so on. The values of the severity of quench, H are listed in Table 7.1 for some common quenching media. The H value for quenching in still water is set at 1.0, as a standard against which all other values are reported. Note that the severity of quench for still air is only 0.02, while that for oil is 0.3, and that for brine is 2.0. Brine is water in which salt is dissolved. The severity of quench can be increased by either agitating the work piece and/or by circulating the quenching fluid. Accordingly, the severity of quench, H for water increases from 1.0 for non-circulating fluid to 4.0 when there is violent circulation.

It is extremely difficult to accurately predict the hardenability of steels based on fundamental principles alone. Therefore, experiments are conducted to actually determine the hardenability. Several different tests are available to determine hardenability, but two of them are popular—Grossman test and the Jominy end-quench test. Of these two, the Jominy end-quench test is simpler and the most common. However, for the sake of completion, we will describe the details of both the tests.

7.5 GROSSMANN* TEST

The *Grossmann method* provides an accurate measure of the hardenability of steels. But, this is also an involved test. Similar to the method used to experimentally determine the severity of quench, in the Grossmann method also, a series of cylindrical steel bars of a given composition, but of different diameters, are quenched in a chosen medium. Transverse sections of the quenched bars are prepared for metallographic examination using the procedures described earlier (Chapter 3: Optical Microscopy), and their microstructures are observed. Using the criterion that the steel is considered hardened if it contains 50% martensite, the diameter of the bar which contains just 50% martensite at the center is determined. This diameter of the

---

* Marcus Aurelius Grossmann (1890–1952) was born and raised in Youngstown, Ohio, where the steel mills lured him into metallurgy. He was interested in "pure" research and made eight trips across the Atlantic to keep abreast of continental steelmaking methods and metallurgical advances. In his later years he endeavored to strike an even balance between primary fundamental research and practical application. His intense interest in research on steelmaking methods and metallurgical advances won him international recognition and acclaim. Grossman was also the research director for U.S. Steel Corporation. He was awarded ASM's Henry Marion Howe Medal in 1939 and the Albert Sauveur Achievement Award in 1949. He was president of ASM International in 1944. The Marcus A. Grossmann Young Author Award was established by ASM International in 1960 in recognition of his abiding interest in and encouragement of younger metallurgists and materials engineers.

TABLE 7.1
Severity of Quench, *H* for Some Common Quenching Media

Condition	Air	Oil	Water	Brine
No circulation of fluid or agitation of piece	0.02	0.25–0.30	0.9–1.0	2.0
Mild circulation or agitation	—	0.30–0.35	1.0–1.1	2.0–2.2
Moderate circulation	—	0.35–0.40	1.2–1.3	—
Good circulation	—	0.4–0.5	1.4–1.5	—
Strong circulation	0.05	0.5–0.8	1.6–2.0	—
Violent circulation	—	0.8–1.1	4.0	5.0

bar is designated as the *critical diameter*, D_0. Sometimes, this value is also known as the *actual critical diameter*. The critical diameter is traditionally expressed in units of inches (but, they can be easily converted into millimeters using the relationship 1 in. = 25.4 mm).

Alternatively, the critical diameter can also be determined by measuring the hardness of the samples on the transverse section and deciding which bar has the hardness value corresponding to 50% martensite at the center.

Note, however, that the critical diameter is not an absolute value since it depends on the composition of the steel and also the quenching medium used. For a specimen of the same chemical composition, the value of D_0 will be larger for water quenching than for oil quenching. Stated differently, the higher the severity of quench, the larger is the critical diameter. Therefore, the effect of cooling rate as a variable can be avoided, if the critical diameter is referred to a hypothetical *ideal quench*, which has a severity of quench corresponding to ∞. That is, an ideal quenching medium is assumed to be able to remove heat from the surface of the hot sample as soon as it can flow out from within the bar, that is, cooled at an infinitely rapid rate. In other words, the surface of the hot sample reaches the temperature of the quenching medium instantaneously. The critical diameter of the bar when using an ideal quench (or perfect quench for which $H = \infty$) is called the *ideal critical diameter*, D_I. The value of D_I is also expressed in inches. The ideal critical diameter, frequently abbreviated as *ideal diameter*, is a true measure of the hardenability of steel and can be used to compare the hardening response of different steels cooled in the same quenching medium.

Note that in the definitions of ideal diameter, D_I and critical diameter, D_0 of samples subjected to ideal quench and actual quench conditions, the steel bar contains 50% martensite at the center. The difference is that for D_I, the severity of quench is ∞, while for D_0, it is a finite value. Therefore, these two values, for which the bar contains 50% martensite at the center, are different. Recall also that the value of D_0 is different for different quenching media, and that D_I is used to avoid cooling rate as a variable. Obviously, $D_I = D_0$ when $H = \infty$. Thus, the ideal critical diameter is a true measure of the hardenability of a steel of given composition.

The hardness of the steel sample is primarily a function of the carbon content. The alloying additions do not change the hardness of the steel substantially. Further,

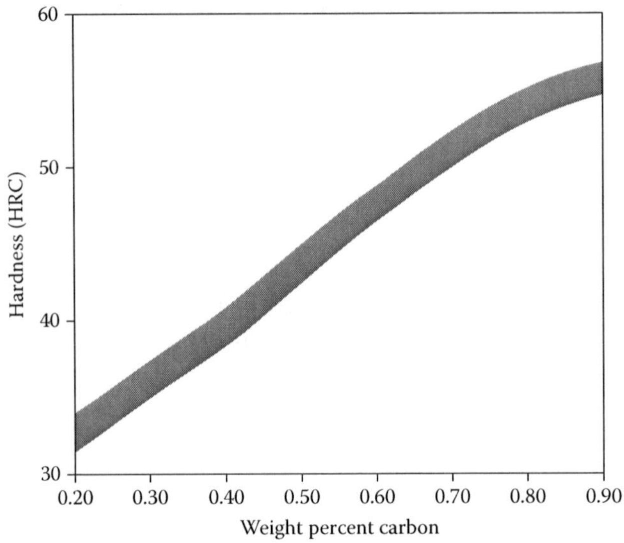

FIGURE 7.4 Variation of hardness corresponding to 50% martensite as a function of the carbon content in quenched samples. (After Grossmann, M. A. and E. C. Bain, 1964. *Principles of Heat Treatment*, 5th edition. Materials Park, OH: ASM International.)

since we require the hardness value of the steel corresponding to 50% martensite, we can have a master plot of hardness corresponding to 50% martensite as a function of carbon content in the steel. Such a plot is reproduced in Figure 7.4. Thus, if we know the carbon content of the steel sample, we can use this master plot to obtain the hardness value corresponding to 50% martensite in the steel.

Since we can only measure the actual critical diameter experimentally and not the ideal critical diameter, we need a way by which we can determine the ideal critical diameter of a given steel sample. Plots, similar to that shown in Figure 7.5 are available relating the actual critical diameter and ideal critical diameter for different severities of quench. From these plots, it may be noticed that the actual critical diameter and ideal critical diameter are the same for a severity of quench $H = \infty$. As the severity of quench decreases, the actual critical diameter for a given ideal critical diameter decreases.

These graphs can be used in two different ways. One is that the ideal critical diameter can be determined if we know the actual critical diameter for a given value of H. Alternately, for a given ideal critical diameter, we can calculate the actual critical diameter for a desired H value. A corollary of this is that if we wish to have an ideal critical diameter, D_I, then we can determine the severity of quench required to obtain 50% martensite in the center of a given diameter steel bar. Thus, the concept of the ideal size permits a rapid estimate of the bar size that will harden to the 50% martensite level in the center in quenches over the entire range of severities. Similar curves between critical plate thickness, ideal plate thickness, and severity of quench have also been developed.

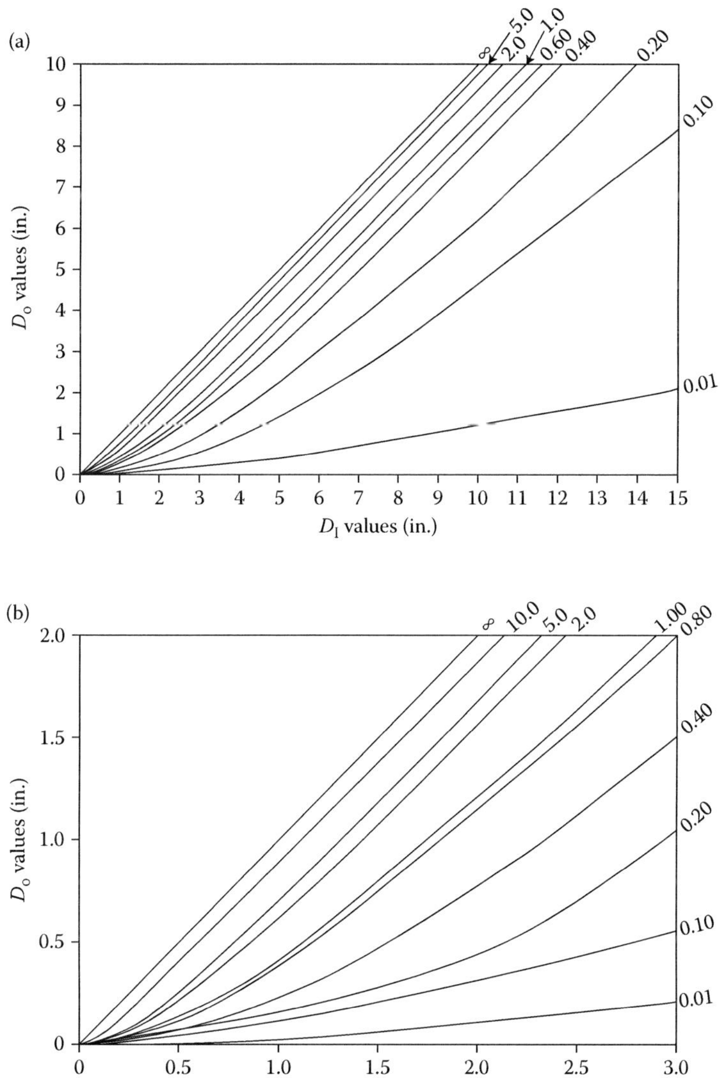

FIGURE 7.5 Relationship between ideal critical diameter, D_I and actual critical diameter, D_o for different severities of quench, H. (a) shows the full range of critical diameters and (b) shows the details at smaller values of the critical diameter. (After Grossmann, M. A. and E. C. Bain, 1964. *Principles of Heat Treatment*, 5th edition. Materials Park, OH: ASM International.)

Example Problem 7.1

a. A series of cylindrical steel rods were quenched into water ($H = 1.0$) and microscopic examination revealed that the rod with 0.8 inches in diameter had 50% martensite in the center. Determine the hardenability of this steel.

b. Determine the hardenability if the same steel sample was quenched in strongly circulated water ($H = 2.0$), and the critical diameter was 1.05 in.

Solution 7.1

a. The hardenability of a steel sample is the ideal critical diameter, D_I for which the steel contains 50% martensite at the center of the sample when it is quenched in an ideal quenching medium. Since Figure 7.5 relates the ideal critical diameter and actual critical diameter for different H values, we can determine the D_I value corresponding to $D_o = 0.8$ in. and for $H = 1.0$. The answer is 1.5 in.
b. Using Figure 7.5 again we note that corresponding to $D_o = 1.05$ in. and $H = 2.0$, the ideal critical diameter, D_I is 1.5 in.

Note: The above example clearly brings out the concept that the ideal critical diameter is a true measure of the hardenability of the steel and that it is fixed for a given steel. It is only the actual critical diameter that is different for different quenching media.

Example Problem 7.2

The ideal critical diameter, D_I of a steel sample was determined to be 2.5 in. What is the actual critical diameter, D_o if the steel is subjected to (a) water quenching with moderate circulation, and (b) oil quenching with strong circulation?

Solution 7.2

a. Referring to Table 7.1, the severity of quench, H for water quenching with moderate circulation will be taken as 1.3. Using this value of $H = 1.3$, D_o corresponding to $D_I = 2.5$ in., can be determined from Figure 7.5 as approximately 1.75 in.
b. Again referring to Table 7.1, the value of H for oil quenching with strong circulation can be taken as 0.8. Using the procedure as in (a), the value of D_I corresponding to $H = 0.8$ in. and $D_I = 2.5$ is determined as 1.55 in.

7.6 JOMINY END-QUENCH TEST

The Grossmann method of determining hardenability of steels is very reliable. But, the method is very complicated, expensive, and time consuming since a large number of steel samples need to be examined. Consequently, this method is most unlikely to be of commercial use. Therefore, the *Jominy End-Quench Test*, among many others, is the one that is practically used in the industry to provide a cost- and time-effective way to determine the hardenability of steels. This end-quench test was developed by Jominy and Boegehold, but is commonly referred to as the Jominy test. The greatest advantage of this method is that a *single test specimen* is used to obtain the hardenability data instead of a series of samples as in the Grossmann method.

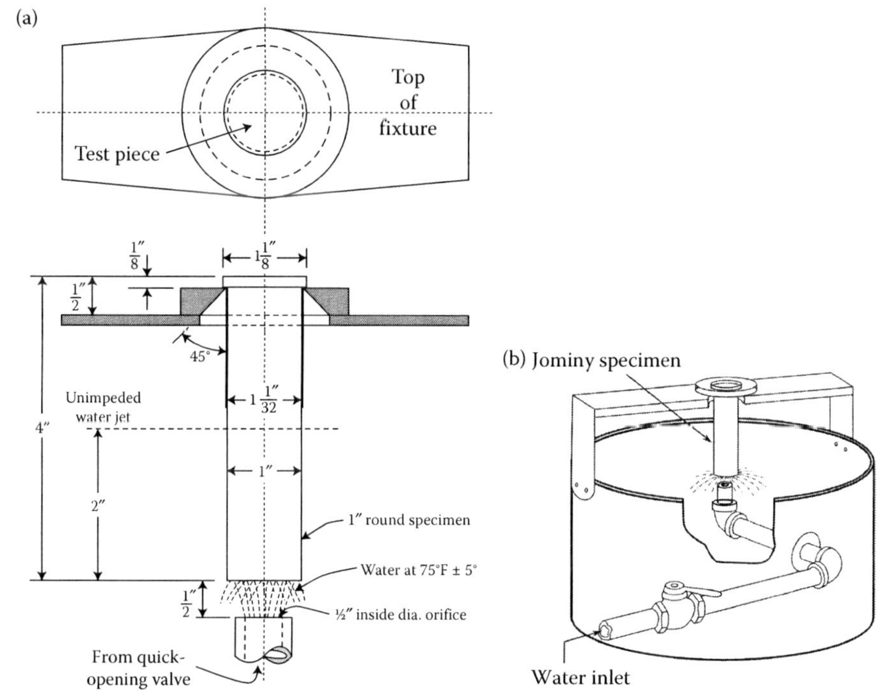

FIGURE 7.6 (a) The size and shape of the specimen used for the Jominy end-quench test. The test specimen is cylindrical in shape and is 1 in. in diameter and 4 in. in length. (b) The fixture used to support the Jominy test specimen during water quenching.

Figure 7.6a shows the shape and dimensions of a Jominy test specimen. The test specimen consists of a cylindrical bar 1 in. in diameter and 4 in. in length. The fully austenitized specimen is subjected to water quenching in a fixture as shown in Figure 7.6b. The specimen is cooled at one end by a column of water, while the other end is cooled by air. Thus, the specimen experiences a range of cooling rates from water quenching to air cooling. After the specimen cools down to room temperature, two parallel flats are ground on opposite sides of the test bar and hardness values are measured along these surfaces every 1/16 inch starting from the quenched end. These hardness values are then plotted as shown in Figure 7.7.

Recall that hardenability has been related to the depth and distribution of hardness in a steel specimen. Therefore, by looking at the Jominy hardness–distance plots of steels, it is possible to grade the steels as low- or high-hardenability steels. If a high hardness value is retained to a larger depth in a sample, then it is considered more hardenable (*deep hardening*) than one for which the hardness falls off more steeply (*shallow hardening*). Figure 7.8 shows the variation of hardness values for shallow- and deep-hardening steel specimens.

As mentioned earlier, hardenability has been defined as the diameter of a steel sample, which contains 50% martensite at the center of the sample when quenched under ideal conditions. That is, the specimen is fully hardened and that there is no

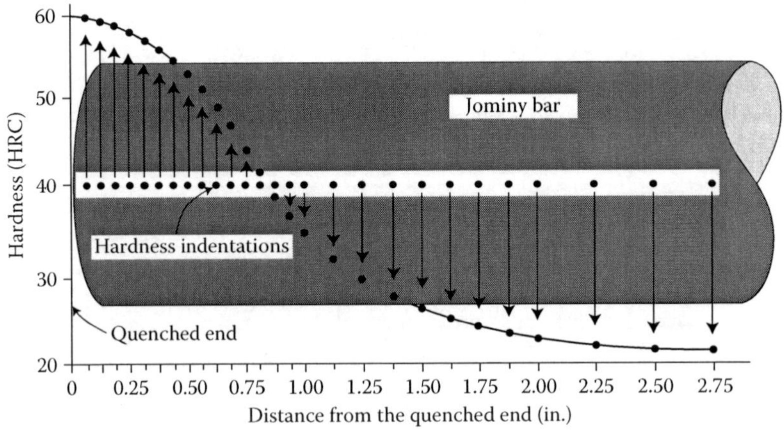

FIGURE 7.7 Variation of hardness with distance from the quenched end in a Jominy end-quench test.

unhardened part in the sample. Therefore, if we have information about the variation of hardness corresponding to 50% martensite as a function of carbon content (Figure 7.4), then the Jominy hardness plots can be used to determine the hardenability of the steel. The actual procedure is as follows. Knowing the carbon content of the steel, we should first determine the hardness value corresponding to 50% martensite content from Figure 7.4. From the experimentally determined Jominy hardness–distance plot, we can determine the distance from the quenched end that corresponds to the hardness value determined from Figure 7.4 for steel with 50% martensite. This will be the critical diameter of the steel for the severity of quench in the actual experiment. Even though we realize that the actual cooling rate varies from point to point in the Jominy bar, we could use the severity of quench value corresponding to the actual experimental conditions, for example, type of quenchant, whether it is agitated or not, and so forth. Using this value of the critical diameter and knowing the

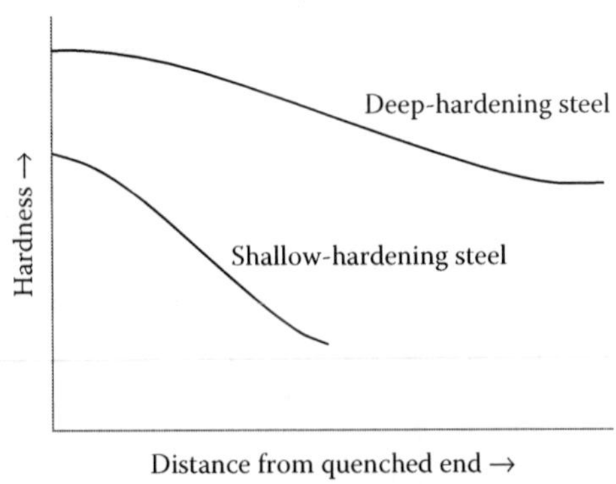

FIGURE 7.8 Variation of hardness with distance from the quenched end in a Jominy test for shallow-hardening and deep-hardening steels.

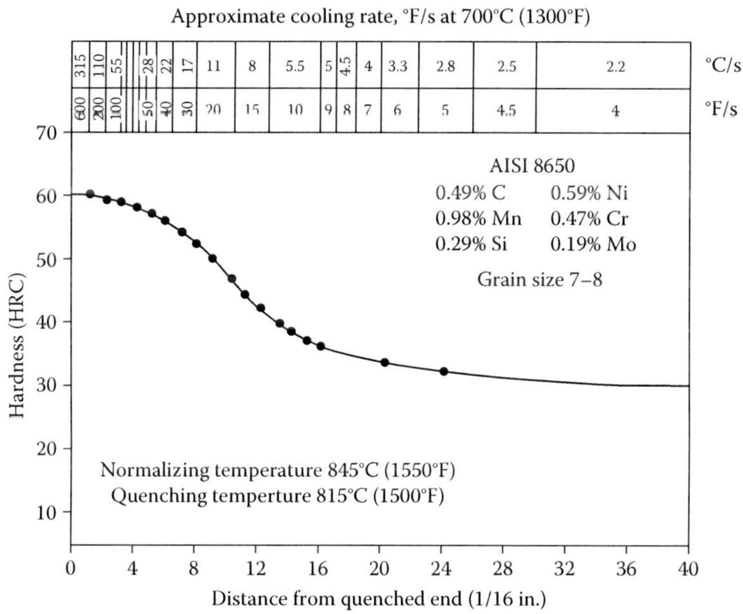

FIGURE 7.9 Conventional method of presenting the Jominy hardness–distance data. The hardness is plotted on the *Y*-axis, the distance from the quenched end on the *X*-axis and the cooling rate on the top. The compositions of the sample and grain size number are also included. (After Krauss, G. 1990. *Steels: Heat Treatment and Processing Principles*, Materials Park, OH: ASM International.)

severity of quench, the ideal diameter can be determined from Figure 7.5 and this ideal diameter is a measure of the hardenability of the steel.

An important point to remember in the Jominy test is that each position of the specimen corresponds to a well-defined cooling rate. Therefore, on a conventional Jominy plot, the approximate cooling rate at that point is represented on the top (horizontal axis) of the diagram as shown in Figure 7.9. Further, note that when the steel is produced, the composition of the steel varies slightly from batch to batch. That is, the composition of the individual elements in the steel is not fixed, but it has a small range of values. Consequently, the hardness of the steel also varies and thus, there is a range of hardness values for the "same" steel composition. Thus, it is practical to expect to see a *hardenability band* for any steel, as shown in Figure 7.10. The upper curve represents the hardness obtained with the upper limit of the composition for the steel, while the lower curve represents the hardness value for the composition at the lower level. The steels designated by the letter *H* (*H*-steels) are guaranteed to meet established hardenabilities. Table 7.2 lists the hardenabilities of different steels. The values are listed as a range of the ideal critical diameters, D_I in inches. The actual chemical compositions of these different steels may be found in standard reference books.

As mentioned earlier, the hardness obtained in the quenched sample is due to the formation of martensite, which again is related to the cooling rate achieved in the

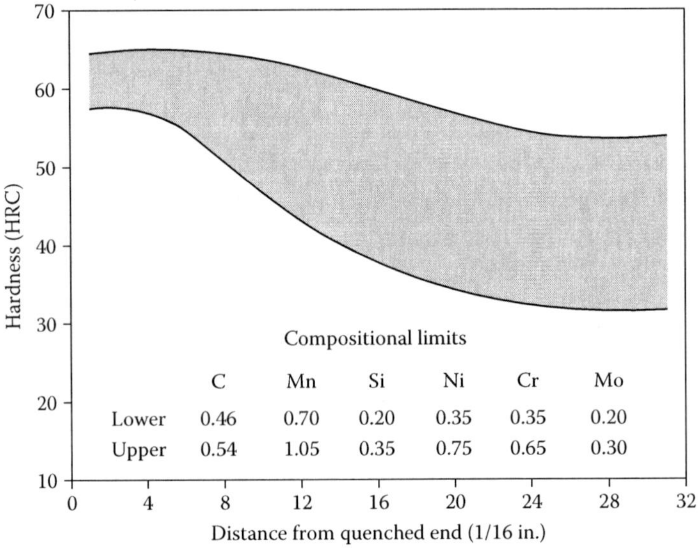

FIGURE 7.10 Hardenability band for 8750 *H* steel. The upper curve represents the hardness for the upper limit of the composition, while the lower curve is that for the lower limit of the composition of the steel. The area between the two curves is referred to as the hardenability band. (After Krauss, G. 1990. *Steels: Heat Treatment and Processing Principles*, Materials Park, OH: ASM International.)

sample. Therefore, if cooling rates obtained at different positions in the specimen of various geometries or sizes (i.e., diameters of the bars) are known, the Jominy data could then be used to plot hardness profiles in the parts. Figure 7.11 shows the equivalent cooling rates for round bars of different diameters quenched in water and oil. These values are indicated for the surface, three-quarter radius, half-radius, and center positions. Note that as the bar diameter increases the cooling rates at the surface and interior points decrease. A similar situation arises when the steel of the same diameter is cooled in water or oil, noting that the cooling rate is lower in oil quenching than in water quenching. Accordingly, the cooling rates differ for (a) different bar diameters, (b) different quenching media, and also (c) for different positions in the cross sections of the bar for a given quenching medium and a given diameter. This may be noted from the top scale of Figure 7.11. These plots could be used to determine the cooling rate and the associated distance from the quenched end of a standard Jominy bar for selected bar diameter, location along the cross section, and for a given quenching medium. Correlations of cooling rates as a function of position in various sizes of bars and plates quenched in various media are also available.

A practical application of these plots will be to obtain hardness profiles of steel samples quenched in different media as follows. Let us assume that a steel specimen of a given diameter is quenched in water. The cooling rates in the specimen could be determined at the surface, three-quarter radius, half-radius, and the center using plots shown in Figure 7.11a. These cooling rates correspond to particular distances from the quenched end of the specimen, as indicated on the bottom scale. Using

TABLE 7.2
Hardenabilities of Some Selected Steels in Terms of the Ideal Critical Diameter, D_I in Inches

Steel	D_I (in.)	Steel	D_I (in.)
1045	0.9–1.3	3140 H	2.6–3.4
1090	1.2–1.6	4140 H	3.1–4.7
		4340 H	4.6–6.0
1320 H	1.4–2.5	4640 H	2.6–3.4
3120 H	1.5–2.3	5140 H	2.2–3.1
4320 H	1.8–2.6	8640 H	2.7–3.7
4620 H	1.5–2.2	8740 H	2.7–3.7
4820 H	2.2–3.2	9440 H	2.4–3.8
5120 H	1.2–1.9		
8620 H	1.6–2.3	5150 H	2.5–3.7
8720 H	1.8–2.4	6150 H	2.8–3.9
		8650 H	3.3–4.5
1330 H	1.9–2.7	8750 H	3.8–4.9
2330 H	2.3–3.2		
3130 H	2.0–2.8		
4130 H	1.8–2.6		
5130 H	2.1–2.9		
8630 H	2.1–2.8		

standard Jominy plots that relate the surface hardness with distance from the quenched end, see, for example, Figure 7.7, the hardness at that surface can be easily determined. The Jominy test is now widely used to determine the hardenabilities of steels in the range of $D_I = 0.4–2.5$ in. These results could also be used to determine the largest diameter round bar which can be fully hardened.

7.7 PARAMETERS AFFECTING HARDENABILITY

We had seen earlier that the hardenability of steel is affected by many parameters, which include the quenching medium, specimen size, and geometry. That is why we used the concept of ideal critical diameter to avoid the effect of quenching rate as a variable. But, in addition to these, other parameters such as carbon content, nature, and amount of alloying elements, and austenitic grain size also are important variables that affect the hardenability of steels. Let us now look at the effect of these variables on hardenability.

7.7.1 Austenitic Grain Size

The grain size of the austenite phase from which the hard martensite phase forms has a significant effect on the hardenability of steels. This is because nucleation of the product phases (e.g., ferrite or pearlite) almost always occurs heterogeneously on the grain boundaries of austenite.

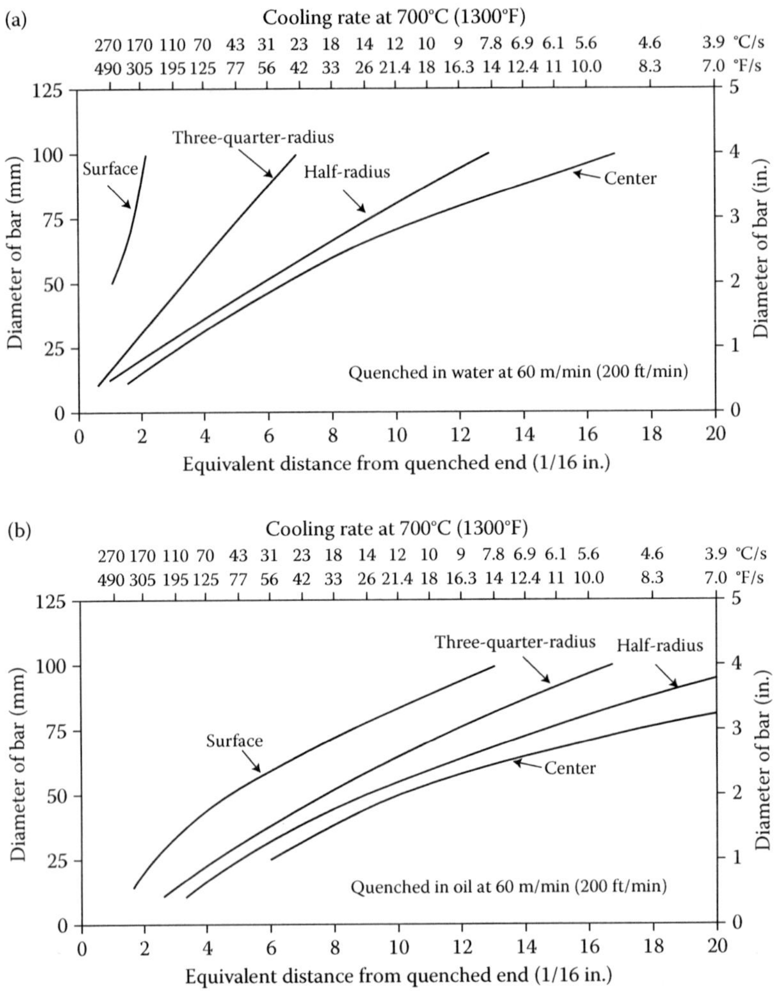

FIGURE 7.11 Equivalent cooling rates for round bars of different diameters quenched in (a) water and (b) oil, as a function of the distance from the quenched end. The data for surface hardness are for "mild agitation" while the other data are for 60 m/min. (After Krauss, G. 1990. *Steels: Heat Treatment and Processing Principles*, Materials Park, OH: ASM International.)

As explained earlier in Chapter 3, the most common way of designating the austenite grain size is by using the ASTM (American Society for Testing and Materials) grain size number, N. This is defined by the simple relationship

$$n = 2^{N-1} \tag{7.1}$$

where n is the number of grains per square inch when observed at a magnification of 100× and N is the ASTM grain size number. Note that as the ASTM grain size

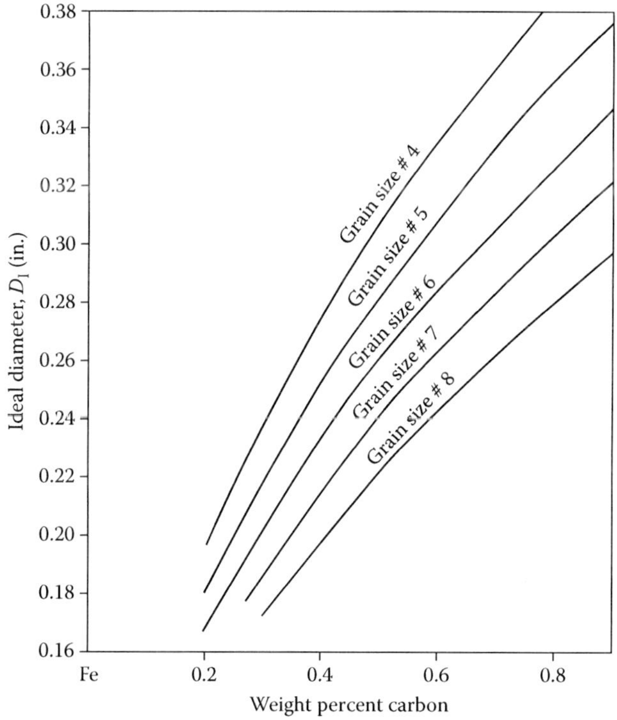

FIGURE 7.12 Variation of the ideal critical diameter as a function of austenitic grain size and carbon content. Note that the hardenability increases with increasing carbon content and increasing grain size (decreasing grain size number).

number, N increases, the number of grains, n increases and consequently, the grain size decreases (see Chapter 3: Optical Microscopy for further details). Most commercial steels have an ASTM grain size number between 1 and 8.

If the ASTM grain size number is large (i.e., the austenite grain size is smaller), then chances of pearlite formation occurring at the grain boundaries are higher and therefore martensite will not form. This reduces the hardenability of the steel. Thus, all other factors remaining constant, the larger is the grain size, the higher is the hardenability of the steel. Figure 7.12 shows how the ideal critical diameter increases with increasing austenite grain size for steels of different carbon content.

A coarse grain size, however, is not an optimum way to increase the hardenability of steels. This is because large grain sizes can be obtained only through high austenitizing heat treatments. The resultant microstructure is quite coarse leading to lower strength and reduced toughness. Further, the tendency to quench cracking increases with increasing grain size. That is why the route of increasing the grain size to increase the hardenability of steels is not followed. Instead, addition of alloying elements to increase the hardenability (to be discussed later) is more desirable.

7.7.2 Carbon Content

The hardenability of steel can be greatly increased by increasing the carbon content. Referring to Figure 7.12, it is clear that the higher the carbon content, the higher is the hardenability. This is again related to the fact that the time–temperature–transformation (*T–T–T*) diagrams are shifted to the right (i.e., longer times) with increasing carbon content. That is, the pearlite transformation takes longer to start and finish. Consequently, the austenite phase will have more time to completely transform to martensite, even at relatively slower cooling rates.

Figure 7.13 shows the Jominy hardness–distance plots for four different steels of the 8600 series. The type of alloying elements and their contents are the same in all the steels, but the carbon concentration is different. The carbon content varies from 0.2 to 0.6 wt%. From these curves it is very clear that the hardenability increases with increasing carbon content. For example, the hardness for any given Jominy distance increases with increasing carbon content.

Again, high carbon content in steel is not always desirable because high-carbon steels lack toughness. Another disadvantage is that high-carbon-quenched steels are likely to cause distortion and cracking during heat treatment and welding. Therefore,

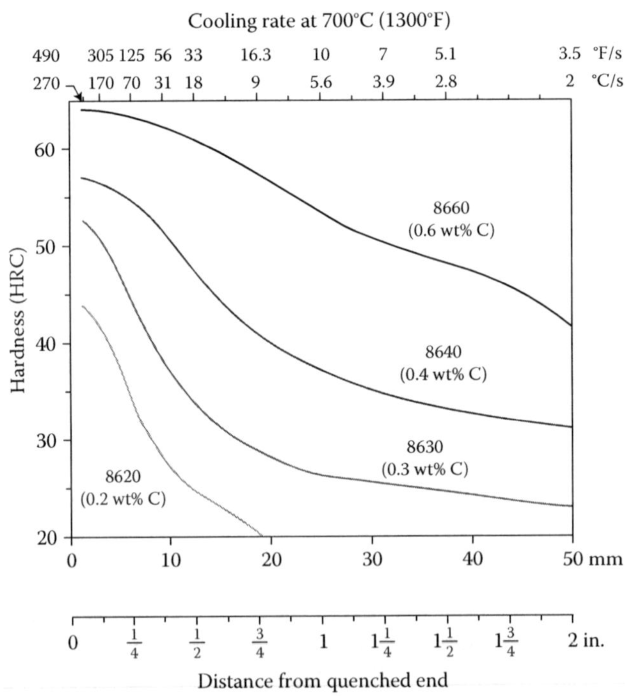

FIGURE 7.13 Hardenability curves for four 8600 series steels of different carbon content. Note that the high hardness is retained until longer distances from the quenched end in steels of higher carbon content. That is, the hardenability is higher for steels with higher carbon content. (After Callister, Jr., W. D. and D. G. Rethwisch. 2010. *Materials Science and Engineering, An Introduction*, 8th edition. New York, NY: John Wiley & Sons, Inc.)

the best way of increasing the hardenability of a plain-carbon steel is to use other alloying elements such as manganese.

Example Problem 7.3

a. Determine the ideal critical diameter for a plain carbon steel containing 0.4 wt% carbon that has an austenitic grain size number of 6.
b. What will be the ideal diameter for the 0.4 wt% C steel with a grain size number of 4?
c. Determine the ideal diameter for a steel containing 0.6 wt% C and an ASTM grain size number of 6.

Solution 7.3

The solutions to all these questions can be obtained by referring to Figure 7.12.

a. It is clear that the ideal critical diameter for this steel with 0.4 wt% C and an ASTM grain size number 6 will be 0.23 in.
b. For the 0.4 wt% C steel with an ASTM grain size number of 4, the ideal diameter will be 0.275 in.
c. For the 0.6 wt% C steel with an ASTM grain size number of 6, the ideal diameter will be 0.285 in.

From the above result it is clear that the hardenability of the steel increases with increasing carbon content and decreasing ASTM grain size number (increasing grain size).

From the above example, it appears that the hardenability of plain carbon steels is very low. If steel bars of about ¼ in. in diameter are quenched with $H = \infty$ they will have 50% martensite in the center. On the other hand, if they are quenched in water, the diameter of the bar that will have 50% martensite in the center will be much smaller. Luckily, however, commercial plain carbon steels contain other alloying elements and therefore larger diameter steel bars could be fully hardened in the center even by water quenching.

7.7.3 Alloying Elements

Like carbon, the addition of other alloying elements to steel delays the pearlite formation since the *T–T–T* diagrams are shifted to longer times. Consequently, the hardenability of the steel is found to increase with increasing amounts of alloying elements. However, the magnitude of increase in hardenability is different for different alloying elements. The efficiency of different alloying elements in increasing the hardenability of steels is evaluated in terms of *multiplying factors*. Table 7.3 lists the concentration and multiplying factors for Ni–Cr steel.

In general, the addition of alloying elements increases the hardenability of steels. Manganese, chromium, and molybdenum are very effective. Boron particularly has a large effect, even in concentrations of the order of 0.001 wt%. Additions

TABLE 7.3
Concentration and Multiplying Factors for Ni–Cr Steel

	Element				
	Carbon	**Manganese**	**Silicon**	**Nickel**	**Chromium**
Concentration (wt%)	0.5	0.8	0.25	1.00	0.28
Multiplying factor	0.24	3.7	1.2	1.4	1.6

of about 0.002–0.003 wt% boron have been found to have an effect equivalent to that of 0.5 wt% Mo addition. The effect of boron is independent of the amount of boron, provided that a sufficient amount is added. The effect of boron is greatest at lower carbon contents and is typically used with lower-carbon steels. Boron can affect the hardenability only when it is in solution. Since boron has strong affinity for oxygen and nitrogen, with which it forms compounds, gettering elements such as aluminum and titanium are added to preferentially react with these elements. Boron would be more widely used if its distribution in steel could be more easily controlled.

The relative effect of the alloying elements on increasing the hardenability (or the ideal critical diameter), that is, the multiplying factor, is plotted in Figure 7.14 as a function of the nature and concentration of the alloying element.

Even though majority of the alloying elements increase the hardenability of steels, there are a few elements which decrease the hardenability. Cobalt is one such

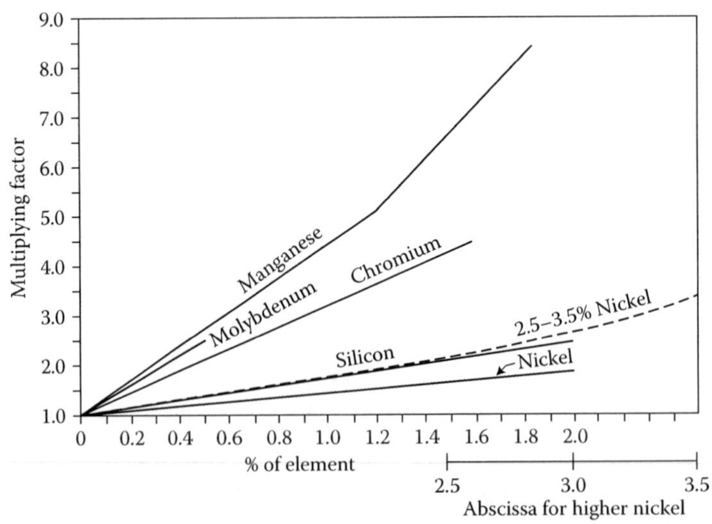

FIGURE 7.14 Multiplying factors as a function of the alloy content for different common alloying elements in alloy steels. (After Grossmann, M. A. and E. C. Bain, 1964. *Principles of Heat Treatment*, 5th edition. Materials Park, OH: ASM International.)

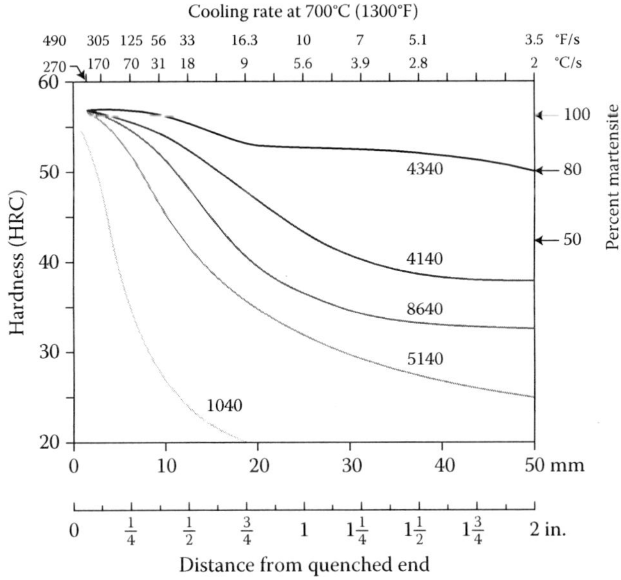

FIGURE 7.15 Hardenability curves for five different steel samples, all having the same carbon content but different amounts of the alloying elements. (After Callister, Jr., W. D. and D. G. Rethwisch. 2010. *Materials Science and Engineering, An Introduction*, 8th edition. New York, NY: John Wiley & Sons, Inc.)

exception. This is because cobalt increases the rate of nucleation and growth of pearlite, thus decreasing the hardenability of the steel.

Figure 7.15 shows the Jominy hardness–distance plots for five different steels, all having the same carbon content (0.4 wt% C), but different alloying contents. The compositions of these five different steels are listed in Table 7.4. It may be seen clearly from this figure again that the hardenability increases with increasing amounts of alloying elements. That is, for a given Jominy distance, the hardness is higher for steels with a larger amount of the alloying elements.

TABLE 7.4
Chemical Compositions of the Five Different Steels for Which the Hardenability Data are Plotted in Figure 7.15

Alloy Steel	Chemical Composition (wt%)							
Designation	**C**	**Ni**	**Cr**	**Mo**	**Mn**	**Si**	**P**	**S**
1040	0.4	—	—	—	—	—	—	—
5140	0.4	—	0.85	—	—	—	—	—
8640	0.4	0.55	0.50	0.20	—	—	—	—
4140	0.4	0.20	1.0	0.19	0.83	0.31	0.18	0.03
4340	0.4	1.85	0.80	0.25	—	—	—	—

7.7.4 Use of Multiplying Factors

A knowledge of multiplying factors of the alloying elements present in steel helps in determining the hardenability of the steel. The way of doing this is as follows:

1. First determine the carbon content of the steel and the ASTM grain size number.
2. From a knowledge of the above two factors (ASTM grain size number and carbon content), determine the ideal critical diameter using Figure 7.12. This will be referred to as the base diameter or *base hardenability*.
3. Knowing the nature and amount of the alloying elements present in the steel, multiply the base hardenability value with the multiplying factor for each of the alloying element (use Table 7.3 and Figure 7.14 for the individual multiplying factor values).
4. The product of all these will give the hardenability of the alloy steel.

Example Problem 7.4

Calculate the hardenability of a 4340 alloy steel which has an ASTM grain size number of 6 and the following composition: 0.4% C, 1.85% Ni, 0.8% Cr, and 0.25% Mo.

Solution 7.4

The base hardenability needs to be first determined based on the carbon content and the ASTM grain size number. Referring to Figure 7.12, it is clear that for a carbon content of 0.4 wt.% and ASTM grain size number 6, the base hardenability is 0.23 in. The multiplying factors for the particular concentration need to be determined next. These values for the alloying elements under consideration are:

Wt.% alloying element	1.85 Ni	0.8 Cr	0.25 Mo
Multiplying factor	1.8	2.6	1.75

Therefore, the ideal critical diameter is found by multiplying the base diameter by the multiplying factors. The value is

$$D_{\text{I}} = 0.23 \times 1.8 \times 2.6 \times 1.75 = 1.88 \text{ in.}$$

Thus, it can be seen that the alloying elements greatly increase the ideal critical diameter, that is, the hardenability is significantly increased by the addition of alloying elements.

This is the diameter of the bar which will have 50% martensite in the center if subjected to an ideal quench. On the other hand, if quenched in still water, $H = 1.0$, the actual diameter of the bar which will have 50% martensite in the center can be determined from Figure 7.5 and it is reduced to 1.2 in.

The multiplying factors have been developed empirically and these have been revised over a period of time. One of the more accepted empirical relationships to evaluate the ideal critical diameter in terms of the multiplying factors is

$$D_I = \text{Base hardenability} \times 2.21\ \text{Mn} \times 1.40\ \text{Si} \times 2.13\ \text{Cr} \times 3.275\ \text{Mo} \times 1.47\ \text{Ni} \quad (7.2)$$

All the compositions in the above relationship are in weight percent of the alloying elements. This relationship appears to be more accurate in practice than the one put forward by Grossmann much earlier.

7.8 JOMINY TESTS AND CONTINUOUS COOLING TRANSFORMATION DIAGRAMS

Recall that a continuous cooling transformation (*CCT*) diagram shows the phases that are present in a specific steel sample on continuous cooling from a high temperature to room temperature. During continuous cooling, the specimen experiences a range of cooling rates. At extremely high cooling rates and on quenching to a temperature below M_f, one will be able to obtain a completely martensitic structure. On the other hand, at very slow cooling rates one will obtain a pearlitic structure (and also the proeutectoid ferrite or cementite depending on the composition of the steel). At intermediate rates of cooling, the phases present could be mixtures of the above or completely new microstructures depending on the chemical composition of the steel.

As seen earlier in this chapter, the Jominy sample also experiences a range of cooling rates—the highest at the water-quenched end and the lowest at the air-cooled end. It is also possible to obtain/calculate the cooling rate at any given point in the Jominy sample. Further, we know the hardness at any point in the Jominy bar. Therefore, it should be possible to superimpose the cooling rate from any point in the Jominy bar on the *CCT* diagram. Consequently, by combining these two plots (Jominy hardness–distance plot and the *CCT* diagram), it is possible to obtain information about the hardness, microstructure (and phases) present at any point in the Jominy sample. That is, it should be possible to obtain the hardness of any particular structure by the combination of these two plots. Figure 7.16 shows the *CCT* diagram for eutectoid composition steel. Four cooling curves corresponding to four positions in the Jominy bar (A, B, C, and D) are superimposed on this *CCT* diagram. The combination of these two suggests that the steel cooled at a rate corresponding to A will have a microstructure consisting of completely martensite and that it will have a hardness of HRC 60 (Rockwell hardness of 60 on the C scale). On the other hand, the steel cooled corresponding to the curve D will have a completely pearlitic structure and that its hardness is HRC 28. Thus, the combination of Jominy plot and *CCT* diagrams can provide full information about the microstructure, phases, and hardness of the material at any given point in the Jominy sample.

7.9 HARDNESS TESTER TO BE USED

Opinions vary regarding the best way to measure the hardness values of the quenched bar. Typically, hardness measurements have been made using a Rockwell

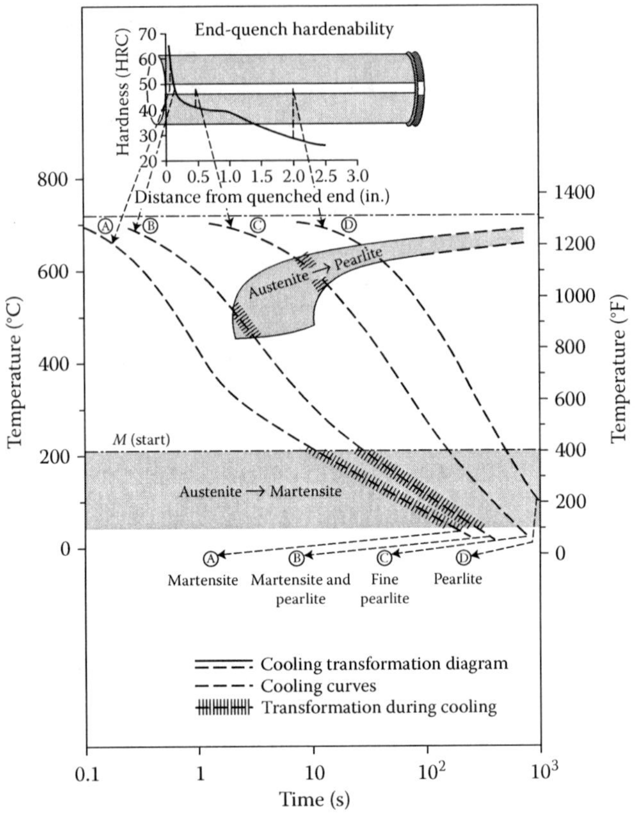

FIGURE 7.16 Correlation of hardenability and continuous cooling transformation diagram data for eutectoid composition steel. From such compositions it is possible to obtain information about the microstructure and hardness of the sample at any given point in the Jominy bar. (From ASM International, Materials Park, OH. With permission.)

hardness tester and the C scale. But, Brinell and Vickers hardness testers have also been used. The choice of measurement depends to a great extent on the specimen dimensions, its hardness, and the precision required. Rockwell C is used if the hardness is greater than 20 HRC. Brinell hardness testing is used if the hardness is below 20 HRC. Vickers hardness is chosen if a microhardness profile is required. Standard precautions such as minimum distance between indentations must be strictly followed.

It has been reported that Vickers hardness measurements provide very good reproducibility and repeatability. It also allows more data to be recorded because the size of the indentation is small and therefore they could be spaced much closer together. The precision of the test can also be improved by the use of a calibrated positioning device that holds the specimen firmly in position. An important advantage of using the Vickers hardness tester is that the measurement can be readily rechecked by measuring the diagonals again or by placing an additional indentation next to the disputed reading.

The repeatability of the standard Rockwell test is typically 2 HRC. Using a 15 kg load, the repeatability of the Vickers hardness measurement is about half that of the Rockwell test. It was also reported that accurately determined hardenability curves often show that the test introduces the scatter rather than the steel specimen.

7.10 JOMINY TEST FOR NONFERROUS ALLOYS

Traditionally, the concept of hardenability and the use of Jominy end-quench test have been applied to ferrous alloys since they can be hardened through the martensitic transformation. However, occasionally, the Jominy test has also been applied to some nonferrous alloys, for example, alloys based on copper, titanium, and aluminum. Even though the purpose in the case of nonferrous alloys is not to investigate the depth of hardening as in the case of steels, the Jominy test has been used to obtain a variety of cooling rates in one sample. The microstructural evolution has been investigated as a function of the cooling rate. For example, in the case of copper-based alloys, the Jominy test has been used to evaluate the critical cooling rate for the formation of a supersaturated solid solution at room temperature. The variation of microstructure as a function of cooling rate was investigated in the case of titanium- and aluminum-based alloys.

7.11 SOME COMMENTS

High hardenability is not always required, nor is it desirable; it is required for many tools and machine parts made of steel. The desirable combination of properties for such components is a hard wear-resisting surface combined with a tough core. That is, we need a low-hardenability steel, in which the hardening takes place to only a limited depth. In such a situation, the formation of martensite on the surface produces a condition wherein the core of the steel develops a tensile internal stress while the surface becomes stressed in compression. This combination is best to prevent fatigue cracks from nucleating and propagating (see Chapter 11: Fatigue Testing for details).

An important use of the Jominy end-quench test is in selecting steels of just the right hardenability for a given required hardness distribution. Let us assume that we have a large number of steels to choose from. Depending on the hardness requirements, we will be able to choose a steel that has the right alloy content that has sufficient hardenability. Further, by choosing an alloy steel of high hardenability, we will be able to achieve the desired hardness distribution by moderate quenching. This will be of great benefit because high quench rates, required to achieve the desired hardness distribution in leaner steels, result in increased residual stresses and also lead to quench cracking.

Alloying elements are added to steels to achieve high strength in the required shapes and sizes. The section thicknesses can be very large, up to a meter in diameter, in the case of large shafts and rotors. Therefore, to achieve full hardening in the whole cross section, alloying elements are added. However, it is important to realize that alloying elements should be added at the optimum level. That is, one could use the minimum amount required to achieve the right hardenability. This is because alloying elements are much more expensive than iron.

One criticism of the Jominy test is that it is not discriminatory when applied to steels of low hardenability. Although the Jominy end-quench test is easy to perform, the procedure must be followed exactly to obtain good results. The single most important aspect of the test is consistency. Each test must be done under identical conditions.

7.12 EXPERIMENTAL PROCEDURE

The Jominy end-quench test is conducted according to the ASTM Standard A 255 "Standard Test Method for End-Quench Test for Hardenability of Steel". As mentioned earlier, the specimen is a cylinder 1 in. in diameter and 4 in. in length. It has a lip at the top and it is 1.25 in. in diameter. The test specimen is first austenitized at a temperature of about 65°C above the Ac_3 point (the GS line in the Fe–Fe_3C phase diagram in Chapter 5 representing the temperature at which the transformation of ferrite to austenite is completed during heating) and cooled in air to eliminate differences in microstructure due to previous hot working. The specimen is then machined to remove any decarburized layer and obtain the correct dimensions of the specimen. From now onward, follow the procedure outlined below:

1. Heat the specimen to about 30°C above the Ac_3 point in a closed container that has a layer of cast-iron chips at the bottom.
2. Hold the specimen at this temperature for about 30 min so that the whole specimen reaches the desired temperature. Following this, transfer the specimen immediately to the hardenability fixture. One should not take more than 5 s to transfer the specimen from the furnace to the quenching fixture. Otherwise, the specimen will cool to a temperature below the Ac_3 temperature and some equilibrium transformations may start occurring.
3. Make sure that the specimen is held 1/2 in. above the water opening in order that a column of water may be directed against the bottom of the test piece.
4. The water opening should be 1/2 in. in diameter and adjustment is made in such a way that before the specimen is placed over it, the column of water is 2.5 in. high from the opening. The water temperature is kept at 75 ± 5°F (24 ± 3°C) and a condition of still air is maintained around the specimen during cooling.
5. Allow the test piece to completely cool down to room temperature in the fixture. Alternately, allow the specimen to be in the fixture for at least 10 min and then quench it in cold water.
6. After the specimen is completely cooled down to room temperature, prepare two flat surfaces 180° apart and 0.015-in. deep, along the whole length of the specimen, to remove the decarburized material. Make sure that the specimen does not get excessively hot during the preparation of the flat surfaces.

7. Measure the hardness of the specimen, starting from the quenched end, at intervals of 1/16 in. for the first inch, then at intervals of 1/8 in. for the next inch, and at an interval of 1/4 in. for the remainder of the bar. The operator error during hardness testing is minimized by using semiautomatic hardness-measuring instruments. If it is difficult to hold the specimen properly on the hardness machine stage, make sure that the specimen is in a fixture that ensures accurate positioning of the specimen and that the specimen is held firmly. List the hardness values as shown in the Table below.
8. Plot the hardness of the specimen against distance from the quenched end. The hardness is plotted on the Y-axis and the distance from the quenched end on the X-axis.

7.13 RESULTS

Hardness values of the Jominy end-quenched sample.

Material:
Chemical composition:
Hardness tester (and scale) used:

#	Distance from the Quenched End (in in.)	Hardness Value	#	Distance from the Quenched End (in in.)	Hardness Value
1	1/16		17	18/16	
2	2/16		18	20/16 (1.25)	
3	3/16		19	22/16	
4	4/16 (0.25)		20	24/16 (1.5)	
5	5/16		21	26/16	
6	6/16		22	28/16 (1.75)	
7	7/16		23	30/16	
8	8/16 (0.5)		24	32/16 (2.0)	
9	9/16		25	36/16 (2.25)	
10	10/16		26	40/16 (2.5)	
11	11/16		27	44/16 (2.75)	
12	12/16 (0.75)		28	48/16 (3.0)	
13	13/16		29	52/16 (3.25)	
14	14/16		30	56/16 (3.5)	
15	15/16		31	60/16 (3.75)	
16	16/16 (1.0)		32	64/16 (4.0)	

Calculate the hardenability of the steel sample.

7.14 ADDITIONAL EXPERIMENT

Different batches of students could be given steel samples of the same carbon content, but different alloying elements. All the results could be pooled together to demonstrate the effect of alloying elements on hardenability of steels.

EXERCISES

7.1 What is hardenability?

7.2 Differentiate between hardness and hardenability.

7.3 Draw schematic curves of hardness distribution across the cross section of a large steel sample when quenched in water.

7.4 Draw schematic curves of hardness distribution across the cross section of a large steel sample when quenched in oil.

7.5 Describe the differences between the hardness distributions between water-quenched and oil-quenched samples.

7.6 What will be the nature of the microstructure on the surface and in the interior of a large steel sample quenched into water?

7.7 Between water quenching and oil quenching, which is more efficient?

7.8 What will happen to the severity of quench on agitating the quenching medium?

7.9 Grade the severity of quench (from high to low) for the quenchants: air, water, and oil.

7.10 When is a steel specimen considered hardened? Give your answer in terms of the martensite content.

7.11 What is the rationale in considering a steel sample hardened when it contains 50% martensite at the center?

7.12 What is critical diameter?

7.13 What is ideal diameter?

7.14 What is the difference between critical diameter and ideal diameter?

7.15 Using Figure 7.5 as a resource, calculate the hardenability of a 1 in. diameter steel sample that shows 50% martensite at the center when quenched in a quenchant that has a severity of quench, $H = 0.8$.

7.16 Using Figure 7.5 as a resource, calculate the critical diameter of a steel sample that has an ideal diameter of 2.5 in., when quenched in a quenchant with $H = 2.0$.

7.17 What are the factors on which hardenability depends?

7.18 What is the effect of austenite grain size on the hardenability of steels? Explain.

7.19 What influence does the presence of alloying elements (other than carbon) have on the shape of a hardenability curve?

7.20 What is base hardenability?

7.21 What are multiplying factors?

7.22 What will be the effect of alloying elements on the hardenability of steels?

7.23 Why are some alloying elements more effective in increasing the hardenability of steels than others?

7.24 Do all alloying elements increase the hardenability of steels?

7.25 Calculate the hardenability of an alloy steel which has an ASTM grain size number of 5 and the composition: 0.4% C, 1% Ni, 0.5% Cr, and 0.3% Mo.

7.26 What tests are used to determine the hardenability of steels?

7.27 Why is the Jominy specimen water quenched only from one end?

7.28 Can the Jominy test be used to determine the hardenability of nonferrous alloys?

7.29 How can one conclude that the hardenability of one steel is higher than another steel?

7.30 Between a plain carbon steel and an alloy steel which will have a higher hardenability?
7.31 Why is the hardenability represented as a band?
7.32 Draw a typical hardenability plot (and label the axes).
7.33 Show low and high hardenabilities in a typical hardenability plot and label them.
7.34 What tests (other than Jominy) can one perform to obtain hardenability data?
7.35 What factors determine the use of a specific hardness test method to evaluate hardenability?
7.36 Why are hardenability data required?

FURTHER READING

ASM Handbook. 1991. *Heat Treating*, Vol. 4. Materials Park, OH: ASM International.

Brooks, C. R. 1996. *Principles of the Heat Treatment of Plain Carbon and Low Alloy Steels*. Materials Park, OH: ASM International.

Callister, Jr., W. D. and D. G. Rethwisch, 2010. *Materials Science and Engineering, An Introduction*, 8th edition. New York, NY: John Wiley & Sons, Inc.

Doane, D. V. and J. S. Kirkaldy (eds.), 1978. *Hardenability Concepts with Applications to Steel*. Warrendale, PA: AIME.

Grossmann, M. A. and E. C. Bain, 1964. *Principles of Heat Treatment*, 5th edition. Materials Park, OH: ASM International.

Krauss, G. 2005. *Steels: Heat Treatment and Processing Principles*. Materials Park, OH: ASM International.

Siebert, C. A., D. V. Doane, and D. H. Breen, 1977. *The Hardenability of Steels*. Materials Park, OH: ASM International.

8 Hardness Testing

8.1 INTRODUCTION

Hardness measurement is a very simple and useful technique to characterize the mechanical properties of materials. It provides an accurate, rapid, and economical way of determining the resistance of materials to plastic deformation, usually by penetration. The hardness of a material is related to its strength, and also to its wear and fatigue resistance. In spite of its importance, hardness is an unusual property in that it is measured in an essentially arbitrary manner. It is not possible to give it a precise definition in terms of the fundamental SI units of mass, length, and time. Compared with other methods of measuring the strength of materials, hardness tests are mostly nondestructive in nature. Additionally, hardness measurements do not require any special size and shape of specimens and the testing equipment is simple and is available in almost every laboratory. Further, other mechanical properties, for example, tensile strength, may be estimated from hardness data. Consequently, hardness measurement is an important means of quality control in manufacturing operations.

The term *hardness* has been defined in many different ways depending on who defines it. A materials scientist defines it as the resistance of a material to penetration. A machinist defines it as the resistance of a material to cutting. A geologist, on the other hand, defines it as the resistance of a material to scratching, and to a lubrication engineer it means resistance to wear, while a design engineer considers it as a measure of the flow stress. People also normally associate hardness primarily with brittleness, that is, the harder a material is the more brittle it is. Even though all the above definitions appear to be significantly different from each other, they are all related to the plastic flow stress of the material. The hardness of a material is a manifestation of several related properties, which may well include a combined effect of yield strength, tensile strength, ductility, work-hardening characteristics, resistance to abrasion, and so on. Depending on the way the hardness is measured, different combinations of these properties are effective in influencing the hardness. Stated differently, hardness is a measure of the resistance of a material to localized plastic deformation (e.g., a small dent or a scratch). Note that hardness is a characteristic of a material and not a fundamental physical property of the material. It depends on a complex set of properties or attributes of the material, for example, microstructure.

Many of the above definitions of hardness are qualitative in nature and therefore a more accurate measure is required to compare the hardness values of different materials and rank them. The hardness of a material can be measured on a macroscopic or microscopic level accurately and reproducibly using modern hardness testers. Hardness tests can also be conducted on thin films, wires, rods, or bulk components.

8.2 TYPES OF HARDNESS MEASUREMENTS

The hardness of a material may be manifested in one or more of the following forms:

1. *Cutting hardness*—resistance to various cutting operations
2. *Abrasive hardness*—resistance to wear when subjected to rotating or sliding motion
3. *Tensile hardness*—strength in terms of yield or tensile strength
4. *Rebound hardness*—resistance offered to strike and rebound, which principally is a measure of the elastic properties
5. *Indentation hardness*—resistance offered to indentation, which is a measure of plasticity and density
6. *Deformation hardness*—measure of the distortion properties, which is of primary importance in the case of sheet materials

Of the different important methods of hardness testing, the following three may be considered the most common:

1. *Scratch hardness measurement:* In this method, used mostly by mineralogists for a qualitative comparison, the materials are ranked on their ability to scratch another material. A harder material is judged as capable of scratching a softer material.
2. *Rebound hardness measurement:* This is a dynamic hardness test based on the energy absorbed under impact loading conditions. In this method, an object of standard dimensions and weight is bounced off the work piece. The height of the rebound is a measure of the hardness; the higher the rebound, the higher is the hardness of the material.
3. *Static indentation hardness measurement:* In this test, a ball, a pyramid, or a cone is forced at right angles into the surface of the material being tested with a known amount of force. The ratio of the force to the area or depth of indentation provides a measure of the hardness of the material. Indentation hardness value is obtained using one of over 12 different test methods; but there are three principal standard test methods—Brinell, Rockwell, and Vickers.

We will now briefly describe the different methods of measuring the hardness of materials by different methods.

8.3 SCRATCH HARDNESS MEASUREMENT

The early hardness tests were based solely on the ability of one material to scratch another that was softer. Such scratch hardness tests were primarily of interest to mineralogists. The testing is done in a qualitative and somewhat arbitrary manner using the hardness indexing scheme known as the *Mohs scale*. The Mohs hardness scale for minerals was devised by the German mineralogist Friedrich Mohs* and has been in use since 1822. This scale is arranged in increasing hardness of the different minerals from 1 on the soft end for talc to 10 on the hard end for diamond. The minerals, which represent the 10 different hardness numbers, are:

1. Talc
2. Gypsum
3. Calcite
4. Fluorite (Fluorspar)
5. Apatite
6. Orthoclase (Feldspar)
7. Quartz
8. Topaz
9. Corundum
10. Diamond

According to this scale, a mineral with a higher number can scratch another mineral with a lower number. Thus, gypsum can scratch talc but not calcite or any other mineral with numbers above 3. Similarly, topaz can scratch quartz or any other mineral with numbers up to 7, but not corundum or diamond. Obviously, diamond can scratch every mineral but not vice versa. Diamond is the hardest material known to man! Figure 8.1 shows the Mohs scale along with the name and composition of the mineral, a photograph of each of the minerals, and their hardness numbers on the Brinell scale.

The steps in the above Mohs scale are not of equal value and the difference between 9 and 10 is much greater than that between 1 and 2. The hardness is determined by finding which of the standard minerals the test material will scratch or not scratch. The hardness of the test material will lie between two points on the scale—

* Friedrich Mohs (1773–1839) was a German geologist/mineralogist, and is famous in the gem world as the creator of the Mohs scale of mineral hardness, which is still widely used today. He was born on January 29, 1773 in Gernrode, Germany, studied chemistry, mathematics, and physics at the University of Halle and also studied at the Mining Academy in Freiberg, Saxony. After assuming the position of a foreman at a mine in 1801, Mohs moved in 1802 to Austria, where he was employed in trying to identify the minerals in a private collection of a banker.

Using the same reasoning as botanists, who sorted plants and animals with similar physical characteristics into genus groups, Mohs began examining the minerals for common properties (more specifically the scratch hardness) which would allow him to group them together for identification. He knew that some of the minerals could scratch others and that if a mineral could make a scratch on another, it must be harder than the mineral that was scratched. He found that all minerals could scratch the mineral talc, so it must be very soft. He also found that diamonds were so hard that no other minerals could scratch them.

In 1812, Mohs became a professor of mineralogy in Graz where he finalized his work on hardness. Creating a scale of 1–10, he assigned each mineral a value, which was to become the Mohs' Scale of Hardness. In 1818, he became a professor in Freiberg, Saxony and in 1826, professor in Vienna. Mohs ended his remarkable career as a mining advisor at the Mining University in Leoben and died on September 29, 1839, aged 66, while holidaying in Agordo near Belluno in Italy.

Mohs hardness number	Name of the mineral	Chemical composition	Brinell hardness	Photograph of the mineral
1	Talc	$Mg_3Si_4O_{10}(OH)_2$	3	
2	Gypsum	$CaSO_4{\cdot}2H_2O$	12	
3	Calcite	$CaCO_3$	53	
4	Fluorite (Fluorspar)	CaF_2	64	
5	Apatite	$Ca_5(PO_4)_3(OH-, Cl-, F-)$	137	
6	Orthoclase (Feldspar)	$KAlSi_3O_8$	147	
7	Quartz	SiO_2	178	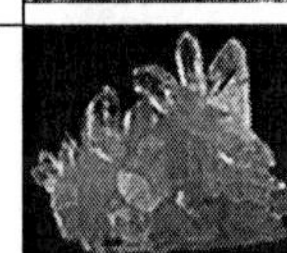
8	Topaz	$(Al_2SiO_4)(OH-, F-)$	304	
9	Corundum	Al_2O_3	667	
10	Diamond	C	—	

FIGURE 8.1 Mohs scale, the name and chemical composition of the mineral, the actual hardness numbers of the minerals, and their photographs.

the first point being the mineral which is scratched by the test material and the next point being the mineral which is able to scratch the test material. That is, the test material will not scratch the mineral representing the second point. On this scale, a finger nail has a hardness value of about 2, annealed copper between 2 and 3, iron with up to 0.1 wt% C between 3 and 4, and tool steel between 7 and 8.

Mohs hardness measurement is a simple test, but is not exactly quantitative and the standards are purely arbitrary. The materials scientist will not find much use for the Mohs scale due to its nonquantitative nature. Further, most hard metals and alloys have hardness values between 4 and 8 on this scale. Pure metal lead has a value of 1.5, gold 2.5, cobalt 5.0, and case-hardened steel 8.0. Consequently, this scale is not suitable for metals and alloys in general since the intervals are not sufficiently widely spaced in the high hardness range. Even though it is possible to subdivide the scale and some derived methods are still commonly used, this is not a very popular method to determine the hardness values of engineering materials.

Similar methods of relative hardness assessment are still commonly used today. An example is the *file test* where a file tempered to a desired hardness is rubbed on the test material surface. If the file slides without biting or marking the surface, the test material would be considered harder than the file. If the file bites or marks the surface, the test material would be considered softer than the file. The file test is useful as a rapid and portable qualitative test for hardened steels, where conventional hardness testers are not available or practical. Files can be tempered back to give a range of known hardness and then used in a similar fashion to the Mohs method to evaluate hardness values and rank the materials.

These relative hardness tests are limited in practical use and do not provide accurate numeric data or scales, especially for modern-day advanced materials.

8.4 REBOUND HARDNESS MEASUREMENT

The *Shore Scleroscope*, invented by Albert F. Shore in 1907, was the first commercial hardness tester produced in the United States, and was used extensively before and during World War I. Even now this test is frequently used when a number of quality checks need to be conducted on large samples. This is essentially a dynamic hardness test, operating on the principle of elastic rebound.

The Shore Scleroscope test consists of dropping a diamond-tipped hammer, which falls inside a glass tube under the force of its own weight from a fixed height, onto the surface of the specimen being tested. The standard hammer is approximately 1/4 in. in diameter, 3/4 in. long, and has a diamond striking tip rounded to 0.01 in. radius. The Shore Scleroscope measures the hardness in terms of the elasticity of the material. The height of the rebound travel of the hammer is a measure of the hardness of the material. The scale is graduated into 100 divisions, the position 100 corresponding to the average rebound height from a quenched and untempered water-hardened tool (high-carbon) steel sample. The scale is usually continued up to 140 units, so that the hardness of still harder materials can also be measured.

For obtaining accurate and consistent values of the hardness, the test specimen surface must be polished to a smooth finish, ensuring that overheating and/or excessive cold working of the surface are avoided. Consequently, soft metals require only a minimum finish, while hard metals are finely ground and polished.

Hardness measurement using the Shore Scleroscope is done in the following manner:

1. The instrument is leveled using the leveling screws so that the glass tube is perfectly vertical.
2. The glass tube containing the hammer is brought into intimate contact with the specimen by lowering the clamp.
3. The hammer is allowed to fall onto the specimen by turning the knob at the side. This operation automatically resets the dial hand to zero.
4. After rebound, the hardness value is read off the dial by turning this knob back to the original position. The scale is set behind the tube and is visible through it.

The smaller the fraction of the kinetic energy of the hammer that is used in plastically deforming the test specimen, the greater will be the height of the rebound, and thus higher is the hardness. It is important to remember that (1) the specimen surface should be flat and smooth and free from grease marks, and (2) a number of readings must be taken to get a fairly good average hardness value. Note that the hammer should not be dropped on the same spot of the specimen more than once.

In normal use, the Shore Scleroscope does not mark the material under test. A further advantage of this method is its portability. In general, this method of hardness testing is used for large workpieces and also in such circumstances when visible or sharp impressions on the surfaces of specimens cannot be tolerated.

8.5 THE DUROMETER TEST

The durometer scale was defined by Albert F. Shore, who developed a measurement device called a durometer in the 1920s. The term durometer is often used to refer to the measurement, as well as the instrument itself. The Durometer is a popular instrument for measuring the indentation hardness of rubber and rubber-like materials such as polymers and elastomers. Widely used in the plastics and rubber industries, this test has evolved from a coarse hand-held measurement to today's repeatable test method. The most popular testers are the Model A used for measuring the hardness of softer materials and the Model D for harder materials. Figure 8.2 shows the photographs of two commercially available durometers used to measure the hardness of nonmetallic materials.

There are several scales of durometer, used for materials with different properties. The two most common scales, using slightly different measurement systems, are the ASTM D2240 type-A and type-D scales. The A scale is for softer plastics, while the D scale is for harder ones. However, the ASTM D2240–00 testing standard calls for a total of 12 scales, depending on the intended use; types A, B, C, D, DO, E, M, O, OO, OOO, OOO-S, and R. Each scale results in a value between 0 and 100, with higher values indicating a harder material.

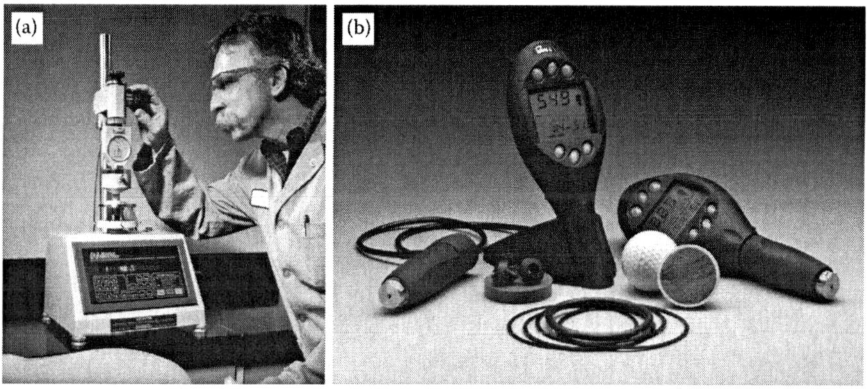

FIGURE 8.2 (a) Photograph of a durometer used to measure the hardness of nonmetallic materials. (Photo courtesy of Newage Testing Instruments, Inc. With permission.) (b) Another type of durometer. (Photo courtesy of Instron Corporation. With permission.)

Durometer, like many other hardness tests, measures the depth of an indentation in the material created by a given force on a standardized presser foot. The operation of the tester is quite simple. The material is subjected to a predetermined test force by a spherical- or conical-shaped indenter. The indenter for scale A is a hardened steel rod 1.1–1.4 mm in diameter, with a truncated 35° cone, 0.79 mm diameter. For scale D, the indenter is a hardened steel rod 1.1–1.4 mm in diameter, with a 30° conical point, 0.1 mm radius tip. The depth of indentation is translated into a hardness value by means of a dial gage or electronically. Test forces range from 822 g (A scale) to 4550 g (D scale).

The material under test should be a minimum of 6.4 mm (0.25 in.) thick. The final value of the hardness depends on the depth of the indenter after it has been applied for 15 s on the material. If the indenter penetrates 2.5 mm or more into the material, the durometer is 0 for that scale; that is, the material is very soft. If it does not penetrate at all, then the durometer is 100 for that scale; that is, the material is very hard. Durometer is a dimensionless quantity, and there is no simple relationship between a material's durometer in one scale, and its durometer in any other scale, or by any other hardness test.

Nonstandardized "micro" scales are also available from many manufacturers. These scales permit testing on thinner and narrow specimens. The hardness number is designated as A/50/15 meaning that the hardness number is 50 using Durometer of type A and 15 s duration.

In the A scale, a rubber band has a hardness number of 25, a door seal 55, a soft skateboard wheel 75, and ebonite rubber 100. The hardness of harder materials such as solid truck tires and hard hats are measured on a D scale and have values of 50 and 75, respectively.

Some semiempirical relationships between the Shore hardness and the Young's modulus for elastomers have been derived.

The use of IRHD, or International Rubber Hardness Degrees, has increased considerably in North America since its origin in Europe. It provides a reproducible result on rubber parts of all shapes and sizes, especially in the determination of the hardness of rubber O-rings. The method is somewhat similar to the Rockwell hardness testing (to be described below). A preliminary test force is applied to the specimen through the indenter. The test is zeroed at this position, and then the total test force is applied. The difference in the distance between these two forces is then measured and converted to an IRHD value. The preliminary (minor) test forces are 8.46 g for microscales and 295.74 g for regular scales. Total test forces are 15.7 g for microscales and 597 g for regular scales.

8.6 INDENTATION HARDNESS MEASUREMENT

The most common way of measuring the hardness of materials in a quantitative way is by indentation testing. In these tests, an indenter is pressed at right angles into the surface of a material to be tested. The forcing of the indenter is done under controlled conditions of load and rate of application. The indenter is usually a ball, cone, or pyramid made of a material that is much harder than the test specimen. The softer the material is, the larger and deeper is the indentation and lower is the hardness number. The depth or size or area of the resulting indentation is measured, which in turn is related to a hardness number. The resistance of the material to yielding or plastic deformation, under the action of the indenter, depends on friction, elasticity, viscosity, and the intensity and distribution of plastic strain produced during indentation. The measured hardness values are only relative (rather than absolute), and care should be exercised when comparing values determined by different techniques.

Several different methods of indentation hardness testing have been developed over the years. Of these, three have gained the most common acceptance and these are:

1. Brinell hardness testing
2. Rockwell hardness testing
3. Vickers hardness testing

Let us now discuss the details of these three hardness tests.

8.7 BRINELL HARDNESS TESTING

8.7.1 Principle

The first report of a method to determine the hardness by measuring the impression made by a steel ball forced into the metal under definite static loads was by Dr. Johan August Brinell (1849–1925) of Sweden in 1900. In the Brinell hardness testing, a spherical indenter is pressed into the smooth flat surface of a test specimen. The load (the term "force" is technically correct, but manufacturers and users commonly use the term "load") applied and the dwell time for the load are different depending on the type of specimen being tested (Table 8.1). The indenter is then removed and the diameter of the resultant recovered round indentation is measured using a special low-power microscope, utilizing a millimeter scale that is etched on the eyepiece.

TABLE 8.1
Test Conditions for Brinell Hardness Testing

Material	Indenter Diameter, D (mm)	Load, P (kg)	Full Load Dwell Time (s)	P/D^2
Steel and cast iron	10	3000	10–15	30
Copper, copper alloys, and aluminum, aluminum alloys	10	500	30	5
Lead, tin, and their alloys	10	100	30	1

The scale is graduated in millimeters from 0 to 7 mm, which permits measurements to be estimated to 0.05 mm accuracy. For more precise measurements, as is required when the impression is very small in diameter, portable precision microscopes are available, which provide an accuracy of 0.001 mm. Figure 8.3 shows a schematic of the Brinell hardness test. A commercial Brinell hardness tester is shown in Figure 8.4. The diameter of the indentation is calculated as the average of at least two diameters, measured usually at right angles to each other. The Brinell hardness number (BHN) can be determined by either measuring the diameter of the impression (the most common method) or the depth of impression made on the test specimen. These two values, however, differ from each other, due to the errors associated with each method of measurement. The BHN is obtained as the ratio of the load applied to the surface area of the indentation, using the formula:

$$\text{BHN} = \frac{P}{(\pi D/2)\left[D - \sqrt{D^2 - d^2}\right]} \tag{8.1}$$

where P is the load applied in kg, D is the diameter of the ball indenter in mm, and d is the diameter of the indentation in mm. BHN has the units of kg mm^{-2}. But, most frequently the units for hardness numbers are omitted.

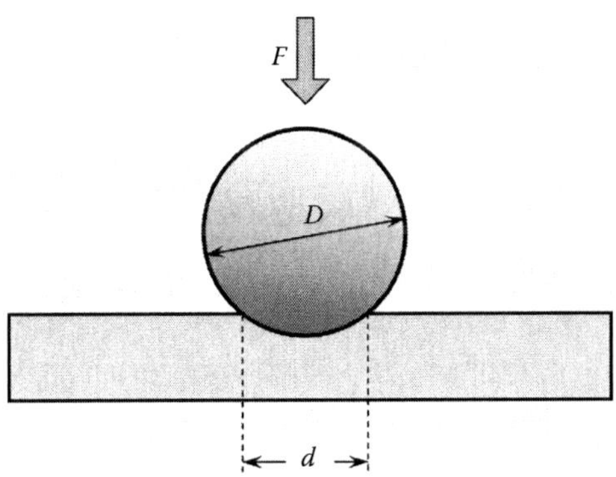

FIGURE 8.3 Schematic of Brinell indentation. The diameter of the indenter (D) and the diameter of the indentation (d) are indicated.

FIGURE 8.4 Photograph of a commercial Brinell hardness tester. (Photo courtesy of Newage Testing Instruments, Inc. With permission.)

Instead of calculating the hardness number every time a test is conducted, standard charts, relating the hardness number to the diameter of the indentation for a given load, could be used to read the hardness value once the average diameter of the indentation is measured. This is possible only when standard loads and indenters are used. Table 8.2 lists the BHN for different indentation sizes and loads when a ball indenter of 10 mm diameter is used. For example, if the diameter of the indentation was measured as 2.2 mm when a load of 500 kg was applied, then the BHN value is 130 (Table 8.2).

Example Problem 8.1

If the size of the indentation during a Brinell hardness test is 3 mm in diameter calculate the hardness number, assuming that the load applied is 3000 kg and the size of the steel ball indenter is 10 mm in diameter.

Solution 8.1

The BHN can be calculated using Equation 8.1:

$$\text{BHN} = \frac{P}{(\pi D/2)\left[D - \sqrt{D^2 - d^2}\right]}$$

Since it has been mentioned that the load applied, $P = 3000$ kg, diameter of the indenter, $D = 10$ mm, and that the size of the indentation, $d = 3$ mm, we can substitute these values into the above equation, and calculate the BHN.

TABLE 8.2
Brinell Hardness Number Calculated as a Function of the Diameter of the Indentation and the Load Applied Using Equation 8.1

Brinell Indentation Diameter (mm)	Load (kg)					
	500	**1000**	**1500**	**2000**	**2500**	**3000**
2.00	158	316	473	632	788	945
2.20	130	260	390	520	650	780
2.40	109	218	327	436	545	653
2.60	92.6	185	278	370	462	555
2.80	79.6	159	239	318	398	477
3.00	69.1	138	207	276	346	415
3.20	60.5	121	182	242	303	363
3.40	53.4	107	160	214	267	321
3.60	47.5	95	142	190	238	285
3.80	42.4	84.8	127	170	212	255
4.00	38.1	76.2	114	152	191	229
4.20	34.4	68.8	103	138	172	207
4.40	31.2	62.4	93.6	125	156	187
4.60	28.4	56.8	85.2	114	142	170
4.80	25.9	51.8	77.8	104	130	156
5.00	23.8	47.6	71.3	95.2	119	143
5.20	21.8	43.6	65.5	87.2	109	131
5.40	20.1	40.2	60.3	80.4	101	121
5.60	18.6	37.2	55.7	74.4	92.5	111
5.80	17.2	34.4	51.5	68.8	85.8	103
6.00	15.9	31.8	47.7	63.6	79.5	95.5
6.20	14.8	29.6	44.3	59.2	73.8	88.7
6.40	13.7	27.4	41.2	54.8	68.8	82.5

Note: The Diameter of the Indenter is 10 mm.

Thus,

$$\text{BHN} = \frac{3000\text{ kg}}{(\pi \times 10\text{ mm}/2)\left[10\text{ mm} - \sqrt{(10\text{ mm})^2 - (3\text{ mm})^2}\right]}$$

$$= \frac{3000\text{ kg}}{\pi \times 5\text{ mm}\left[10\text{ mm} - \sqrt{(100-9)\text{ mm}^2}\right]}$$

$$= \frac{3000\text{ kg}}{\pi \times 5\text{ mm}\left[10\text{ mm} - 9.54\text{ mm}\right]}$$

$$= 415\text{ kg mm}^{-2}$$

Example Problem 8.2

Calculate the size of the indentation for a steel specimen with a hardness number of 300. Assume that the diameter of the indenter used is 10 mm and that a load of 3000 kg is applied.

Solution 8.2

Since a load of 3000 kg is used, using the equation for BHN (Equation 8.1), one gets

$$300 \text{ kg mm}^{-2} = \frac{3000 \text{ kg}}{(\pi \times 10 \text{ mm}/2)\left[10 \text{ mm} - \sqrt{(10 \text{ mm})^2 - d^2}\right]}$$

or

$$10 \text{ mm} - \sqrt{100 \text{ mm}^2 - d^2} = \frac{3000 \text{ kg}}{(300 \text{ kg mm}^{-2}) \times \pi \times 5 \text{ mm}} = 0.637 \text{ mm}$$

Therefore,

$$\sqrt{100 \text{ mm}^2 - d^2} = 10 \text{ mm} - 0.637 \text{ mm} = 9.363 \text{ mm}$$

or

$$d^2 = 100 \text{ mm}^2 - (9.363 \text{ mm})^2 = 12.334 \text{ mm}^2$$

Thus, diameter of the indentation, $d = 3.51$ mm.

Comparing the solutions to the above two Example Problems 8.1 and 8.2, it is clear that, maintaining all other parameters the same, when the hardness of the material is higher the smaller is the diameter of the indentation.

8.7.2 Indenters and Loads

The standard indenter for Brinell hardness testing is a spherical hardened steel ball of 10 mm diameter. The diameter of the ball indenter should not deviate by more than 0.005 mm and it should have a hardness of at least 850 Vickers hardness number (VHN) (see later in the chapter for details of Vickers hardness testing). For hard metals like steels, a load of 3000 kg is applied for 10–15 s. On the other hand, a load of 500 kg is applied for 30 s for soft metals like aluminum or copper.

The principle of the Brinell tester is based on the fact that with indenters of different sizes, similar geometrical impressions having proportional diameters are produced which give the same BHN, provided that the loads applied are proportional to the square of the ball diameters. These ratios of P/D^2 are 30 for iron and steel and

5 for brass, bronze, and other nonferrous metals and alloys. These ratios assume special importance when ball indenters of nonstandard diameters are used.

Example Problem 8.3

Consider a case where the hardness of a steel specimen has to be measured using a steel ball of 7 mm diameter. Calculate the load that has to be used, if the hardness value obtained should be the same as when a standard 10-mm diameter steel ball is used.

Solution 8.3

Since 7-mm diameter steel ball is not a standard size for Brinell hardness testing, we need to make sure that the ratio of P/D^2 is maintained at 30 for steel specimens. From the above relationship

$$\begin{aligned} P &= 30\ D^2 \\ &= 30\ \text{kg mm}^{-2} \times (7\ \text{mm})^2 \\ &= 1470\ \text{kg} \end{aligned}$$

Therefore, a load of 1470 kg has to be used.

Note, for example, that for a BHN of 85, the indentation size will be 6.3 mm when a steel ball indenter of 10-mm diameter and a load of 3000 kg are used (the standard test). However, for the same hardness number, the indentation size will be 4.4 mm when a (nonstandard) steel ball indenter of 7 mm diameter and a load of 1470 kg are used. Thus, the BHN is independent of the ball diameter and test loads used, provided the two are used in the proper ratio. Similarly, if a ball diameter of 5 mm is used, then the load to be used will be 750 kg.

Hardened steel balls of other diameters [e.g., 1–3 (±0.0035) mm, 3–6 (±0.004) mm, or 6–10 (±0.0045) mm] and other loads (1000, 1500, 2000, and 2500 kg) also could be used. But, when nonstandard loads and/or indenters are used, the test report should specify the load and ball size. The diameter of the indentation is expected to be in the range of 25–60% of the diameter of the indenter. Thus, the test load used should be adjusted according to the expected hardness of the material (i.e., the size of the impression), specimen thickness, and test surface. The specimen thickness should be at least 10 times the depth (not the diameter) of the indentation, and these minimum thickness values are specified in the literature. Table 8.3 lists these minimum thickness values. The depth of indentation during Brinell hardness testing can be calculated using the equation

$$\text{Depth (mm)} = \frac{P}{\pi D \times (\text{BHN})} \tag{8.2}$$

where the symbols have the same meaning as for Equation 8.1.

TABLE 8.3
Minimum Thickness Requirements for Brinell Hardness Specimens

Minimum Thickness of Specimen		Minimum Hardness for which the Brinell Test May be Made Safely		
mm	**Inches**	**3000 kg Load**	**1500 kg Load**	**500 kg Load**
1.6	0.0625	602	301	100
3.2	0.125	301	150	50
4.8	0.1875	201	100	33
6.4	0.25	150	75	25
8.0	0.3125	120	60	20
9.6	0.375	100	50	17

Example Problem 8.4

Calculate the depth of the indentation corresponding to a BHN of 350 of a steel specimen measured with a load of 3000 kg and a hardened steel ball indenter of 10 mm diameter.

Solution 8.4

The depth of the impression is given by Equation 8.2:

$$\text{Depth (mm)} = \frac{P}{\pi D \times (\text{BHN})}$$

and it is given that $P = 3000$ kg, $D = 10$ mm, and BHN = 350 kg mm^{-2}. Substituting these values into the above equation, we get

$$\text{Depth} = \frac{3000\ \text{kg}}{\pi \times 10\ \text{mm} \times 350\ \text{kg mm}^{-2}} = 0.27\ \text{mm}$$

Therefore, the depth of indentation is 0.27 mm. Note, however, that this depth is different from the diameter of the indentation.

Hardened steel ball indenters can be used to test materials up to a BHN of 444. That means, the hardness of highly hardened steels and other very hard materials cannot be measured using hardened steel ball indenters. Under these situations, the hardness values measured will be in error due to the possible flattening and permanent deformation of the steel ball. Tungsten carbide ball indenters are used to test samples with Brinell hardness values of 444–627. This highest value of hardness that can be measured using a Brinell test corresponds to approximately 60 HRC (a hardness number of 60 using a Rockwell C scale; see later sections for details). Use of a tungsten carbide ball should again be stated in the test report. The diameter of the tungsten carbide ball can also be different depending on the requirement; the more common diameters are 2.5, 5.0, and 10.0 mm.

8.7.3 Hardness Designation

The Brinell hardness number has been traditionally designated as BHN, even though in recent times it is more commonly designated as HB. A well-structured BHN reveals the test conditions, and will appear as "200 HBS 10/3000/15," which means that a BHN of 200 was obtained using a 10 mm diameter hardened steel ball indenter with a 3000 kg load applied for a period of 15 s. The letter S indicates that a steel ball indenter is used. If a tungsten carbide ball is used to measure the hardness, the hardness number may appear as "500 HBW 1/30/20" which means that a BHN of 500 was obtained using a tungsten carbide ball of 1 mm diameter with a 30 kg load and 20 s dwell time.

8.7.4 Advantages and Disadvantages

Compared to other hardness test methods, the Brinell indenter makes the deepest and widest indentation. On the one hand, this is an advantage because the test averages the hardness over a larger amount of material and is thus representative of the bulk or macrohardness of a material, particularly of those materials with heterogeneous structures. An additional advantage of this method of testing is that it is relatively insensitive to surface irregularities, that is, a satisfactory surface may be obtained by grinding or filing with a relatively fine file. Polishing of the surface is not necessary. Nevertheless, it is advisable to ensure that the specimen is free from pits, oxide scale, or grease before the test. However, because of the large size of the indentation, the surface of the specimen may get damaged; this assumes added concern if the specimen is a finished product.

8.7.5 Precautions

Certain precautions need to be taken to obtain accurate and reproducible hardness numbers using the Brinell hardness tester.

1. Since the hardness number is dependent on the size of the indentation, the diameter of the indentation needs to be measured carefully to an accuracy of 0.01 mm. The larger the diameter of the indentation, the lower is the hardness of the material.
2. The maximum range of Brinell hardness values is 16 HB for very soft aluminum to 627 HB for hardened steels. Brinell hardness testers work best for testing softer metals and medium-hard steels.

8.7.6 General Observations

It is important to realize that the size of the indentation will be different if different loads and/or different sizes of the indenters are used. Consequently, the hardness number will be different for the same material. However, the BHN will remain constant for a given material only when P/D^2 is a constant; geometrically similar indentations are produced when this condition is satisfied. The value of the constant depends on the hardness of the test material. Important values are listed in Table 8.1.

Equation 8.1 would be applicable and correct hardness numbers are obtained only if the steel indenter was in no way deformed during the test. However, it is always possible that some deformation of the ball indenter occurs, especially when very hard metals are tested. Because of the attending elastic recovery of the indentation when the load is released, the indentation is not perfectly spherical. This introduces errors in the hardness values calculated using Equation 8.1. Notwithstanding these difficulties, the hardness values calculated using Equation 8.1 are accepted as valid for all practical purposes, provided that standard conditions of testing are maintained.

8.8 ROCKWELL HARDNESS TESTING

As mentioned above, the Brinell hardness testing is slow, not useful on fully hardened steel, and leaves too large an impression to be considered nondestructive. Therefore, alternate and more efficient hardness testers are required.

Rockwell hardness testing is the most widely used method for determining the hardness of materials because it is rapid (measurements are done in a matter of seconds), simple to perform (no microscopic measurements are required as in the Brinell or Vickers methods), and does not require highly skilled operators. Figure 8.5 shows a commercial Rockwell hardness tester.

Rockwell hardness testing is conducted by applying two loads (a minor and a major load) to the specimen and measuring the difference in the depth of penetration between these two load applications.* The minor load is applied to eliminate errors that could be caused by specimen surface irregularities. The minor load is 10 kg when used with 60, 100, or 150 kg major loads and 3 kg on the superficial tests where the major loads are 15, 30, or 45 kg. The minor load should be applied and maintained for a duration that shall not exceed 3 s. The increase in force from the minor to the major load should occur in no less than 1 s, nor greater than 8 s. The force is then maintained for a duration of 4 ± 2 s. The principle of the test is described below.

8.8.1 Principle

The Rockwell indenter is forced into the test material under a preliminary minor load, F_0. When equilibrium has been reached, the relative positions of the indenter and specimen surface will be as shown in Figure 8.6. This is considered as the zero position. With the minor load still in position, an *additional* major load, F_1, is applied resulting in increased penetration. The major load, which is usually much larger than the minor load, is held for a predetermined amount of time and then released, but with the minor load still applied. Thus, the total load applied onto the specimen

---

* The concept of differential depth hardness measurement first appeared in the book *Die Kegelprobe* (crudely, the cone trial) by Professor Paul Ludwig of Vienna in 1908. The differential-depth method subtracted out the errors associated with the mechanical imperfections of the system, such as backlash and surface imperfections of the specimen. The Rockwell hardness tester, a differential-depth machine, was co-invented by Connecticut natives Hugh M. Rockwell (1890–1957) and Stanley P. Rockwell (1886–1940), for which a patent was granted in 1919.

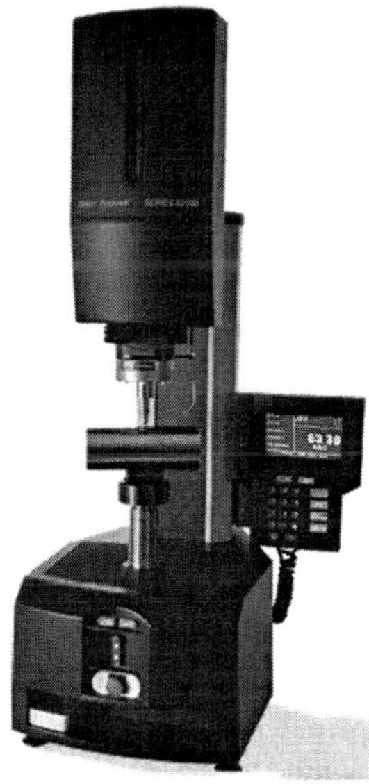

FIGURE 8.5 Photograph of a commercial Rockwell hardness tester. (Photo courtesy of Instron Corporation. With permission.)

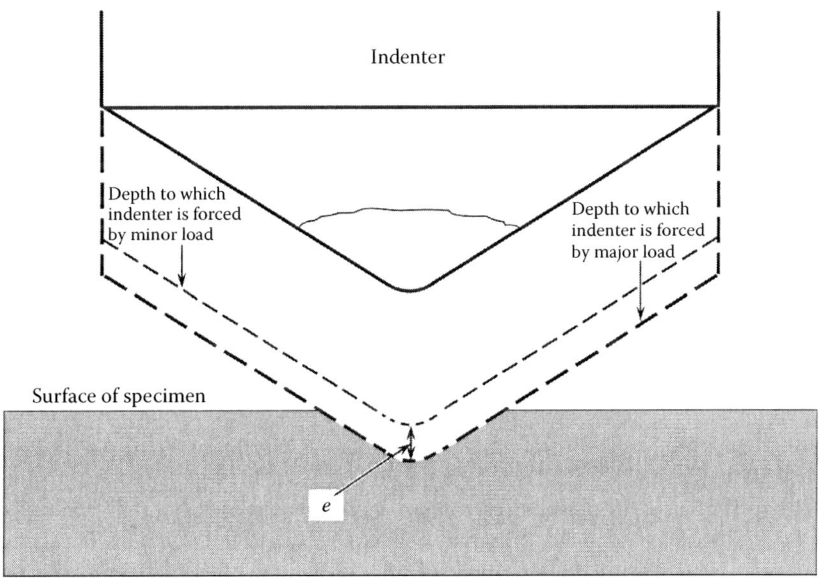

FIGURE 8.6 Principle of the Rockwell hardness testing. The figure shows the relative positions of the indenter and the specimen surface on application of minor and major loads and on removal of the major load.

surface is $F = F_o + F_1$. For example, in the Rockwell B scale (to be explained later), the minor load (F_o) is 10 kg and the additional load (F_1) is 90 kg and the total final load is 100 kg. Similarly, for the C scale, the minor load is 10 kg and the additional load is 140 kg, and so the total load is 150 kg. (In majority of the cases in the literature, this final load has been incorrectly referred to as the major load.) When equilibrium is again reached, the position of the indenter with respect to the specimen surface will be as shown in Figure 8.6. Removal of the major load allows partial (elastic) recovery and reduces the depth of penetration. It is this permanent increase in depth of penetration, resulting from the application and removal of major load, which is used to calculate the Rockwell hardness number. The Rockwell hardness number, HR is given by

$$\mathrm{HR} = E - e \tag{8.3}$$

where E = a constant depending on the nature of the indenter (100 units for the diamond indenter, scale C, and 130 units for the steel ball indenter, scale B), and e = permanent increase in depth of penetration due to major load, F_1, measured in units of 0.002 mm. In the case of superficial hardness measurement, one hardness number represents an indentation depth of 0.001 mm, regardless of the type of indenter used.

The "differential depth" measurement method is of significance because errors due to mechanical defects of the system, such as backlash, are eliminated, as well as errors resulting from slight surface imperfections in the specimen.

Alternately, the depth of indentation can be determined from the hardness values. The Rockwell hardness number reported on the C scale should be subtracted from 100 and that on the B scale should be subtracted from 130. The remainder should then be multiplied by 0.002 mm and that will be the depth of the indentation. For example, if the hardness measured using the diamond indenter is reported as HRC 65, then the depth of the indentation is $(100 - 65) \times 0.002$ mm = 0.07 mm. Similarly, if the hardness is measured with a hardened steel ball indenter and the value reported is HRB 65, then the depth of the indentation is $(130 - 65) \times 0.002$ mm = 0.13 mm. Thus, one can see that for the same hardness number, the depth of penetration is higher in the B scale than in the C scale. Therefore, it is very important to specify the scale used to measure the hardness of a material. However, such computations of the depth of penetration are not required in practice.

Example Problem 8.5

The hardness of a steel specimen was measured using the C scale of a Rockwell hardness tester. Calculate the Rockwell hardness number if the depth of the indentation was measured as 0.09 mm.

Solution 8.5

Using Equation 8.3, the Rockwell hardness number can be calculated as HR = $E - e$, where e is the depth of the indentation, measured in units of 0.002 mm. Since the depth of penetration was given as 0.09 mm, then $e = 0.09$ mm/0.002 mm = 45. Since the "C" scale is used, $E = 100$. Therefore, the Rockwell hardness number of the steel sample, HRC = 100 − 45 = 55.

Example Problem 8.6

One is interested in measuring the Rockwell hardness number of a copper specimen using the B scale. If the depth of indentation was found to be 0.15 mm, calculate the Rockwell hardness number of the specimen.

Solution 8.6

Using Equation 8.3 again, HR = $E - e$. Since the depth of indentation was reported as 0.15 mm, then $e = 0.15$ mm/0.002 mm = 75. Since the "B" scale was used to measure the hardness of the copper sample, $E = 130$. Thus, the Rockwell hardness number HRB = 130–75 = 55.

Example Problem 8.7

Calculate the depth of penetration of the diamond indenter in a Rockwell hardness tester if the hardness of the steel specimen was measured as HRC 60.

Solution 8.7

Since the hardness was measured using the diamond indenter and the "C" scale, the value of E in Equation 8.3 is 100. Further, HRC was measured to be 60. Therefore,

$$60 = 100 - e, \text{ or}$$
$$e = 40$$

The depth of penetration can be calculated as 40×0.002 mm = 0.080 mm.

8.8.2 Indenters and Loads

Basically two different types of indenters are used in the Rockwell hardness tester. One is a sphero-conical diamond indenter, called Brale, which is used only for hard materials such as cemented carbides, shallow case-hardened steels, nitrided steels, or hard-cast irons. It has a 120° diamond cone with a 0.2 mm radius spherical tip (Figure 8.7a). The second indenter is a 1/16″ diameter hardened steel ball (Figure 8.7b) that is used for measuring the hardness of softer materials such as nonferrous alloys. But, depending on the application, other sizes of the hardened steel balls, for example, 1/8″, 1/4″, and 1/2″, are also used.

The ASTM has revised the E18–03 (Standard Test Methods for Rockwell Hardness and Rockwell Superficial Hardness of Metallic Materials) to allow the use of

(a) (b)

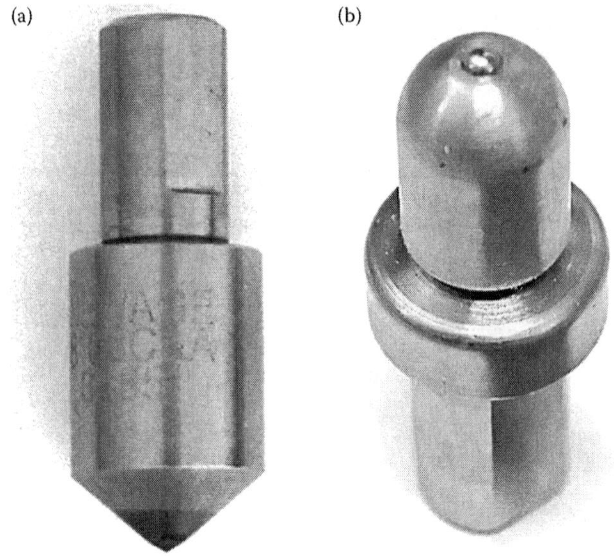

FIGURE 8.7 The two types of indenters used in Rockwell hardness testers. (a) The Brale 120° sphero-conical diamond indenter and (b) 1/16″ hardened steel ball indenter. (Photos courtesy of Newage Testing Instruments, Inc. With permission.)

tungsten carbide balls in addition to hardened steel balls. ASTM plans to eliminate the use of hardened steel balls completely. During the transition period, it is recommended that the Rockwell hardness values are reported with the scale designation followed by the letter "S" to indicate the use of a steel ball or the letter "W" to indicate the use of a tungsten carbide ball.

Rockwell hardness tests comparing the use of steel and tungsten carbide balls have shown different results. For example, depending on the material tested and its hardness level, Rockwell B scale tests using a tungsten carbide ball indenter have given results up to one Rockwell point lower than when a steel ball indenter is used. National Institute of Standards and Technology (NIST) maintains the national Rockwell hardness standards for the United States.

8.8.3 Rockwell Scales

Since the Rockwell hardness tester is used for measuring the hardness of a variety of materials, a large number of scales are employed. The important differences amongst these scales are the major loads and the type of indenter used. Based on this combination, the scales have been classified as listed in Table 8.4. Even though as many as 15 scales have been listed, only two of them, B and C, are most commonly used. For example, the Rockwell C scale is used for hard metals such as heat-treated steels. The hardness of a steel file will be about R_C 65, a steel spring R_C 45, a good knife between R_C 52 and 58.

TABLE 8.4
Type of Indenter, Major Load, and Typical Applications of the Different Rockwell Scales

Scale Symbol	Indenter	Major Load (kg)	Dial Figures	Typical Applications
A	Brale	60	Black	Cemented carbides, sheet steel, shallow case-hardened steel
B	1/16″ dia steel ball	100	Red	Copper alloys, aluminum alloys, annealed low-C steels, malleable iron
C	Brale	150	Black	Titanium alloys, hardened steels, deep case-hardened steel, cast irons, and materials harder than R_B 100
D	Brale	100	Black	Thin but hard steel, medium case-hardened steel
E	1/8″ dia steel ball	100	Red	Cast iron, aluminum and magnesium alloys, bearing metals
F	1/16″ dia steel ball	60	Red	Annealed copper alloys, thin soft sheet materials
G	1/16″ dia steel ball	150	Red	Malleable cast irons, beryllium copper, phosphor bronze, gun metals
H	1/8″ dia steel ball	60	Red	Soft materials, high-pearlitic alloys, aluminum, lead, zinc
K	1/8″ dia steel ball	150	Red	Aluminum and magnesium alloys
L	1/4″ dia steel ball	60	Red	Thermoplastics
M	1/4″ dia steel ball	100	Red	Thermoplastics
P	1/4″ dia steel ball	150	Red	Thermosetting plastics
R	1/4″ dia steel ball	60	Red	Very soft plastics and rubbers
S	1/2″ dia steel ball	100	Red	—
V	1/2″ dia steel ball	150	Red	—

Note: The Minor Load is always 10 kg (for All the Scales); Brale—120° sphero-conical diamond indenter; 1″ = 25. 4 mm; For scales K through V, use the smallest ball size and heaviest load that will not give an anvil effect, that is, the anvil, on which the specimen is placed, also contributes to the resistance of penetration of the indenter.

Diamond indenters are not calibrated for values below 20 on the C scale and therefore the C scale should not be used when the readings fall below this level; the B scale should be used for these softer materials. As an example, soft iron may read from 0 to 10 on the C scale, but would read from 40 to 50 on the B scale. If the 1/16″ diameter steel ball is used on materials harder than R_B 100, there is a danger that the ball will be flattened. If values above 100 are obtained, the next heavier load or next smaller indenter should be used.

There is no upper limit to the hardness of a material that can be tested with a diamond indenter. However, Rockwell C scale should not be used on tungsten carbide, because the material will fracture or the diamond indenter life will be reduced considerably. The Rockwell A scale is the accepted scale in the carbide products industry.

In summary, it is important to use the correct type of scale (load and indenter) for the given material. For example, a B scale (with 100 kg load) should not be used for materials harder than R_B 100 and the C scale (with 150 kg load) should not be used for materials softer than R_C 20.

If one is not clear about what scale to use, then a decision has to be made based on the type of material (use Table 8.4 to choose the appropriate scale), specimen thickness (use superficial hardness scales for very thin specimens), and scale limitations discussed above. In the worst scenario, one could use a scale that appears to be right and, based on the test result, decide if it is really right for the type of material under consideration.

8.8.4 Superficial Testing

The above-described load and indenter combinations cannot be used for all types of section thickness of the samples. For example, examine the underside of a thin section of a specimen after you conduct the hardness testing using one of the above combinations of load and indenter. If a bulge or an impression of the indenter is noticed, then it suggests that the anvil has also contributed to the resistance of penetration of the indenter and therefore the measured hardness value is not the correct hardness of the specimen. In such a situation, the superficial Rockwell hardness test should be used. If the impression can still be seen after the superficial test, then a lighter load should be used. A minor load of 3 kg and a major load of 30 kg are recommended for most superficial testing. Superficial testing can also be used for case-hardened and nitrided steel having a very thin case. (It may be quickly realized that the Brinell hardness testing is not suitable to measure the hardness of thin sections of samples.)

In all cases of superficial testing, a minor load of 3 kg is used; the major load may be 15, 30, or 45 kg. A Brale indenter is used for superficial testing of hard materials (similar to C scale); but, it is marked N for thin sections of specimens and the loads used are different. Scales marked T are used for softer materials (similar to B scale) and scales marked W, X, and Y are used for very soft materials. The different scales used in superficial testing are listed in Table 8.5.

8.8.5 Hardness Designation

The Rockwell hardness number has no units. The hardness numbers in Rockwell testing are expressed as a combination of a hardness number and a scale symbol. Once the scale is identified, the load and indenter used are also known. Thus, a well-structured Rockwell hardness number looks like HRB 80, which means that a Rockwell hardness number of 80 was obtained using the B scale. Once the B scale is mentioned, the minor and major loads are known as 10 and 90 kg, respectively and

TABLE 8.5
Type of Indenter, Major Load, and Typical Applications of the Different Rockwell Superficial Scales

Scale Symbol	Indenter	Major Load (kg)	Typical Applications
15 N	Brale	15	Similar to C scale, but for thin sections
30 N	Brale	30	Similar to C scale, but for thin sections
45 N	Brale	45	Similar to C scale, but for thin sections
15 T	1/16″ dia steel ball	15	Similar to B scale, but for thin sections
30 T	1/16″ dia steel ball	30	Similar to B scale, but for thin sections
45 T	1/16″ dia steel ball	45	Similar to B scale, but for thin sections
15 W	1/8″ dia steel ball	15	Used for very soft materials such as thin sheets, foils, coatings, surface heat treatments, polymers, including their surface treatments
30 W	1/8″ dia steel ball	30	
45 W	1/8″ dia steel ball	45	
15 X	1/4″ dia steel ball	15	
30 X	1/4″ dia steel ball	30	
45 X	1/4″ dia steel ball	45	
15 Y	1/2″ dia steel ball	15	
30 Y	1/2″ dia steel ball	30	
45 Y	1/2″ dia steel ball	45	

Note: The Minor Load is always 3 kg (for all the scales).

that 1/16″ diameter steel ball is the indenter. In other words, complete details of the test conditions are known. Sometimes, people use other ways of designating the Rockwell hardness number, for example, R_C 65. This means that the hardness obtained was 65 using the C scale. Thus, HRB and HRC or R_B and R_C, followed by the hardness number, are the common ways of designating the Rockwell hardness numbers. Since a number of scales are used in the Rockwell tester, it is very important to designate the scale that was used to determine the hardness number.

The superficial hardness number has been designated in the literature in a slightly different way than the regular hardness number. The superficial hardness number is prefixed by the major load and the letter representing the scale. For example, 30N78 represents a superficial hardness number of 78 measured using the N scale and a load of 30 kg. Other superficial scales, such as T, W, X, and Y are also prefixed with the major load when recording the hardness. In this connection, it may be noted that the 1/16″ diameter hardened steel ball indenter should not be used on materials harder than 30T82. However, in order to avoid confusion in the way the hardness number is designated in the regular and superficial methods, it may be a good practice to designate the superficial hardness number also as, for example, 80HR30N. This means that the superficial Rockwell hardness number is 80 measured using a load of 30 kg and the N scale.

8.8.6 Anvils

Depending on the size and shape of the test specimen, which cannot be placed in a stable condition on the test platform in the Rockwell hardness tester, different types of anvils are used to support the test specimens (Figure 8.8). Flat anvils are the most common, since the specimens most frequently have a flat surface. V-type anvils are used for round specimens. A spot anvil is used when the tester is being checked on a Rockwell test block. The diamond spot anvil (similar to the spot anvil, except that a diamond ground and polished to a flat surface is set into the spot) is used only with the superficial hardness tester and only in conjunction with the steel ball indenter for testing soft metals. Special anvils can be fabricated to accommodate specific work piece configurations. The most important criterion in selecting the anvil design is that the test specimen should remain rigid and prevent movement of the specimen during testing.

A jack rest is used for supporting long, heavy parts such as shafts. It consists of a separate elevating screw and anvil support similar to that on the tester. Without adequate support, overhanging work can damage the indenter and produce inaccurate readings.

8.8.7 Precautions

If the surface of the anvil is rough or gouged, or it is worn, or ground to a taper, then the measured hardness values are likely to be in error. One must ensure that there are also no dust particles, grease, or other impurities on the surface, which will cause deflection of the specimen sitting on it. If so, the measured hardness value is again wrong.

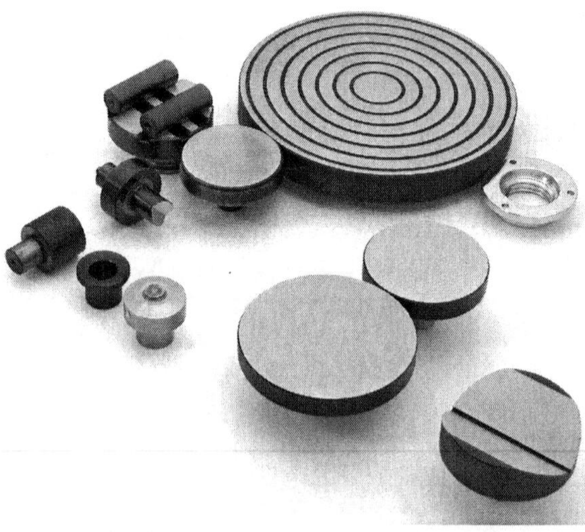

FIGURE 8.8 Different types of anvils used to support specimens during Rockwell hardness testing. (Photo courtesy of Newage Testing Instruments, Inc. With permission.)

Accurate and reproducible hardness values can be obtained only when the specimen surface is smooth and flat. A rough or ridged surface caused by coarse grinding will not produce results that are as reliable as from a smooth surface. An often overlooked point is the surface preparation of the underside of the test specimen; it is necessary that the underside is also flat.

Accurate and consistent test results can only be obtained when the test surface is shiny. If you are only interested in knowing roughly how hard a part is, then a rough surface will work. Even though the Rockwell method begins its hardness measurement beneath the surface of the part, the inherent variability of a rough surface can and will cause inconsistent results. Scale and decarburized surfaces also yield erroneous results. In these cases, the surface scale must be removed and the specimen must be ground to below the decarburized layer before conducting a hardness test. Errors can also result from testing curved surfaces. This effect may be eliminated by grinding a small flat spot on the specimen. (This technique will be useful, e.g., in measuring the hardness on a round Jominy end-quenched sample.)

8.8.8 Advantages and Disadvantages

The advantages of the Rockwell hardness method are that the test can be performed rapidly and that the hardness number can be directly read out. It is also free from personal errors involved in microscopic measurement of the indentation size as in a Brinell (or the Vickers) hardness test. Due to the small size of the indentation, the finished products can generally be tested without damage. The disadvantages include many arbitrary nonrelated scales and possible effects from the specimen support anvil. Brinell and Vickers methods do not suffer from this effect.

8.9 VICKERS HARDNESS TESTING

The Vickers hardness test was developed in 1924 by Smith and Sandland at Vickers Ltd as an alternative to the Brinell method to measure the hardness of materials. Since then, this method has been the most popular among scientific community. Because of the shape of the indenter employed and the wide latitude available in the applied loads, this method can be used to measure the hardness of massive samples of the softest and hardest metals and alloys as well as thin sheets or surface hardness of case-hardened materials or coatings. The Vickers instrument is a precision unit that normally requires care in operation, rather than skill on the part of the operator. The hardness tester has been semiautomatic in operation, in that after the specimen surface is brought into contact with the indenter, the preset load is applied for some definite length of time, after which the load is automatically removed. Present-day instruments (Figure 8.9) are a little more automated in that the size of the indentation and the hardness numbers can be directly read off on an LCD screen and the results printed.

FIGURE 8.9 Photograph of a commercial Vickers hardness tester. (Photo courtesy of Newage Testing Instruments, Inc. With permission.)

8.9.1 Principle

In this method, a square-based diamond pyramid indenter is pressed into the smooth polished surface of the specimen. The load is applied for a predetermined time and removed. The average value of the length of the two diagonals of the square indentation is measured and the Vickers hardness number (VHN) calculated as the ratio of the applied load to the surface area of the indentation, using the equation

$$\text{VHN} = \frac{P}{D^2/\left[2\sin\left(136^\circ/2\right)\right]} = 1.854\frac{P}{D^2} \tag{8.4}$$

where P is the applied load in kilograms and D is the arithmetic mean of the two diagonal lengths of the indentation in millimeters. Standard conversion tables relating the average diagonal length and hardness number for a given load are available. Part of such a table is reproduced in Table 8.6.

Note that the size of the indentation in the Vickers hardness test is a couple of hundred micrometers, depending on the hardness of the material. It is usually less than a millimeter. But, in the Brinell hardness test, the size of the indentation is a few millimeters. That is, the indentation size is an order of magnitude larger in the Brinell hardness test.

Equation 8.4 is theoretically valid only when the impression has been made with a perfect indenter—a circumstance that is never fully realized in actual practice. However, the error attending the use of Equation 8.4 when the indenter is not perfect is insignificant, provided the dimensions of the indenter are within the tolerances specified for such indenters.

TABLE 8.6
Vickers Hardness Numbers as a Function of the Average Diagonal Size Calculated using Equation 8.4 for Different Loads

Diagonal of Indentation (μm)	Load				
	1 g	10 g	1 kg	10 kg	25 kg
1	1854	18,540	—	—	—
5	74.17	741.7	—	—	—
10	18.54	185.4	18,540	—	—
15	8.242	82.42	8242	—	—
20	4.636	46.35	4635	—	—
25	2.967	29.67	2967	—	—
30	2.060	20.60	2060	—	—
35	1.514	15.14	1514	15,140	—
40	1.159	11.59	1159	11,590	—
45	—	9.157	915.7	9157	—
50	—	7.417	741.7	7417	18,540
60	—	5.151	515.1	5151	12,877
70	—	3.884	388.4	3884	9710
80	—	2.897	289.7	2897	7242
90	—	2.289	228.9	2289	5722
100	—	1.854	185.4	1854	4635
110	—	1.533	153.3	1533	3832
120	—	1.228	122.8	1228	3070
130	—	1.097	109.7	1097	2742
140	—	—	94.6	946	2365
150	—	—	82.4	824	2060
160	—	—	72.4	724	1810
170	—	—	64.2	642	1605
180	—	—	57.2	572	1430
190	—	—	51.4	514	1285
200	—	—	46.4	464	1160
210	—	—	42.0	420	1050
220	—	—	38.3	383	957
230	—	—	35.0	350	875
240	—	—	32.2	322	805
250	—	—	29.7	297	742
260	—	—	27.4	274	685
270	—	—	25.4	254	635
280	—	—	23.6	236	590
290	—	—	22.0	220	550
300	—	—	20.6	206	515

Example Problem 8.8

Calculate the VHN of a material if the measured average length of the diagonals is 250 μm and the load applied is 15 kg.

Solution 8.8

The VHN can be calculated using Equation 8.4:

$$\text{VHN} = 1.854\frac{P}{D^2}$$

where P is the applied load in kg and D is the average length of the diagonals in mm. It is given that $D = 250\ \mu\text{m} = 0.25$ mm, and $P = 15$ kg. Therefore,

$$\begin{aligned}\text{VHN} &= 1.854\frac{15\,\text{kg}}{(0.25\,\text{mm})^2}\\ &= 445\ \text{kg/mm}^2\end{aligned}$$

8.9.2 Indenter

Only one type of indenter is used in the Vickers hardness testers (Figure 8.10). It consists of a diamond in the form of a square-based pyramid with an included angle of 136° between opposite faces of the pyramid. This angle is chosen in order to obtain results approximating to those of the Brinell ball scale, which was well established by that time. It was known that a nearly proportional hardness scale would be obtained in the Brinell test, particularly at high-hardness values, if the load and ball diameter were always adjusted so that the diameter of the indentation would be 0.25–0.50, or an average of 0.375, of the diameter of the ball. If tangents were drawn to the edges of a Brinell indentation whose diameter is 0.375 times the Brinell ball indenter, the angle will be 136° (Figure 8.11), equivalent to the included angle between opposite faces purposely specified in the Vickers indenter.

8.9.3 The Test

The actual test consists of preparing the surface of the test specimen smooth and flat. If one is interested in looking at the microstructure, the specimen could also be etched (but not too deep to cause surface irregularities) to reveal the microstructural features. The prepared specimen is placed on the anvil, which is slowly raised toward the indenter assembly unit (that contains the indenter and microscope objectives of different magnifications) until the specimen surface is in focus at the desired magnification. Once the proper area for indenting has been chosen, the microscope objective turret is moved aside and the indenter is brought into

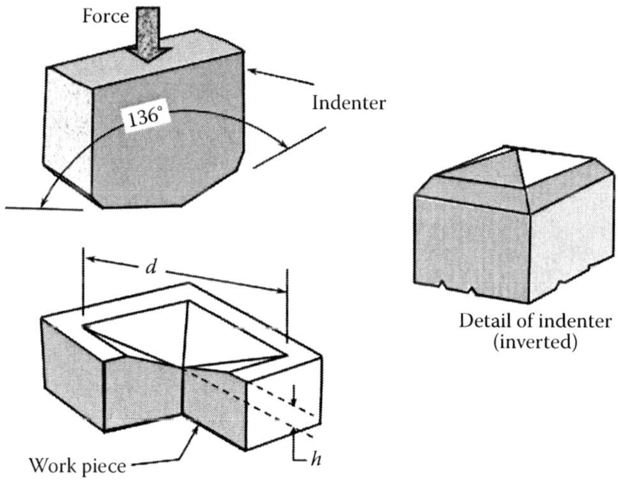

FIGURE 8.10 Sketch of the Vickers indenter showing the shape and the included angles.

position such that it is just above the area of interest. (*Note*: Do not adjust either the position of the specimen or that of the indenter at this stage. The machine is calibrated in such a way that when the load is applied the indenter will directly make the indentation at the chosen place.) Apply the predetermined load for the chosen dwell time. The loads that can be applied range from 1 to 120 kg and the dwell

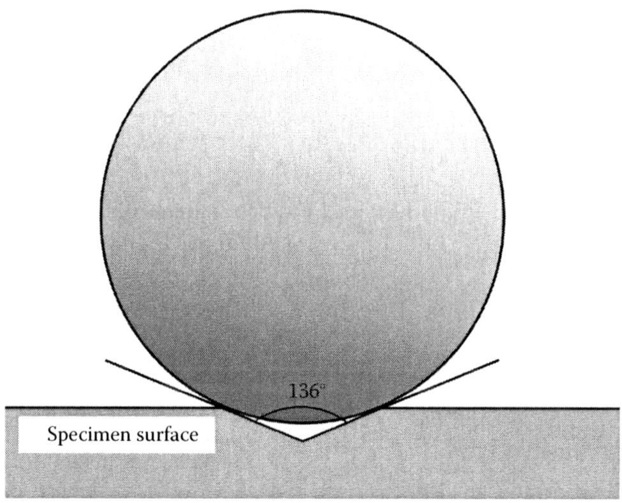

FIGURE 8.11 Geometry of the Brinell indentation showing equivalence of the Brinell and Vickers indentations. When $d/D = 0.375$ (d is the diameter of the indentation and D is the diameter of the indenter) for the Brinell indentation, the included angle of the tangents drawn to the edges of the Brinell indentation is 136°, equivalent to the included angle between opposite faces of the Vickers indenter.

times are about 10–30 s. Once the indentation is made, the specimen supporting anvil (without disturbing the position of the specimen) is lowered by means of a hand wheel. A special measuring microscope, attached to the side of the instrument, is then brought into position. The measuring microscope is focused on the indentation, and by bringing one of the knife edges to coincide with one edge of the diagonal and by moving the other knife edge to the other end of the diagonal, the length of the diagonal is measured, and noted (Figure 8.12). The length of the other diagonal is also measured in a similar way by rotating the microscope eyepiece set normal to the previous position. The average of the two diagonal lengths is calculated and then the VHN is computed using Equation 8.4. Instead of actual calculations, one can conveniently use standard conversion tables relating the average diagonal length and hardness for a given load. A partial list of the Vickers hardness values as a function of the diagonal length is reproduced in Table 8.6. Modern Vickers hardness testers directly provide the hardness values on the screen and the results could also be printed.

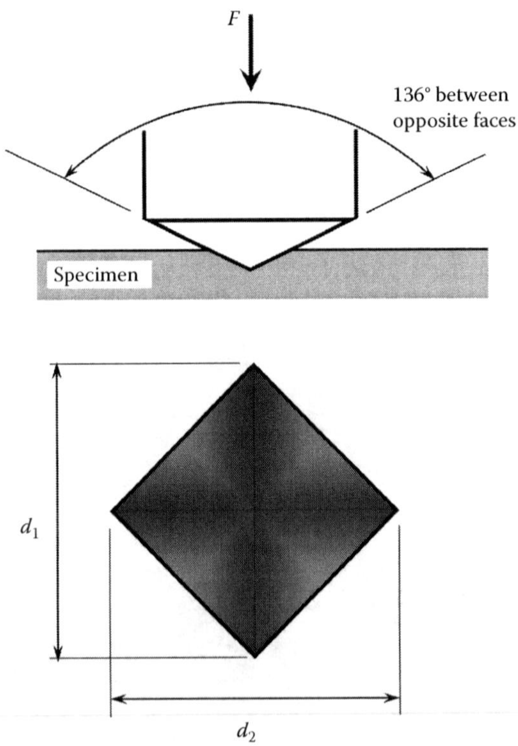

FIGURE 8.12 Technique to measure the length of the diagonal of a Vickers indentation. One of the knife edges is made to coincide with one edge of the diagonal and the second knife edge is then brought to coincide with the other edge of the diagonal. The distance between the two knife edges is then measured using the scale on the eye piece.

8.9.4 Hardness Designation

The Vickers hardness number is usually abbreviated as VHN. In the literature you would also find it referred to as H_V. The most appropriate designation would be as "600HV/10" meaning a VHN of 600 was obtained using a 10 kg force. Another designation some people prefer to use is "HV 5" indicating that the VHN at a load of 5 kg is being reported. A complete description, however, would be like "200 HV 20/15" which means that a VHN of 200 was obtained with a 20 kg load and a dwell time of 15 s. Since a diamond pyramid indenter is used to measure the hardness, this hardness is also referred to as diamond pyramid hardness (DPH) or diamond pyramid number (DPN). It has been noted that practically identical hardness numbers could be obtained on a uniform material even if several different load settings were used, that is, the area of the indentation is proportional to the load applied.

8.9.5 Precautions

Since the size of indentation in Vickers hardness testing is very small (on the order of a couple of hundred micrometers), it is important to have a very smooth surface. Uneven surfaces do not produce perfect diamond-shaped indentations and therefore, the error in the measurement of the two diagonal lengths is large. Figure 8.13 shows the distortions in the diamond pyramid indentations due to elastic effects. Pincushion distortion arises due to sinking of material around the flat faces of the pyramid. On the other hand, barrel distortion occurs due to ridging of the material around the faces of the indenter. Even though the lengths of the diagonals are the same in the two perpendicular directions in all the images, whether distortion had occurred or not (Figure 8.14), the areas of indentation are not the same in the three cases. For example, when pincushion distortion occurred (Figure 8.14a), the area is smaller than it would have been for a perfect indentation (Figure 8.14b). Similarly,

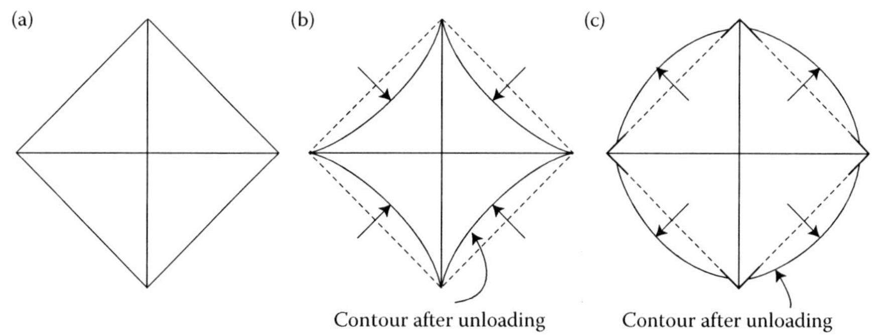

FIGURE 8.13 Distortion of diamond pyramid indentations due to elastic effects. (a) Perfect indentation. (b) Pincushion distortion which occurs due to sinking of the material around the flat faces of the pyramid. (c) Barrel distortion of the indentation, which occurs due to ridging of the material around the faces of the indenter.

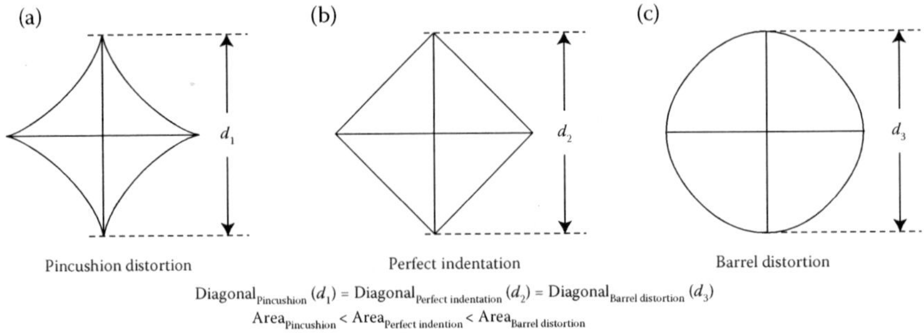

FIGURE 8.14 Distortion in the Vickers indentations due to elastic effects. Even though the lengths of the diagonal are the same in all the three indentations, note that the areas are different. The surface area of the indentation with the pincushion distortion (a) is smaller than when the indentation is perfect (b). The area of the indentation with the barrel distortion (c) is larger than when the indentation is perfect (b).

for the barrel distortion (Figure 8.14c), the area is larger than for a perfect indentation. Even though the hardness number appears to be the same in all the three cases, because the average length of the diagonals is the same, these are not correct values because the surface areas of the indentation are different. This clearly brings out the fact that Equation 8.4 is valid only for a perfect indentation. Surface imperfections and oxide scales should be removed. Decarburized or recarburized or nitrided layers should be removed by grinding, unless they are of primary interest, because hardness measurements on such surfaces will not be representative of the entire specimen. Round or curved surfaces may be tested, but it is desirable to have a flat surface by either filing or grinding.

8.9.6 Advantages and Disadvantages

The main advantages of Vickers hardness testing are that extremely accurate readings can be taken, and just one type of indenter is used for all types of metals and surface treatments. This is different from other tests (Brinell and Rockwell) where different loads, scales and/or indenters are used. The Vickers hardness testing leaves only tiny indentation marks on the surface that are not usually a problem for production items, making the technique suitable for quality control. The Vickers hardness test is used widely in research work because it provides a continuous scale of hardness for a given load from very soft to very hard metals whereas in the Brinell and Rockwell tests, it becomes necessary to change either the indenter and/or the load. However, the Vickers machine used to be a floor-standing unit that is more expensive than the Brinell or Rockwell machines. At present, bench models of the Vickers machine are available. The Vickers test is not suitable for routine testing as it requires careful surface preparation and is slow on account of microscopic measurement of the diagonal lengths. Further, measurement of the diagonal length is also subject to human error.

8.10 MICROHARDNESS TESTING

Most engineering materials are polycrystalline in nature and polyphasic in character, that is, they contain many grains and more than one phase or microconstituent. It is possible that the hardness of these microconstituents varies widely. Conventional hardness tests provide only the average hardness value and do not provide any indication of the hardness of the individual microconstituents/phases. The mechanical behavior, especially the wear resistance, of a material can be rationalized in terms of the hardness of the individual microstructural features.

Microhardness testing is of great importance for heterogeneous materials when the hardness of an individual phase in a multiphase microstructure needs to be measured. Such examples include:

- General multiphase microstructures consisting of phases of different hardness values
- Cast irons (which contain soft graphite dispersed in a harder matrix)
- Powder metallurgy parts (which contain residual porosity)
- Newly developed cellular materials which contain lot of porosity

Microhardness testing is useful in research and development, materials testing, and quality control programs. It is used to measure the hardness of small parts (precision parts, screws, metal bearings, tiny watch gears, hair springs, etc.), thin foils and wires, and for testing materials that cannot withstand heavy loads, for example, plastic sheet and rod, dental materials, lacquers, paints, and other coatings. Microhardness testing is generally carried out using either a Vickers hardness tester or a Knoop (pronounced *nūp*) hardness tester. Figure 8.15 shows the photograph of a commercial microhardness tester.

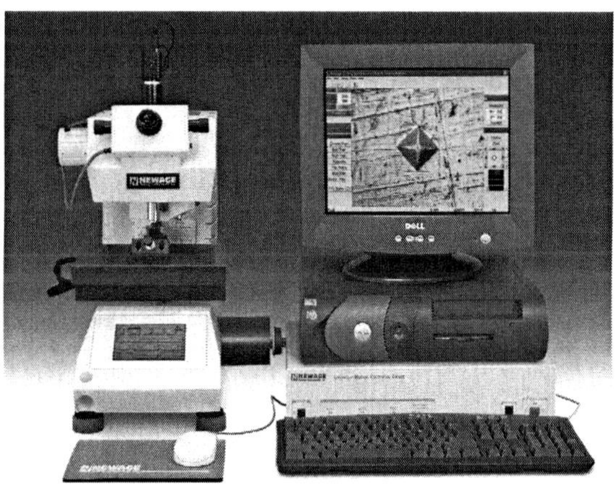

FIGURE 8.15 Photograph of a commercial Vickers microhardness tester. (Photo courtesy of Newage Testing Instruments, Inc. With permission.)

8.10.1 Principle

The term microhardness test usually refers to static indentations made with loads not exceeding 1 kg. In this test also, an indenter is pressed into the polished surface of a specimen and the hardness number is calculated as the ratio of the load applied to the surface area of the indentation. The major difference between a regular hardness test and microhardness test is that the loads used are very small in microhardness testing; consequently the size of the indentation also is small.

The equation used to calculate the microhardness number using the Vickers microhardness tester is the same as that used in the regular hardness measurement, namely, VHN = 1.854 P/D^2. On the other hand, the Knoop hardness number (KHN) is calculated as the ratio of the load applied, P (kg) to the unrecovered projected area of the indentation, A (mm^2), using the equation

$$\text{KHN} = \frac{P}{A} = \frac{P}{CL^2} \tag{8.5}$$

where P is the load applied in kilograms, L is the measured length of the longitudinal diagonal of the indentation in millimeters, and C is the Knoop indenter constant (=0.07028), relating projected area of the indentation (A) to the square of the length of the long diagonal (L^2), that is, $C = A/L^2$. [The projected area of the indentation can be visualized as consisting of 4 triangles, each with a base of $L/2$ and a height of $W/2$, where L and W are the measured long and short diagonals of the indentation. Thus, the projected area of the indentation, A is given by

$$A = 4 \times \frac{1}{2} \times \frac{L}{2} \times \frac{W}{2} = 0.5\,LW \tag{8.6}$$

Since we know from the geometry of the indenter that $W/L = 0.14056$, the projected area is $A = 0.5\ LW = 0.5 \times L \times 0.14056L = 0.07028\ L^2$. Hence, $C = 0.07028$]. Standard tables are available relating KHN and L for a given load. The use of these tables is usually a more convenient way to obtain the hardness values than to actually calculate them using Equation 8.5.

Example Problem 8.9

Calculate the Knoop hardness number of a particular phase in a specimen when the length of the long diagonal was measured as 100 μm and a load of 500 g was applied.

Solution 8.9

The Knoop hardness number can be calculated using Equation 8.5, namely, KHN = P/CL^2.

It has been mentioned that $P = 500$ g = 0.5 kg and $L = 100$ μm = 0.1 mm. The Knoop indenter constant $C = 0.07028$. Substituting these values into the above equation, we get

$$\text{KHN} = \frac{P}{CL^2} = \frac{0.5\text{ kg}}{0.07028 \times (0.1\text{ mm})^2} = \frac{0.5\text{ kg}}{0.0007028\text{ mm}^2} = 711\text{ kg mm}^{-2}$$

8.10.2 Indenters

Two types of indenter are used in microhardness testing. One is the Vickers diamond pyramid (of the same shape as in a regular Vickers hardness tester) or the Knoop-elongated diamond pyramid.

The Knoop indenter is a diamond ground to a pyramidal form that produces a diamond-shaped indentation, having a ratio between the long and short diagonals of L/W as 7.114 (or $W/L = 0.14056$). The depth of indentation is about 1/30 of its length. (In the case of Vickers indentation, the depth of indentation is about 1/7 of the diagonal length.) The shape of the indenter and the included angles of the Knoop indenter are shown in Figure 8.16.

8.10.3 The Test

The procedure of testing is very similar to that of the standard Vickers hardness test, except that it is done with smaller loads on a microscopic scale with higher-precision instruments. The loads used are in the range of 1–2000 g. The surface being tested

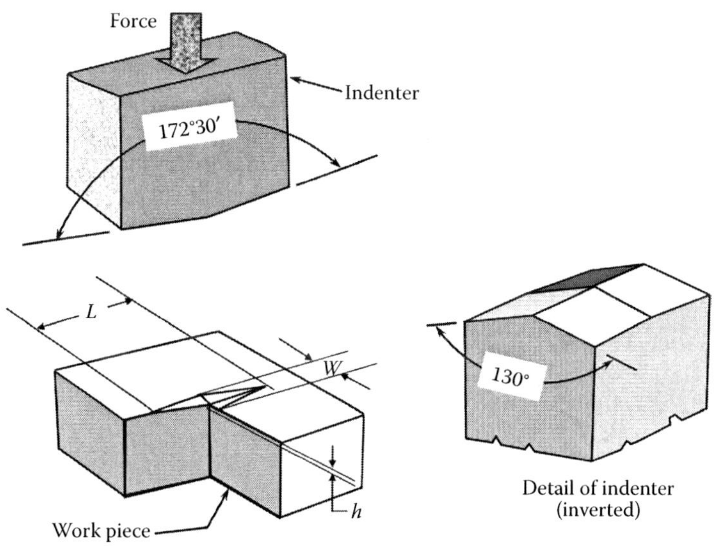

FIGURE 8.16 Schematic of the Knoop indenter used for microhardness testing. The shape of the indenter and the included angles are shown.

generally requires a metallographic finish with the microstructural features clearly visible. It should be noted that the smaller the load used, the higher is the surface finish required. A variety of fixtures are available for holding and clamping the work pieces for microhardness testing.

The proper phase (or area), whose hardness has to be measured, that is, where the indentation has to be made, is first chosen. From this point onward, the same procedure employed for regular Vickers hardness testing is followed. Precision microscopes are employed to measure the size of the indentations; these usually have a magnification of around 500× and measure to an accuracy of ±0.5 μm. Also with the same observer, differences of ±0.2 μm can usually be resolved. But, considerable care and experience are necessary to obtain this accuracy. As an example, Figure 8.17 shows the Knoop indentations made in a D2 tool steel sample. Note that the indentation in the carbide particle (white diamond-shaped object) is smaller in size than that in the matrix, suggesting that the hardness of the carbide phase (1930 KHN) is much higher than that of the matrix (801 KHN).

The main difference between the Vickers and Knoop microhardness measurements is that while both the diagonals of the indentation are measured and their average calculated in the Vickers test, only the long diagonal length is measured in the Knoop test. Once the diagonal length is measured, Equation 8.5 can be used to calculate the Knoop hardness number, or Equation 8.4 to calculate the VHN. Use of Tables of relating hardness to the diagonal length is usually a more convenient way to look up the KHN or VHN values from the measurements.

Depending on the load, the microhardness readings may be affected by indenter shape, vibrations, friction within the tester, surface preparation, cold working, indentation size, elastic recovery, bulge formation, and other indentation characteristics.

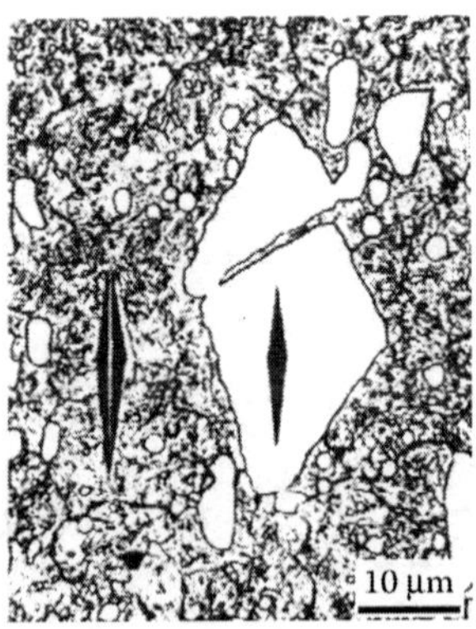

FIGURE 8.17 Microstructure of a D2 tool steel containing the Knoop indentations made both in the Cr-V-carbide precipitate (white constituent) and the matrix phase (darker constituent). Both the indentations were made with a 50 g load. Note that the indentation in the harder carbide phase is smaller than that in the softer matrix phase. The measured hardness for the carbide and the matrix phases are 1930 KHN and 801 KHN, respectively. (The specimen was polished and etched in Vilella's reagent.) (Photo courtesy of ASM International, Materials Park, OH.)

8.10.4 Hardness Designation

Like in the Vickers test, the Knoop hardness number is also expressed fully as "200 HK 500/15" which means that the hardness value measured is 200 with a Knoop tester using a load of 500 g and 15 s dwell time. A common way of expressing the number is with the prefix KHN.

8.10.5 Comparison between Vickers and Knoop Microindentation Tests

Comparing the indentations made with Knoop and Vickers diamond pyramid indenters for a given load and test material, it may be noted that:

- Vickers indenter penetrates about twice as deep as the Knoop indenter.
- Vickers indentation diagonal is about 1/3 of the length of Knoop long diagonal.
- Vickers test is less sensitive to surface conditions than the Knoop test.
- Vickers test is more sensitive to measurement errors than the Knoop test.
- Vickers test is the best for small rounded areas.
- Knoop test is the best for small elongated areas.
- Knoop test is good for very hard and brittle materials and very thin sections.

8.10.6 Precautions

The measured hardness is dependent on the applied load, especially when the loads are very low. However, from a practical standpoint, as along as a single load is used throughout a series of tests, the load dependence is not significant. Therefore, if any comparison of hardness values is to be made, it is important to consider the load dependence. As a general rule, Knoop hardness numbers with loads of less than 500 g and VHN with loads of less than 1000 g are invalid unless load dependence is considered.

8.11 NANOINDENTATION TESTING

The idea of nanoindentation arose from the realization that an indentation test is an excellent way to measure the mechanical properties of small volumes of materials. In principle, if a very sharp indenter is used, the size of the indentation is very small and consequently the volume of the material tested is also small. Under these conditions, the size of the indentation is so small that it will be difficult to see it without the help of a microscope. This problem is overcome by depth-sensing indentation methods in which the load and displacement of the indenter (vertically down into the specimen) are recorded during the indentation process. These data are analyzed to obtain the contact area, and from that the mechanical properties, without the necessity of actually seeing the indentation.

Nanohardness testers are precision depth-sensing instruments. They measure not only the hardness but also other mechanical properties such as Young's modulus, time-dependent creep behavior, and fracture toughness of the sample material. These properties can be obtained on any type of material—soft, hard, brittle, or ductile.

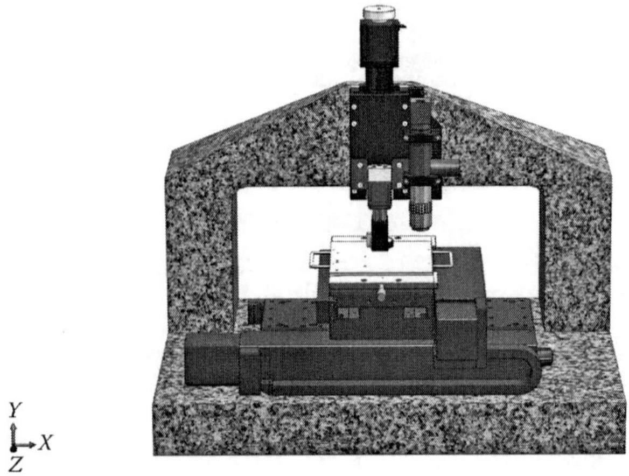

FIGURE 8.18 Photograph of TI 950 commercial nanoindenter. (Photo courtesy of Hysitron, Inc. With permission.)

These machines are relatively new and a number of commercial instruments are available. Figure 8.18 shows a photograph of a commercial nanoindenter.

8.11.1 Indenters and Loads

Three different types of indenters are used in these hardness testers—spherical, Vickers, and Berkovich. A spherical indenter is used to characterize the stress–strain behavior of materials. This type of indenter is especially useful when the specimen is homogeneous and an average response of the specimen needs to be obtained (since the size of the indentation is relatively large). The Vickers indenter is similar to the regular indenter used in a Vickers hardness tester and is a square-based diamond pyramid with an included angle of 136°. The hardness can be calculated in exactly the same way as in regular Vickers hardness testing by measuring the average length of the two diagonals and using Equation 8.4 or by looking up the relevant tables. The standard Berkovich indenter also is a pyramidal indenter (like the Vickers indenter) but with only three sides and an included angle of 141.9°. These two types of (pointed) indenters (Vickers and Berkovich) are used when information from a highly localized area needs to be obtained. In both the Vickers and Berkovich indentations, the ratio of the actual surface area to indentation depth is the same. Thus, one could use either a sharp indenter or a spherical indenter depending on the purpose.

The loads used in nanoindentation are relatively low. They range from a few milliNewtons up to 1 N. The loading rate also plays an important role; usually it is about 100 mN/s.

8.11.2 The Test

In this test, a preset load is continuously applied to an indenter tip, placed normal to the specimen surface. The load is then gradually withdrawn until partial or full relaxation of the sample has occurred. The load and displacement are continuously

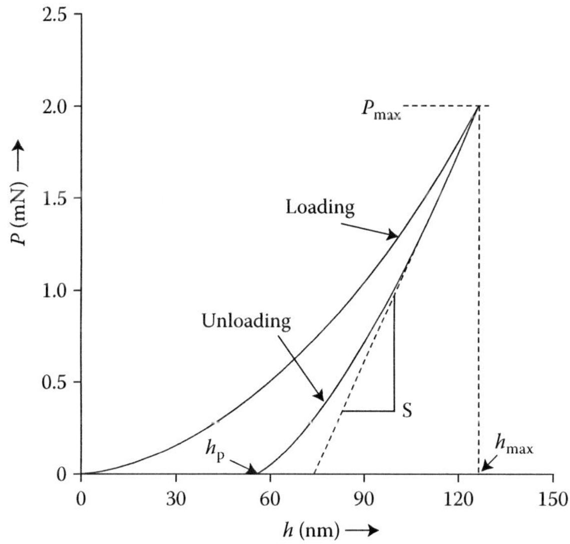

FIGURE 8.19 Typical load (P)–displacement (h) curve obtained during nanoindentation testing. The different parameters used in the analysis of the results are also indicated in the figure.

recorded throughout this process to produce a load–displacement curve, as shown in Figure 8.19.

One could also measure the hardness of a material by allowing the indenter to penetrate into the specimen surface and measuring the resistance offered by the specimen to plastic deformation. The hardness number is automatically calculated using the appropriate formula.

8.11.3 Analysis of Data

Referring to Figure 8.19, it may be noted that it consists of a loading curve that characterizes the resistance of the material against penetration of the indenter and the initial part of the unloading curve that describes the elastic recovery of the indent. When the load is removed, the material tries to regain its original shape but due to plastic deformation, its complete recovery is not possible. Thus, if the specimen undergoes only elastic deformation, the unloading curve retraces the loading curve (Figure 8.20a). Materials such as diamond and quartz exhibit such a behavior. On the other hand, Figure 8.20b shows the load–displacement curve for a fused quartz specimen which undergoes plastic deformation at the loads applied.

In Figure 8.19, P_{max} is the maximum load applied at the indenter tip, h_{max} is the maximum displacement of the indenter into the material at P_{max}, S is the slope of the initial part of the unloading curve, and h_p is the plastic displacement. The mean contact pressure is taken as the indentation hardness (H) of the material and is calculated by dividing the load applied at the pointer by the projected area of contact. The elastic modulus determined from the initial slope of the unloading part of the P–h curve is termed as indentation modulus of the material.

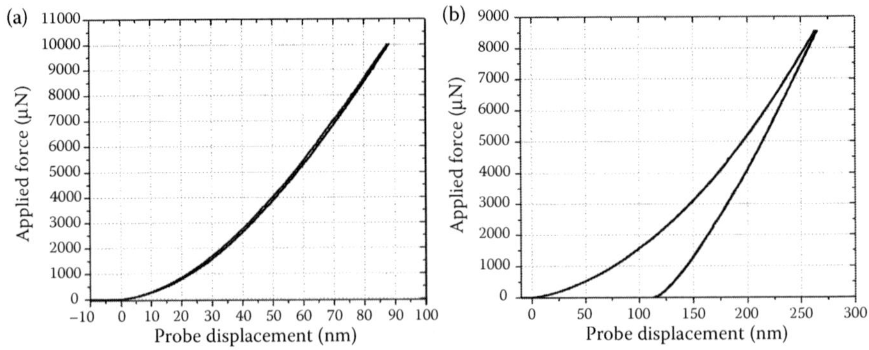

FIGURE 8.20 Load–displacement curves during nanoindentation (a) when the specimen undergoes only elastic deformation (e.g., diamond film) and (b) when the specimen undergoes elastic-plastic deformation (e.g., fused quartz). (Images courtesy of Hysitron, Inc. With permission.)

As mentioned above, one could use either a sharp indenter or a spherical indenter during nanoindentation experiments. For the case of a sharp indenter, the slope of the initial part of the unloading curve is used to determine the unloading stiffness, and the extrapolated depth, h_p, to calculate the area of contact. This is known as the Oliver–Pharr analysis.

For the case of a spherical indenter, the elastic contact between the indenter and the material surface (considered to be of infinite radius of curvature) is taken into account. From this, the relation between the load (P) and the depth of indentation (h) can be calculated as

$$P = Ch^{3/2} \tag{8.7}$$

where

$$C = \frac{(2^{3/2} E^* D^{1/2})}{3} \tag{8.8}$$

Here D is the diameter of the indenter and E^* is the reduced modulus defined as

$$E^* = \left[\left[\frac{(1 - \nu^2)}{E} \right] + \left[\frac{(1 - \nu_i^2)}{E_i} \right] \right]^{-1} \tag{8.9}$$

In Equation 8.9, E and ν represent the elastic modulus and Poisson's ratio, respectively. The subscript i refers to the indenter and the symbols without the subscript refer to the material whose properties are being measured. This is known as the Hertzian analysis. (In the case of a Berkovich indenter, e.g., Equation 8.7 will be modified as $P = Ch^2$.)

The general analysis of the curve begins with fitting the experimental loading portion of the *P–h* curve with a 3/2 power to calculate the constant *C* in Equation 8.7. Once the value of *C* is obtained, the value of E^* in Equation 8.8 can be calculated knowing the value of *D*, the diameter of the indenter. This value of E^* is then subsequently used to calculate the value of the modulus of elasticity of the material, *E* knowing the values of E_i and ν_i, the modulus and Poisson's ratio of the indenter, respectively, using Equation 8.9. Poisson's ratio of the material is usually assumed to be about 1/3.

8.12 GENERAL OBSERVATIONS

Table 8.7 summarizes the details of the indenters, loads used, and the typical indentation sizes during hardness measurement of different materials by different hardness testers. The following comments are applicable to all hardness test methods described above.

8.12.1 Reproducibility

The hardness measurements can be in error either due to improper calibration of the instrument or improper testing methods. Unfortunately, it is not practical to be able to reproduce or repeat the test result accurately. This is because the microstructure and/or material could be different, on a microscopic level, from point to point. Further, it is not possible to retest exactly the same area because an indentation has already been made and the properties of the material at that point have been altered from the original situation due to cold working of the material. The only way to check the repeatability and reproducibility of results on a hardness tester will be to study highly consistent test blocks and thus minimize the material and operator variations.

8.12.2 Surface Preparation

It is important that the sample has a smooth and flat surface and that it has parallel faces to achieve accurate and reproducible hardness values. The level of preparation depends on the magnitude of the load applied, with the caveat that the smaller the load the higher is the required surface smoothness. Thus, for microhardness and nanohardness measurements, the surface must be extremely smooth.

The surface to be tested should always be firmly supported prior to the application of loads. Only rarely does holding the part by hand suffice. Appropriate specimen supports (or anvils) must be used for specimens that are large or those which do not have flat parallel faces. In the nanohardness tester, for example, the specimen is placed on an automatic stage, which moves in the *X*- and *Y*-directions, according to a predetermined pattern. It should also be remembered that the specimen surface should not deviate from the horizontal by more than 5°. Remember that each regular Rockwell point is equal to 0.002 mm in depth and that each Superficial Rockwell point is equal to 0.001 mm.

TABLE 8.7
Summary of Details for Hardness Measurement

Hardness Test	Typical Indenters	Indent: Diagonal Length or Diameter	Indent: Depth	Load	Method of Measurement	Remarks
Brinell	10 mm diameter hardened steel ball	2.5–6.0 mm	Up to 1 mm	3000 kg for ferrous materials and 500 kg for nonferrous metals	Measure the diameter of the indentation under a microscope and read hardness from tables	Size of the indentation is large and surface may be damaged. Cannot be used for very thin samples
Rockwell	120° diamond sphero-cone or 1/16″ diameter hardened steel ball	0.1–1.5 mm	25–375 μm	Major 60–150 kg; Minor 10 kg	Read hardness directly from dial or digital display	Measure depth of penetration and not diameter. Number of scales and indenters. Superficial hardness can be measured. Most common in industry
Vickers	136° diamond pyramid	100–300 μm	0.03–0.1 mm	1–120 kg	Measure indent diagonal with a microscope; Read hardness from tables	Widespread use in scientific research. Small indent but high local stress
Microhardness	136° diamond pyramid or Knoop	40 μm	1–4 μm minimum	1–1000 g	Measure indent diagonal with a microscope; Read hardness from tables	Laboratory test to measure hardness of individual microconstituents
Nanohardness	Berkovich, Vickers, Spherical	Up to a few μm		Up to 1 N	Automatic	

8.12.3 Minimum Thickness of Test Section

The material immediately surrounding the indentation is cold worked, and the extent of cold working depends on the work hardening characteristics of the material. It has been reported that the depth of the material affected is at least 10 times the depth of the indentation. Therefore, for obtaining accurate hardness values, the minimum thickness should be at least 10 times the depth (not the diameter) of the indentation. However, this "minimum thickness" ratio of 10:1 should be regarded only as an approximation or as a guideline. If a doubt exists regarding the accuracy of the hardness value (because of insufficient specimen thickness), then the side directly beneath the indentation (the bottom surface, i.e., the surface away from the surface on which the indentation is made) should be examined. If the surface is damaged or a bulge exists (known as the "anvil effect"), then the hardness obtained is not accurate and the specimen is not sufficiently thick for the applied load. Then a lighter load should be used. It should also be remembered that the minimum thickness varies with the load applied to measure the hardness. Minimum thickness values for different hardness tests are listed in the Test Specifications.

8.12.4 Minimum Spacing between Indentations

The material surrounding the indentation gets work hardened. Therefore, if another indentation is placed within this work-hardened area, the hardness value obtained will be higher than the real value (that is obtained outside this cold-worked area). Accordingly, a minimum spacing equal to 2.5–3 times the size (not the depth) of the indentation should be maintained between two indentations. Further, an indentation should not also be placed too close to the edge of the specimen. If an indentation is made too close to the edge of the specimen, the specimen edge will bulge and the hardness number obtained will be lower than the real value. Again, a minimum distance of 2.5–3 times the size of the indentation should be maintained from the edge of the specimen.

8.13 CORRELATIONS

Since the hardness of a material represents its resistance to plastic deformation, it should be, in principle, possible to relate the hardness measured by different techniques. Further, it is sometimes necessary to test the specimen in one scale and report the hardness value in another scale, especially if comparisons have to be made with materials whose hardness was measured on a different scale. However, exact and mathematical relationships cannot be developed for such comparisons. This is essentially because different hardness measurement techniques use different loads and different shapes of indenters. This problem is exacerbated by the inherent mechanical properties of the material (elastic and cold-working behavior). In spite of these difficulties, tables and charts have been generated relating the hardness values obtained by different hardness testers. For example, the hardness values obtained by the Brinell and Rockwell testers could be related to each other, as listed in Table 8.8. The hardness values of several engineering materials measured on different scales are compared in Figure 8.21.

TABLE 8.8
Comparison of the Brinell and Rockwell Hardness Numbers

Brinell		Rockwell		Brinell		Rockwell	
Indentation Diameter (mm)	BHN[a]	B	C	Indentation Diameter (mm)	BHN[a]	B	C
2.25	747		65.3	3.75	262	(103.0)	26.6
2.30	712		—	3.80	255	(102.0)	25.4
2.35	682		61.7	3.85	248	(101.0)	24.2
2.40	653		60.0	3.90	241	100.0	22.8
2.45	627		58.7	3.95	235	99.0	21.7
2.50	601		57.3	4.00	229	98.2	20.5
2.55	578		56.0	4.05	223	97.3	(18.8)
2.60	555		54.7	4.10	217	96.4	(17.5)
2.65	534		53.5	4.15	212	95.5	(16.0)
2.70	514		52.1	4.20	207	94.6	(15.2)
2.75	495		51.6	4.25	201	93.8	(13.8)
2.80	477		50.3	4.30	197	92.8	(12.7)
2.85	461		48.8	4.35	192	91.9	(11.5)
2.90	444		47.2	4.40	187	90.7	(10.0)
2.95	429		45.7	4.45	183	90.0	(9.0)
3.00	415		44.5	4.50	179	89.0	(8.0)
3.05	401		43.1	4.55	174	87.8	(6.4)
3.10	388		41.8	4.60	170	86.8	(5.4)
3.15	375		40.4	4.65	167	86.0	(4.4)
3.20	363		39.1	4.70	163	85.0	(3.3)
3.25	352	(110.0)	37.9	4.80	156	82.9	(0.9)
3.30	341	(109.0)	36.6	4.90	149	80.8	
3.35	331	(108.5)	35.5	5.00	143	78.7	
3.40	321	(108.0)	34.3	5.10	137	76.4	
3.45	311	(107.5)	33.1	5.20	131	74.0	
3.50	302	(107.0)	32.1	5.30	126	72.0	
3.55	293	(106.0)	30.9	5.40	121	69.8	
3.60	285	(105.5)	29.9	5.50	116	67.6	
3.65	277	(104.5)	28.8	5.60	111	65.7	
3.70	269	(104.0)	27.6				

[a] BHN values above 500 are for tungsten carbide ball; below 500 for standard steel ball. The values in parentheses are not reliable.

8.13.1 Brinell versus Vickers Hardness Numbers

Since the geometry of the indenters is designed on the same principle for both Brinell and Vickers hardness testers, the hardness value obtained should be the same in both the cases. In fact, the Vickers and Brinell hardness numbers are practically identical

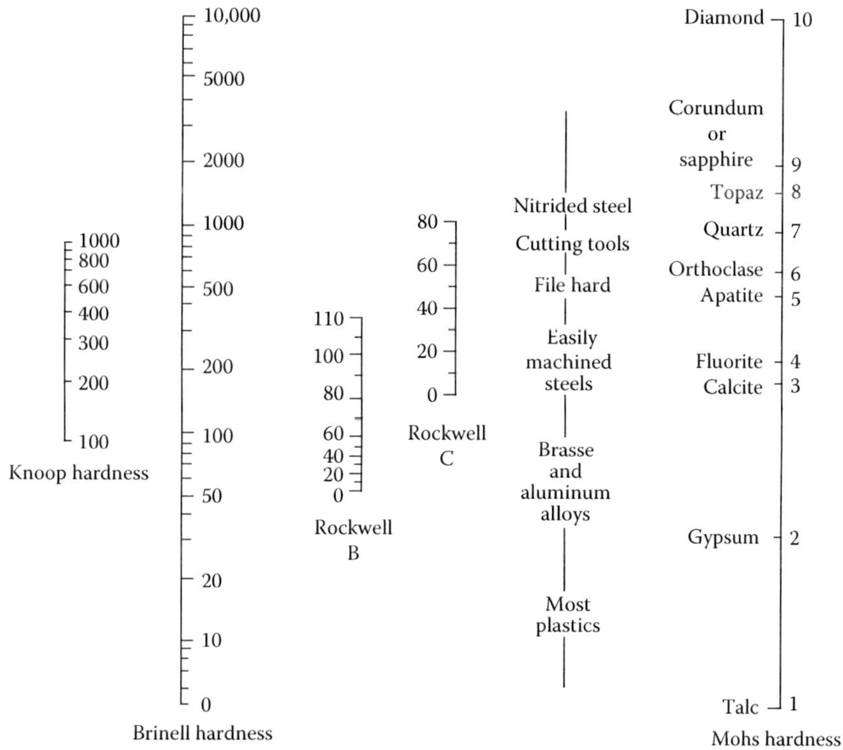

FIGURE 8.21 Comparison of the hardness values of materials measured on different scales and different types of hardness testers.

up to about 300 Brinell. Beyond this hardness value, however, the two curves tend to diverge (Figure 8.22) and to rather serious proportions at about 600 Brinell. This departure of the Brinell curve from proportionality is related to deformation of the hardened steel ball indenter at high hardness values, whereas the Vickers curve is practically straight owing to the insignificant amount of distortion of the diamond indenter at these high hardness values. With the introduction of the tungsten carbide ball for measurement of hardness using the Brinell tester, the two scales again may be close to each other. This is again related to the very little deformation experienced by the tungsten carbide ball indenter. For all practical purposes, the Vickers hardness measurements should be considered the only reliable evaluation of hardness at hardness values above about 600 Brinell.

8.13.2 Relationship between Strength and Hardness

It has been mentioned in the Introduction that hardness is a measure of the resistance of a material to plastic deformation. The strength of a material is also defined as the resistance of a material to plastic deformation. Thus, hardness and strength should be related to each other in some direct way. The relationship between hard-

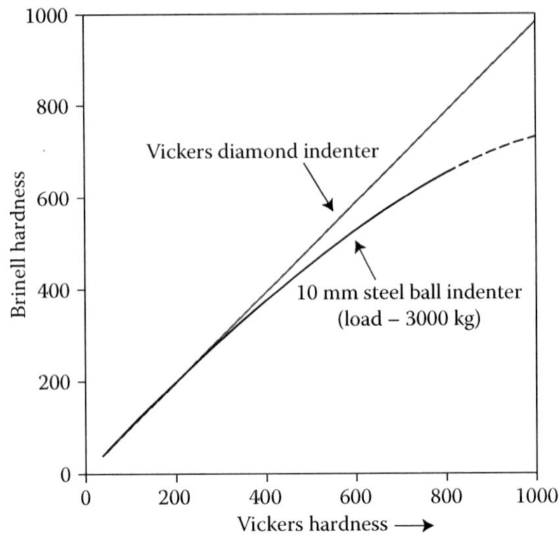

FIGURE 8.22 Comparison of Brinell and Vickers hardness numbers. Note that the two values are identical up to about 300 Brinell. At higher values, there is divergence and its magnitude increases at higher hardness numbers. The BHNs are lower because of the flattening of the hardened steel ball indenter.

ness and strength is different for different metals and alloys. Usually, the hardness and strength for a given alloy can be correlated only over a rather narrow range of conditions. Also the range of correlation is different for different methods of hardness determination.

It has been empirically shown that the VHN of any material is approximately three times the yield strength value.

$$\text{VHN} \approx 3 \times \text{yield strength} \tag{8.10}$$

Thus, the yield strength of a material can be estimated by measuring the hardness value. Such an estimate will be especially useful in those cases where new materials are being developed and when these are not available in sufficient quantities for a regular tensile testing experiment.

There is a close relationship between the tensile strength of a material and its BHN. This relation is a direct one, but the constants used in the correlation are different for different metals and alloys and depend to a great extent on the structural condition of the metal. For example, for steels, the constants of proportionality vary from 461 to 515. An approximate relationship between the tensile strength and BHN of steels is

$$\text{Tensile strength}\left(\text{in psi}\right) \cong 500 \times \text{BHN}\ (\text{kg}\,\text{mm}^{-2}) \tag{8.11}$$

Materials which have similar work hardening characteristics exhibit good correlation between hardness and tensile strength.

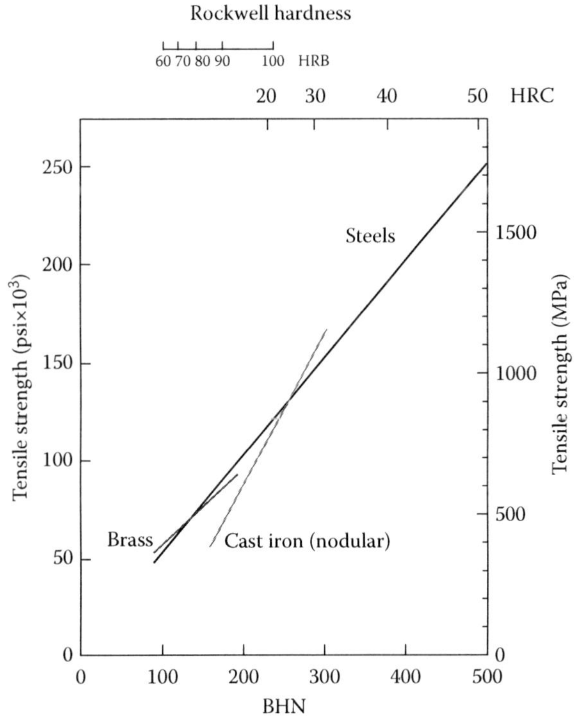

FIGURE 8.23 Comparison of the Rockwell and Brinell hardness numbers with the tensile strength values of brass, steel and nodular cast iron samples. Note that the proportionality is different for different materials.

All tables and charts are only approximate equivalents, particularly when converting to a method or scale which is physically not possible for the particular test material and thus cannot be verified. Unless an actual correlation has been completed by testing in different scales, established conversions may or may not provide reliable information.

Figure 8.23 compares the Brinell and Rockwell hardness numbers of brass, steel, and nodular cast iron specimens with their tensile strength values. This figure clearly indicates that the same proportionality relationship is not valid for all types of materials.

Table 8.9 lists the hardness values of some of the engineering materials in their annealed condition. It may be noted form this Table that the hardness value increases with increasing alloying content and that ceramic materials are much harder than metallic materials.

8.14 EXPERIMENTAL PROCEDURE

Most of the laboratories will have different types of hardness testers. The actual operation of the instrument will depend on the manufacturer and therefore it is advised that the student gets familiar with the operation of the different instruments

TABLE 8.9
Hardness Values of Some Typical Engineering Materials

Material	Hardness	Scale
	Aluminum Alloys	
1060 Al (99.6 Al min)—Annealed	19	Brinell
1100 Al (99.0 Al min, 0.12 Cu)—Annealed	23	Brinell
2024 Al (4.4 Cu, 1.5 Mg, 0.6 Mn)—Annealed	47	Brinell
2024 Al (4.4 Cu, 1.5 Mg, 0.6 Mn)—T6 temper	125	Brinell
7075 Al (5.6 Zn, 2.5 Mg, 1.6 Cu, 0.23 Cr)—Annealed	60	Brinell
Cu–Be (97.9 Cu, 1.7 Be)—Annealed	45–78	Rockwell B
	Steels	
0.2 wt% C steel—Annealed	111	Brinell
0.4 wt% C steel—Annealed	149	Brinell
0.9 wt% C steel—Annealed	197	Brinell
302 stainless steel—Annealed	147	Brinell
4340 Alloy steel—Annealed	217	Brinell
	Copper Alloys	
99.99 Cu	10	Rockwell B
Cartridge Brass (Cu–30 Zn)	60	Rockwell B
	Metallic Glasses	
$Pd_{77.5}Cu_6Si_{16.5}$	500	Vickers
$Fe_{80}B_{20}$	1100	Vickers
$W_{40}Re_{40}B_{20}$	2400	Vickers
	Ceramic Materials	
Cemented Carbides	1100–2000 88–94	Vickers Rockwell A
WC-6 wt% Co	1700	Knoop
Al_2O_3	1365	Vickers
SiC	2700	Knoop
Cubic boron nitride	4500	Knoop
Diamond	7000–10,000	Knoop

Note: The compositions indicated are in weight percent except for metallic glasses, where the subscripts indicate the atomic percentage of the element.

before starting to use them. In general, the Vickers- and Rockwell-type hardness testers are available in most laboratories. The Knoop microhardness tester is not commonly available in most laboratories. The nanoindenter is still a research-type instrument and therefore every undergraduate student may not have access to it. But, if it is available, one could measure the hardness and modulus of materials such as thin films, coatings, or a bulk material.

Take about 1/2″ diameter and 1/2″ high cylindrical samples of aluminum, copper, brass, and steel. One can take other samples also; but these samples were chosen to provide a range of hardness values. It is important to have the samples in a well-annealed state.

1. Make sure that the samples have a smooth surface and that the two flat surfaces are parallel to each other.
2. If necessary, prepare the surface of the specimen to be smooth and shining. If Vickers hardness testing and/or microhardness testing are planned, then it is important that the polished specimens are also etched (lightly) to reveal the microstructural features.
3. Take each sample and measure the hardness value using different hardness testers.
4. Measure the hardness at least in six different locations to ensure that you will get the same value everywhere, which indicates that the material is homogeneous.
5. Remember that the scales, indenters and loads are different for soft and hard materials in a Rockwell hardness tester. Remember to use the appropriate load and indenter and read off the correct scale.
6. If you are not sure of which scale to use in the Rockwell hardness tester, start from the B scale using 100 kg load and 1/16″ diameter steel ball. If you get an unrealistic or unreasonable value, for example, if you get a hardness number of >100, then go the C scale and use the 150 kg load and Brale indenter.
7. There is scope for personal error in measuring the diagonal lengths in the Vickers hardness testing. Take a few measurements on the same indentation to get an accurate value of the length of the diagonal.
8. When interested in measuring the microhardness value, make sure that the appropriate phase is chosen under the microscope and that the indentation is made in that area.
9. Compare the hardness values you get with the standard values reported in the literature. Table 8.9 lists the hardness values of some of the more common engineering materials in their annealed state.

8.15 OBSERVATIONS

Even though in many instances one directly reads off the hardness numbers on the machine itself, it may be a good practice, at least in the early times, to note down the measurements, for example, diameter of the indentation in a Brinell hardness test or the length of the diagonal in a Vickers hardness test. Using the formulae provided in the text, one could calculate the hardness numbers.

Report the hardness numbers measured in a tabular form as below:

Hardness Values Measured

Instrument used: Rockwell hardness tester (mention the Make and Model)

Material	Major Load	Indenter	Scale	Hardness Number	Literature Value
Aluminum					
Copper					
Brass					
Steel					

Instrument used: Vickers hardness tester (mention the Make and Model)
Load used: ...... kg
Dwell time: ...... s

Material	Diagonal Length I (μm)	Diagonal Length II (μm)	Average Diagonal Length (μm)	Hardness Number	Literature Value
Aluminum					
	Repeat the measurements on the same diagonal				
Copper					
Brass					
Steel					

From the hardness measurements you would notice that aluminum is the softest and steel the hardest material, amongst those chosen for this experiment. It would also be clear that brass is harder than copper due to solid solution strengthening effect.

The hardness and strength of materials depend on the microstructural features. Further, the microstructure can be altered by cold working, heat treatment, and so on. By cold working, the material gets work hardened and therefore it is harder. Similarly, the effect of heat treatment was apparent in the section on Heat Treatment of Steels, where it was shown that steel, of the same composition, can exhibit different types of microstructures and therefore different hardness values, depending on the heat treatment. For example, annealed steel will be the softest, normalized steel harder, and the quenched steel (containing martensite) will be the hardest. That is the reason why one should take the materials in the annealed state so that the effects of microstructural changes due to cold working and/or heat treatment are excluded/minimized.

8.16 ADDITIONAL EXPERIMENTS

Hardness measurement is so simple and yet very useful that it is used in a number of research investigations. In addition to the simple experiments described above, the students could do additional experiments, depending on the time available. The following is a brief description of the different experiments that could be conducted.

8.16.1 Influence of Cold Working

Cold working (plastic deformation of materials at low temperatures, typically <0.4 T_m, where T_m is the melting point of the material) results in increase in strength (and hardness) of a material. This is essentially due to introduction of a high density of crystal imperfections, the most important of them being the dislocations. The magnitude of increase in strength is a function of the amount of cold work introduced into the material. Some details about this behavior will be explained in Chapter 9. Therefore, a neat experiment for students to conduct will be to cold work a simple metal like copper or aluminum to different extents and measure the hardness as a function of the percentage of cold work. A plot of hardness versus percentage of cold work shows that the hardness increases with increasing amount of cold work. Cold working can be carried out by rolling the metal if it is in the form of a rod or sheet.

8.16.2 Influence of Annealing on a Cold-Worked Metal

As mentioned in the above experiment, cold working of a metal increases the hardness (and strength) of a metal. If the metal is worked extensively, the hardness increases so much that it becomes brittle. Further working of the metal becomes difficult. Since the hardness increase is due to an increase in the dislocation density, the material can be made softer and more deformable by annealing this cold-worked material. The annealing process reduces the dislocation density. Therefore, a neat experiment could be for the students to anneal the cold-worked metal (1) either at different temperatures for a fixed length of time (isochronal experiments) or (2) at a fixed temperature as a function of time (isothermal experiments). These experiments can be repeated for different times and at different temperatures. Annealing can be done by putting the cold-worked samples in a furnace maintained at the desired temperature.

If the sample is annealed at different temperatures, the grain size of the material (it is simpler to use a single-phase material like a pure metal or a solid solution alloy for this purpose) increases and this also has an influence on the hardness of a material. So, using microscopic methods (described in Chapter 3), the grain size of the metal can be measured and related to its hardness to determine if the Hall–Petch equation $H = H_o + Kd^{-1/2}$, where H is the measured hardness, d is the grain size, and H_o and K are constants is satisfied.

8.16.3 Influence of Alloying Additions

In addition to cold working and grain size reduction, another effective way of increasing the hardness of an alloy is to add alloying elements that form solid solutions with the host metal. Hardness measurement can be an effective tool to evaluate this solid solution strengthening behavior of materials. Therefore, the students can be asked to measure the hardness of materials with different types and amounts of solute elements. From such measurements, which could be distributed among a number of groups of students, it should be possible to establish the relative efficiencies of different solute elements in contributing to solid solution strengthening.

8.16.4 Microhardness of Different Phases

In a multicomponent alloy, the different phases will have different hardness values. This can be easily demonstrated using a microhardness tester, especially when the size of the different phases is reasonably large. Take an alloy, for example, a eutectic Al–Si alloy in which the large-sized second-phase Si is distributed in an Al matrix. By measuring the hardness of the individual phases, pure Al and Si, using a microhardness tester, it can be easily shown that Si is harder than Al. A similar experiment can also be done using a tool steel sample, where the carbide precipitate phase is harder than the matrix solid solution phase.

8.16.5 Influence of Aging on the Hardness of a Supersaturated Solid Solution Alloy

Precipitation hardening is a common form of heat treatment given to nonferrous alloys to increase their strength and hardness. In this process, a two-phase alloy that can exist as a single-phase alloy at higher temperatures (but still in the solid state) is chosen for study. On heating the alloy to a temperature above the solvus temperature and maintaining the alloy at that temperature, it will exist as a single-phase alloy. This part of the heat treatment is known as solution heat treatment since the second phase dissolves in the matrix. On quenching this alloy from this elevated temperature to room temperature, the alloy will remain as a supersaturated solid solution phase in a metastable state. However, on aging this alloy, the second phase will come out in the form of a fine precipitate (usually this is preceded by the formation of Guinier–Preston zones and some transition phases) and increase its strength. The increase in its strength can be monitored by measuring the hardness as a function of time at the aging temperature.

A number of such additional experiments may be devised to gain practice of hardness measurements. It should be realized that hardness is measured in a very large number of experiments to monitor different processes. This is mainly because hardness measurement is simple, the values are easy to interpret, and hardness measurement machines are available in almost every laboratory.

EXERCISES

8.1 What is hardness?

8.2 What are the different ways in which hardness can be measured?

8.3 What is Mohs scale of hardness?

8.4 What is the softest mineral in the Mohs scale?

8.5 What is the hardest material in the Mohs scale?

8.6 Why is Mohs scale not suitable to measure the hardness of different metals and alloys?

8.7 Name the different methods in which indentation hardness can be measured.

8.8 Why are a minor load and a major load used in the Rockwell hardness testing?

8.9 What scale in the Rockwell hardness tester would you use to determine the hardness of copper alloys?
8.10 What scale in the Rockwell hardness tester would you use to determine the hardness of case-hardened steel?
8.11 What does R_C65 (or HRC65) signify?
8.12 Brinell Hardness Testers generally use a 10 mm diameter steel ball indenter and a 3000 kg load to measure the hardness of steels. What load should we use if we change the size of the indenter to 5 mm diameter ball?
8.13 What are the units of BHN?
8.14 What is the type of indenter used in Vickers Hardness Testing?
8.15 What are VHN and DPN and do they represent the same thing?
8.16 What is the specific advantage of the Vickers Hardness Tester over other systems?
8.17 Why is Vickers hardness measurement not used frequently in the industry?
8.18 What is the qualitative relationship between the indentation size and the hardness?
8.19 How can we measure the hardness of individual phases in a microstructure?
8.20 What is the difference between macrohardness and microhardness?
8.21 What is the ratio of the two diagonals in a Knoop indentation?
8.22 What is the minimum thickness of test section to obtain reliable values of hardness?
8.23 What should be the minimum spacing between two indentations to obtain reliable hardness data?
8.24 Both BHN and VHN have the units of load/area (e.g., kg mm^{-2}). Are these two the same?
8.25 Draw a plot relating the VHN and BHN of a given material.
8.26 What is the relationship between hardness and strength of a metal?
8.27 What is nanohardness?
8.28 What will be the magnitude of loads used in a nanoindenter?
8.29 How is the hardness of a material determined by a hardness-testing machine?
8.30 What type of indenters are used in (a) Rockwell C hardness and (b) Rockwell B hardness measurements?

FURTHER READING

Courtney, T. H. 2000. *Mechanical Behavior of Materials*, 2nd edition. New York, NY: McGraw-Hill.

Fischer-Cripps, A. 2010. *Nanoindentation*, 2nd edition. New York, NY: Springer-Verlag.

Dieter, G. E. 1986. *Mechanical Metallurgy*, 3rd edition. Boston, MA: McGraw-Hill.

Dowling, N. E. 2007. *Mechanical Behavior of Materials*, 3rd edition. Upper Saddle River, NJ: Pearson: Prentice-Hall.

Hertzberg, R. W. 1996. *Deformation and Fracture Mechanics of Engineering Materials*, 4th edition. New York, NY: John Wiley & Sons, Inc.

Meyers, M. A. and K. K. Chawla, 2008. *Mechanical Behavior of Materials*, 2nd edition. Cambridge, UK: Cambridge University Press.

9 Tensile Testing

9.1 INTRODUCTION

The mechanical properties of materials are very important in determining their applications. For example, one may be interested in choosing a material for load-bearing applications. In that case, a material with high strength is required. Alternately, if one is interested in a material for easy fabrication, then the material should have high ductility. Thus, for different applications, an engineer will have to select a material with an appropriate combination of properties. Even though people talk of materials being strong or weak, a quantitative comparison of the strength properties is essential for an intelligent choice of suitable material.

The strength of a material can be understood as its resistance to *plastic deformation*. It may be measured by applying a static load uniformly over a surface cross section and monitoring the dimensional changes of the specimen. The load may be applied in four basically different ways—*tension*, *compression*, *shear*, and *torsion*. These four different modes are schematically shown in Figure 9.1. In tension, a load is applied in such a way that the load pulls the specimen apart and the specimen expands in the direction of application of load. In compression, the load is applied so as to contract the specimen in the direction of application of load. In shear, the load is applied in opposite directions on two parallel faces of a specimen. In torsion, which is a variation of pure shear, a specimen is twisted by a torque, which produces a rotational motion about the longitudinal axis of one end of the specimen relative to the other. Even though the load may be applied in many different ways, it is the shear force which causes plastic deformation in a material. Consequently, irrespective of the way in which the load is applied, it is the shear component which needs to be calculated. Unless a critical shear stress is exceeded, plastic deformation will not occur in the material. This critical shear stress value is different for different materials, and for the same material, it is different in different orientations of the crystal.

9.2 MEASUREMENT OF STRENGTH

The strength of a material can be determined when the load is applied in any of the above four different modes. Consequently, one measures the tensile properties, compressive properties, shear properties, and torque properties. But, the engineering *tensile test* (also referred to as a *tension test*) is most common because:

- It is easy to perform.
- The specimen design is so simple that it can be directly cast or machined from parts.

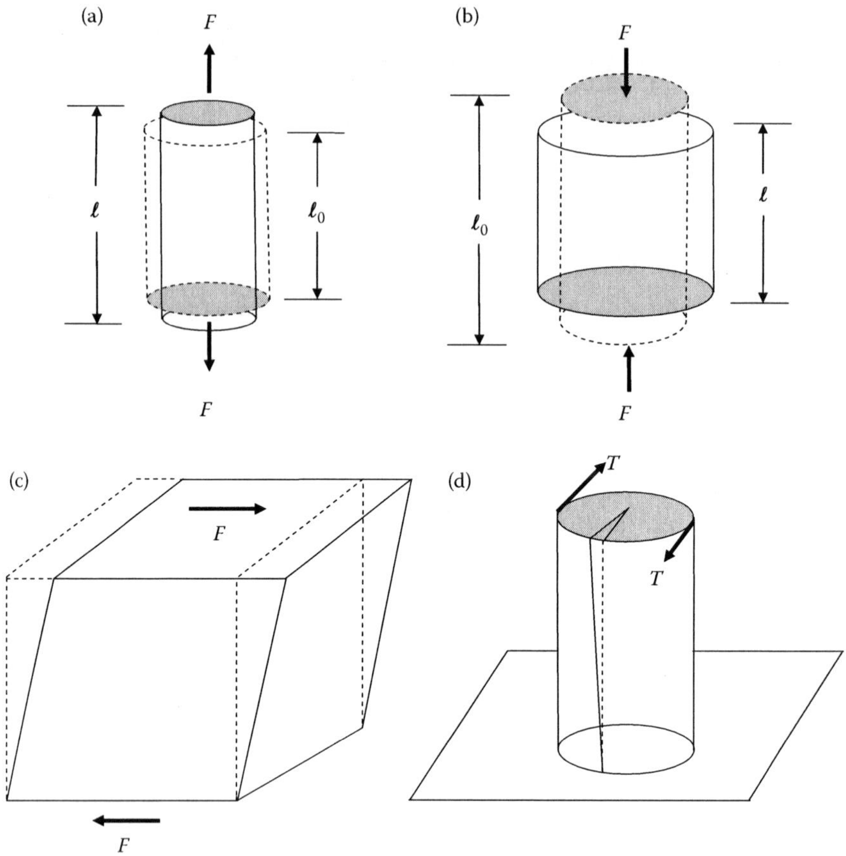

FIGURE 9.1 Schematic illustration of the different ways in which a load can be applied to a material: (a) tension, (b) compression, (c) shear, or (d) torsion.

- The test can be carried out in a few minutes.
- Measurements can be taken with equipment that is available in most material laboratories.

Therefore, the tensile test is the most widely used method to provide basic design information on the strength of materials and also as an acceptance test for the specification of materials. Further, the tensile properties are often measured during development of new materials and processes so that different materials and processes can be compared. Finally, *tensile properties* are often used to predict the behavior of a material under forms of loading other than uniaxial tension.

The compression test is conducted if the specimen experiences such forces in service. Examples are materials in manufacturing applications. Compression tests are also conducted to obtain the strength properties of materials which behave in a brittle manner in tension, for example, ceramics and some intermetallic compounds. Examples of torsion are found for machine axles and drive shafts, and also for twist drills. In this chapter, however, we will focus only on tensile testing of materials.

In the tensile test a specimen is subjected to a continually increasing uniaxial tensile force while simultaneous observations are made of elongation of the specimen. Note that the force (also frequently referred to as load) is applied slowly in comparison to the Impact test where the loads are applied at a fast rate.

Before going into details of how to conduct a tensile test and discuss what information and properties can be obtained from such a test, it would be useful to understand some basic definitions and the terms we will be using in this chapter. Brief information about the background on the science of deformation behavior of materials will also be provided.

9.3 BASIC DEFINITIONS

As mentioned above, on application of a load, the specimen changes its dimensions. For a particular applied load, the change in length would, however, be different for different cross-sectional areas of the specimen. But, the results of a single test could be used for all sizes and cross-sectional areas of the specimens, by normalizing the results of a test obtained on a material, when the loads and elongations (changes in length) are expressed in terms of stress and strain. The concepts of *stress* and *strain* eliminate the necessity of knowing the size and shape (dimensions) of the specimen.

9.3.1 STRESS

Stress, σ is defined as the load applied per unit area of cross section. Thus,

$$\sigma = \frac{F}{A_0} \tag{9.1}$$

where F is the load (force) applied perpendicular to the specimen surface, and its cross-sectional area is A_0 before any load is applied. The load may be measured in units of Newtons (N), pounds (lb), or kilograms (kg). The area of cross section may be measured in units of square millimeters (or square meters) or square inches. Thus, the units of stress are N m^{-2} (also referred to as Pascal, Pa), lb in.$^{-2}$, psi (pounds per square inch), or kg mm^{-2}. The most frequently used unit for stress is MPa (MegaPascals), which is 10^6 N m^{-2} or N mm^{-2}. MPa is the SI unit whereas psi is the customary US unit. The conversion factors for these units are as follows:

1 lb = 4.448 N = 0.454 kg
1 kg = 2.204 lb = 9.803 N
1 N = 0.225 lb = 0.102 kg
1 MPa = 10^6 Pa = 10^6 N m^{-2} = N mm^{-2}
1 GPa (Giga Pascal) = 1000 MPa = 10^9 N m^{-2}
1 psi = 6895 Pa = 0.006895 MPa = 0.000704 kg mm^{-2}
1 ksi = 1000 psi = 6.895 MPa

1 MPa (Mega Pascal) = 0.145 ksi = 145 psi = 0.102 kg mm^{-2}
1 kg mm^{-2} = 1.420 ksi = 9.797 MPa

Example Problem 9.1

A cylindrical aluminum specimen, 0.5 in. in diameter, is pulled in tension at a load of 2000 lb. (a) Calculate the resultant stress. (b) Calculate the stress if the same load is applied on a cylindrical specimen whose diameter is 0.4 in.

Solution 9.1

Since stress is defined as the load per unit area of cross section, we need to calculate the area of cross section of the specimen. The specimen is cylindrical in shape, and therefore its cross section is circular and the area of cross section can be calculated as

$$A = \pi r^2 = \pi \frac{d^2}{4}$$

where r and d are the radius and diameter of the specimen, respectively. Also, $r = d/2$.

a. Since the diameter $d = 0.5$ in., $A = \pi \times (0.5 \text{ in.}/2)^2 = 0.196$ sq. in. Therefore, the resultant stress, $\sigma = 2000$ lb/0.196 sq. in. = 10,204 psi.
b. If the specimen diameter, $d = 0.4$ in., then the area of cross section, $A = \pi \times (0.4 \text{ in.}/2)^2 = 0.126$ sq. in. The stress in this case would be, $\sigma = 2000$ lb/0.126 sq. in. = 15,873 psi.

From the above example, it is easy to realize that even though the load applied is the same, the stress will be different depending on the area of cross section on which the load is being applied. The smaller the area of cross section on which the load is applied, the larger will be the stress, for the same applied load.

9.3.2 Strain

Strain, ε is defined as the ratio of the change in length to the original length of the specimen. Thus,

$$\varepsilon = \frac{\Delta l}{l_0} \tag{9.2}$$

where Δl is the change in length ($= l_f - l_0$). Here l_0 is the original length (before any load is applied) and l_f is the final length of the specimen (after the test is completed). The length may be measured in units of meters or inches; but, since strain is the ratio of the change in length to the original length, it is dimensionless. However, units such as inch/inch or meter/meter are occasionally used. Sometimes strain is also expressed as a percentage, in which case the strain value is multiplied by 100.

Example Problem 9.2

A tensile test was conducted on an aluminum specimen that is 2 in. in length. (a) If the length of the specimen after the test is 2.1 in. calculate the strain. (b) Calculate the strain if the change in length is 0.1 in., but the original length of the specimen was 3 in. (c) Calculate the percentage strain in both the cases.

Solution 9.2

Strain is defined as the ratio of the change in length to the original length, that is, $\varepsilon = \Delta l/l_o$.

a. The change in length of the specimen is, $\Delta l = 2.1 - 2.0$ in. = 0.1 in. Since the original length of the specimen is 2 in., then the strain, $\varepsilon = 0.1$ in./2.0 in. = 0.05 (in./in.).
b. In this case, $\Delta l = 0.1$ in. and $l_o = 3.0$ in. Therefore, $\varepsilon = 0.1$ in./3.0 in. = 0.033 (in./in.).
c. The percentage strain in case (a) is $0.05 \times 100 = 5\%$ and in case (b) it is $0.033 \times 100 = 3.3\%$.

Thus, one can see that for the same elongation (change in length), the strain is smaller the longer the specimen is and it is larger the shorter the specimen is.

9.3.3 True Stress and True Strain

The values of stress and strain calculated in the above manner are referred to as *engineering stress* and *engineering strain*, respectively, since the original dimensions (length and diameter or cross-sectional area) of the sample are used. These are the values most commonly reported. But, it should be realized that the specimen dimensions (both length and diameter) change during the test. For example, during a tensile test, the length increases and the diameter decreases (to conserve the volume of the material). Thus, for a truly representative calculation of stress and strain, the instantaneous values of the length and diameter need to be considered and not the original dimensions. These values of stress and strain calculated this way will be referred to as *true stress* and *true strain*, respectively. If the results of tensile testing are to be used to predict how a material will behave under other loading conditions, it is necessary to plot the data in terms of true stress and true strain and not in terms of engineering stress and engineering strain.

True stress, σ_t is defined as the load divided by the instantaneous area of cross section, A_i, that is, the cross-sectional area at the time the applied load is F. This is expressed as

$$\sigma_t = \frac{F}{A_i} \tag{9.3}$$

In terms of the engineering stress, the true stress can be expressed as

$$\sigma_t = \sigma(1 + \varepsilon) \tag{9.4}$$

True strain, ε_t is defined as the change in length divided by the instantaneous length, l_i, of the specimen. This can be expressed as

$$\varepsilon_t = \frac{\delta l}{l_0} + \frac{\delta l}{l_1} + \frac{\delta l}{l_2} + \frac{\delta l}{l_3} + \cdots = \sum_i \left(\frac{\delta l}{l_i} \right) \tag{9.5}$$

where $l_1 = l_0 + \delta l$, $l_2 = l_1 + \delta l$, ... When expressed in a differential form, Equation 9.5 becomes exact, that is,

$$d\varepsilon_t = \frac{dl}{l} \tag{9.6}$$

On integrating Equation 9.6 from $l = l_0$ to $l = l_i$, the true strain, ε_t, can be expressed in terms of the original length, l_0 and instantaneous length, l_i as

$$\varepsilon_t = \ln\left(\frac{l_i}{l_0} \right) \tag{9.7}$$

This relationship arises from taking an increment of true strain, $d\varepsilon$, as the incremental change in length, δl, divided by the length, l_i, at the time, that is, $d\varepsilon = \delta l / l_i$. Integration of this expression results in Equation 9.7.

As long as deformation is uniform along the gage length of the specimen, true stress and true strain can be calculated from engineering stress and engineering strain. Assuming uniform deformation and constancy of volume, that is, $A_0 \times l_0 = A_i \times l_i$, we will have

$$\frac{A_0}{A_i} = \frac{l_i}{l_0} \tag{9.8}$$

(The material volume remains constant during plastic deformation. This is not so for elastic deformation. The sample extension along the tensile axis is accompanied by lateral contractions along the two axes normal to the tensile axis.) Combining Equations 9.2 and 9.8, we get

$$\frac{A_0}{A_i} = \frac{l_i}{l_0} = 1 + \frac{l_i - l_0}{l_0} = 1 + \frac{\Delta l}{l_0} = 1 + \varepsilon \tag{9.9}$$

Therefore, Equation 9.3 can now be rewritten as

$$\sigma_t = \frac{F}{A_i} = \frac{F}{A_0} \times \frac{A_0}{A_i} = \sigma(1 + \varepsilon) \tag{9.10}$$

Since the engineering stress σ is defined as F/A_0, then the true stress, σ_t and engineering stress, σ are related to each other through the equation $\sigma_t = \sigma\,(1 + \varepsilon)$, where

ε is the engineering strain. Similarly, we can derive an expression for true strain in terms of engineering strain, ε by substituting $1 + \varepsilon$ for l_i/l_0 into the expression for true strain, that is, Equation 9.7. Thus,

$$\varepsilon_t = \ln(1+\varepsilon) \tag{9.11}$$

At very low strains, the differences between true stress and engineering stress and true strain and engineering strain are very small. The difference continues to increase with increasing strain.

The true strain can be expressed as

$$\varepsilon_t = \int_{l_0}^{1} \frac{\Delta l}{l_i} = \ln\left(\frac{l_i}{l_0}\right) = \ln\left(\frac{A_0}{A_i}\right) \tag{9.12}$$

In the above equation, A_i is the instantaneous area of cross section. The true strain can be calculated either in terms of the length or the area of cross section before *necking* (to be explained later) occurs in the specimen. But, only Equation 9.12 can be used after necking takes place in the specimen.

Unless specified otherwise, stress and strain always refer to the engineering stress and engineering strain, respectively.

9.4 DEFORMATION BEHAVIOR

Materials can basically deform in two different ways—elastic and plastic. When a load is applied to a material, it will change the dimensions of the specimen.

9.4.1 Elastic Deformation

When a solid material is subjected to small tensile loads (or stresses), the specimen expands in length (elongates) and when the load is removed, the material returns to its original size and shape. This reversible deformation is called *elastic deformation*. Elastic deformation is temporary and recoverable in the sense that if the load applied to the specimen is removed the specimen will go back to its original size and shape. The deformation of a rubber band is entirely elastic. On an atomic scale, interatomic bonds between atoms are stretched on loading. When the load is removed, the bonds relax and the specimen goes back to its original shape. If the bonds stretch, then the strain is tensile, and if the bonds contract, then the strain is compressive. In case of polymers, however, the long-chain molecules may exist in a tangled arrangement before deformation. When a tensile load is applied, it tends to straighten these molecules and align them in the direction of applied stress. The stretching of interatomic bonds occurs only after the alignment of the molecules. Consequently, polymers can exhibit very large elastic strains. Figure 9.2a shows a typical stress–strain plot involving elastic deformation of metallic materials. It may be noted that the stress

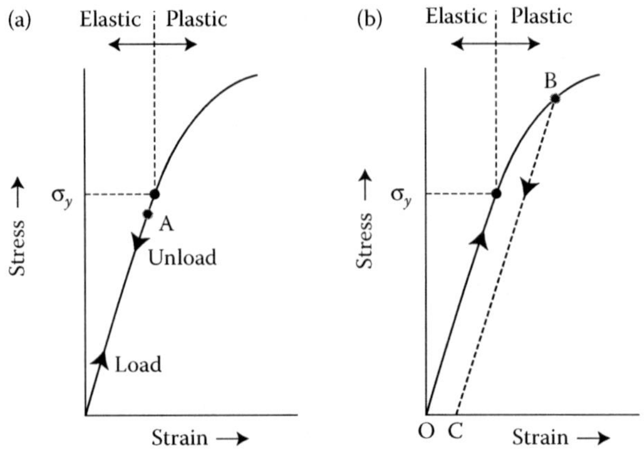

FIGURE 9.2 Schematic stress–strain curve during loading and unloading cycles for (a) elastic deformation and (b) plastic deformation. In the elastic region, the specimen will return to its original shape when the load is removed prior to reaching the yield point. Thus, the strain is completely recovered after the specimen is unloaded from point A. On the other hand, if the specimen is loaded into the plastic region, up to point B, then on unloading the specimen will trace the path BC, that is, parallel to the initial linear portion of the stress–strain curve. The specimen will experience permanent strain corresponding to OC.

and strain are proportional to each other and this relationship is known as *Hooke's law.* (Strictly speaking, Hooke's* law states that the elongation is proportional to the

* Robert Hooke (1635–1703) was an English natural philosopher who played an important role in the scientific revolution, through both experimental and theoretical work. Hooke was born on July 18, 1635 in Freshwater on the Isle of Wight, England. As a youth, he was fascinated by observation, mechanical works, and drawing. In 1653 he was enrolled in Christ Church, Oxford and became an Assistant to Robert Boyle, constructing, operating, and demonstrating Boyle's air pump. Hooke obtained the Master of Arts in 1662 or 1663. During 1663 and 1664, Hooke produced his microscopical observations, subsequently collated in *Micrographia* in 1665. On March 20, 1664, he was appointed Gresham Professor of Geometry. He also served as Surveyor to the City of London after the Great Fire of London. He received his doctorate degree in December 1691. Hooke is known for his law of elasticity (Hooke's law), his book *Micrographia*, and for first applying the word "cell" to describe the basic unit of life. Much of Hooke's scientific work was conducted as curator of experiments of the Royal Society. He has been recognized as one of the most important scientists of his age and was characterized as "England's Leonardo."

In 1660, Hooke discovered the law of elasticity which bears his name and which describes the linear variation of tension with extension in an elastic spring. He first described this discovery in the anagram "ceiiinosssttuv," whose solution he published in 1678 as "Ut tensio, sic vis" meaning "As the extension, so the force." His work on elasticity culminated, for practical purposes, in his development of the balance spring or hairspring, which for the first time enabled a portable timepiece—a watch—to keep time with reasonable accuracy. His name has been commemorated through 3514 Hooke, an asteroid (1971 UJ) and craters on the Moon and Mars, named in his honor. The British Society of Cell Biology awards annually The Hooke Medal to an emerging leader in cell biology.

Hooke's reputation suffered after his death and this is popularly attributed to a dispute with Isaac Newton over credit for his work on gravitation (that the gravitating power decreases with square of the distance). Newton, as President of the Royal Society, did much to obscure Hooke, including, it is said, destroying (or failing to preserve) the only known portrait of the man. Robert Hooke died on March 3, 1703 in London.

load applied. But, a similar linear relationship exists between stress and strain also.) When the applied stress increases, the strain also increases and the rate of increase of strain is always proportional to the rate of increase of stress. However, when you unload the specimen, say from point A, the specimen goes back to its original size and shape and the strain in the specimen is zero. That is, the deformation in the specimen is nonpermanent. This is true when the applied loads are small or the material has a very high strength. This type of elastic deformation occurs in metallic and ceramic materials. In the case of polymeric materials, however, the specimen may not return to zero strain condition on unloading of the specimen, even in the elastically deformed condition. This is because the stress employed to align the long chain molecules cannot be recovered since the chains cannot be made to go back to the tangled arrangement, that is, this is a nonreversible process.

9.4.2 Plastic Deformation

When the applied loads are high, beyond the point up to which Hooke's law is obeyed, the deformation experienced by the specimen is permanent, that is, *plastic deformation* occurs in the specimen. (Note that the term "plastic deformation" means nonelastic deformation and does not mean that the deformed material is plastic, i.e., a polymeric material.) It means that if you unload the specimen, it will not go back to its original size and shape. Figure 9.2b shows the typical tensile stress–strain plot into the plastic region. If the specimen is now loaded up to point B, the specimen is plastically deformed. If the load is now removed and brought back to zero, the specimen will still have some strain, indicated by point C. That is, the specimen experiences a permanent elongation (or strain) and this cannot be recovered (unless you anneal the specimen at an elevated temperature).

On an atomic scale, permanent deformation can be viewed as breaking of bonds with the original atom neighbors and then reforming of bonds with new neighbors. A very large number of atoms are involved in this process. When the load is removed, the atoms do not come back to their original positions. Such a process occurs through movement of crystal defects known as dislocations in crystalline materials. The movement of dislocations is difficult in ceramic materials and therefore the amount of plastic deformation experienced by ceramic materials is much smaller than in metals. Consequently, ceramic materials and some intermetallic compounds do not exhibit much of plastic deformation.

9.4.3 Elastic Strain Recovery

Consider the early stages of a tensile test. With reference to Figure 9.3, let us assume that the specimen is loaded up to point B. The strain corresponding to this point is ε_B. Since the specimen has undergone plastic deformation, the strain consists of both *elastic strain* that can be recovered and the plastic strain that cannot be recovered. If the load is now released from this point B, the stress–strain curve will follow the path BC, which is parallel to the initial elastic portion of the stress–strain plot. The strain corresponding to point C, ε_C, represents the permanent (nonrecoverable) strain experienced by the specimen. Thus, the difference between ε_B and ε_C ($=\varepsilon_B-\varepsilon_C$)

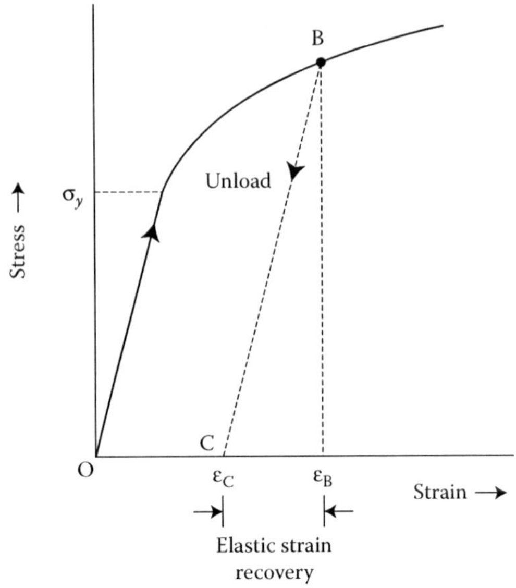

FIGURE 9.3 Schematic tensile stress–strain plot showing the phenomenon of elastic strain recovery. Part of the total strain in the specimen is recovered after unloading the specimen. In this figure, the difference between ε_B and ε_C, that is, $\varepsilon_B - \varepsilon_C$, represents the elastic strain that has been recovered.

represents the elastic strain that has been recovered due to unloading of the specimen. The elastic strain can always be recovered on unloading the specimen.

9.4.4 Strain Hardening

From the typical stress–strain curve it may be noticed that the slope of the curve in the plastic deformation region decreases with increasing elongation. This is because continued plastic deformation requires higher and higher loads. In other words, the material becomes harder (stronger) during plastic deformation and therefore it becomes more difficult to plastically deform the specimen. This phenomenon, known as *strain hardening* or *work hardening*, can be easily explained with reference to Figure 9.4.

Let us assume that the specimen is loaded up to point X in the plastic region for the material. Also note that the yield strength (the stress above which plastic deformation occurs) of the material is σ_0. If the load is now removed, the stress will come back to zero position along the line XY, which is parallel to the initial elastic (straight line) portion of the stress–strain curve. But, note that the specimen did not go back to its original size, that is, the specimen experiences some permanent strain, OY on unloading. If the specimen is now reloaded up to point X, it is noticed that the deformation is entirely elastic. That is, the new yield strength is σ_1. Yielding will now occur at a load level higher than where the previous unloading began. In other words, the new yield strength is σ_1, and clearly σ_1 is higher than σ_o.

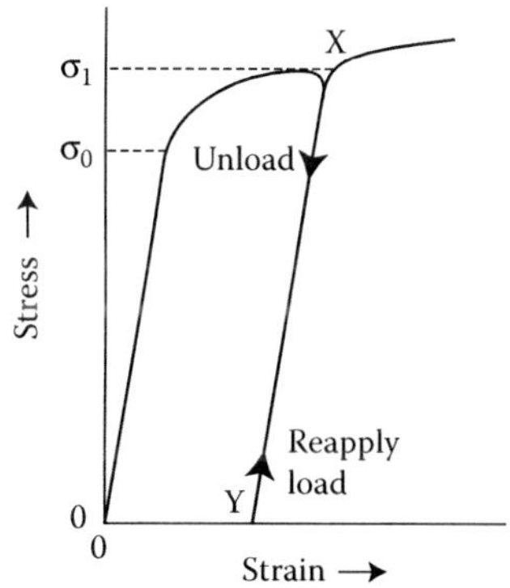

FIGURE 9.4 Schematic tensile stress strain curve to explain the concept of strain hardening. The initial yield strength is indicated as σ_0. If the specimen is loaded up to point X in the plastic region, and then unloaded, the specimen will return to point Y, where the specimen experiences some permanent (plastic) deformation. If the specimen is now reloaded from point Y, then elastic deformation continues up to point X, beyond which the deformation is plastic. In other words, the yield strength now is σ_1, which is higher than σ_0. This phenomenon of increased yield strength on continued straining is called strain hardening.

A practical example of the phenomenon of work hardening is that if you try to deform a metal or alloy at room temperature, the ease with which the specimen could be deformed decreases with continued deformation. In other words, it becomes more and more difficult to work the material, that is, more stress is required to deform the material. Thus, the phenomenon of strain hardening is also known as work hardening. In fact, the material may become so hard that it may have to be softened by annealing (heating to an elevated temperature) before it could be fabricated into the final shape.

9.4.5 Anelasticity

It has been assumed so far that the elastic deformation is independent of time. It means that an instantaneous elastic elongation is produced on application of load and also that the elongation becomes zero on removal of load, that is, complete recovery occurs. In most engineering materials, a time-dependent elastic elongation component may also exist. That is, elastic deformation will continue after the application of load and that some finite amount of time is required for complete recovery of elongation, after removal of the load. This time-dependent elastic behavior is known as *anelasticity* (Figure 9.5). For metals, the anelastic component is very small and is often neglected. But, this component is quite significant for some polymeric materials; in this case, it is called *viscoelastic behavior.*

9.5 THE TENSILE TEST

The tensile test can be conducted on different geometries of the specimens—cylinders (with a circular cross section), sheets (with a rectangular cross section), or wires (with circular cross section, but of small diameters). The choice of the specimen geometry

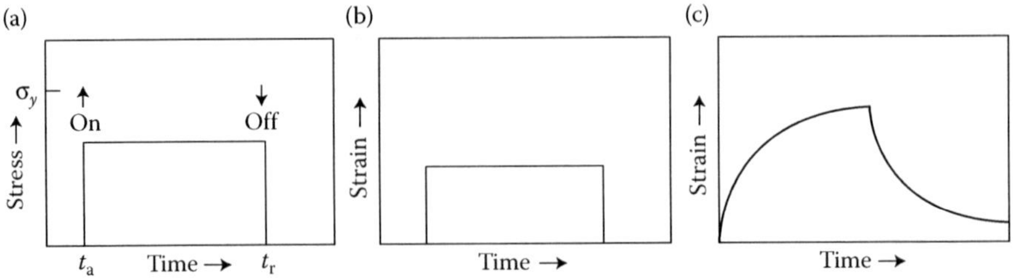

FIGURE 9.5 (a) Load vs. time, where the load is applied instantaneously at time t_a and released at time t_r. The strain vs. time responses for a totally elastic and viscoelastic material are represented in (b) and (c), respectively.

and size are frequently determined by the product form in which the material will be used and/or the amount of material available for samples. A flat specimen geometry is preferred when the end product is a thin plate or sheet. Specimens with circular cross section are preferred for products such as extruded bars, forgings, and castings.

Standard tensile specimens are usually tested in two common geometries—cylinders and sheets, as recommended by the American Society for Testing and Materials (ASTM). Figure 9.6 shows the dimensions of a standard round tensile specimen. In the more common cylindrical specimen, the standard diameter of the specimen is 0.505 in. and the gage length (the region over which measurements are made and is centered within the reduced section; this value is used in ductility computations) is at least four times the diameter. Accordingly, the standard gage length is 2.0 in. In some special cases, when sufficient quantity of the material is not available for standard specimen dimensions, modified specimen dimensions have been used.

The actual tensile testing of the specimens is conducted in a machine that can apply a uniaxial load onto the cross section of the sample and elongate the specimen at a constant slow strain rate (usually about 10^{-4} to 10^{-2} s^{-1}). The applied load and the resultant elongation are simultaneously and continuously measured using a transducer (called load cell) and an extensometer or a strain gage, respectively. The test takes several minutes to perform and is continued until the specimen fractures. Thus, the test is destructive in nature. Machines for tensile testing are available from commercial manufacturers such as Instron, MTS, and others. Figure 9.7a shows a photograph of a modern universal testing machine to perform tensile, compressive, flexural, shear, and tear tests on a variety of materials. Figure 9.7b shows a close-up view of the loading of the tensile specimen in the grips.

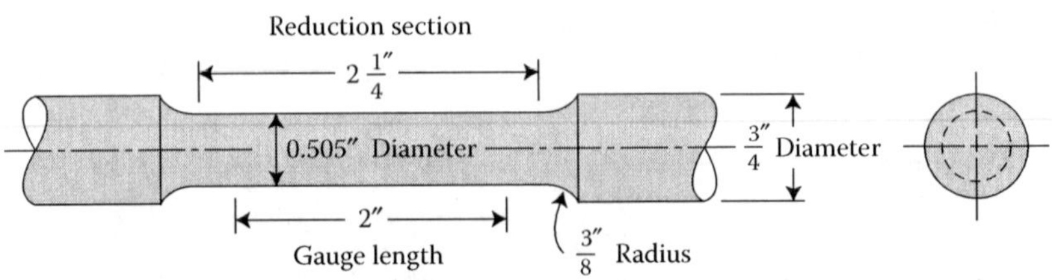

FIGURE 9.6 Dimensions of a standard tensile specimen with a circular cross section.

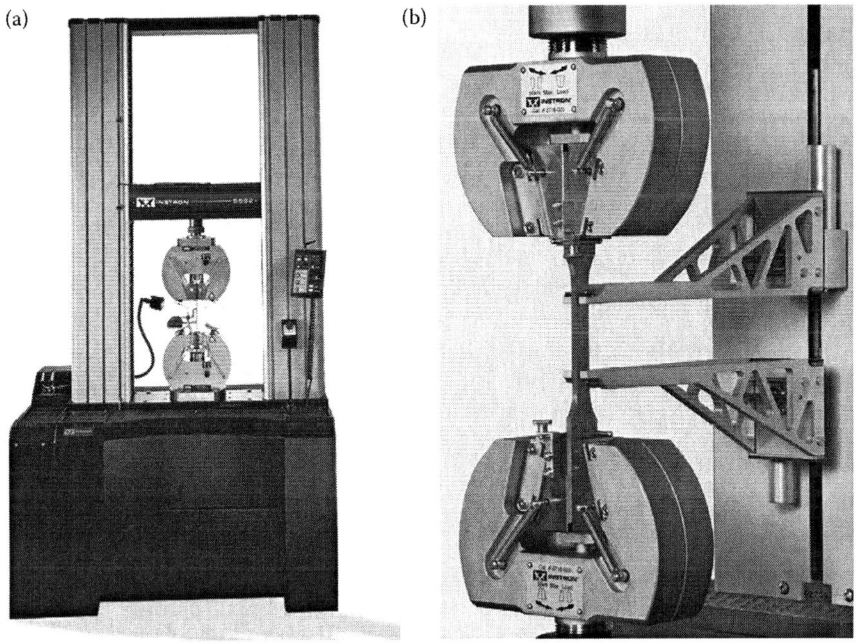

FIGURE 9.7 (a) Photograph of a commercial universal testing machine to perform tensile, compressive, flexural, shear, and tear tests on a variety of materials. (b) Close-up view of the loading of the specimen in the grips. (Photos courtesy of Instron Corporation. With permission.)

In the standard tensile test, the specimen with threads at both the ends is screwed into the threaded grips, each one attached to a crosshead. The most important concern in the selection of a gripping method is to ensure that the specimen can be held at the maximum load without slippage or failure in the grip section. Bending should also be minimized. It is important to realize that the specimen is centered in the grips and that it is also parallel to the direction of load application. One of the crossheads is stationary and the other is moving (Figure 9.8). With appropriate gears, the moving crosshead is moved at slow rates and thus a load is applied to the specimen; consequently the specimen gets elongated.

The load and elongation are recorded on a strip chart (or digitally). Thus, the data recorded in the test are change in length of the specimen (elongation) as a function of load (or force). Since the original gage length and area of cross section of the specimen (before the application of load) are known, the load and displacement (change in length) can be converted into stress and strain, respectively, using Equations 9.1 and 9.2. It is the *stress versus strain curve*, which is the starting point for further analysis to extract information about the mechanical properties of the specimen material. The actual operation of the machine is different for different manufacturers and the brochures or instruction manuals provided by the manufacturer should be consulted for details of operation of the machines.

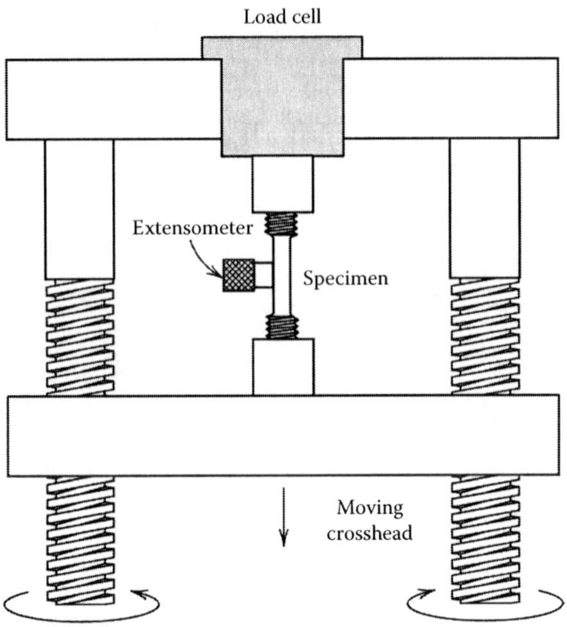

FIGURE 9.8 Schematic representation of the apparatus used to conduct tensile stress–strain test. The specimen is elongated by the moving cross head. The magnitude of the applied load is measured by the load cell and the elongation by the extensometer.

9.6 PROPERTIES OBTAINED FROM THE TENSILE TEST

A typical stress–strain curve from a tensile test on a metallic specimen is shown schematically in Figure 9.9. It may be noted that the strain increases linearly as the stress increases in the initial part of the test. Beyond a particular stress value, this linear relationship is no longer followed. On continued application of stress, the stress reaches a maximum value beyond which a decrease in stress, but continued increase in strain, is noticed. Eventually the specimen breaks. The strain continues

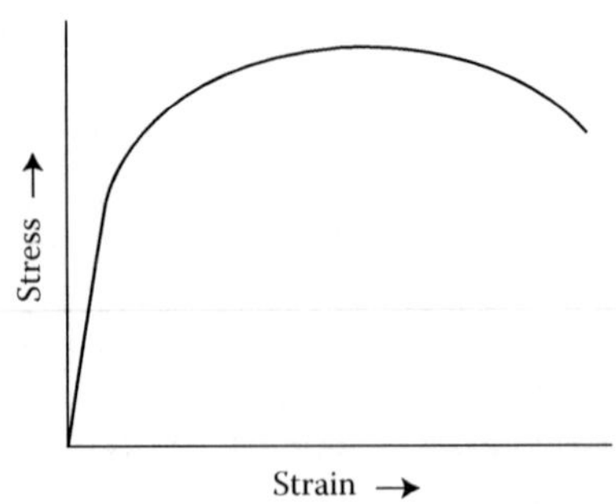

FIGURE 9.9 Typical tensile stress–strain curve of a metallic material.

to increase throughout the test until the specimen breaks. It is possible to obtain several engineering mechanical properties from the tensile test.

Once plastic deformation has begun, there will be both elastic and plastic contributions to the total strain, ε_T. This can be expressed as $\varepsilon_T = \varepsilon_e + \varepsilon_p$, where ε_e is the elastic contribution and ε_p is the plastic contribution to the total strain. The elastic strain is recoverable while the plastic strain is not (see Figure 9.3).

9.6.1 Proportional Limit

The stress up to which a linear relationship exists between the applied stress and resultant strain is called the *proportional limit*, that is, the strain is proportional to stress. Above the proportional limit, the relationship between stress and strain is no longer linear (proportional).

9.6.2 Yield Strength

Yield strength, σ_y is the stress at which plastic deformation or yielding is observed to start, that is, permanent deformation of the material starts at this stress value. The strain corresponding to the yield strength is called the *yield point strain*. Below the yield strength, the deformation is entirely elastic and the strain is fully recoverable. Sometimes the yield strength is also referred to as *yield stress* or the point on the stress–strain curve as the *yield point*. In most materials the yield strength and proportional limit are quite close to each other. In Great Britain, the yield strength is referred to as *proof stress*.

In practice, however, it is very difficult to determine precisely or even detect the onset of yielding in materials. This is especially true in ductile metals with FCC structure (e.g., Cu, Al, ...), since the yield point is not well defined in such materials. Further, the measured values depend on the sensitivity of the strain rate measurements. Therefore, it becomes necessary to devise a simple and reproducible method to estimate the yield strength of materials. The most accepted way of doing this is to measure the stress corresponding to a plastic strain of 0.002 or 0.2%, and this is known as the *0.2% offset yield strength*, designated as $\sigma_{0.2}$. The 0.2% offset yield strength is obtained by drawing a line parallel to the initial elastic (linear) portion of the stress–strain curve from the point of 0.2% (0.002) strain, and identifying the stress corresponding to the point of intersection between this line and the stress–strain curve. This is schematically shown in Figure 9.10. Design engineers normally prefer to use the 0.2% offset yield strength values in their calculations.

Instead of the traditional 0.2% offset strain value, some people use 0.1% offset strain or other values. In Great Britain, the stress corresponding to these values will be referred to as 0.2% proof stress or 0.1% proof stress, and so on.

The transition from elastic to plastic deformation is very abrupt in some materials. This happens in low carbon steels and many linear polymers. In these materials the onset of plastic deformation is denoted by a sudden drop in the stress value. The maximum stress reached before the stress drop is known as the *upper yield point* (Figure 9.11). The sudden drop in the stress value is associated with the appearance

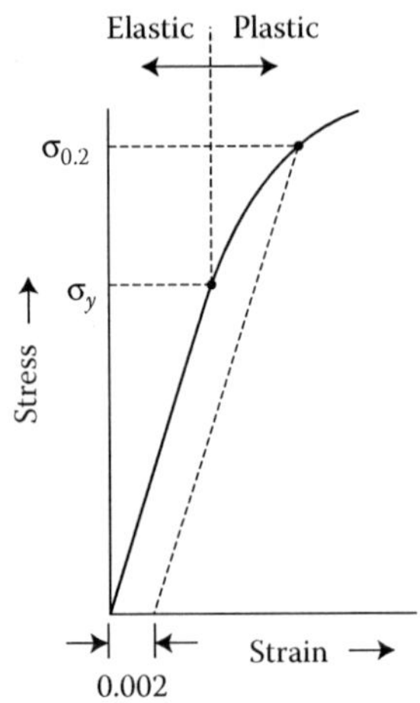

FIGURE 9.10 Method of estimating the 0.2% offset yield stress.

of a Lüders band across the specimen. That is, after the initial maximum, all the deformation at any instance is occurring within a relatively small region of the specimen. Continued elongation of the specimen occurs by propagation of the deforming region (Lüders bands) along the gage section rather than by increased strain within the deforming region. Only after the entire gage section has been traversed by the band does the stress rise again. This constant stress at which the Lüders band propagates (it actually oscillates around an average value) is known as the *lower yield point*. The yield strength in such cases is defined by the lowest stress at which plastic deformation occurs and is identified as the lower yield point. The upper yield point characterizes the stress at which plastic deformation first begins.

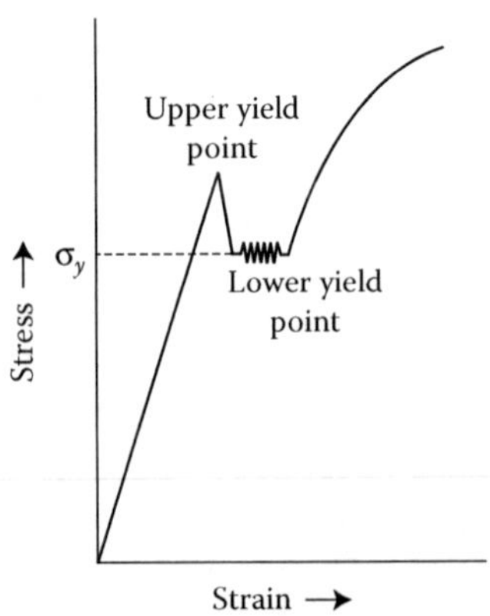

FIGURE 9.11 A stress–strain curve showing the yield point phenomenon. Note the upper and lower yield points. Such a behavior is exhibited by some materials such as low-carbon steels.

The yield point phenomenon is associated with the presence of an extremely small amount of interstitial impurity atoms (of the order of 0.001 at%) such as carbon and nitrogen, which pin the dislocations and prevent their movement. Thus, when the dislocations are locked by the carbon and nitrogen atoms present in the ferrite solid solution phase, they cannot easily move and consequently higher stresses are required to pull the dislocations away from these atmospheres. This stress required to tear the dislocations away from the impurity atmospheres (also known as Cottrell atmosphere) will correspond to the upper yield point. Once the dislocations have been dislodged, the stress required to move these free dislocations is less and thus we get the lower yield point.

9.6.3 Young's Modulus

It was mentioned earlier that, at relatively low stress levels, the stress and strain are proportional to each other, that is,

$$\sigma \propto \varepsilon \quad \text{or} \quad \sigma = E\varepsilon$$

$$\text{or} \quad E = \frac{\sigma}{\varepsilon} \tag{9.13}$$

This relationship is known as Hooke's law and the proportionality constant, E is known as the *modulus of elasticity*, or *Young's modulus**.

As shown in Figure 9.12, the modulus of elasticity can be measured as the slope of the elastic portion of the stress–strain curve. This modulus of elasticity is a measure of the stiffness of the material, or the resistance that a material offers to elastic

* Thomas Young (1773–1829) was an English genius and is famous for having partly deciphered Egyptian hieroglyphs (specifically the Rosetta stone). He made notable scientific contributions to the fields of vision, light, solid mechanics, energy, physiology, musical harmony, and Egyptology. He also developed Young temperament, a method of tuning musical instruments. He was born on June 13, 1773 in Milverton, Somerset and learned Greek and Latin by the time he was 14 years of age. He was also acquainted with several other languages. He began to study medicine in London in 1792, moved to Edinburgh in 1794, and a year later went to Germany and obtained his PhD in Physics in 1796. In 1799, he established himself as a physician in London. Young published many of his first academic papers anonymously to protect his reputation as a physician. He was appointed Professor of Physics at the Royal Institution, but resigned his professorship in 1803, fearing that his duties will interfere with his medical practice. Of his many scientific achievements, the most important was to establish the wave theory of flight.

Young described the characterization of elasticity that came to be known as Young's modulus, denoted as E, in 1807. However, the first use of the concept of Young's modulus in experiments was by Giordano Riccati in 1782. The idea can also be traced back to a paper by Leonhard Euler in 1727. Young's modulus relates the stress in a body to its associated strain, and is independent of the geometry of the component under investigation. Young's modulus allowed, for the first time, prediction of the strain in a component subjected to a known stress and vice versa. Prior to Young's contribution, engineers were required to apply Hooke's law ($F = kx$) to identify the deformation (x) of a body subjected to a known load (F). k is a function of both geometry and material. Young's modulus depends only on the material, and not its geometry, thus allowing a revolution in engineering strategies. He was elected a Fellow of the Royal Society in 1794, and was appointed its foreign secretary in 1802. He was chosen a Foreign Associate of the French Academy of Sciences in 1827 and was elected a foreign member of the Royal Swedish Academy of Sciences in 1828. He died on May 10, 1829 in London.

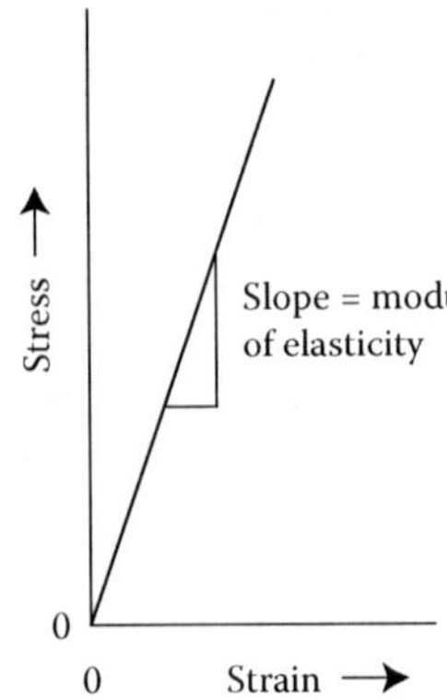

FIGURE 9.12 Schematic showing the way to determine the modulus of elasticity in the elastic portion of the stress–strain curve.

deformation. Thus, for a given stress level, a material with a high modulus of elasticity (that is a stiffer material) will show much less elastic deflection than a material with a lower modulus of elasticity. Thus, one can see in Figure 9.13 that, for a given stress, the amount of deflection in tungsten ($E = 407$ GPa) is much less than in aluminum ($E = 69$ GPa). The modulus of elasticity is an important design parameter for engineers to estimate the amount of dimensional changes a material may show under the application of a given stress.

Example Problem 9.3

Calculate the modulus of elasticity if the strain produced in a material is 0.0005 when a stress of 200 MPa is applied.

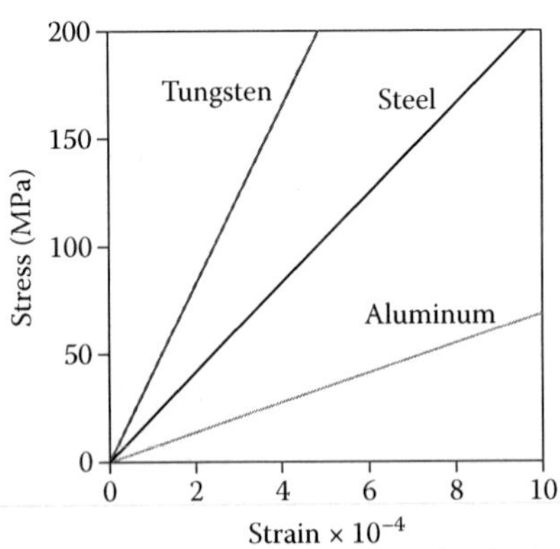

FIGURE 9.13 Comparison of the elastic behavior of three materials with different modulus of elasticity. For a given stress, the material with the lowest modulus of elasticity will exhibit the largest elastic strain.

Solution 9.3

Since the modulus of elasticity is defined as, E = stress/strain, E in the present case can be calculated as

$$E = \frac{200 \text{ MPa}}{0.0005} = 400{,}000 \text{ MPa} \quad \text{or} \quad 400 \text{ GPa}$$

On an atomic level, macroscopic elastic tensile strain in a material is a manifestation of an increase in interatomic spacing or stretching of interatomic bonds. Consequently, a material with a higher modulus of elasticity will result in much smaller elastic strain and therefore stretching of the interatomic bonds will be smaller than in a material with a lower modulus of elasticity. Accordingly, the modulus of elasticity can be taken as a measure of the interatomic bonding forces in a material. The interatomic bonding forces are stronger and the modulus of elasticity higher the higher the melting point of the metal is. Furthermore, the modulus of elasticity is proportional to the slope of the interatomic force–separation curve at the equilibrium atomic spacing, r_0, that is,

$$E \propto \left(\frac{dF}{dr}\right)_{r_0} \tag{9.14}$$

Figure 9.14 shows a force (F)–distance (r) plot for two different materials. The distance of separation between the atoms at $F = 0$ corresponds to the equilibrium atomic spacing, r_0. The tangent to the F–r curve at r_0, that is, $(dF/dr)_{r_0}$ represents a

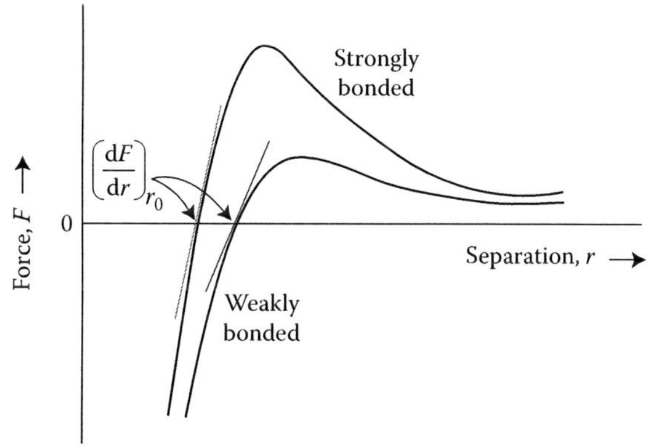

FIGURE 9.14 Force–interatomic distance curve for two materials showing the relationship between atomic bonding and modulus of elasticity. A steep dF/dr slope is an indication of a high modulus.

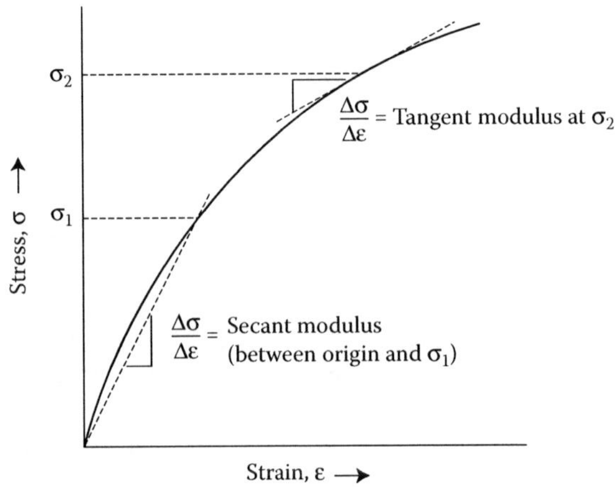

FIGURE 9.15 Schematic stress–strain diagram showing nonlinear elastic behavior and how the tangent and secant moduli are determined.

measure of the modulus of elasticity of the material. The steeper the slope the higher is the modulus of elasticity.

In some materials such as gray cast iron, concrete, and many polymers, the stress is not proportional to strain even at low stresses, that is, the elastic portion of the stress–strain curve is not linear. Therefore, the modulus of elasticity in such materials cannot be measured as the slope of the "linear" portion of the stress–strain curve. Instead, when the stress–strain curve is nonlinear in the elastic portion, either tangent or secant modulus is normally used. The tangent modulus is taken as the slope of the stress–strain curve at some specified level of stress, while the secant modulus represents the slope of a secant drawn from the origin to some given point of the stress–strain curve (Figure 9.15).

The modulus of elasticity varies from as low as 45 GPa for magnesium to as high as 407 GPa for tungsten. Since the value of the modulus of elasticity of a material is determined by the interatomic bond strength, its microstructure does not play any role. (We will see later that microstructure, especially the grain size, plays a very important role in determining the strength properties of materials). The modulus of elasticity, however, depends on the direction in which it is measured in a single-crystal specimen, that is, it exhibits the phenomenon of *anisotropy*. The modulus of elasticity increases from polymeric materials (due to the nature of secondary bonds) to metals to ceramic materials (due to the high strength of the ionic and covalent bonds). Accordingly, compared to metals, the modulus of elasticity is typically lower for polymers and higher for ceramic materials. It is because of the higher modulus of elasticity of the ceramic materials that they are introduced into metallic matrices to produce metal-matrix composites (MMCs) for critical applications as structural materials. However, it should be realized that the modulus of elasticity in ceramic materials depends on the extent of porosity in the sample.

9.6.4 Poisson's Ratio

When a tensile stress is imposed on a metal specimen, say along the z-direction, the specimen expands in the z-direction. That is, there is a positive strain along the direction of applied stress. Since the volume of the material needs to be conserved, there will be a corresponding contraction along the other two directions, x and y, perpendicular to the direction of the applied stress (Figure 9.16). If the applied stress is truly uniaxial and the material is isotropic, then the compressive strain along the x- and y-directions will be equal to each other, that is, $\varepsilon_x = \varepsilon_y$. The ratio of the lateral and axial strains is called *Poisson's ratio*, ν and is expressed as

$$\nu = -\frac{\varepsilon_x}{\varepsilon_z} = -\frac{\varepsilon_y}{\varepsilon_z} \tag{9.15}$$

The negative sign is included in the expression to suggest that ν will always be positive, since ε_x (or ε_y) and ε_z will always be of opposite sign. Theoretically, for isotropic materials, $\nu = 0.25$ and the maximum value of ν (for which there is no net volume change) is 0.5. Typically, for many metals and alloys, the Poisson's ratio

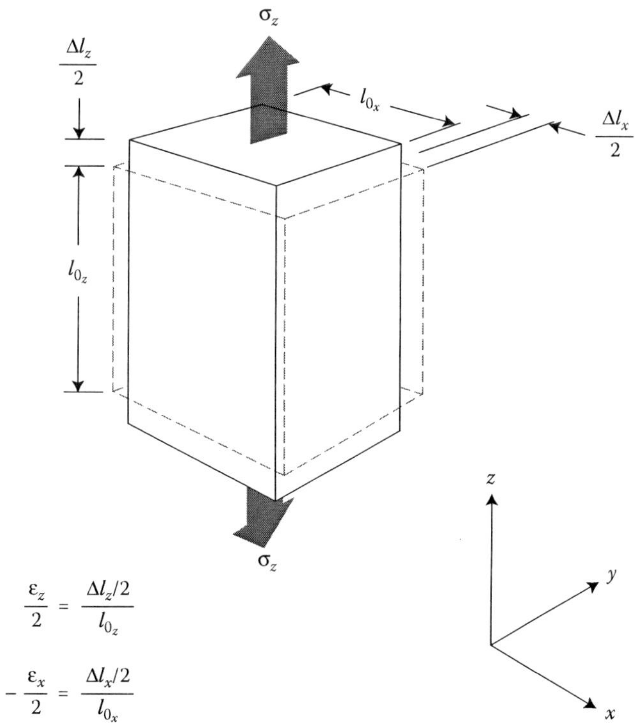

FIGURE 9.16 Axial (z) elongation (positive strain) and lateral (x and y) contractions (negative strains) in response to an imposed tensile stress. Solid lines represent dimensions after stress application; dashed lines, before.

varies between 0.25 and 0.35. Values of the Poisson's* ratio for some materials are included in Table 9.1. Note that the value of the Poisson's ratio cannot be obtained from a routine tensile test. Since elastic strains are usually very small, reasonably accurate measurement of Poisson's ratio in a tensile test requires that the strain is measured very accurately with a very sensitive extensometer. Strain gages should be used for lateral strains.

9.6.5 Ultimate Tensile Strength

The stress corresponding to the maximum applied force (load) is called the *ultimate tensile strength*, σ_{uts} or simply the *tensile strength*, σ_{ts}. The strain corresponding to the tensile strength is called the *uniform strain*, ε_u because up to this point the strain is uniformly distributed throughout the gage section of the specimen. (Sometimes the upper yield strength of low-carbon steels is higher than the subsequent maximum. In such cases, some prefer to define the tensile strength as the subsequent maximum instead of the initial maximum, which is higher.)

The tensile strength is the maximum stress that a structural member could sustain during tensile loading. If this stress is applied and maintained, fracture will result. Even though this is the maximum stress that a material can withstand, this value is not of any engineering significance. This is because by the time the material experiences the stress corresponding to the tensile strength, it has experienced so much of plastic deformation that it becomes useless. The stress that is of engineering significance is the yield strength.

9.6.6 Necking

Deformation in ductile materials may not occur uniformly; consequently, one region deforms more than the other. As a result, a small constriction or neck begins to form in the specimen at some point. This phenomenon is known as *necking*. Since the area of cross section of the specimen is smaller at the point of formation of the neck, a lower load (resulting in a higher stress) is sufficient to continue to plastically deform the specimen. This is known as *strain localization*. Subsequently, deformation is confined to the neck area and this is referred to as *strain accumulation*. Therefore, the strain is nonuniform from the point of necking onward. Necking occurs only during tension and not during compression testing; in the latter case, the specimen bulges.

* Siméon Denis Poisson (1781–1840) was a French mathematician, geometer, and physicist. He was born on June 21, 1781 in Pithiviers, Loiret, France. He entered the École Polytechnique in Paris in 1798 and one of his papers was published in Recueil des savants étrangers, an unprecedented honor for a youth of 18. He was appointed a full professor at the École Polytechnique in 1806 succeeding Jean Baptiste Joseph Fourier. Subsequently, he occupied many important positions culminating in his appointment as geometer to the Bureau des Longitudes succeeding Pierre-Simon Laplace in 1827. He was elected a foreign member of the Royal Swedish Academy of Sciences in 1823. As a teacher of mathematics Poisson is said to have been extraordinarily successful. As a scientific worker, his productivity has rarely, if ever, been equaled. Notwithstanding his many official duties, he found time to publish more than 300 works, several of them extensive treatises. Poisson is well-known for Poisson's ratio, Poisson equation, and Poisson distribution in probability theory. In 1831, he derived the Navier–Stokes equations independently of Claude–Louis Navier. He died on April 25, 1840 at Sceaux, Hauts-de-Seine in Paris.

TABLE 9.1
Typical Room Temperature Mechanical Properties of Some Metals and Alloys in the Annealed Condition

Metal/Alloy	Yield Strength (MPa)	Tensile Strength (MPa)	Ductility, % El. (in 50 mm (2 in.) gage length)	Modulus of Elasticity (GPa)	Shear Modulus (GPa)	Poisson's Ratio
Aluminum	35	90	40	69	25	0.33
Copper	69	200	45	110	46	0.34
Brass (Cu-30 wt% Zn)	75	300	68	97	37	0.34
Iron	130	262	45	207		0.27
Nickel	138	480	40	207	76	0.31
1020 Steel (Fe-0.2 wt% C)	180	380	25	207	83	0.30
Titanium	450	520	25	107	45	0.34
Molybdenum	565	655	35			
Tungsten				407	160	0.28
Magnesium				45	17	0.29
Lead				14		0.45
Al_2O_3		207		380		0.26
Si_3N_4				303		0.24
SiC (dense)		170		470		
PVC		41	2–30	2.8		

Every metallic tensile specimen has some inhomogeneity along its length (e.g., porosity or small inclusions inside the specimen or machine marks, taper, or other marks on the surface). Strain is localized in these regions, which leads to a locally grater reduction in area. The increased flow stress due to work hardening, for strains less than ε_u, more than compensates this incipient instability. However, the rate of work hardening decreases as deformation continues, that is, the increase in flow stress per unit strain becomes less with increasing deformation. Hence, it becomes more difficult to sufficiently remove the incipient instability. At strains corresponding to ε_u, the work-hardening capacity has decreased to a level that an instability once formed continues to develop. This instability takes the form of a neck and this process is referred to as "necking." The criterion for necking can be quantitatively expressed by realizing that at this point, the stress (or load) has reached the maximum, that is, $dF = 0$. Using the definition of true stress (Equation 9.3),

$$dF = 0 = \sigma_t \, dA_i + A_i \, d\sigma_t \tag{9.16}$$

$$\text{or} \quad \frac{d\sigma_t}{\sigma_t} = -\frac{dA_i}{A_i} \tag{9.17}$$

In other words, the onset of necking is characterized by the fractional increase in the flow stress (as defined by $d\sigma_t/\sigma_t$ being exactly balanced by the fractional decrease in load-bearing area (as measured by dA_i/A_i)). Prior to necking $|(d\sigma_t/\sigma_t)| > |(dA_i/A_i)|$ and incipient instabilities are removed. The strain in the specimen after necking is not uniformly spread all over the gage length; instead it is focused in the neck region. Consequently, the strain after necking is referred to as *nonuniform strain*.

With ductile materials, the tensile strength corresponds to the point at which deformation starts to localize, forming a neck. Less ductile materials fracture before they neck. In this case, the fracture strength is the tensile strength. Very brittle materials (such as glass at room temperature) do not yield before fracture. Such materials have tensile strengths but not yield strengths.

The yield and tensile strengths of metallic materials vary widely anywhere from 30 MPa for commercially pure metals up to 3000 MPa for high-strength steels. Table 9.1 lists the mechanical properties (yield strength, tensile strength, ductility, and others) of some selected materials. It may be noted that while ultrahigh-purity metals have very low yield strengths (1–10 MPa), the yield strengths of alloys are much higher (10–1000 MPa). The yield strength of plastics and elastomers is usually lower (0.2–100 MPa) than that of metals and alloys. Ceramic materials are usually not tested in tension due to the presence of surface flaws and their inherent brittleness; instead they are tested in compression. Their compressive strength values are very high and can be over 1 GPa.

9.6.7 Fracture Strength

The stress at which the material fractures is known as the *fracture strength* and the strain corresponding to this stress is known as *fracture strain*. The fracture strength of a material is also not normally specified for engineering design purposes. The fracture strength is less than the tensile strength in ductile materials because of necking. In completely brittle materials the (yield strength or) tensile strength and fracture strength are the same. Some brittle materials may show a small amount of plastic deformation, with a strain of usually <5%. In that case, the tensile strength and fracture strength are the same.

Apart from these strength properties that can be directly read off from the stress–strain curve, it is possible to obtain other mechanical properties of the material from the tensile test. But, these properties need to be calculated. The method of obtaining these properties will be described in the following sections.

9.6.8 Ductility

This is another important mechanical property of a material. People normally talk of a material being ductile or brittle and this is only a qualitative way of classifying materials. But, a better (and quantitative) method of classification of materials would be to measure the degree of plastic deformation that has been sustained by the specimen at fracture. A material that experiences lot of plastic deformation before fracture is referred to as *ductile* while one which experiences very little or no plastic deformation is referred to as *brittle*.

Knowledge of ductility of materials is important, firstly in indicating to a designer the extent of plastic deformation the material will experience before fracture and secondly, in specifying the degree of allowable plastic deformation during fabrication operations.

Ductility may be quantitatively expressed as either percent elongation or percent reduction in area. The *percent elongation*, %EL is the percent plastic strain at fracture, and is defined as

$$\%\text{EL} = \left(\frac{l_f - l_0}{l_0}\right) \times 100 \tag{9.18}$$

where l_f is the gage length of the specimen after it had fractured and l_0 is its original length. Since the specimen breaks into two pieces after fracture, l_f is measured by putting the two broken pieces together appropriately. Percent elongation, as a measure of ductility, has the disadvantage that it is really composed of two parts: the uniform elongation that occurs before necking, and the localized elongation that occurs during necking. The second part is sensitive to the specimen size. If the gage length of the specimen is large, the necking elongation converted to percent is very small. On the other hand, with a small gage length (relative to its diameter), the necking elongation can be very large and account for most of the elongation. In other words, %EL depends on the length-to-diameter ratio for the specimen. Thus, it is customary to specify the initial gage length of the specimen when %EL is reported.

The percent reduction in area, %*RA* is defined as

$$\%\,RA = \left(\frac{A_0 - A_f}{A_0}\right) \times 100 \tag{9.19}$$

where A_0 is the original area of cross section of the specimen and A_f is the final area of cross section at the point of fracture (in the necked region). The main advantage of using %*RA* to report as a measure of ductility is that it is independent of the gage length and the length-to-diameter ratio. Percent reduction in area, as a measure of ductility, however, has the disadvantage that with very ductile materials it is often difficult to measure the final cross-sectional area at fracture. This is particularly true of sheet specimens. For a given material, the magnitudes of %EL and %*RA* will be usually different.

To eliminate the effect of gage length on the magnitude of %EL and provide a measure of ductility that correlates with forming operations in which the gage length is very short, it is possible to determine the *zero-gage-length elongation*, ε_0. From the constancy of volume relationship for plastic deformation, that is, $A_f l_f = A_0 l_0$ and the definition of %*RA*, it can be shown that

$$\frac{l_f}{l_0} = \frac{A_0}{A_f} = \frac{1}{1 - RA} \tag{9.20}$$

$$\varepsilon_0 = \frac{l_f - l_0}{l_0} = \frac{A_0}{A_f} - 1 = \frac{1}{1 - RA} - 1 = \frac{RA}{1 - RA} \tag{9.21}$$

where *RA*, the reduction in area, is defined as $[(A_0 - A_f)/A_0]$.

Example Problem 9.4

A copper specimen, 10 mm in diameter and 50 mm long has been subjected to a tensile test. After the test the final length was measured to be 54.25 mm and the final diameter as 9.2 mm. Calculate the ductility of this material.

Solution 9.4

Since ductility may be defined in terms of either percent elongation or percent reduction in area, we calculate both.

$$\%\,\text{Elongation} = \left(\frac{l_f - l_0}{l_0}\right) \times 100 = \frac{54.25\text{ mm} - 50.00\text{ mm}}{50.00\text{ mm}} \times 100$$

$$= \frac{4.25\text{ mm}}{50.00\text{ mm}} \times 100 = 8.5\%$$

$$\%\text{Reduction in area} = \left(\frac{A_0 - A_f}{A_0}\right) \times 100$$

$$= \frac{\pi \times (10\text{ mm}/2)^2 - \pi \times (9.2\text{ mm}/2)^2}{\pi \times (10\text{ mm}/2)^2} \times 100 = 15.4\%$$

Note that the % elongation and % reduction in area are different.

9.6.9 Toughness

Toughness may be defined as the ability of a material to absorb energy in the plastic region prior to fracturing. This quantity, occasionally referred to as tensile toughness and designated by the symbol U, may be calculated as the area under the true stress–true strain curve up to the point of fracture. It is calculated as

$$U = \int_0^{\varepsilon_f} \sigma \, d\varepsilon \tag{9.22}$$

The units for toughness are stress × strain, that is, force per unit area × length per unit length = (force × length) per unit volume = energy per unit volume. For SI units, these values are expressed as joules per cubic meter, J/m^3, equivalent to Pa and in customary US units, they are expressed as inch-pounds force per cubic inch, equivalent to psi.

Since toughness is a product of stress and strain, high toughness values can be obtained when a material exhibits high tensile strength or high ductility or both. A ductile material exhibits high ductility but its strength is low. But, its toughness could be reasonably high. On the other hand, a brittle material has a very high strength but

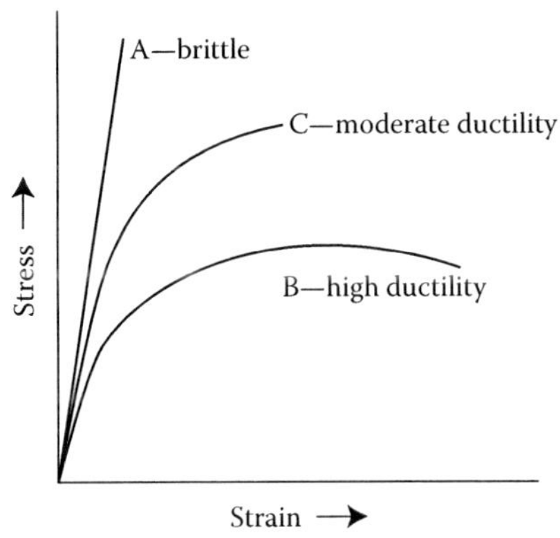

FIGURE 9.17 Toughness evaluation of materials. High toughness values can be obtained either by having a high yield strength or large elongation or both.

its ductility is extremely low. Consequently, its toughness is low. Thus, it is usually observed that ductile materials are usually tougher than brittle materials.

Figure 9.17 shows the schematic true stress–true strain plots for three different materials. Material A has high yield strength but low strain while material B has low yield strength but high ductility. Thus, high toughness may be achieved by having high yield strength and/or high strain. A high toughness can also be obtained when both yield strength and strain are reasonably high (material C).

9.6.10 Resilience

Resilience may be defined as the capacity of a material to absorb energy when it is deformed elastically and then, upon unloading, to have this energy recovered. The associated property is the *modulus of resilience*, designated by the symbol U_r, which is the strain energy per unit volume required to stress the material from zero stress up to the point of yielding (Figure 9.18). U_r may be evaluated as the area under the stress–strain curve up to the point of yielding and calculated as

$$U_r = \int_0^{\varepsilon_y} \sigma \, d\varepsilon \tag{9.23}$$

Assuming a linear elastic region,

$$U_r = \frac{1}{2}\sigma_y \varepsilon_y \tag{9.24}$$

where σ_y is the yield strength and ε_y is the yield point strain. The units of resilience are those of energy and are the same as that of toughness, namely, Pa or psi.

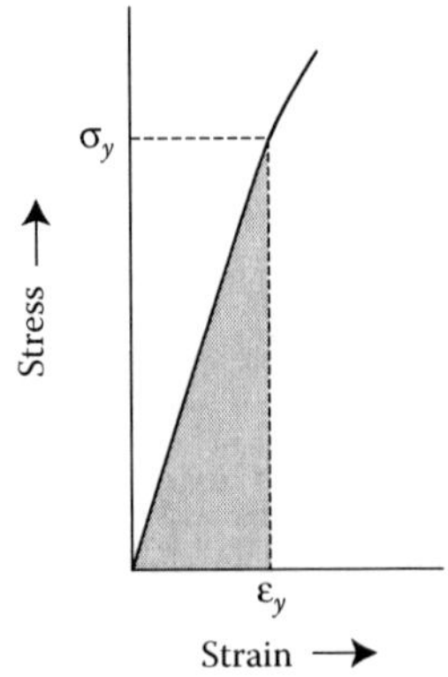

FIGURE 9.18 Calculation of resilience from the stress–strain curve.

Since the yield point strain can also be expressed in terms of the yield stress and modulus of elasticity, $\varepsilon_y = \sigma_y/E$, where E is the modulus of elasticity, substituting this value into Equation 9.24 yields

$$U_r = \frac{1}{2}\sigma_y\varepsilon_y = \frac{1}{2}\sigma_y\left(\frac{\sigma_y}{E}\right) = \frac{\sigma_y^2}{2E} \tag{9.25}$$

Thus, the higher the yield strength, higher is the resilience. That is, resilient materials have high yield strength and/or low modulus of elasticity. Such materials would be useful in spring applications and also as materials for golf club heads. (The newly developed Zr-based bulk metallic glassy alloys are used for producing the new generation golf club heads because of the ability to recover the elastic energy stored in them.)

9.7 TRUE STRESS VERSUS TRUE STRAIN CURVE

We had earlier defined the terms true stress and true strain and mentioned that these values truly represent the instantaneous stress and strain experienced by the specimen. Thus, when we calculate the true stress and plot it against true strain, the curve will appear as shown in Figure 9.19. For comparison purposes, the engineering stress versus strain curve is also included. From this plot, it may be noted that the two curves are identical up to the yield point. Once plastic deformation has set in, there is a permanent decrease in the specimen diameter and consequently, the true stress is higher than the engineering stress. (This is because the engineering stress is calculated based on the original area of cross section while the true stress is calculated on the basis of the instantaneous area of cross section of the specimen.) This trend continues on further loading and the true stress continues to increase until fracture of the specimen occurs. That is, the difference between the true and engineering stresses and strains increases with the amount of plastic deformation

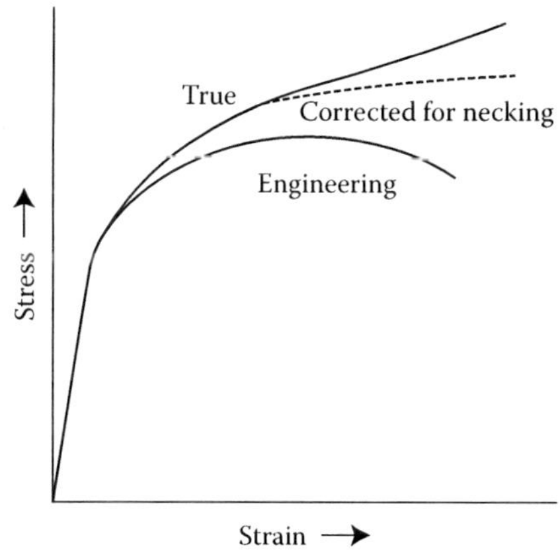

FIGURE 9.19 Comparison of the engineering stress vs. strain and true stress vs. true strain curves obtained for the same material and from the same test.

experienced by the sample. Therefore, one can understand that the "apparent" decrease in *engineering stress* beyond the point of necking is not real and that it is only a consequence of the definition of engineering stress in terms of the original cross-sectional area of the specimen.

As shown in Figure 9.9, the engineering stress reaches a maximum value at the strain value, ε_u, that is, the specimen experiences uniform elongation up to this point. Continued deformation beyond this point requires an ever-decreasing applied force due to the onset of necking. On this basis, it might be inferred that the material has exhausted its work hardening capacity at ε_u, and has begun to "work soften" for strains greater than this. This point is associated with a geometrical instability.

During necking the stress state is no longer uniaxial, that is, other stress components also exist, and therefore, the correct axial stress within the neck is slightly lower than the stress computed from the applied load and the cross-sectional area of the neck.

For many metals and alloys, the region in the true stress versus true strain curve from the point of yielding up to necking can be approximated by the equation

$$\sigma_t = K\varepsilon_t^n \tag{9.26}$$

where K and n are constants, whose values vary from material to material.

By taking logarithms on both sides of the above equation, we get

$$\log \sigma_t = \log K + n \log \varepsilon_t \tag{9.27}$$

Thus, by plotting $\log \sigma_t$ against $\log \varepsilon_t$, one gets a straight line, whose slope is n and the intercept is $\log K$, from which the value of K can be calculated.

K is called the strength coefficient (with units of stress) and the parameter n is often referred to as the *strain-hardening exponent*, a measure of the material's

TABLE 9.2
K and n Values for Some Metals and Alloys

Material	n	K (MPa)
Low-carbon steel (annealed)	0.21 (0.26)	600 (530)
4340 Low-alloy steel (annealed)	0.15	640
4340 Low-alloy steel (tempered at 315°C)	0.12	2650
304 Stainless steel (annealed)	0.44 (0.45)	1400 (1275)
Copper (annealed)	0.44 (0.54)	530 (315)
Aluminum (annealed)	0.20	180
Naval brass (annealed)	0.21	585
2024 Aluminum alloy (heat treated—T3)	0.17	780
AZ-31B Magnesium alloy (annealed)	0.16	450

work-hardening behavior. This correlates with a material's resistance to necking. The strain-hardening exponent may have values from $n = 0$ (a perfectly plastic solid) to $n = 1$ (an elastic solid). Thus, higher values of n represent improved ability to be deformed plastically during the shaping process without excessive thinning or fracture of the piece. Values of n for most metals and alloys at room temperature are between 0.02 and 0.50. Table 9.2 lists the K and n values for some metals and alloys.

9.8 GENERAL OBSERVATIONS

Even though we have so far discussed about how to obtain the mechanical properties of materials by conducting a tension test, similar properties can be obtained under compression and shear tests also. In such cases, the properties will be referred to as compressive yield strength, shear strength, compressive strain, shear strain, shear modulus, and so on. It is the shear modulus which we use to evaluate the parameters for plastic deformation. For example, it is the magnitude of the shear stress that determines if a metal crystal is going to be plastically deformed or not. We use the shear modulus of a metal to calculate the energy of a dislocation, the stacking-fault energy of a material, and so on.

For most metals and alloys, the tensile and compressive strengths will be similar. In case of some metallic alloys, however, the tensile and compressive strengths are different. In the case of ceramic materials, the compressive strength is much higher than the tensile strength, essentially due to the presence of cracks and porosity in ceramic materials.

As mentioned earlier, we can have tensile modulus, shear modulus, bulk modulus, etc. These different modulus values are different for different materials; but, these are related to each other. For example, in isotropic materials, the shear modulus G and the modulus of elasticity E are related to each other through the Poisson's ratio, ν as

$$E = 2G(1+\nu) \tag{9.28}$$

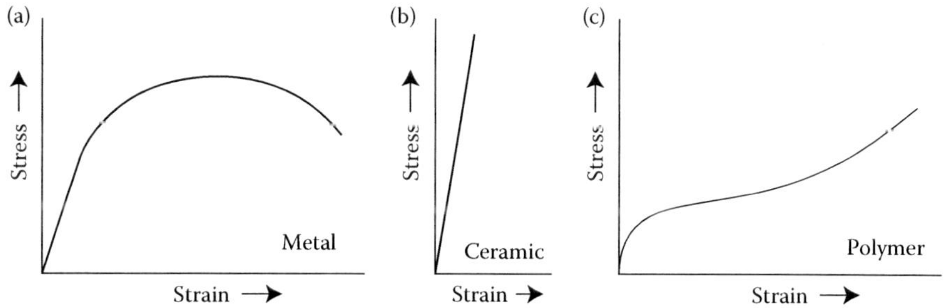

FIGURE 9.20 Typical stress–strain curves of different types of materials (metals, ceramics, and polymers) tested in tension. (a) metals, (b) ceramics, and (c) polymers.

Further, it has also been shown that a small value of the ratio of shear modulus/bulk modulus indicates a high ductility of the material.

Most of the above discussion pertains to metals and alloys. When polymeric and ceramic materials are tested in tension, their stress–strain curves look different from those of metals and alloys. Figure 9.20 presents the typical stress–strain diagrams for metals, ceramics, and polymers. It may be recalled that ceramic materials are rarely tested in tension; they are almost always tested in compression. This is essentially due to the difficulty of machining samples into shapes necessary for tensile testing, especially the threading and the reduced cross-sectional area in the gage length portion. Further, most of the ceramic materials are so brittle that they fail even during mounting of the specimens in the grips of the tensile testing machine. In comparison to metals, ceramics (1) have a higher modulus of elasticity, (2) rarely exhibit plastic deformation, and (3) often have a higher fracture stress (if the ceramic is devoid of flaws, such as cracks and porosity).

9.9 INFLUENCE OF VARIABLES ON TENSILE PROPERTIES

The mechanical properties of materials vary with different parameters. Some of the important parameters that affect the tensile properties of materials are:

1. Chemical composition
2. Microstructure
3. Temperature
4. Heat treatment
5. Prior history of plastic deformation
6. Strain rate
7. Anisotropy

9.9.1 Chemical Composition

Depending on the relative atomic sizes of the solvent and solute atoms and some other considerations, addition of solute elements into a solvent forms a solid solution.

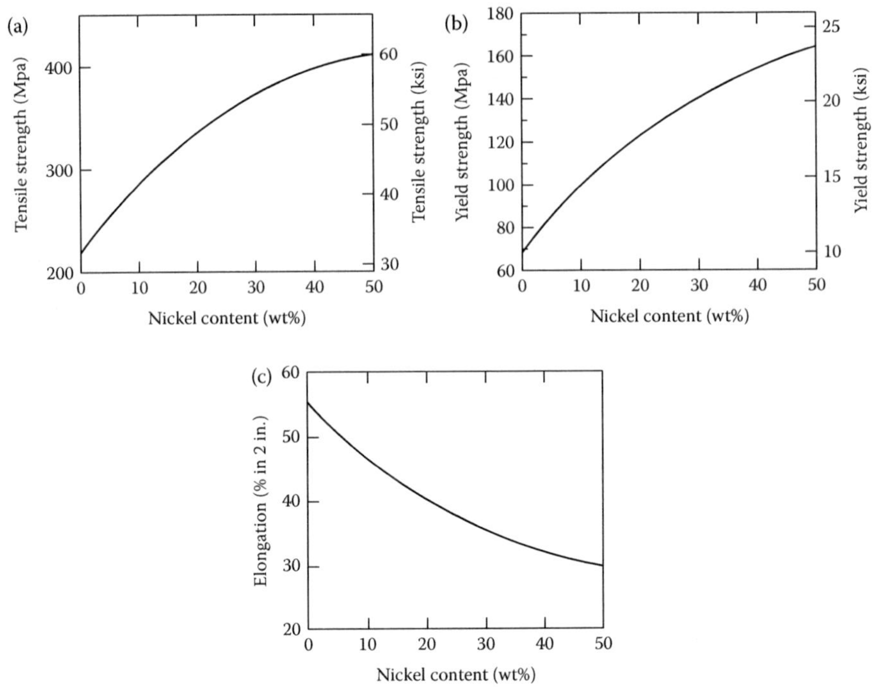

FIGURE 9.21 Effects of solid solution strengthening. The yield strength and tensile strength increase and the %elongation decreases with increasing amount of solute present in the solid solution.

Since the atomic sizes of the solute and solvent atoms are different, the strain caused by the introduction of solute atoms of a different size into the solvent matrix inhibits the movement of dislocations. This results in the strengthening of the alloy, and this phenomenon is known as *solid solution strengthening*. Consequently, solid solution alloys are always stronger (and harder) than pure metals. Even if the two elements do not form a solid solution, but an intermetallic or a mixture of two phases is formed, the alloy is still stronger than the pure metal. Figure 9.21 shows that solid solution alloys are stronger than the pure metals. When a solid solution forms, the solvent atoms experience the misfit stress due to the differences in size between the solvent and solute atoms. The rate of increase of strength per unit solute content is different for different solute elements. Both theoretical estimates and experimental results indicate that the strength measured either as the critical resolved shear stress (the minimum amount of stress required to initiate plastic deformation in a material) or the yield stress increases with $\sqrt{c}$, where c is the solute content. Figure 9.22 shows the increase in strength of solid solution alloys as a function of $\sqrt{c}$.

Addition of solute atoms can also have other strengthening effects. For example, the solute atoms can form insoluble dispersoids and their presence in the metal matrix can lead to dispersion strengthening, and the high strength of such materials can be retained up to high temperatures.

9.9.2 Microstructure

The general microstructural features, especially the grain size, have a very significant effect on the yield strength of a material—the smaller the grain size, the higher is the yield strength. This empirical relationship, known as the *Hall–Petch relationship*, can be expressed as*

$$\sigma_y = \sigma_0 + kd^{-1/2} \tag{9.29}$$

where σ_y is the observed yield strength, d is the average grain size, and σ_0 and k are constants. σ_0 may be regarded as the yield strength of the material with an infinitely large grain size (effectively, a single crystal). Grain boundaries act as barriers to the movement of dislocations. Thus when a dislocation approaches the grain boundary, which has higher yield strength than the matrix itself, the dislocation will experience a repulsive force near the grain boundary, and dislocations will pile up along the slip plane behind the grain boundary. When the stress concentrations around the tip of the slip band cause the grain boundary to yield, deformation is transferred to the next grain. This is the reason why a fine-grained material has higher yield strength than a coarser-grained material. Thus, as the grain boundary area in a fine-grained material is large and this resists the movement of dislocations, the strength of the material increases with a decrease in grain size. Note that

---

* Norman James Petch (February 13, 1917–December 9, 1992) was trained in chemistry at Queen Mary College in London, in metallurgy at Sheffield, and in physics at the Cavendish Laboratory, University of Cambridge. He was Professor of Metallurgy at Leeds University, Cochrane Professor of Metallurgy at Newcastle upon Tyne University, Pro-Vice-Chancellor, and Professor of Metallurgy at Strathclyde University. While at Cambridge, he determined the positions of carbon atoms in the complicated crystal structure of cementite in steel. He also related the cementite carbon atom positions to those in martensite, contributing to the establishment of a fundamental crystal structure basis for understanding the physical metallurgy of iron and steel. His meticulous work on the influence of polycrystal grain size on the plastic yielding and fracturing behavior of iron and steel materials at low temperatures, that was complementary to studies of the yield point behavior of steel by his Cambridge colleague, E. O. Hall, led to the well-established Hall–Petch equation for the stress dependence on reciprocal square root of average grain diameter; perhaps, the most widely quoted equation in materials science. He was awarded the American Society for Metals (ASM) Gold Medal. He was also a Fellow of the Royal Society of London.

Eric Ogilvie Hall, while at the Cavendish Laboratory in Cambridge, UK, wrote three papers in 1951, which appeared in volume 64 of the *Proceedings of the Physical Society B*. In his third paper, Hall showed that in mild steels of different grain sizes ranging from 6 to 40 μm, the difference between the lower yield strength of the steel and that of the single crystal ($\sigma_y - \sigma_0$) was inversely proportional to the square root of the grain size. Hall concentrated on the yielding properties of mild steels. This relationship followed the theoretical calculations by Eshelby, Frank, and Nabarro.

On the basis of his experimental work, N. J. Petch published a paper in 1953 independent from Hall's. Petch's paper concentrated more on brittle fracture. By measuring the variation in cleavage strength with respect to ferritic grain size at very low temperatures, Petch found a relationship exact to that of Hall's. Thus this important relationship is named after both Hall and Petch. The discovery of Hall–Petch effect has proven to be a powerful tool for providing "cheap" increased strength, and simultaneously making the material tougher.

The Hall–Petch relation was experimentally found to be an effective model for materials with grain sizes ranging from 1 μm to 1 mm. However, experiments on many nanocrystalline materials demonstrated that if the grains reached a small enough size, typically less than 10–30 nm, the yield strength would either remain constant or decrease with decreasing grain size.

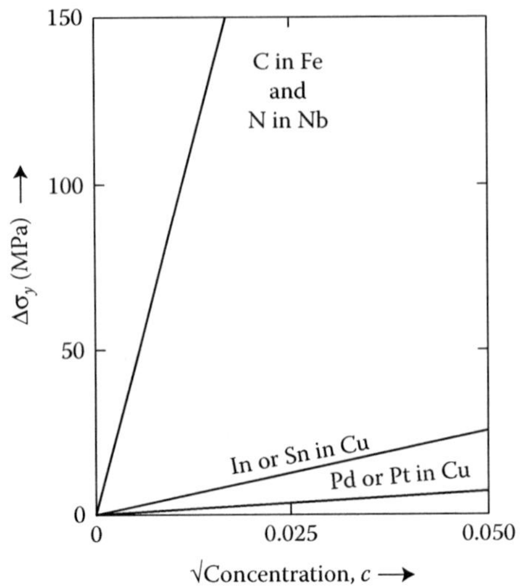

FIGURE 9.22 Effect of the nature and amount of solute on the strength of solid solution alloys. (After Ralls, K. M., T. H. Courtney, and J. Wulff, 1976. *Introduction to Materials Science and Engineering*, New York, NY: John Wiley & Sons, Inc.)

nanostructured materials (with a grain size of less than about 100 nm) have a strength of 4–5 times that of coarse-grained materials of the same composition. When the yield strength, σ_y is plotted against the reciprocal of the square root of the grain size, $d^{-1/2}$, a straight line is obtained. Thus, it can be clearly seen that the strength of a material increases with a decrease in grain size (Figure 9.23), in accordance with Equation 9.29. The slope of this straight line is k and the intercept corresponding to $d^{-1/2} = 0$ is σ_o.

Other microstructural features that could increase the strength of a material are the presence of fine dispersoid or precipitate particles, and increase of dislocation density due to cold-working operations.

9.9.3 Temperature

The mechanical properties of materials change significantly with temperature. Relative to ambient temperature, materials become stronger, but less ductile, at lower temperatures and weaker, but more ductile, at elevated temperatures. For example, the yield strength, tensile strength, and modulus of elasticity decrease with an increase in temperature. But, ductility increases with temperature (Figure 9.24). This is the reason why materials are easily fabricated at higher temperatures; one needs to apply less force to deform the material.

At low homologous temperatures (ratio of the temperature, T at which deformation is occurring and T_m, the melting point of the material, both expressed in degrees Kelvin) ($<0.3\ T/T_m$), dislocation interactions and pile-ups increase and recovery is minimal. Therefore, continued deformation requires higher stresses. On the other hand, at higher temperatures (between 0.3 and 0.5 times T/T_m), thermally activated

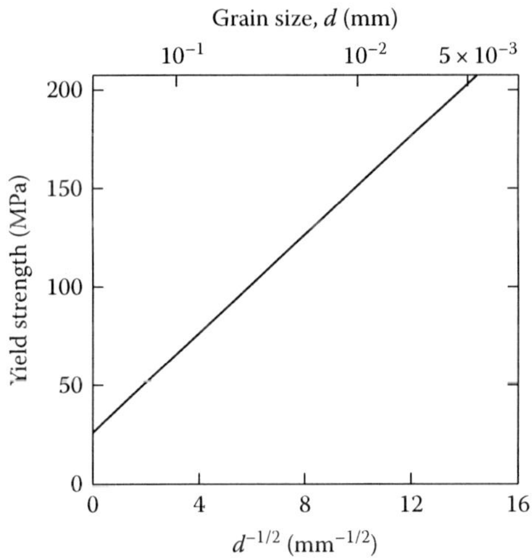

FIGURE 9.23 A Hall–Petch plot showing the linear relationship between yield strength, σ_y and reciprocal of the square root of the grain size, $d^{-1/2}$.

processes allow local stresses to be relieved and therefore strength is decreased. At still higher temperatures (>0.5 T/T_m), diffusional processes become important, recovery, recrystallization and grain growth can reduce the dislocation density, and consequently the strength of the material decreases further.

At higher temperatures some materials may undergo phase transformations. Second phase particles may dissolve, and other changes may take place. Some materials, especially those with a BCC structure may be ductile at higher temperatures, but become brittle at temperatures below the ductile–brittle transition temperature (see Chapter 10: Impact Testing for more details).

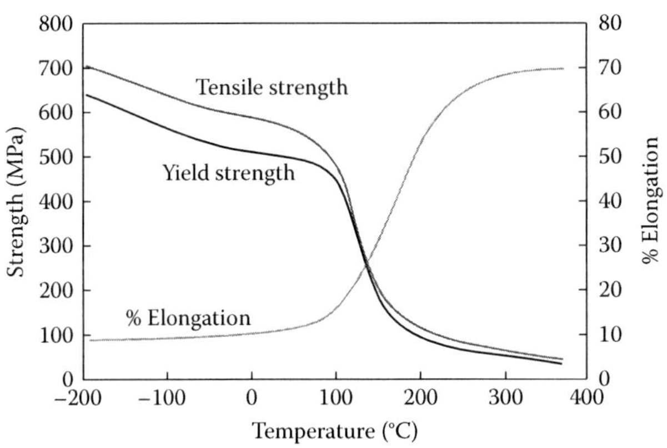

FIGURE 9.24 The effect of temperature on the mechanical properties of materials. While the yield strength and tensile strength decrease with an increase in temperature, the ductility increases.

9.9.4 Heat Treatment

Heat treatment of alloys can result in altering the nature, number, and proportion of phases in an alloy system. Consequently, the properties of materials will also be different. For example, precipitation of fine second-phase particles in a solid solution matrix can increase the strength of the alloy. This is referred to as precipitation hardening. Quenching of an austenitized steel sample can produce a martensitic structure and increase its hardness and strength. Annealing heat treatment of a cold-worked metal or alloy can decrease its strength and enable further mechanical working of the material. Thus, heat treatment methods provide an effective way of altering the mechanical properties of materials without changing their composition. The variation of mechanical properties of steels in different heat-treated conditions was discussed earlier (Chapter 6: Heat Treatment of Steels) including the reasons why heat treatment operations change the microstructure and mechanical properties.

9.9.5 Prior History of Plastic Deformation

As mentioned earlier, a material deformed in the plastic region will experience strain hardening. Thus, a prior cold-worked sample will have a higher strength than an annealed metal or alloy. Further, the higher the degree of cold work, the harder and stronger the material is, unless the specimen cracks to relieve the internal stresses. On the other hand, a specimen annealed after cold working will go through the processes of recovery, recrystallization, and grain growth and the strength decreases. Thus, the amount of cold work and/or annealing experienced by the material will determine the actual strength of the alloy. Figure 9.25 shows the increase in strength and decrease in ductility of different materials as a function of the amount of cold work.

9.9.6 Strain Rate

Strain rate, or the rate at which a specimen is deformed, can have significant effect on the mechanical properties of a material. For most materials, the strength properties tend to increase with the strain rate. Figure 9.26 shows the stress–strain curves of niobium single crystals tested at different strain rates. Notice that the yield strength of the material has increased as the strain rate has increased. The rate of strain hardening, however, decreases with increasing strain rate, according to the equation

$$\sigma_t = K'\left(\dot{\varepsilon}_t\right)^m \tag{9.30}$$

where $\dot{\varepsilon}_t$ is the true strain rate, m is the strain-rate sensitivity, and K' is a constant indicating the material flow stress at a true strain rate of 1. An important effect of strain rate is seen during impact testing of materials. The high rate of application of load causes an otherwise ductile material to fail in a brittle manner. (See Chapter 10: Impact Testing for further details and the material behavior under these test conditions.)

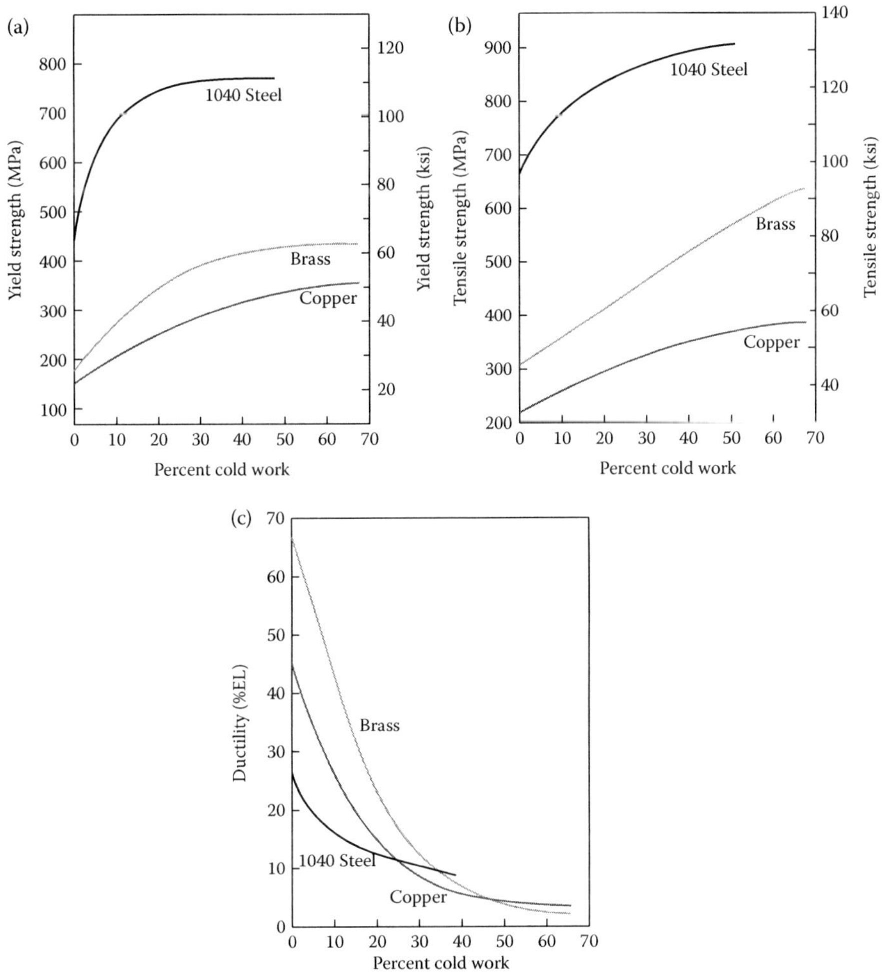

FIGURE 9.25 Effect of cold work on the strength properties of copper, brass and 1040 steel (Fe-0.4 wt% C). While the yield strength and tensile strength increase, ductility decreases with increasing percent cold work.

The strain-rate sensitivity parameter, m can be obtained by plotting log σ_t against log $\dot{\varepsilon}_t$, when a straight line is obtained. The slope of the straight line is m and the value of K' can be obtained as the intercept on the Y-axis. m has a direct influence on the ductility of materials. The higher the value of m, the higher is the ductility. The tendency to form a neck (which is responsible for the eventual failure of the materials) can be reduced at higher values of m. In fact, a large value of m (>0.3 but usually higher than 0.5) is a necessary prerequisite for observing superplastic deformation in metallic materials. Superplasticity is the ability of certain alloys to undergo extensive, neck-free, tensile deformation prior to fracture. The highest tensile elongation reported to-date is 8000% in a Cu-10 wt% Al-based alloy.

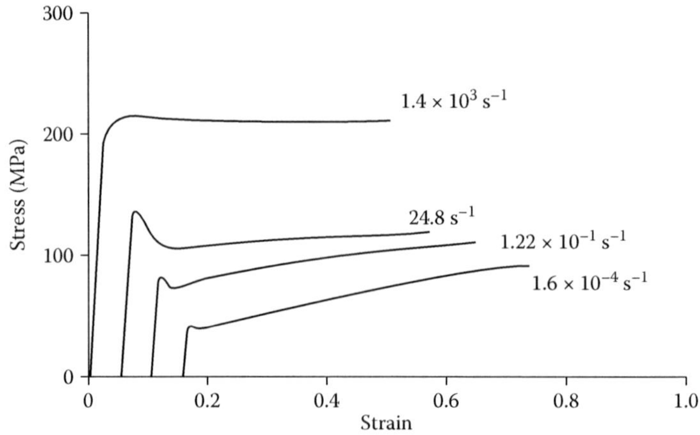

FIGURE 9.26 Stress–strain curves for single crystals of niobium at different strain rates. Note that the strength increases with increasing strain rate.

9.9.7 Anisotropy

Anisotropy means that the properties are different when measured in different directions in a single crystal specimen. However, in a polycrystalline material, the large number of grains oriented randomly and differently average out the effects and so it is not possible to detect anisotropy in a polycrystalline material. (Materials that have been either cold-rolled or deformed uniaxially may show elongation of grains in the direction of deformation. In this case also, the properties are different along the direction of rolling and in perpendicular directions. But, this phenomenon in a polycrystalline material will be referred to as texture. Similarly in fiber-reinforced composite materials, the strength properties of the composite will be higher in the longitudinal direction than in the transverse direction).

Mechanical properties such as modulus of elasticity and strength will be different depending on the direction in which the property is measured. Table 9.3 lists the values of Young's modulus, E for different materials. Note that the modulus for metallic materials is low in the <100> direction than in the <111> direction.

The effect of anisotropy on the yield strength is even stronger. Anisotropy has also an effect on magnetic and other physical properties. A common example of exploitation of anisotropy is development of cold-rolled grain-oriented (CRGO) Fe–Si alloys for use in transformer cores. The steel is cold-rolled in such a way that majority of the grains are oriented along the <100> direction, along which magnetization can be easily achieved.

Having discussed the basic definitions and deformation behavior of materials and the different properties that could be obtained from a tensile test, let us now see how a tensile test can be conducted and the way of deriving the information from the raw data.

TABLE 9.3
Modulus of Elasticity (*E*) Values (in GPa) for Polycrystalline and Single Crystals in Different Orientations of Different Materials with a Cubic Structure

Material	$E_{polycrystal}$	$E_{<111>}$	$E_{<100>}$
Aluminum	70	76	64
Copper	121	192	67
α-Iron	209	276	129
Gold	78	117	43
Tungsten	411	411	411
Diamond	—	1200	1050
Titanium carbide		429	476
Magnesium oxide	310	343	247
Sodium chloride	37	32	44

9.10 EXPERIMENTAL PROCEDURE

Properly machined specimens of commercially pure aluminum, copper, brass, and steel, suitable for tensile testing, will be provided to you. You are expected to conduct the tensile tests and obtain the different mechanical properties of these materials. For the sake of completion, let us assume that all the properties listed above (both that can be directly read off from the stress–strain curve and those that need to be calculated) will be determined.

If you have access to more than one tensile testing machine, use the one with which you are most familiar. Otherwise, read the operating instructions or get properly trained on the machine before you use the machine. An Instron or MTS type machine is most commonly available in majority of the laboratories.

Perform the test according to the sequence of steps listed below.

1. The first step involving determination of mechanical properties of the material is to accurately measure the dimensions of the sample. Mark the gage length on the specimens. Measure the diameter and gage length (sometimes, the total length of the specimen is also measured) of each of the specimens, using a calipers. It is a good practice to measure these dimensions at three or four places to ensure that the dimensions are uniform. If you find that there is a small variation in the dimensions, take the average value for both the diameter and gage length.
2. Insert any one of the specimens into the grips of the test machine and make sure that there is no play in the specimen and that the specimen is sitting properly. Make sure that the specimen is centered in the grip and that it is parallel to the direction of application of load.
3. Begin to apply the load and record the load versus elongation data. The load is plotted on the *Y*-axis and the elongation on the *X*-axis. If you are

using a strip chart to record the load versus elongation data, make sure that you record the positions of the gears, since these determine the speed with which the chart and the crosshead are moving. Also make sure that you choose the proper load cell (the maximum load that can be applied to the specimen) and that the full scale on the chart corresponds to the maximum load. Most of the modern-day machines provide the data digitally.

4. Note that the load rises rapidly in the initial stages. When necking is approaching (which you could notice by watching that the specimen cross section is decreasing in the center of the specimen), there is a large amount of extension for a small increase in the applied load. Further, the time gap between start of necking and fracture of the specimen is short. You can hear the specimen break into two pieces with a loud thud, and the pen on the chart will come back to the zero load position.
5. From the positions of the gears, you would be able to calculate the relative speeds with which the crosshead and the chart are moving. This will help in calculating the elongation of the specimen.
6. Since the original (gage) length and diameter of the specimen are known, convert the load versus elongation chart to the stress versus strain plot. Remember that stress is load/area of cross section and that strain is change in length/original length.
7. By identifying the value of 0.002 on the strain axis, that is, 0.2% strain, draw a line parallel to the initial linear portion of the stress–strain curve and calculate the 0.2% offset yield strength. Note the tensile strength and fracture strength as the maximum stress on the plot and the stress at which the specimen has fractured, respectively.
8. Calculate the modulus of elasticity as the slope of the initial linear portion of the stress–strain curve.
9. By putting the broken two halves of the specimen together, measure the final (gage) length and diameter at the point of fracture. Calculate the percentage elongation and the percentage reduction in area from these and the original dimensions of the specimen.
10. Calculate the modulus of resilience and toughness of the specimen.
11. Plot a true stress versus true strain curve by noting the instantaneous dimensions of the specimen during the test. Use about 20 points on the engineering stress versus strain curve to obtain data for the true stress versus true strain curve. Notice that while the stress continues to rise with the application of load in the true stress versus true strain curve, it appears to decrease beyond the ultimate tensile strength in the engineering stress versus strain curve.
12. Obtain values of the constants (K and n) in the power law using Equation 9.26.
13. Repeat the experiment and calculations for other specimens also.
14. Tabulate your observations and results.
15. Compare the mechanical properties you have obtained with the standard values listed in the literature. It will also be useful to visually observe the fracture surface of the specimens to determine whether it has fractured in a

ductile or brittle manner. A more accurate method will be to observe these fractured surfaces in the scanning electron microscope.

As an interesting observation, make a notch in one of the spare tensile specimens and conduct the tensile test. You will notice that the specimen fails at a much lower stress than an unnotched specimen.

9.11 OBSERVATIONS

a. *Specimen Dimensions*

Dimensions	Aluminum	Copper	Brass	Steel
Original length (in.)				
Original diameter (in.)				
Final length (in.)				
Final diameter (in.)				

b. *Load versus Displacement and Stress versus Strain Data*

Measured		Calculated			
Load (lb)	Elongation (in.)	Engineering Stress (psi)	Engineering Strain (in./in.)	True Stress (psi)	True Strain (in./in.)

Note: It is not necessary that you have to express your stress values only in units of psi. They can be expressed as kg mm^{-2} or MPa also.

9.12 RESULTS

Property	Aluminum	Copper	Brass	Steel
Proportional limit, psi				
0.2% Offset yield strength, psi				
Ultimate tensile strength, psi				
Fracture strength, psi				
% Reduction in area				
% Elongation				
Modulus of resilience, inch-pounds force/cubic in. (psi)				
Toughness inch-pounds force/cubic in. (psi)				
K, psi				
n				

9.13 ADDITIONAL EXPERIMENT

It is important to remember that the surface of the tensile specimen should be very smooth and should not contain any scratches or dents. If such surface defects are present, then they act as stress raisers and as a result the strength values measured will be lower than for a perfectly smooth specimen. To demonstrate this concept, the students could be asked to perform an additional tensile test. Take a regular tensile test specimen of standard dimensions. Using a steel file, introduce a deep scratch on the surface, preferably somewhere in the middle of the gage section of the specimen. Perform the tensile test on this specimen and note that the strength values are lower than for a perfect material without any scratches or dents on the surface of the specimen.

EXERCISES

Define the following terms:

9.1 Stress.
9.2 Strain.
9.3 True stress.
9.4 True strain.
9.5 Young's modulus.
9.6 Yield strength.
9.7 Ultimate tensile strength.
9.8 Fracture strength.
9.9 Percentage elongation.
9.10 Percentage reduction in area.
9.11 Toughness.
9.12 Resilience.
9.13 Ductility.
9.14 Poisson's ratio.

Distinguish between

9.15 Engineering stress and true stress.
9.16 Engineering strain and true strain.
9.17 Elastic deformation and plastic deformation.
9.18 Upper yield point and lower yield point.

A steel specimen, with a diameter of 0.5″ and a gage length of 2″ is subjected to uniaxial tension. Calculate

9.19 The stress if it is subjected to a load of 5000 lb.
9.20 The strain if it is extended to 2.15″.
9.21 The yield strength when the corresponding load is 7000 lb.
9.22 The ultimate tensile strength if the maximum load is 10,000 lb.
9.23 Fracture strength if the load at which the specimen broke was 9000 lb.
9.24 Percentage elongation if the final gage length of the specimen is 2.45″.
9.25 Predict the microstructure of the above specimen after the test.
9.26 Predict the type of fracture for the above specimen.
9.27 Why do mild steel samples exhibit an upper yield point?
9.28 What will happen to the strength of the alloy at high temperatures?
9.29 What is uniform elongation?
9.30 What is strain hardening?
9.31 Briefly explain the phenomenon of solid solution strengthening.
9.32 What is the effect of grain size on the yield strength of a material?
9.33 Why is it easy for materials to be plastically deformed at higher temperatures?

9.34 What is the effect of strain rate on the strength of a material?
9.35 What is specific strength?
9.36 What conditions of the stress–strain diagram produce high toughness values?
9.37 What are upper and lower yield points?
9.38 What type of materials exhibit serrated yielding?
9.39 Compare the microstructure of a pure aluminum sample before and after the tensile test.
9.40 What is superplastic deformation?
9.41 Why is the shape of the engineering stress versus engineering strain curve different from that of the true stress versus true strain curve?
9.42 List three factors that determine the shape of the stress–strain curve during tensile testing.

FURTHER READING

Callister, Jr., W. D. and D. G. Rethwisch. 2010. *Materials Science and Engineering, An Introduction*, 8th edition. New York, NY: John Wiley & Sons, Inc.

Courtney, T. H. 2000. *Mechanical Behavior of Materials*, 2nd edition. New York, NY: McGraw-Hill.

Dieter, G. E. 1986. *Mechanical Metallurgy*, 3rd edition. Boston, MA: McGraw-Hill.

Dowling, N. E. 2007. *Mechanical Behavior of Materials*, 3rd edition. Upper Saddle River, NJ: Pearson: Prentice-Hall.

Hertzberg, R. W. 1996. *Deformation and Fracture Mechanics of Engineering Materials*, 4th edition. New York, NY: John Wiley & Sons, Inc.

Meyers, M. A. and K. K. Chawla 2008. *Mechanical Behavior of Materials*, 2nd edition. Cambridge, UK: Cambridge University Press.

Ralls, K. M., T. H. Courtney, and J. Wulff, 1976. *Introduction to Materials Science and Engineering*, New York, NY: John Wiley & Sons, Inc.

10 Impact Testing

10.1 INTRODUCTION

The fracture behavior of a material is different when it is tested at low strain rates (i.e., the rate at which the specimen is strained, $d\varepsilon/dt$ where ε is the strain and t is the time) or high strain rates. The strain rate is low during conventional tensile testing and is typically about 10^{-4}–10^{-2} s^{-1}. A material may exhibit *ductile fracture* under these conditions. However, if the same material is subjected to a sudden, intense blow, that is, when loading occurs at high strain rates, it may behave in a much more brittle manner than under low strain rates. Such a behavior can be seen very clearly in the case of plastics. Let us take a plastic such as polyethylene or Silly Putty and subject it to stretching slowly. The polymer molecules will have time to disentangle or for the chains to slide past each other and cause large amounts of elongation. On the other hand, if the loading is done suddenly, that is, at high strain rates, the molecules will not have sufficient time to disentangle and therefore the specimen will experience *brittle fracture*, that is, the specimen will break suddenly. Thus, the rate of loading plays an important role in determining the fracture behavior of a material. Accordingly, an *impact test* is frequently used to ascertain the brittleness of a material. During impact loading, the strain rates are typically high, about 10^3 s^{-1}, that is, 5–7 orders of magnitude faster than during a conventional tensile test.

Just as hardness is the analog of strength measured by the tensile test, *impact energy*, the energy necessary to fracture a standard test piece under an impact load, is the analog of toughness. Prior to the advent of *fracture mechanics* as a scientific discipline in the 1960s, impact testing techniques were the only methods used to evaluate the fracture behavior of materials. But, impact testing is done at present also to evaluate the fracture behavior of materials.

Brittle fracture is of concern in equipment designed for use at low temperatures. For example, sinking of the Royal Mail Ship "Titanic" in the Atlantic Ocean in 1912 during its maiden voyage is attributed to the brittle failure of the hull at the very low temperatures of the seawater encountered in the ocean (–2°C). The fracture of a large number of car axles in Alaska during the winter of 1988–1989 was attributed to the severe weather conditions.

10.2 IMPACT-TESTING TECHNIQUES

Impact test conditions are chosen to be most severe so as to easily cause brittle fracture. The three basic factors that contribute to a brittle-cleavage-type fracture are:

1. High strain rate, that is, rapid rate of deformation
2. Stress concentration
3. Triaxial stress state, which may be introduced by the presence of a notch

Any one or all of these factors can be individually (or together) responsible for the brittle behavior of materials; all three of these factors need not be present at the same time to produce brittle fracture. Metallic specimens lose their strength at elevated temperatures. Conversely, they have a higher strength and lower percentage elongation to fracture at lower temperatures. Therefore, testing of specimens at low temperatures enhances the brittle behavior of materials. Many steels tend to lose toughness as very low (subfreezing) temperatures are approached. The notched-bar impact test provides a means of determining the embrittling temperature range for different compositions.

Many test procedures have been designed to evaluate the impact energy of materials. Of these, two have gained acceptance by the materials community. These are the *Charpy test* and the *Izod test*. The test specimen may be either notched or unnotched. The use of a V-notch in the specimens will be better to measure the resistance of the material to crack propagation. While the Charpy test can be used for most of the metallic materials, the Izod test is frequently used for nonmetallic (polymeric) materials. Further, while the Charpy test is used more commonly in the United States, the Izod test was favored in Great Britain, but is rarely used nowadays. We will, however, describe both the testing methods.

10.2.1 Charpy Impact Test

Figure 10.1 shows the dimensions of the specimen used for the Charpy impact test. The specimen is in the shape of a bar of square cross section, into which a V-notch has been machined. The specimen has a cross section of 10 mm × 10 mm and a length of 55 mm. A 45° V-notch, 2-mm deep with a 0.25 mm root radius, is made half-way across the length of the specimen. The V-notch is cut to specifications with a special milling cutter. Other types of notch have been used, but they are not popular. The function of the notch is to insure that the specimen will break as a result of

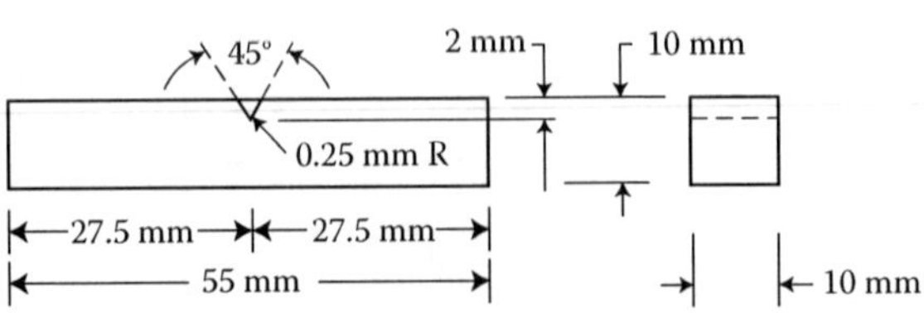

FIGURE 10.1 Dimensions of the Charpy test specimen.

the impact load to which it is subjected. Without the notch, many alloys would simply bend without fracturing, and it would therefore be impossible to determine their ability to absorb energy.

The Charpy specimen is held in a rigid vise and supported as a beam in a horizontal position and loaded behind the notch by the impact of a heavy swinging pendulum (Figure 10.2). This can be considered a three-point loading condition. The pendulum is heavy and the impact velocity of the pendulum is approximately 16 ft s^{-1} (4.5 m s^{-1}). Typically, it is expected to be between 10 and 20 ft s^{-1} (3–6 m s^{-1}). A schematic of the Charpy impact testing setup is shown in Figure 10.3.

The testing is relatively simple and the principle is straightforward. The specimen is positioned as shown in Figures 10.2, with the notch facing *away* from the direction of impact of the pendulum. The load is applied as an impact blow from the heavy pendulum, which is released from a cocked position, at a fixed height, h. Upon release, the pendulum swings through its arc, and a knife-edge mounted on the pendulum strikes and fractures the specimen at the notch. The notch acts as a point of stress concentration for this high-velocity impact blow. Thus, the net test result is to subject the sample to elastic deformation, plastic deformation, and fracture in rapid succession. The pendulum continues its swing after impacting the specimen, rising to a maximum height of h', which is lower than h. The impact energy (i.e., energy absorbed by the specimen during failure) is calculated from the difference in potential energy of the pendulum before and after fracture of the specimen. Thus, the impact energy is given by

$$\text{Impact energy} = mg(h - h') \tag{10.1}$$

where m is the mass of the pendulum and g is the acceleration due to gravity. The impact energy is usually expressed in foot-pounds (ft-lb) or joules (J) (1 J = 0.737 ft-lb

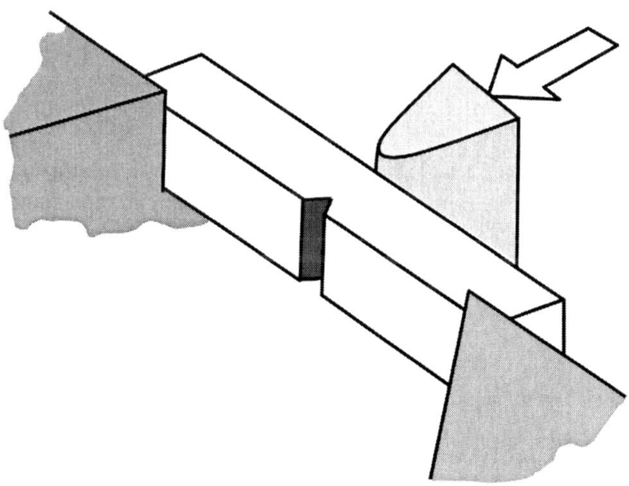

FIGURE 10.2 Placement of the specimen for the Charpy impact testing.

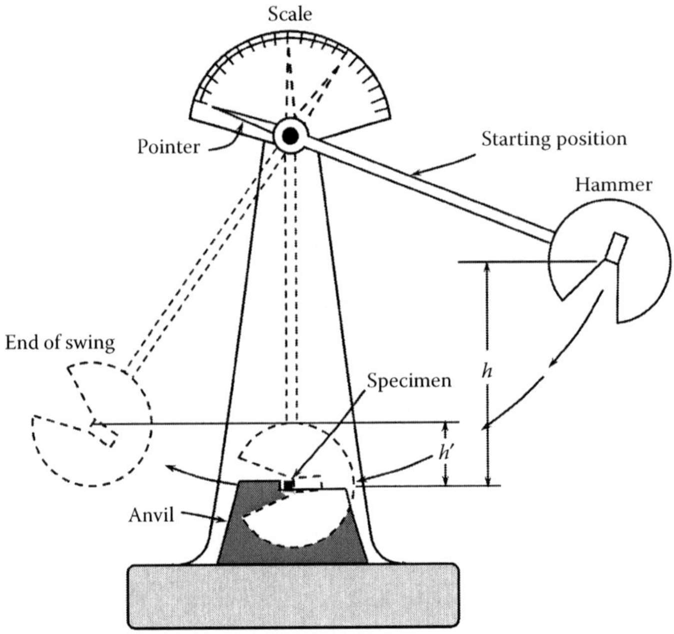

FIGURE 10.3 A schematic of the Charpy impact testing setup.

or 1 ft-lb = 1.356 J). Modern machines are equipped with scales and pendulum-actuated pointers, which yield direct readings of energy absorption. A typical commercial instrument used for standard Charpy and Izod impact tests is shown in Figure 10.4.

10.2.2 Izod Impact Test

The Izod test differs from the Charpy test in the size of the specimen and the way in which the specimen is mounted. The Izod test specimen has a square cross section of 10 mm × 10 mm and a length of 75 mm with a 2 mm deep 45° notch. But, the notch is not in the center of the specimen. Specimens with a circular cross section have also been used; but, these are not very common. The specimen is placed vertically and the impact hammer strikes the specimen above the notch and breaks it. Thus, this type of loading can be considered a cantilever-beam loading. Note that the notch faces the striking hammer in the Izod test whereas in the Charpy test, the pendulum strikes the surface away from the notch. Like in the Charpy test, the hammer continues its swing and the impact energy (energy absorbed by the specimen) is measured in exactly the same way as in the Charpy test. The dimensions of the Izod test specimen and positioning of the specimen with respect to the pendulum striking direction are shown in Figures 10.5 and 10.6, respectively.

The Izod impact energy is expressed in units of foot-pounds per inch (ft-lb/in) or joules per meter (J/m).

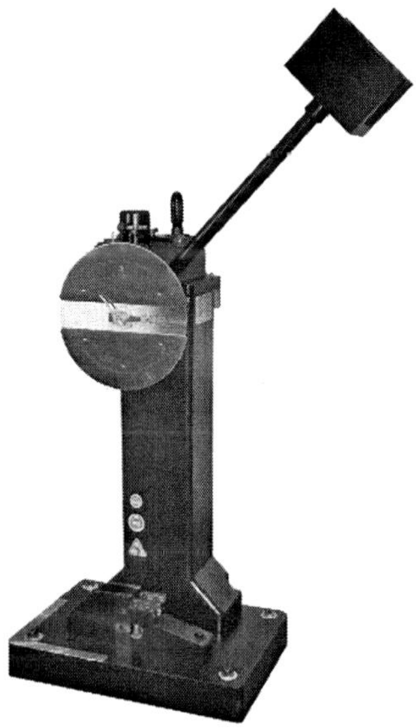

FIGURE 10.4 A commercial pendulum-style instrument used for standard Charpy and Izod impact tests. (Photo courtesy of Instron Corporation. With permission.)

10.3 DUCTILE–BRITTLE TRANSITION

Some materials fail in a ductile manner at high temperatures and in a brittle manner at low temperatures. One of the primary functions of the Charpy or Izod tests is to determine whether or not a material experiences the ductile-to-brittle transition with decreasing temperature, and if so the temperature range over which this occurs. For this purpose, a number of samples are tested at different temperatures and the energy absorbed by the specimen at these different temperatures is determined.

We had seen earlier (Chapter 9: Tensile Testing) that ductile fracture of materials occurs with lot of energy absorbed by the specimen. In contrast, brittle fracture occurs with little or almost no energy absorbed. Since the specimen absorbs energy during fracture of the material, this absorbed energy could be related to the fracture behavior of the material. Thus, the higher the energy absorbed by the specimen, the more ductile is the fracture.

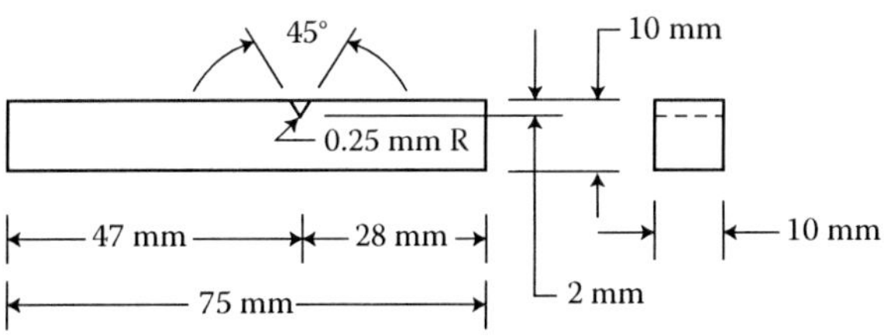

FIGURE 10.5 Dimensions of the Izod test specimen.

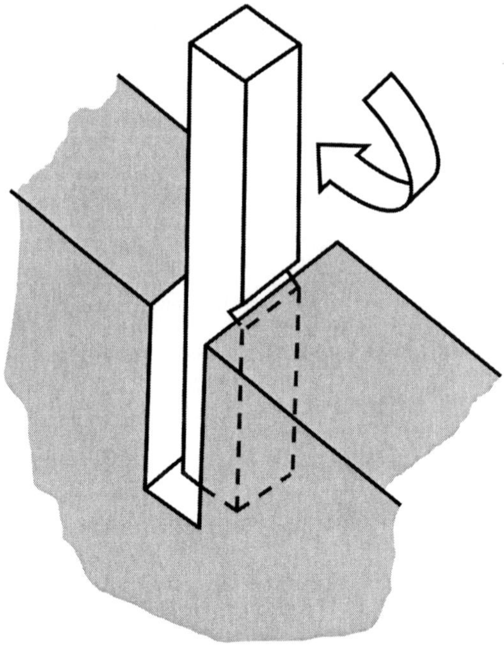

FIGURE 10.6 Positioning of the specimen for the Izod impact testing.

When the energy absorbed by the specimen is plotted against temperature, one obtains a plot, as shown in Figure 10.7. Note that the energy absorbed is constant and very high at high temperatures, in conformity with a ductile fracture. This value is referred to as the *upper-shelf energy* or value in the literature. With decreasing temperature, the energy absorbed starts to drop suddenly over a relatively narrow temperature range. Below this temperature range, the energy again remains constant but with a small value (referred to as the *lower-shelf energy*), due to the brittle fracture the specimen experiences at these low temperatures. The energy determined this way is referred to as the *Charpy V-Notch (CVN) energy.*

10.4 DETERMINATION OF DUCTILE–BRITTLE TRANSITION TEMPERATURE

As mentioned above, the specimen fractures in a ductile fashion at high temperatures and in a brittle manner at low temperatures. Ductile fracture occurs slowly while brittle fracture occurs suddenly. That is, there will not be any warning for the user or operator when brittle failure occurs in the material. But, one could watch the specimen deform and change shape slowly for ductile materials. Thus, in practice it is desirable that a material is used above a temperature at which the material will fracture in a ductile manner and not in a brittle manner. This temperature, above which the material fractures in a ductile manner, and below which it fractures in a brittle manner, is referred to as the *ductile–brittle transition temperature* (DBTT). Generally,

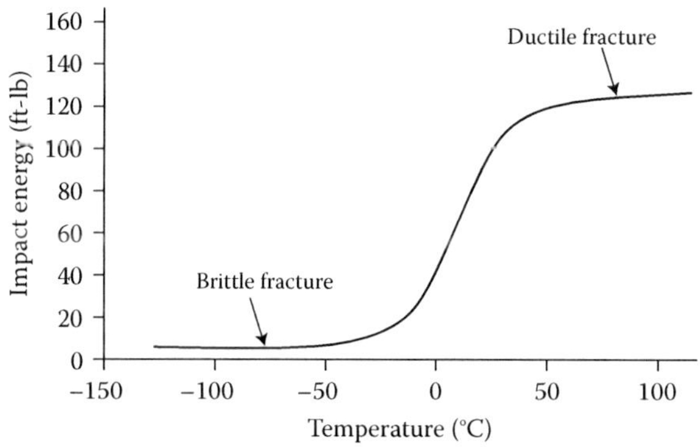

FIGURE 10.7 Energy absorbed by the test specimen as a function of test temperature.

DBTT is not a well-defined temperature; the transition occurs over a range of temperatures. The DBTT can be conveniently measured as *the temperature (or a narrow temperature range) at which the absorbed energy shows a rapid decrease with decreasing temperature*. However, for many materials, the ductile-to-brittle transition occurs over a wide range of temperatures. It is then difficult to specify a single DBTT and no specific criteria have also been established to define this temperature. For most metallic materials (especially steels), the DBTT is in the range of −100°C to +100°C, depending on the alloy composition and test conditions.

The DBTT may be determined or estimated in different ways and a brief description of the different ways of doing this is provided below.

1. The most conservative criterion is to select the upper-shelf temperature at which the fracture is entirely ductile. This transition temperature criterion is called the *fracture transition plastic* (FTP). The probability of brittle fracture is negligible above the FTP. The use of FTP is, however, very conservative and its use in many applications may be impractical.

   Another well-defined criterion is to consider that temperature at which the fracture is 100% brittle. This temperature is known as the *nil ductility temperature* (NDT). Below the NDT, the probability of ductile fracture is negligible.
2. The point of intersection of the tangent of the steep drop portion of the impact energy with the high constant value at high temperatures is called the *upper transition temperature*. Similarly, the point of intersection of the tangent of the steep drop portion of the impact energy with the low constant value at low temperatures is called the *lower transition temperature* (Figure 10.8a). The average of these two temperatures may be taken as the DBTT.
3. Another method of estimating the DBTT is to identify the temperature at which the absorbed energy is 15 ft-lb (or 20 J). This value was chosen because it was found that, based on extensive tests on World War II steel ship plates, brittle fracture would not initiate if the impact energy was 15 ft-lb at the test temperature. The temperature so chosen is frequently

referred to as the *ductility transition temperature*. The ductility transition temperature has become an accepted criterion for low-strength ship steels. However, for other materials, this 15 ft-lb value has no special significance.

This method is, however, not recommended since the temperature thus estimated is usually much lower than the safe operating temperature for the material (Figure 10.8b).

4. The third method is to designate that temperature as the DBTT, at which the impact energy is the average of the highest and lowest values (Figure 10.8c). For example, if the highest impact energy measured is 100 ft-lb and the lowest is 5 ft-lb, then the average energy is 52.5 ft-lb. The temperature corresponding to the energy of 52.5 ft-lb is then designated as the DBTT.
5. One could observe the fracture surface and determine whether the specimen had fractured in a ductile or brittle manner. The surface of the fractured specimen is fibrous if the fracture is ductile. That is, the surface produces features with a fine silky appearance, and evidence of deformation caused

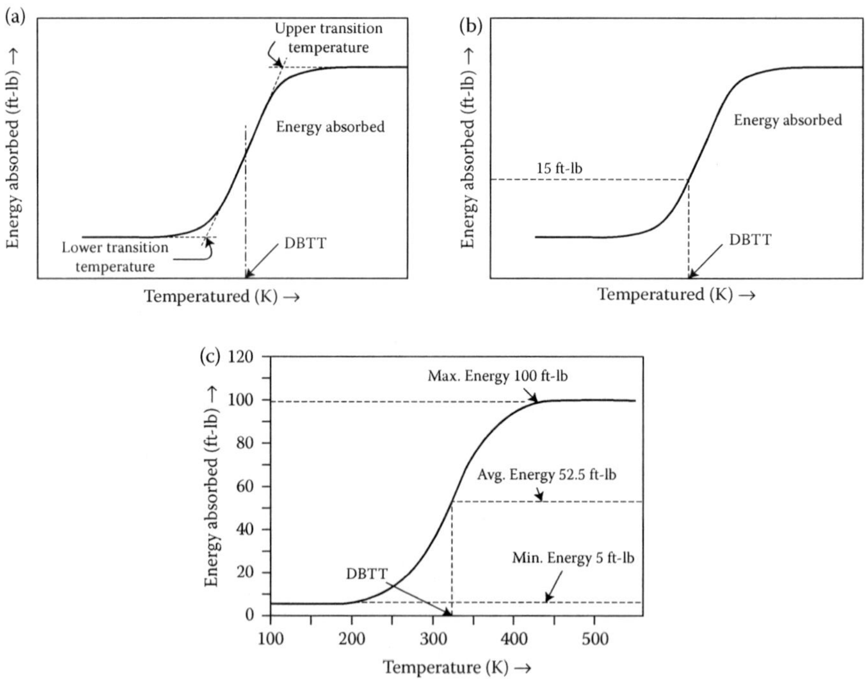

FIGURE 10.8 Methods of evaluating the DBTT. (a) Energies corresponding to points of intersection of the tangents to the steep portions of the curve with the upper and lower constant energy values are known as the upper transition and lower transition temperatures, respectively. The average of these two temperatures is taken as the DBTT. (b) Temperature corresponding to an absorbed energy of 15 ft-lb (or 20 J). (c) Temperature corresponding to the average of the highest and lowest energies absorbed by the specimen.

by bending before fracture. Dull or rough surfaces are other adjectives used to describe the ductile behavior. On the other hand, brittle fracture results in a surface which appears coarsely crystalline (granular) and shows no evidence of deformation by bending. The different modes of failure can readily be distinguished even without observing them at a high magnification. The flat facets of cleavage (brittle) fracture provide a high reflectivity and bright appearance. On the other hand, the dimple surface of a fibrous ductile fracture provides a light-absorptive surface and dull appearance. The fibrous (ductile) fracture appears first around the outer surface of the specimen where triaxial constraint is the least (Figure 10.9). Alternatively, if the nature of fracture is not clear, the fracture surfaces could be observed at high magnifications in an SEM and the nature of fracture (ductile or brittle) and the extent of brittle (or ductile) fracture could be determined. One could then use the fracture appearance as a measure of the DBTT. The most conservative estimate of DBTT would be to evaluate the temperature at which the fracture is 100% ductile. Another method is to evaluate the temperature at which the fracture is 50% brittle and 50% ductile. This temperature, at which the fracture is 50% ductile and 50% brittle, is referred to as *fracture appearance transition temperature* (FATT). In most cases, the DBTT and FATT are nearly the same. Figure 10.10 shows the method of determining the FATT and Figure 10.11 shows the fracture appearance of some steel specimens tested at different temperatures.

Another quantitative measure of the notch toughness is the degree of ductility or deformation (lateral expansion or notch root contraction). For example, if the amount of lateral expansion on the compression side of the bar is measured, it is found that the lateral expansion increases with increasing temperature. It is small at low temperatures and large at high temperatures. Figure 10.12a shows how the lateral expansion is measured by placing together the fractured surfaces and Figure 10.12b

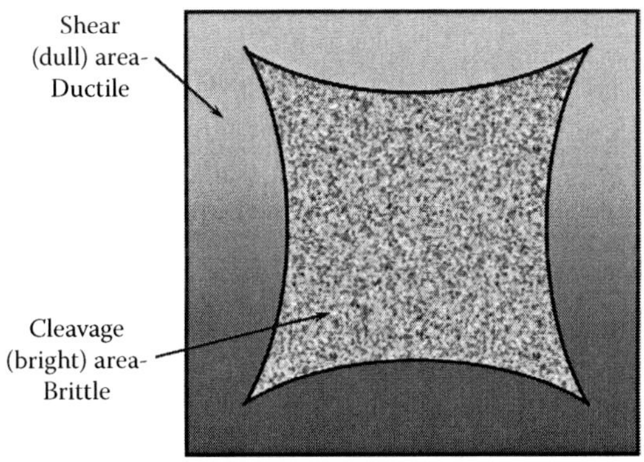

FIGURE 10.9 Schematic of the appearance of a fracture surface. The dull outer portion represents ductile fracture, while the bright interior portion represents brittle fracture of the specimen.

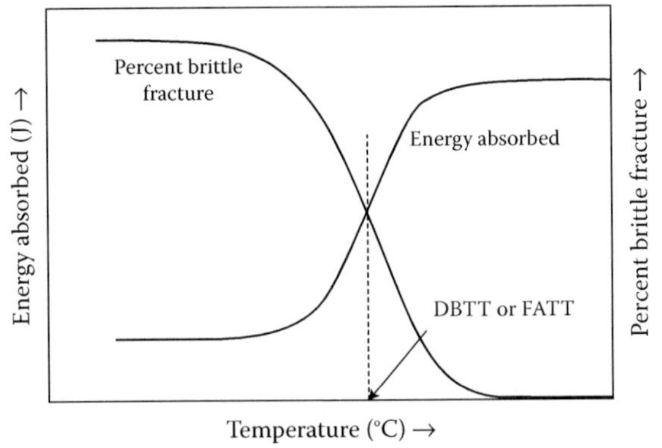

FIGURE 10.10 DBTT measured based on the appearance of the fracture surface.

shows the temperature dependence of the lateral expansion showing the transition behavior.

In reality, the DBTT measured by these different methods may be different. In such cases, the most conservative transition temperature is that at which the fracture is 100% ductile.

Table 10.1 lists the impact energy data of different materials determined by the Charpy V-Notch method.

Note: Sometimes, practitioners estimate the upper transition temperature and lower transition temperature. Although not common for metallic materials, the upper and lower transition temperatures seem to acquire importance for polymeric materials. In the glassy state (i.e., at temperatures below the glass transition temperature, T_g), molecular motion is essentially frozen and therefore the material behaves in a brittle fashion.

A serious drawback of the conventional pendulum-type testers (Charpy or Izod, as shown in Figure 10.4) described above is that they provide only one value—the total impact energy—and nothing else. The conventional tests cannot provide

FIGURE 10.11 Variation in fracture surface appearance of A36 steel specimens tested at different temperatures. The temperatures indicated are in °C. (After Hertzberg, R. W. *Deformation and Fracture Mechanics of Engineering Materials*, 4th edition. John Wiley & Sons, Inc., New York, NY, 1996. With permission.)

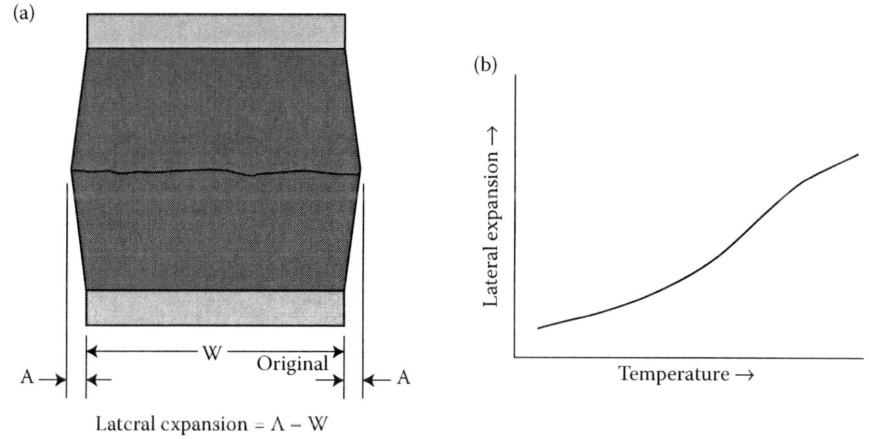

FIGURE 10.12 (a) Method to determine the lateral expansion at the compression side of the Charpy bar. (b) Temperature dependence of the lateral expansion showing the transition behavior.

additional information on the ductility, dynamic toughness, fracture, and yield loads or the behavior of the specimen during the entire impact event. Therefore, instrumented impact testers have been developed. In these instruments, available commercially (Figure 10.13), a free-falling dart (or tup) is dropped from a specified height onto a test specimen. The force is measured continuously while the specimen is penetrated. The data are used to determine the type of failure and maximum load, in addition to the amount of energy required to fracture the specimen. By studying the shape of the load-time or load-deflection curve, the type of failure can be analyzed, and important information about its performance in service can be gathered. The new piezoelectric-equipped strikers offer increased sensitivity, allowing testing of a variety of materials including foam containers for eggs and ultrathin films used in packaging industry.

10.5 EFFECT OF VARIABLES ON IMPACT ENERGY

The impact energy is affected by several material variables and the specimen design features. These include specimen size and shape, notch configuration and depth, chemical composition, and grain size. Further, differences in the Charpy properties from batch to batch can be really large. Let us now discuss in some detail the effect of these variables on the impact energy of materials.

10.5.1 Notch Sensitivity

Notches caused by poor machining, fabrication, or design concentrate stresses and reduce the toughness (and impact energy) of materials. The notch sensitivity of a material can be evaluated by comparing the absorbed energies of notched and unnotched specimens. The absorbed energies are much lower in notched specimens if

TABLE 10.1
Impact Energies of Different Types of Materials

Material	Impact Energy [J (ft-lb)]
Metallic Materials	
Mild steel	203 (150)
1040 Carbon steel	180 (133)
4340 alloy steel, Hardened	18 (13)
4340 alloy steel, Annealed	106 (78)
8630 Low-alloy steel	55 (41)
410 Stainless steel	34 (25)
Monel 400 (Nickel alloy)	298 (220)
50:50 Solder (Lead alloy)	21.6 (16)
Aluminum	27 (20)
Aluminum bronze (Cu-9 wt%Al alloy)	48 (35)
CP titanium (99.2% pure)	43 (32)
Ti-5.2 wt%Al-2.5 wt%Sn	26 (19)
Ti-6 wt% Al-4 wt% V	19 (14)
Polymeric Materials	
High-density polyethylene	0.4–14
Polyvinyl chloride (PVC)	0.4–20
Polycarbonates	12–16
Polystyrene	0.2–0.5
Polystyrene (high impact)	0.5–4
Epoxy	0.2–3.0
Epoxy (fiberglass reinforced)	2–30

Note: The Impact Energy is Determined by the Charpy V-Notch Method for Metallic Materials and by the Izod Method for Polymeric Materials. Impact energy values for the metallic materials are from the notched Charpy tests. For the polymeric materials, they are from the notched Izod tests and are in units of ft-lb/in.

the material is notch sensitive. As an illustrative example, if the tensile strength of two specimens of a material (of identical composition and microstructure) is measured—one without a notch and the other with a notch—the strength will be substantially lower in the notched specimen. Such a behavior can be easily demonstrated by taking two standard tensile specimens, putting a notch in the gage section of one of the specimens, and measuring the tensile strengths of both the specimens. (This could be done as an additional experiment, or just for curiosity, in the Tensile Testing section.)

10.5.2 Configuration of the Notch

A surface crack permits lower absorbed energies than a V-notch. The DBTT is also dependent on the shape of the notch in the specimen. For identical materials, the sharper the notch, the higher is the apparent transition temperature.

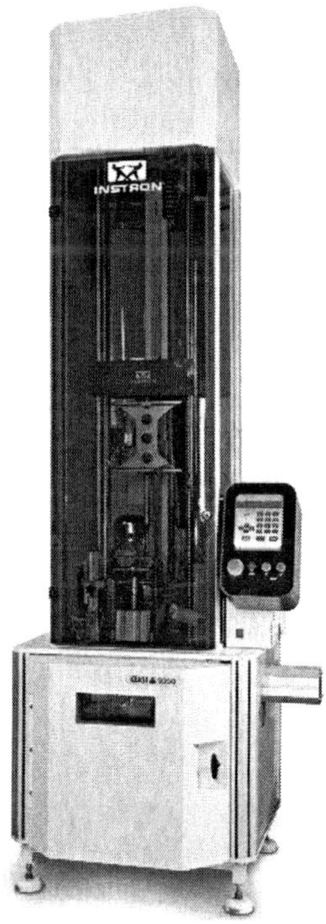

FIGURE 10.13 A commercial drop weight tester. (Photo courtesy of Instron Corporation. With permission.)

10.5.3 Size of Specimen

Since it is more difficult for a thick material to deform, smaller energies are required to break thicker materials; thus lower energies are absorbed by them. Sometimes, it is also possible that the thicker specimens contain cracks which are completely inside and these could reduce the impact energy of the specimen. Stated differently, the transition temperature increases with increasing thickness of the sample (Figure 10.14).

10.5.4 Rate of Application of Energy

Higher rate of application of energy to the specimen reduces the absorbed energy and increases the DBTT. However, it was reported that the greatest shift in the transition temperature due to increased strain rate was found in the low-strength steels. No apparent strain-rate sensitivity was present in alloys with strengths in excess of 825 MPa.

10.5.5 Crystal Structure of Material

Generally, materials with an FCC crystal structure fracture in a ductile fashion, while materials with a HCP structure fail in a brittle manner. Figure 10.15 shows that the impact energy is high for FCC materials and low for HCP materials. However,

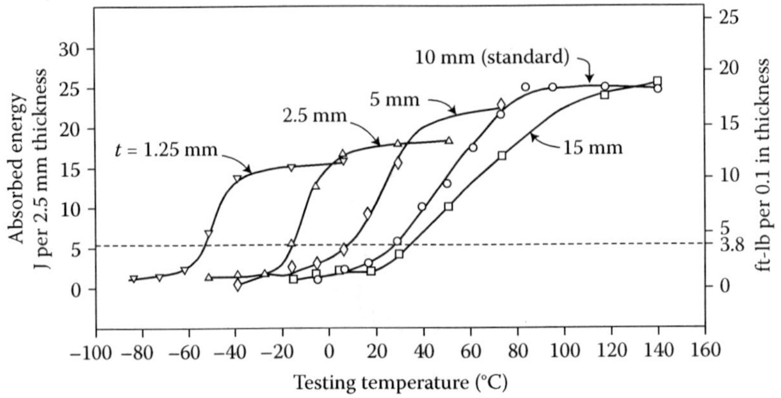

FIGURE 10.14 Absorbed energy vs. temperature curves for 38-mm-thick plate of A283 steel tested with Charpy V-notch specimens of various thicknesses. (After Hertzberg, R. W. *Deformation and Fracture Mechanics of Engineering Materials*, 4th edition. John Wiley & Sons, Inc., New York, NY, 1996. With permission.)

materials with a BCC structure show a change in the fracture behavior from ductile to brittle as the temperature decreases. This ductile-to-brittle transition in BCC alloys can be attributed to the slower dislocation mechanics for these alloys compared to that for FCC or HCP alloys. Increasing yield strength, combined with decreasing dislocation velocities, at decreasing temperatures is responsible for the brittle failure of these alloys.

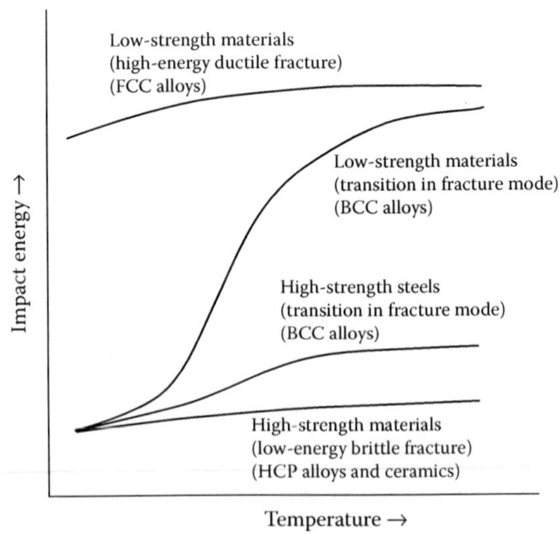

FIGURE 10.15 Influence of crystal structure on the impact energy as a function of temperature. Whereas metals with the FCC structure do not show any ductile–brittle transition, metals with a BCC structure do exhibit the transition.

10.5.6 Chemical Composition

The chemical composition of steel has a major role in determining the DBTT. Most alloying elements in steel raise the transition temperature, although the rate of increase is different for different alloying elements. The transition temperature increases by about 25°F for each increase of 0.1 wt% C and decreases by about 10°F for each increase of 0.1 wt% Mn. A Mn:C ratio of at least 3:1 is required for satisfactory notch toughness.

Phosphorus also has a strong effect in raising the transition temperature. For each increase of 0.01 wt% P, the transition temperature increases by about 13°F. The presence of nitrogen is also considered detrimental. Si (in amounts in excess of 0.25 wt%) and Mo also raise the transition temperature, while Cr has little effect. But, Ni is generally accepted to be beneficial.

It was reported that oxygen, in excess of 0.003 wt% produced intergranular fracture in high-purity iron. When the oxygen content was raised from 0.001 to the high value of 0.057 wt%, the DBTT increased from 5 °F to 650°F. That is why the steel deoxidation practice has an important effect on the DBTT.

Figure 10.16 shows the variation of Charpy impact energy with temperature for two different types of steels. Figure 10.16a shows the variation of DBTT as a function of carbon content in plain carbon steels and Figure 10.16b shows the effect of Mn content in Fe–Mn–0.5C steel. It may be noted that the DBTT decreases with decreasing carbon content and increasing Mn content.

10.5.7 Grain Size

The DBTT is found to decrease with a decrease in grain size in steels. For example, an increase of one ASTM number in the ferrite grain size (i.e., decrease in the ferrite grain size in case of plain carbon steels or austenite grain size in case of high-alloy steels) results in a decrease in the DBTT by about 30°F for mild steel. (The concept of ASTM grain size number was discussed in Chapter 3: Optical Microscopy and Chapter 7: Hardenability of Steels.) It was reported that decreasing the grain size from ASTM grain size #5 to #10 resulted in the decrease of DBTT from 70°F to 60°F.

Addition of small amounts of Nb and V results in the formation of fine NbC or VC precipitates during transformation. This dispersion strengthening raises the strength of the steel and at the same time retards grain growth. The impact resistance is consequently improved. Thus, it is reasonable to expect that ultrafine grained and nanocrystalline materials exhibit a very low DBTT. Therefore, these materials are expected to be reasonably ductile at temperatures much lower than their strength properties suggest.

10.5.8 Irradiation

Owing to irradiation in nuclear reactors, pressure vessel steels lose part of their toughness. The DBTT is found to increase markedly to higher temperatures and the upper-shelf value is noted to decrease as a result of irradiation (Figure 10.17).

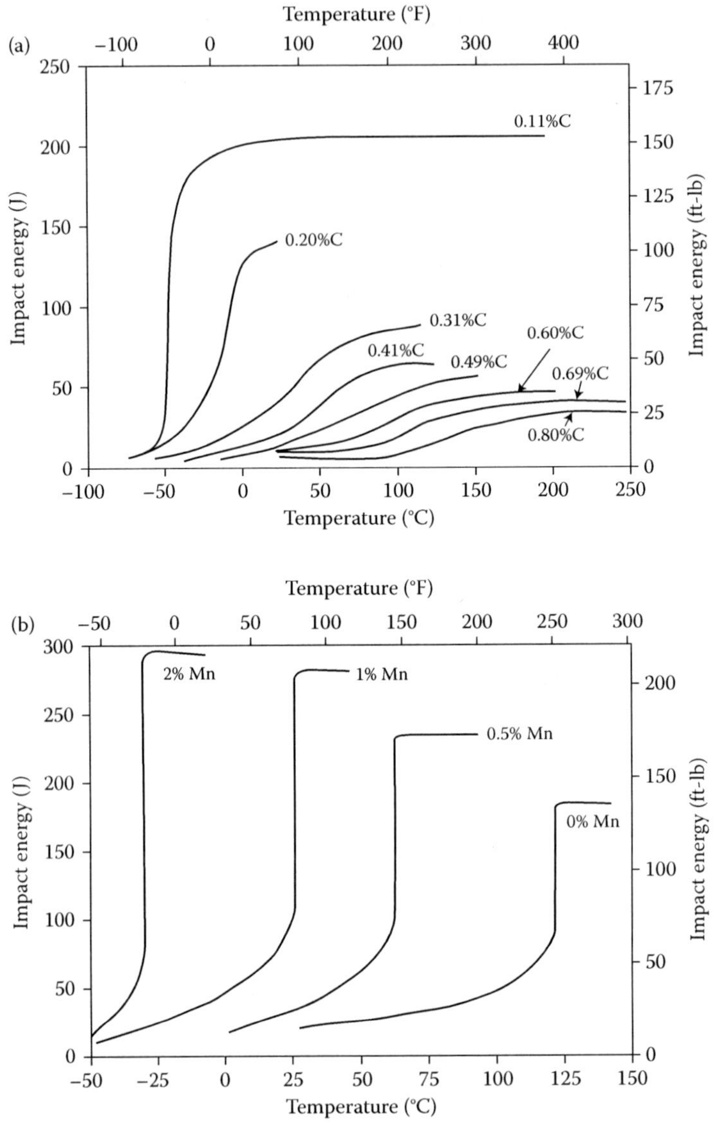

FIGURE 10.16 Variation of impact energy as a function of temperature for (a) different carbon contents of plain carbon steel specimens and (b) different Mn contents for an Fe–Mn–0.5C steel specimens. All compositions are in weight percentage.

10.5.9 Heat Treatment

It is generally known that the tensile properties of tempered martensites of the same hardness and carbon content are alike, irrespective of the amount of other alloying additions. But, this generalization is true only for room-temperature impact

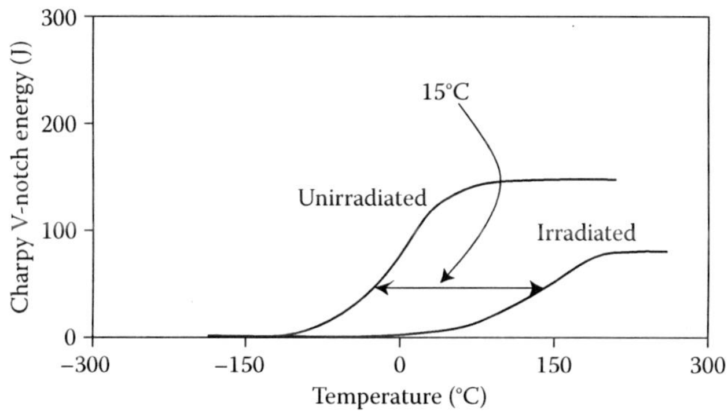

FIGURE 10.17 Effect of irradiation on the Charpy impact energy for a nuclear pressure vessel steel.

resistance of heat-treated steels; but not true for the variation of impact resistance with temperature. The maximum variation in DBTT could be as much as 200°F.

A quenched and tempered steel sample (containing martensite) will have a good combination of strength and toughness and consequently exhibit a higher impact energy than a normalized and tempered steel sample (containing pearlite) (Figure 10.18). Thus, the DBTT will be lower for the quenched and tempered steel sample.

Figure 10.19 summarizes the effect of different variables on the DBTT.

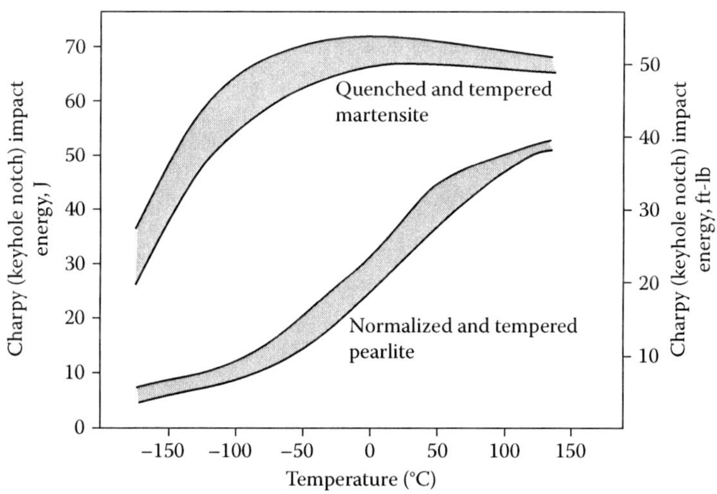

FIGURE 10.18 Comparison of the impact of energy of quenched and tempered vs. normalized and tempered steel samples.

↑

- High rate of load application
- Increasing carbon content
- Increasing thickness of samples
- Irradiation of samples
- Notches and notch sharpness
- Surface cracks

Ductile brittle transition temperature (DBTT)

- Decreasing grain size
- Increasing manganese content
- Materials with FCC crystal structure

↓

FIGURE 10.19 Summary of the effect of different variables on the DBTT.

10.6 PRECAUTIONS

Since the impact tests have to be conducted at different temperatures, the specimens have to be either heated or cooled. Therefore, it is important that the specimen is accurately positioned on the anvil support within 5 s of removal from the heating (or cooling) medium.

Sometimes the impact energy has to be determined for specimens in different heat-treated conditions. Since the specimens have to be machined to size and shape, any heat treatment of the specimens should be performed prior to final machining.

Even though a large number of tests need to be conducted to get a reasonably accurate DBTT, in practice, the tests will be conducted at a limited number of temperatures (because of the cost of the specimens and testing).

Considerable scatter is noticed in the notched-bar impact tests, particularly in the region of the transition temperature. This is partly due to the local variations in the properties of the steel and partly due to the difficulty in preparing perfectly reproducible notches. As mentioned above, both notch shape and depth are critical. Proper placement of the specimen in the impact machine is also important.

10.7 CORRELATIONS WITH OTHER MECHANICAL PROPERTIES

The energy absorbed by the material during impact testing (i.e., *impact toughness*) is not always directly related to the tensile toughness (i.e., area under the tensile true stress–true strain curve). (Remember that engineers mostly consider the area under the *engineering* stress–strain curve to represent the tensile toughness.) High tensile toughness is obtained when a material exhibits both high strength and high ductility (Figure 10.20). Thus, most metals exhibit reasonable toughness values (see Table 10.1). But, even though ceramic materials and intermetallic alloys exhibit high strength values, they have low tensile and impact toughness values due to their limited ductility. In contrast, bulk metallic glasses (multicomponent metallic alloys with a noncrystalline structure that can be cast into thick sections of a few millimeters) in diameter exhibit high strength and reasonable strain (elastic strains of up to about 2%) and therefore they also have high toughness.

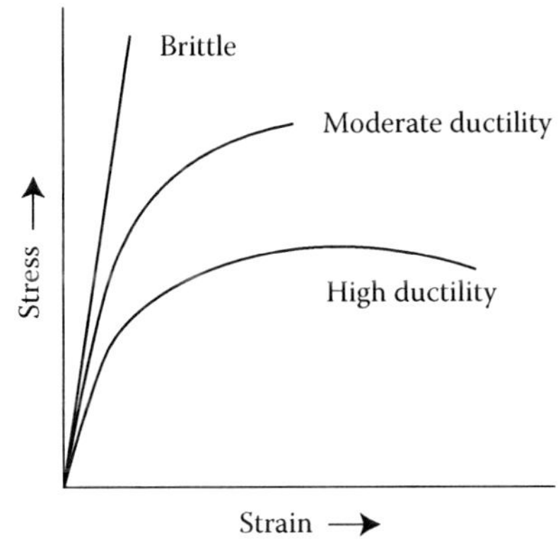

FIGURE 10.20 Conditions under which a high tensile toughness is obtained. The material should have a high yield strength and/or high ductility.

A high tensile toughness is not always translated as high impact toughness. Metals that show excellent tensile toughness may show a brittle behavior under high strain rates, that is, poor impact toughness. Thus, strain rate can shift the DBTT significantly.

As mentioned earlier, prior to the advent of fracture mechanics as a scientific discipline, impact testing techniques were established to ascertain the fracture characteristics of materials. The toughness of a material is measured quantitatively using the fracture mechanics approach. Fracture toughness measures the ability of a material containing a flaw to withstand an applied load. Thus, both *plane strain fracture toughness* (when the thickness of the specimen is much larger than the flaw size responsible for failure), K_{Ic} and impact tests determine the fracture properties of materials. While fracture toughness is quantitative in nature (since it determines the K_{Ic} values), results of the impact tests are more qualitative in nature. Thus, they are of little use for design purposes. Further, the impact energy is dependent on several parameters and it may not be always possible to control these parameters. Consequently, the impact test is best used for comparison purposes and for selection of materials. It is also important to realize that impact tests are relatively easy to perform while the plane strain fracture toughness measurements require expensive equipment and well-designed specimens. Thus, impact test is a quick, convenient, and inexpensive way to compare different materials. Thus, the Charpy V-notch (CVN) test is used as a screening test for evaluating notch toughness changes influenced by chemical composition (alloying and impurity elements, including gases) and physical (microstructure and grain size) and mechanical properties (hardness and yield properties) of materials.

Attempts have been made to correlate K_{Ic} and CVN energies, but only with limited success. Most of these correlations are dimensionally incompatible, ignore differences between the two measures of toughness (in particular loading rate and notch acuity), and are valid only for limited types of materials and ranges of data. Additionally, these correlations can be widely scattered.

10.8 DBTT IN NONMETALLIC MATERIALS

Apart from metallic materials (with a BCC structure), ceramics and polymeric materials also experience the ductile-to-brittle transition. For ceramic materials the transition occurs at elevated temperatures, usually above 1000°C, while polymers show this behavior at very low temperatures. Thus, natural rubber is superelastic at room temperature, but when cooled to liquid nitrogen temperature (77 K), it becomes brittle. The explosion of the space shuttle Challenger in 1986 was attributed to the embrittlement of rubber seals in the booster rocket because of cold weather. Most commonly, the glass transition temperature, T_g, is taken as the DBTT for polymeric materials. The impact energies of some polymeric materials also are presented in Table 10.1.

10.9 EXPERIMENTAL PROCEDURE

Get familiar with the type of impact testing machine(s) you have. These machines are usually very simple and easy to operate. The most important precaution you have to take while using the machine is not to stand in the way of the swinging pendulum. The pendulum should hit the specimen but not you!

The objective of the experiment is to determine the DBTT for the given low-carbon steel specimens by measuring the impact energy at different temperatures. Conduct the test according to the following procedure.

1. Take a set of steel specimens of known composition and heat treatment. Make sure that you have at least 10 specimens. These specimens need to have a notch according to the specifications of the test (Charpy or Izod).
2. Depending on the type of machine you have (Charpy or Izod), mount the specimen in the proper orientation (notch away from the pendulum side if it is Charpy and toward the pendulum if it is the Izod test).
3. Measure the height, h, of the pendulum before the test is conducted.
4. Note down the weight of the pendulum.
5. Release the hammer such that it hits the specimen, fractures it and swings to the other side.
6. Measure the height, h', up to which the pendulum has risen after fracturing the specimen.
7. Usually, both the initial and final heights can be read off on a dial in the machine.
8. You should also be able to change the temperature of the specimen over a reasonably wide range. These different temperatures (above and below room temperature) could be easily obtained in the laboratory by choosing different baths: boiling water (100°C), room temperature (~25°C), ice water (0°C), dry ice (–78°C), and liquid nitrogen (–196°C). Have these different baths ready before you are ready to start the experiment.
9. Repeat the above test at different temperatures depending on the heating and/or cooling media you have.
10. Ensure that the tests are done at reasonably short temperature intervals, especially near the DBTT. Since we do not know the DBTT yet, it may be a

good idea to do a test at room temperature and another at the lowest temperature one could do the test. Then, the remaining tests could be conducted at some intermediate temperatures.

11. Make sure that the time lapse between the time the specimen is taken out of the bath and it is mounted in the machine is short (<5 s).
12. Measure the height of the pendulum after impact at each temperature.
13. Calculate the impact energy (energy absorbed) as the product of the mass of the pendulum times the height difference between the values before and after impact.
14. Plot the data of impact energy as a function of temperature and estimate the DBTT.
15. Look at the fracture surface of the specimen in the SEM at each test temperature and determine whether the fracture is ductile or brittle in nature. Estimate the percentage brittle/ductile fracture and plot this as a function of temperature.
16. Just for fun, take a rubber hose and try to hammer it at room temperature. You would notice that nothing happens because it is highly elastic. Then dip one end of the rubber hose in liquid nitrogen for a few minutes (until it reaches the liquid nitrogen temperature) and try to hammer it. The rubber will now shatter since it is brittle at low temperatures, such as liquid nitrogen temperature. Make sure that you wear safety glasses when you do this experiment since the broken hard rubber pieces can fly in different directions.
17. List the impact energy values measured in a tabular form as follows:

10.10 OBSERVATIONS

Machine used:
Specimen (type, composition, and condition):
Pendulum weight: ----- lb
Initial pendulum height, h: ----- ft.

Test Temperature (°C)	Pendulum Height After Impact, h' (ft)	Energy Absorbed = Mass of Pendulum × Height Difference (ft-lb)

Estimate the DBTT of the tested material using the different methods discussed in the text.

Remember a few important points.

1. Use at least two specimens for each test temperature so that a reliable value is obtained for each temperature.
2. Allow the specimens to reach the bath temperature before the test is conducted.

3. Transfer the specimens from the bath to the test equipment as fast as possible. Otherwise the temperature of the specimen will change too much, especially when the specimens were maintained at lower or higher temperatures than room temperature.
4. It is also important to remember that you should never come in direct contact with liquid nitrogen. Contact with liquid nitrogen or with objects at liquid nitrogen temperature will freeze your skin instantaneously and cause blisters which heal very slowly.
5. Transfer the specimens from the baths using tongs, making sure that the tongs are dipped inside the bath for a sufficient time such that the tongs reach the bath temperature.

EXERCISES

10.1 What is impact?

10.2 What is the difference in the strain rates between regular tensile testing and impact testing?

10.3 Name three conditions that promote brittle-cleavage type of fracture.

10.4 Hardness is the analog of strength. What is the analog of impact energy?

10.5 Distinguish between ductile and brittle fracture.

10.6 What are the factors that contribute to brittle fracture?

10.7 What are the two standardized tests for Impact testing?

10.8 What is the difference between Izod and Charpy impact tests?

10.9 What is meant by DBTT?

10.10 How can one determine the DBTT?

10.11 What are the different ways by which you can estimate the DBTT?

10.12 What is the mode of fracture above the DBTT?

10.13 What is the mode of fracture below the DBTT?

10.14 How can you lower the DBTT?

10.15 Do all metals exhibit the DBTT?

10.16 Do all type of materials exhibit the DBTT?

10.17 What is the effect of specimen size on the DBTT?

10.18 What is the effect of carbon content on the DBTT?

10.19 What is the effect of grain size on the DBTT?

10.20 Is it safe to use a material below the DBTT?

10.21 Rubber is usually elastic at room temperature. Will it continue to be so at liquid nitrogen temperature?

10.22 Ceramics are usually brittle. Can they be made ductile?

10.23 Name two nonmetallic type of materials that exhibit the DBTT.

10.24 Draw a typical plot showing the variation of impact energy with temperature.

FURTHER READING

Courtney, T. H. 2000. *Mechanical Behavior of Materials*, 2nd edition. New York, NY: McGraw-Hill.

Dowling, N. E. 2007. *Mechanical Behavior of Materials*, 3rd edition. Upper Saddle River, NJ: Pearson: Prentice-Hall.

Hertzberg, R. W. 1996. *Deformation and Fracture Mechanics of Engineering Materials*, 4th edition. New York, NY: John Wiley & Sons, Inc.

McEvily, A. J. 2002. *Metal Failures: Mechanisms, Analysis, Prevention*. New York, NY: John Wiley & Sons, Inc.

11 Fatigue Testing

11.1 INTRODUCTION

Fatigue is a form of failure in structures subjected to dynamic and fluctuating stresses. Fatigue failures have been detected in different metallic and polymeric materials, especially in load-bearing components that have been subjected to repetitive stresses. Some typical examples are bridges, machine components, springs, gas turbine disks and blades, train wheels and axles, pump impellers, crankshafts, and some of the aircraft components. Even surgical prostheses implanted into the human body have been shown to exhibit *fatigue failure* in service.

Fatigue is such a common feature in the animate and inanimate world that a scientific and rigorous definition has been often overlooked. Fatigue may be defined as the progressive, localized, permanent structural change that occurs in materials subjected to fluctuating stresses and strains that may result in cracks or fracture after a sufficient number of fluctuations.

Fatigue failure normally occurs after a lengthy period of repeated stress or strain cycling. Fatigue constitutes approximately 90% of all metallic failures. Fatigue failure is catastrophic, occurring suddenly and without warning. There is very little gross plastic deformation associated with fatigue failure and so is brittle like. Fatigue failure is known to occur in all types of materials—metals, polymers, and ceramics, even though ceramics are less likely to fail by fatigue fracture.

The phenomenon of "fatigue" was discovered in the 1800s when several investigators noted that bridge and railroad components were cracking when subjected to repeated loading. With continuing industrialization and increased use of materials, more and more failures of components subjected to repeated loads were recorded. It is very easy to demonstrate the phenomenon of fatigue. Take a metal paper clip and bend it forward and backward repeatedly. After a few repetitions, the paper clip will break into two pieces. The paper clip fractures even though the stress experienced by it is much below the yield strength of the material. Thus, three factors need to be simultaneously present for fatigue failure to occur—a cyclic stress (to initiate the crack), a tensile stress (to propagate the crack), and plastic strain (for the material to fail). It is also important to remember that the cyclic stress employed is much below the static yield strength of the material.

Fatigue failure occurs in three distinct stages. First, a small crack is initiated, usually at the surface of the sample. Unless the metal surface is carefully prepared, it is likely to contain scratches, sharp corners, notches, pits, or other surface irregularities and these could act as nucleating sites for crack initiation. We will see later that it is possible to increase the fatigue life of a material by either removing these surface defects or increasing the surface hardness to make it difficult to form the defects. The second stage is the propagation of the crack that has already been initiated. This occurs gradually as the stress on the specimen is repeated. The last stage is the sudden failure of the sample due to the inability of the material to support the stress.

11.2 DEFINITIONS

One can easily understand some of the basic definitions used in fatigue testing with reference to Figure 11.1, which shows a schematic of cyclic load application.

Here S_{max} is the maximum level of stress that is applied. (Note that we are using the symbol S to represent engineering stress, instead of the conventional symbol, σ. This is because people in the industry use the symbol S and the fatigue plots are most popularly referred to as *S–N* plots.) By convention, positive stresses are termed tensile in nature and negative stresses compressive. S_{max} is always positive and its magnitude is generally smaller than the static yield strength of the material. The maximum applied stress may be above the yield strength when one conducts a *low-cycle fatigue* test or if the material under investigation has an appreciable work hardening rate. Further, the applied stress may be axial (tension–compression), flexural (bending), or torsional (twisting).

S_{min} is the minimum level of stress. The nature of this stress may be the same or different from that of S_{max} in the sense that S_{max} may be tensile while S_{min} can be either tensile or compressive.

The *range of stress*, ΔS is the difference between the maximum and minimum levels of stress and is given by

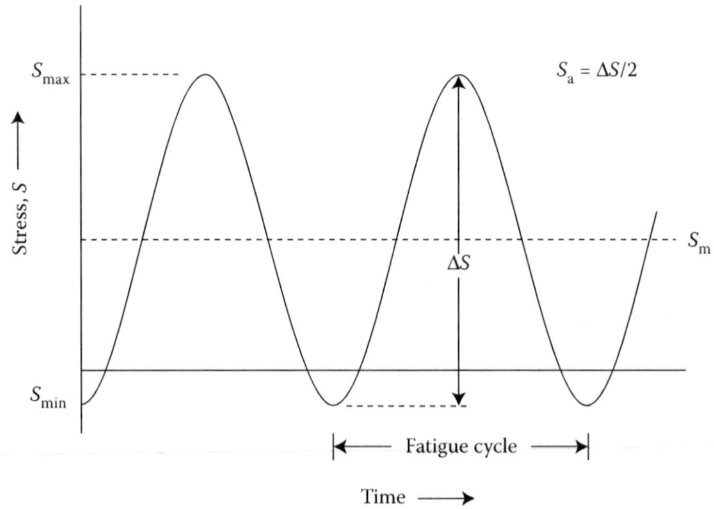

FIGURE 11.1 A schematic of a cyclic stress application to define the different parameters used in fatigue testing.

$$\Delta S = S_{max} - S_{min} \tag{11.1}$$

The compressive stress will not allow the cracks to propagate and so the compressive stress is taken as equivalent to a zero stress, irrespective of its actual value.

The *mean stress*, S_m is the algebraic mean of the maximum and minimum stress levels, that is,

$$S_m = \frac{(S_{max} + S_{min})}{2} \tag{11.2}$$

The *stress amplitude*, S_a is half the range of stress, that is,

$$S_a = \frac{\Delta S}{2} = \frac{S_{max} - S_{min}}{2} \tag{11.3}$$

The *stress ratio* is defined as the ratio of the minimum to maximum stress level, that is,

$$R = \frac{S_{min}}{S_{max}} \tag{11.4}$$

A *fatigue cycle* is defined as the time between two successive maxima (or minima) in the stress. The number of fatigue cycles per second is defined as the cyclic frequency.

Let us now look at a couple of examples of calculating the above parameters in two different situations—one when both the maximum and minimum are tensile stresses and the other when the maximum stress is tensile and the minimum stress is compressive in nature. It is also possible that the magnitudes of the maximum and minimum stresses are the same, but of opposite sign, and then the mean stress is zero. Figure 11.2 shows the different ways in which a cyclic stress may be applied to the material. In Figure 11.2a, the S_{max} and S_{min} values are equal and opposite in magnitude. That is, the stress alternates from a maximum tensile stress to a maximum compressive stress and therefore this is referred to as *completely reversed stress cycle*. The mean stress is zero in this case. On the other hand, Figure 11.2b represents the *repeated stress cycle*, when the maximum and minimum stress levels are asymmetrical relative to the zero stress level. Figure 11.2c represents the random stress cycle, where there is no well-defined maximum or minimum stress level.

Example Problem 11.1

Calculate the different parameters used in fatigue testing for two different cases when (i) S_{max} is 100 MPa and S_{min} is 20 MPa and (ii) S_{max} is 80 MPa and S_{min} is –20 MPa (compressive in nature).

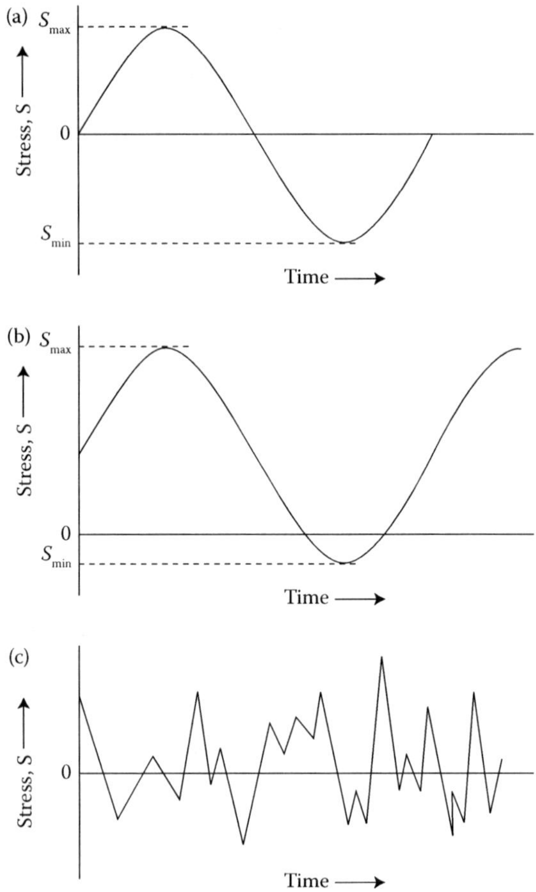

FIGURE 11.2 Three different ways in which cyclic stresses may be applied to a material. These figures represent the variation of stress with time under (a) reversed stress cycle, (b) repeated stress cycle, and (c) random stress cycle.

Solution 11.1

The important parameters we need to calculate are (a) range of stress, (b) mean stress, and (c) stress amplitude.

For case (i), when the S_{max} and S_{min} are 100 and 20 MPa, respectively:

a. Range of stress, $\Delta S = S_{max} - S_{min}$ = 100 MPa – 20 MPa = 80 MPa
b. Mean stress, $S_m = (S_{max} + S_{min})/2$ = (100 MPa + 20 MPa) /2 = 60 MPa
c. Stress amplitude, $S_a = \Delta S/2$ = 80 MPa/2 = 40 MPa

For case (ii), when the S_{max} and S_{min} are 80 and –20 MPa (compressive), respectively:

a. Range of stress, $\Delta S = S_{max} - S_{min}$ = 80 – 0 MPa = 80 MPa

Recall that the compressive stress is always taken to be zero.

b. Mean stress, $S_m = \frac{(S_{max} + S_{min})}{2} = \frac{100\,\text{MPa} + (-20\,\text{MPa})}{2} = 40\ \text{MPa}$

c. Stress amplitude, $S_a = \Delta S/2 = 80\ \text{MPa}/2 = 40\ \text{MPa}$

11.3 FATIGUE TESTING

Even though there are many different types of tests available to determine the fatigue behavior of materials, the most commonly used method is to use the rotating-beam test. A schematic of the reversed bending fatigue machine is shown in Figure 11.3. Figure 11.4 shows a photograph of R. R. Moore-type rotating beam fatigue system while Figure 11.5 shows a servohydraulic machine for conducting high-cycle fatigue testing of metallic and composite materials used in airplane fuselages. The specimen for fatigue testing is of different design from a standard tensile test specimen. Figure 11.6 shows the sketch of a rotating-beam fatigue specimen. The surface of the specimen is polished to a smooth surface finish and carefully tapered toward the center. In this test, the specimen is subjected to alternating tensile and compressive stresses of equal magnitude. During the test, the specimen experiences a tensile stress on the lower surface and a compressive stress on the upper surface by the weights attached in the center of the apparatus. The revolution counter measures the number of stress cycles the specimen experiences. The test is generally continued until the specimen breaks.

The normal way of conducting the test is according to the ASTM Standard E466 "Standard Practice for Controlling Constant Amplitude Axial Fatigue Tests of Metallic Materials," and ASTM Standard E468 "Standard Practice for Presentation of Constant Amplitude Fatigue Test Results for Metallic Materials." The specimen is loaded at a given stress, S (usually below the yield strength of the material) and the number of stress cycles, N_f it takes for the material to fail is determined. The test is then repeated at different stress levels and the number of stress cycles to failure is again determined at each stress level. From these data, an S–N curve is plotted, with the stress, S plotted on the Y-axis and the number of cycles,

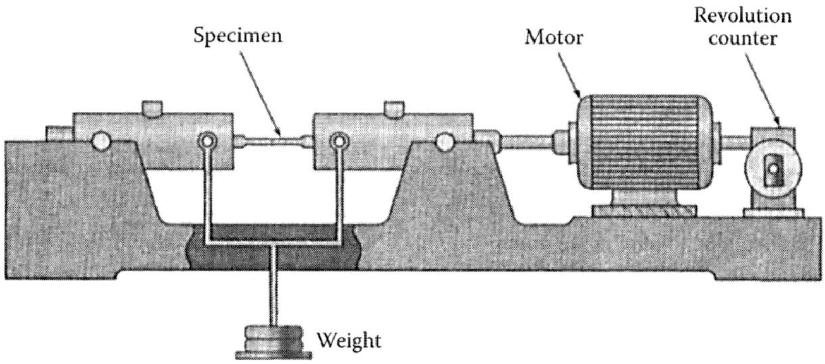

FIGURE 11.3 Schematic diagram of a R. R. Moore reversed bending fatigue machine.

FIGURE 11.4 Photograph of the R. R. Moore rotating-beam fatigue-testing system. The specimen can be rotated up to 9999 complete stress cycles per minute with bending stress applied by means of dead weights. (Photo courtesy of Instron Corporation. With permission.)

N_f at which failure occurs plotted on the X-axis. Note that the number of cycles to failure is plotted on a logarithmic scale since the number is very large. This curve is also known as the Wöhler curve after the German scientist August Wöhler,* who pioneered studies on developing ways to continuously measure the fatigue behavior of metals to determine general construction principles for building of locomotives and wagons, rail profiles, axles, tires, and brakes. Normally, the stress on the Y-axis corresponds to the stress amplitude defined above as $\Delta S/2$. A typical S–N curve is shown in Figure 11.7.

* August Wöhler (1819–1914), a German engineer, best remembered for his systematic investigations of metal fatigue, was born in the town of Soltau on June 22, 1819. After graduating from the Technische Hochschule in Hannover, Germany, he was recruited in 1840, to the Borsig works in Berlin where he worked on the manufacture of rail tracks. By 1847, Wöhler was chief superintendent of rolling stock on the Lower Silesia-Brandenberg Railroad. His growing reputation led to his appointment in 1852 by the Prussian minister of commerce to investigate the causes of fracture in railroad axles, work that was to occupy Wöhler over the next two decades. His work on fatigue marks the first systematic investigation of S–N curves, also known as Wöhler curves, to characterize the fatigue behavior of materials. Such curves can be used to minimize the problem of fatigue by lowering the stress at critical points in a component. Wohler showed clearly that fatigue occurs by crack growth from surface defects until the product can no longer support the applied load. He developed apparatus for repeated loading of railway axles, mainly because many accidents were caused by sudden fatigue fracture. The presentation of his work at the Paris Exposition in 1867 brought it to a wide international audience. In 1901, Wöhler was one of the first people to be awarded an honorary doctorate from the Technische Hochschule in Berlin. He died in Hannover on March 21, 1914.

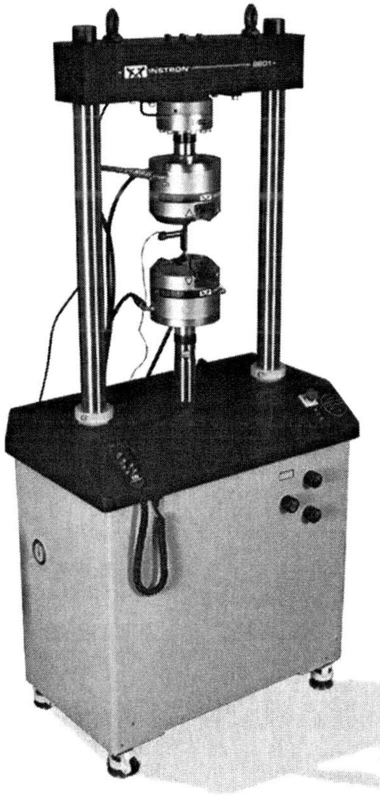

FIGURE 11.5 100 kN servohydraulic machine for high-cycle fatigue testing of metallic and composite materials used in airplane fuselages. (Photo courtesy of Instron Corporation. With permission.)

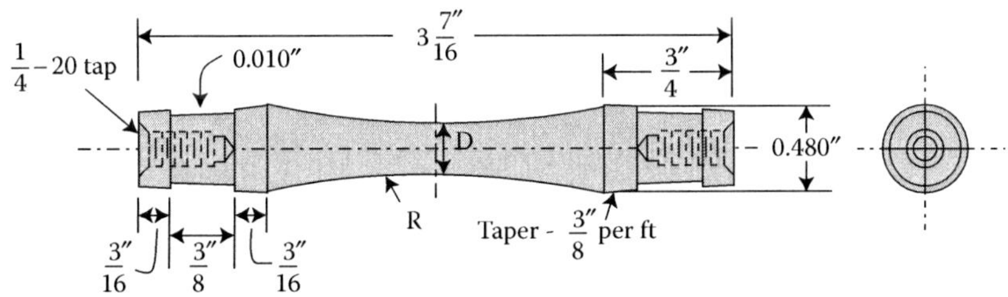

D = 0.200–0.400 in. selected on basis of ultimate strength of material
R = 3.5–10 in.

FIGURE 11.6 A schematic of the rotating-beam fatigue-testing specimen.

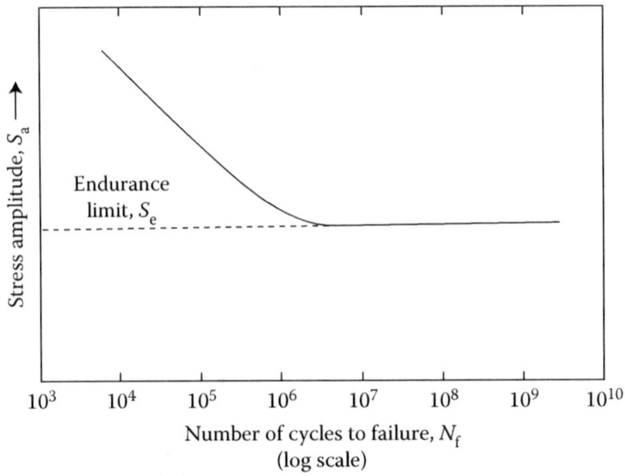

FIGURE 11.7 A typical *S–N* curve. The stress is actually the stress amplitude, calculated as half of the stress range, $\Delta S = S_{max} - S_{min}$. The *S–N* plot exhibits a well-defined endurance limit, S_e. Note that as the stress level is decreased, the number of cycles to failure increases. But, when the specimen is subjected to a stress level below S_e, the endurance limit, the specimen will not fail irrespective of the number of cycles to which the specimen is subjected.

A few important points are worth noting in Figure 11.7. The number of cycles to failure increases as the applied stress is lowered. In other words, the stress required to cause fatigue failure decreases as the number of cycles is increased. But, there is a limiting stress, S_e below which the specimen will not fail. Stated differently, even if the sample is subjected to a very large number of cycles below this limiting stress, fatigue failure does not occur in the specimen. This limiting stress, S_e, or the horizontal part of the *S–N* curve, is known as the *fatigue strength* or *endurance limit* of the material. That is, the stress to cause failure first decreases as the number of cycles is increased, and then there is a plateau in the *S–N* curve. This situation will prevail irrespective of the number of cycles the stress is applied. The value of S_e is a function of the applied stress amplitude, S_a and it is known that S_e decreases as S_a increases, as shown in Figure 11.8. The number of cycles to failure is also referred to as the *fatigue life* of the material. The number of cycles to failure of materials exhibiting the endurance limit is generally between 10^6 and 10^{10}.

Some materials do not exhibit a well-defined endurance limit. In these cases, the number of cycles to failure continues to increase as the cyclic stress on the specimen decreases. Alternately, the stress to cause failure is lower as the number of cycles is increased. A typical example is shown in Figure 11.9a. FCC metals such as aluminum and copper or polymeric materials such as nylon, epoxy, and polyethylene are typical examples of materials exhibiting this type of fatigue behavior. In such cases, the endurance limit is taken as the stress corresponding to 10^7 cycles to failure in the case of metals and 10^6 cycles in the case of polymeric samples. As noted in Figure 11.9b, the fatigue life (the number of cycles to failure at a given stress level) is longer at lower stress levels, and the fatigue strength is higher at shorter number of cycles to failure.

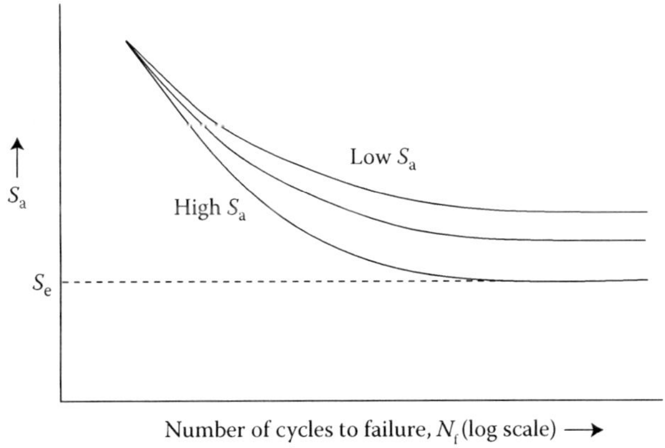

FIGURE 11.8 Variation of the endurance limit, S_e as a function of the stress amplitude, S_a. Even though the number of cycles to failure may not be significantly different, S_e decreases with increasing value of S_a.

Figure 11.10 shows the *S–N* data for various metals and polymers. Even though the data are presented in such a way that there is not much scatter in the data, in practice there is significant scatter. Therefore, a more realistic representation may be to present the *S–N* data in the form of scatter bands.

11.4 SOME TYPICAL EXAMPLES OF FATIGUE FAILURE

Fatigue failure involves (i) formation of a small crack at some point of stress concentration on the surface of the specimen, (ii) growth of the crack, and (iii) final failure once the crack has reached a critical size. Since both crack initiation and crack growth require time, careful examination of the fracture surface of the material can reveal features that are characteristic of fatigue failure.

Cracks associated with fatigue failure always originate at a point of stress concentration on the surface of the specimen. These points of stress concentration could be sharp corners, notches, machine markings, or metallurgical inclusions. Once nucleated these cracks grow and propagate across the specimen under the repeated stress cycles.

As mentioned before, majority of the engineering components fail by fatigue processes. Further, many of the failures of aircraft components also occur by fatigue. Figure 11.11 shows a photograph of the Aloha Airlines Boeing 737 aircraft after its fuselage was ripped apart during its flight 243 from Hilo International Airport to Honolulu in the Hawaiian Islands on April 28, 1988. As the aircraft reached its normal flight altitude of 24,000 ft, a small section on the left side of the roof ruptured. The resulting explosive decompression tore off a large section of the roof, consisting of the entire top half of the aircraft skin extending from just behind the cockpit to the fore-wing area. But, the flight was able to land safely at Kahului airport in Maui. Detailed investigations proved that this accident was caused by metal fatigue and stress fractures exacerbated by crevice corrosion. It was later revealed that a

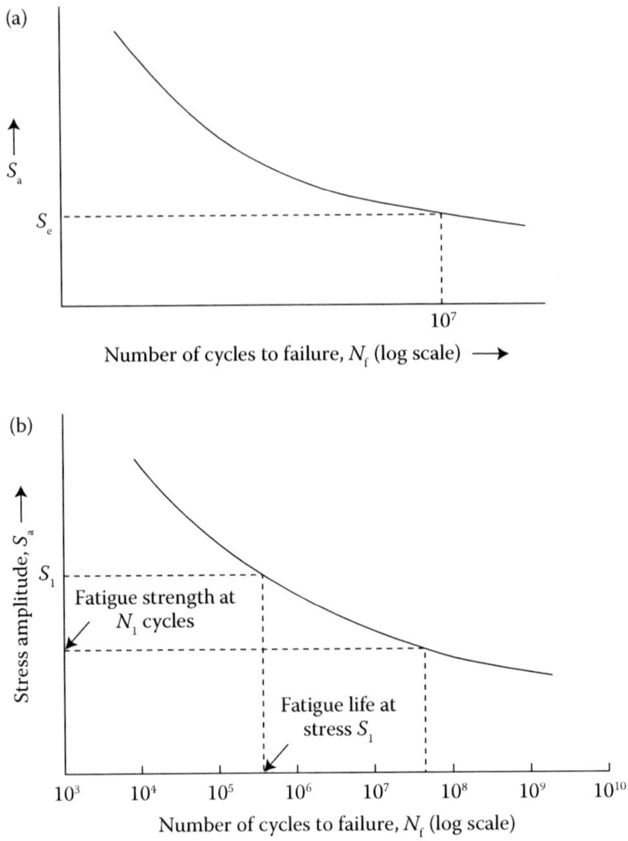

FIGURE 11.9 (a) An *S–N* curve that does not exhibit a well-defined endurance limit. Such a behavior is shown by metals having an FCC structure and also by polymers. The operational endurance limit in such cases is taken as the stress at which failure of the specimen occurs in 10^7 cycles. (b) Shows that the fatigue life increases at lower stress levels and that the fatigue strength is higher at shorter number of cycles to failure.

passenger had noticed a crack in the fuselage upon boarding the aircraft prior to the ill-fated flight but did not notify anyone of this. The crack was located aft of the front portside passenger door, and was probably due to metal fatigue related to the 89,680 compression and decompression cycles experienced in the short hop flights by Aloha. Microscopic and/or macroscopic examination can easily help identify features characteristic of fatigue failure. The point at which fatigue failure originates is always smooth. The roughness of the surface increases as the crack grows in size. Very distinctive microstructural features can be observed during this process. These are characterized by clamshell or "beach" marks, as shown in Figure 11.12 or striations as shown in Figure 11.13.

"Beach" marks appear as concentric ridges, frequently in a circular or semicircular pattern, the spacing of which increases as we move away from the crack initiation

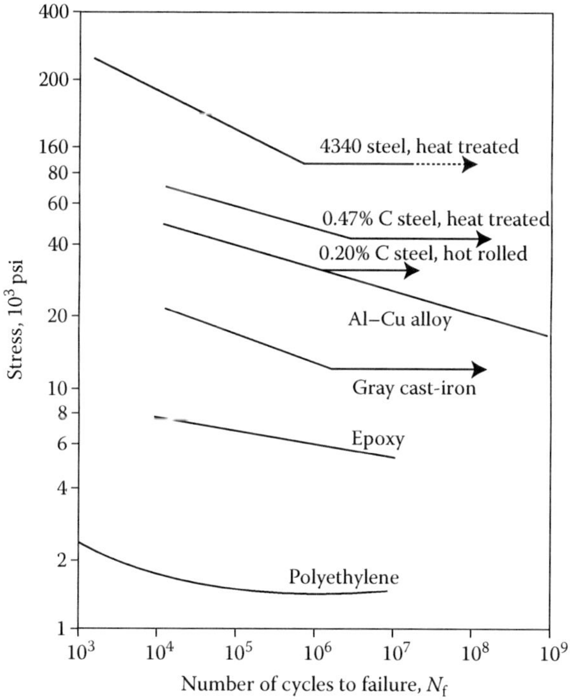

FIGURE 11.10 *S–N* data for various metals and polymers.

FIGURE 11.11 Photograph of the fuselage of the Aloha Airlines Boeing 737 aircraft after its roof was torn off. The accident was caused by metal fatigue (repeated compression and decompression cycles, numbering 89,680) and crevice corrosion (the plane was primarily operating in saltwater-saturated climate).

FIGURE 11.12 Optical micrograph of the fracture surface of a rotating steel (0.4 wt% carbon) shaft that experienced fatigue failure. "Beach" marks are clearly visible in the micrograph. The fatigue crack initiated at the left bottom corner of the keyway and extended almost through the entire cross section before final fracture occurred. (Photo courtesy of ASM International, Materials Park, OH. With permission.)

site. The spacing between the beach marks is reasonably large and therefore they can be observed using an optical microscope, or sometimes with an unaided eye. Such markings are noticed on components that experience interruptions in loading and consequently there are periods of time when the cracks do not propagate. An example would be when a machine is operated during the day time and shut down during the night time. Each "beach" mark band represents a period of time over which crack growth had occurred.

Fatigue striations (Figure 11.13), on the other hand, are microscopic in size and consequently one uses either an SEM or a TEM to observe these. Each striation represents the distance over which the crack has grown in one cycle. The width of the striations depends on the stress range and increases with increasing stress range. Note that fatigue striations are much finer than "beach" marks and that each "beach" mark may contain literally thousands of striations.

The presence of "beach" marks and striations is a proof that the material had failed through fatigue. But, their absence does not mean that fatigue was not the cause of failure. Another important point to note is that these "beach" marks and striations appear in the region that experienced crack propagation. But, during the last stage of fatigue failure, when the cracks grow very rapidly, these markings are not observed. Depending on whether evidence for plastic deformation was available or not, the failure can be determined to be ductile or brittle.

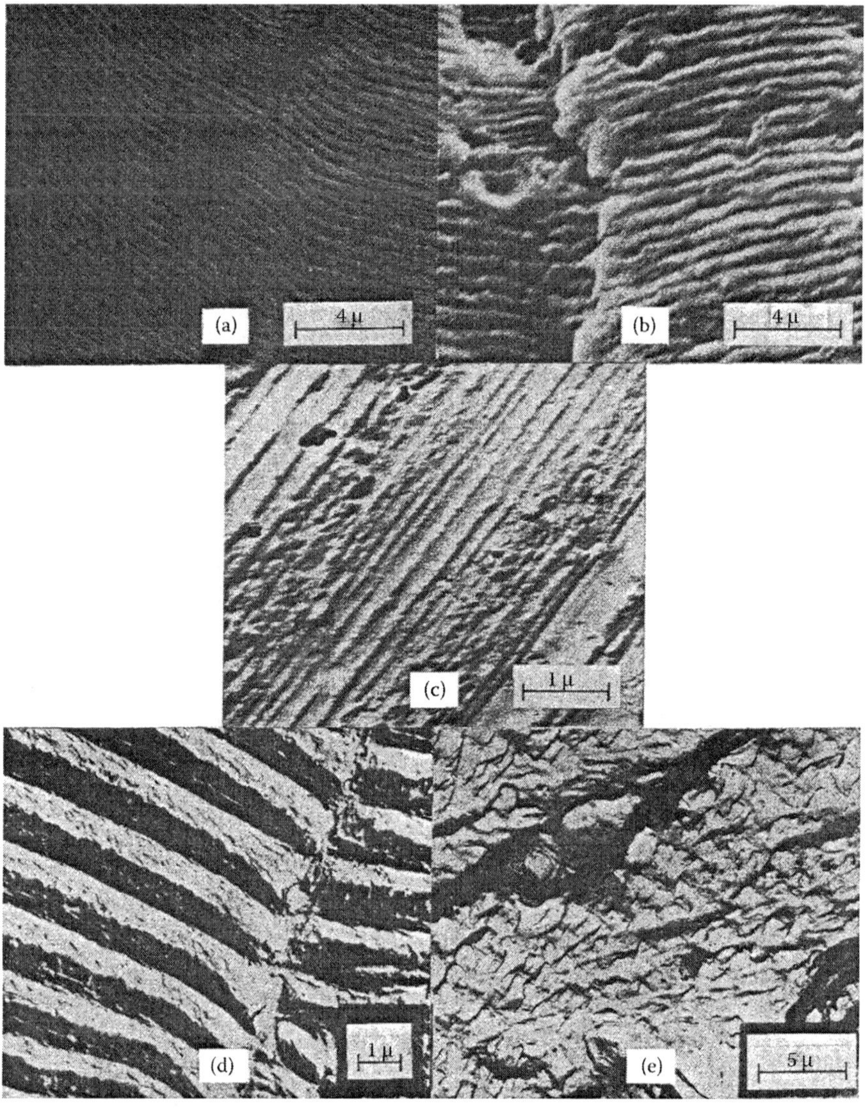

FIGURE 11.13 Electron fractographs showing fatigue striations found on fracture surface and within macroscopic bands. The spacing of the striations corresponds to the distance the crack advances during a stress cycle. (a), (c), (d), and (e) are transmission electron fractographs and (b) is a scanning electron fractograph. (After Hertzberg, R. W. 1996. *Deformation and Fracture Mechanics of Engineering Materials*, 4th edition. New York, NY: John Wiley & Sons, Inc. With permission.)

11.5 FATIGUE FAILURE MECHANISM

Fatigue failure occurs through formation of fatigue cracks and their growth to a large-enough size when the material will not be able to withstand the applied stress. If a ductile material is subjected to stress, it will undergo permanent (plastic) deformation if the applied stress is beyond the yield strength of the material. If the specimen contains stress raisers such as surface cracks, notches, etc., fatigue damage could occur by a localized slip process, since slip is the most common mode of deformation. Plastic deformation in one direction and then in the opposite direction causes surface ridges and grooves, referred to as slipband extrusions and slipband intrusions (Figure 11.14). These intrusions and extrusions will be the nucleating sites for fatigue cracks. When the specimen is subjected to continued cycling of stress, these slip bands cause cracks to form at or near the surface and propagate into the specimen along planes that experience high shear stresses. At this stage, the crack growth rate is very slow, typically about 10^{-10} m cycle^{-1}. On continued cycling, the crack may grow to a small depth and reorient itself in a direction normal to the tensile axis. Under these conditions, the crack starts growing rapidly (on the order of micrometers per cycle) and fatigue striations (Figure 11.13) are created as the crack advances across the cross section of the metal specimen. When the crack reaches a large size and the specimen is not able to support the applied stress, the material fails.

11.6 FACTORS AFFECTING THE FATIGUE STRENGTH OF MATERIALS

Factors other than the chemical composition of the material and its tensile strength affect the fatigue behavior of materials. By knowing the way in which these additional factors affect the fatigue life of materials, it becomes possible to prolong their fatigue life and prevent early fatigue failures.

11.6.1 STRESS CONCENTRATORS

For many common loading conditions, the surface of a specimen experiences the maximum stress. Consequently, most cracks leading to fatigue failure originate on

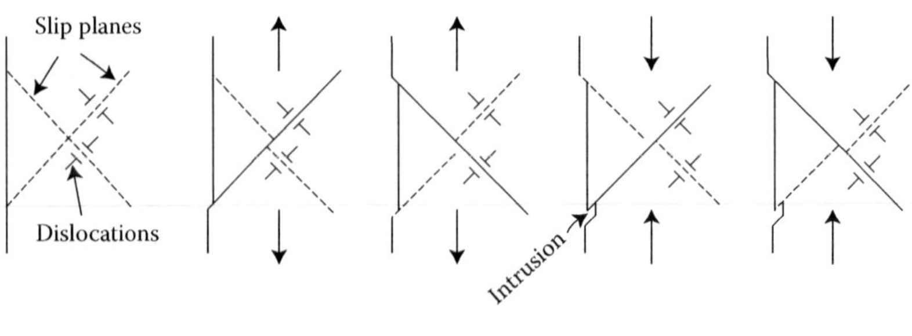

FIGURE 11.14 Mechanism for the formation of slip-band extrusions and intrusions as a result of localized slip deformation on the surface of the material.

the surfaces. Therefore, as mentioned earlier, it is very important to have a smooth surface on the specimen to avoid easy nucleation of cracks. Any surface roughness could act as a stress concentrator. Stress concentrator means that the nominal applied stress gets magnified. In other words, the sample experiences a much higher stress than what is applied and consequently, the number of stress cycles needed for failure is reduced. Keyways on shafts, bolt, and river holes, abrupt changes in cross section, and sharp corners are typical examples of stress raisers. The sharper the surface feature, that is, the smaller the radius of curvature, the more severe is the stress concentration. In fact, failure of the Comet long-distance passenger planes in the 1950s was related to the design defect of square-shaped escape hatch windows that caused stress concentrations. Avoiding these stress concentrators, by proper and careful design, in components subjected to repetitive loading can substantially increase the fatigue life. But, it will not be always possible to avoid the presence of stress concentrators, especially when the component to be designed is complicated.

11.6.2 Rough Surfaces

Small scratches and grooves are almost invariably introduced into samples during machining operations. Crack nucleation can easily occur at these points. Therefore, it is desirable to have as smooth a surface as possible to avoid fatigue failure. It is observed that grinding, honing, and polishing of surfaces remove these sources of potential cracks and increase the fatigue life significantly. For example, a component with a smooth surface obtained by fine grinding will have increased fatigue resistance than the one with a rough surface.

11.6.3 Surface Condition

As mentioned earlier, fatigue cracks originate at the surface of the specimen. Therefore, if the condition of the surface could be suitably modified to prevent easy formation of cracks, the fatigue life of the component can be substantially enhanced.

It is also important to remember that surface cracks propagate easily under a tensile load. On the other hand, compressive loads shut the cracks closed. That is, in the presence of a tensile stress on the surface, less additional load is required to produce equivalent fatigue damage relative to the unstressed surface. Therefore, it is necessary to have a compressive stress state on the surface to increase its fatigue life.

Compressive stress state on the surface can be achieved by *shot peening*. In this process hard particles, having a small size (within the range of 0.1–1.0 mm), are shot at high velocities onto the surface of the component. These particles produce plastic deformation at the surface, provided the material that was treated is ductile. This plastic deformation has two advantages. One is that the material experiences work hardening at the surface and this provides more resistance to plastic deformation. The other advantage is that this process produces compressive residual stresses on the surface, to a depth of one-quarter to one-half of the diameter of the particles impinging on the surface. This surface compressive stress is equivalent to reducing the effective mean stress level during fatigue testing. Also, specimens with compressive residual stresses on the surface will be able to tolerate higher levels of applied tensile stresses.

Surface compressive residual stresses and increased surface hardness to increase the fatigue life of components can also be achieved in ferrous alloys by *case hardening* treatments such as *carburizing* and *nitriding*. In these processes, the steel sample is exposed to a carbon (or nitrogen)-rich atmosphere at a sufficiently high temperature such that the steel is in the austenitic region. Because of rapid diffusion of these interstitial atoms into steel at these elevated temperatures, the carbon (or nitrogen) content on the surface increases substantially and consequently the surface hardness is also increased. This increased hardness provides superior *fatigue resistance*. In addition to the increased surface hardness, these treatments also produce favorable residual compressive stress patterns. *Case carburization* (increase of carbon content on the surface, or case) of automobile engine crankshafts or gears made from carbon steels is a common method to improve their fatigue life.

But, it should be noted, that not all surface treatments are effective in increasing the fatigue life of components. For example, it is reported that nickel or chrome plating for decorative or other purposes decreases the fatigue resistance of ferrous alloys. This is due to the fact that the residual surface stresses are now tensile in nature.

11.6.4 Corrosion and Environmental Effects

These factors normally reduce the fatigue life of components. When a component is exposed to a corrosive medium, localized chemical attack occurs and the metal gets removed from the exposed grains and grain boundaries. This results in the formation of pits and a rough surface is produced. Further, the corrosion products that are formed are left behind. All these act as stress concentrators and therefore cracks can easily nucleate at these sites. Further, due to the corrosive environment, the crack propagation rate also is increased. As a result of all these, the fatigue life is reduced. Failure that occurs by the simultaneous action of corrosion and fatigue is known as *corrosion fatigue*.

Corrosion fatigue can be minimized or prevented by reducing the effects of corrosion and/or fatigue in causing metal failure. For example, the effects of corrosion can be minimized by choosing a more corrosion-resistant material, by applying a protective coating on the surface, or by reducing the corrosiveness of the environment. The effects of fatigue can be minimized by using the methods described above, namely, reducing the applied stress level, imposing a compressive stress on the surface, or increasing the surface hardness. Thus, a component

- Made of a corrosion-resistant alloy
- That has been deposited with a hard coating
- That has a smooth surface
- That has a surface compressive stress
- Operated at low stress levels, is likely to have a long fatigue life

Just like corrosion fatigue, one can also visualize *thermal fatigue*. This can occur even in components that are not subjected to fluctuating stresses. When a material is heated in a nonuniform manner some parts of it get heated to a higher temperature and consequently expand more than the other parts. During cooling, these parts experience more cooling and contract more. As a result of this, the material experiences thermal stresses, if the system is mechanically restrained. Consequently, under these oscillating thermal stresses, the material will fail. The magnitudes of these thermal stresses developed depend on the modulus of elasticity E, coefficient of thermal expansion α_l, and the temperature change ΔT, according to the relation

$$S = \alpha_l E \Delta T \tag{11.5}$$

Such a situation is very common in composites containing a reinforcement of fibers or particulates in a matrix. Since the matrix and the reinforcement have different coefficients of thermal expansion, the rate of expansion during heating (and contraction during cooling) will be different for the matrix and the reinforcement. Consequently, there is a length differential at any given temperature and this leads to shear stresses at the matrix/reinforcement interface and eventually causes fatigue failure. This can be minimized when the coefficients of thermal expansion of the matrix and the reinforcement are close to each other, which is not easy to achieve in practice. Therefore, a practical solution to this problem is to coat the fibers with a compliant coating (with a material having a coefficient of thermal expansion intermediate between those of the matrix and the interface).

Yet another form of failure that occurs especially in ceramics and glasses is termed *static fatigue*. Note, however, that static fatigue does not involve cyclic loading and occurs under static loading conditions. Further, this failure is more due to reasons that are chemical than mechanical. The phenomenon of static fatigue in the case of silicate materials can be explained with reference to Figure 11.15. Since majority of the silicate materials contain Si and O, there exist the Si–O–Si networks. In the presence of water, these Si–O–Si groups react with water and produce two Si–OH groups:

$$\mathrm{Si-O-Si + H_2O \rightarrow Si-OH + Si-OH} \tag{11.6}$$

Since these hydroxyl units are not joined together, there is a break in the silicate network and this could be considered an incipient crack. When such a reaction occurs

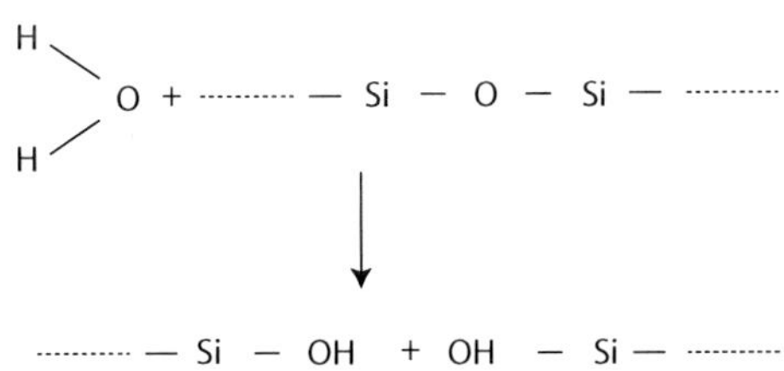

FIGURE 11.15 Interaction of water molecules with the silicate network leading to static fatigue failure in ceramic materials.

at the tip of a surface crack, the effective length of the crack increases and eventually the material fails. With time, the length of the crack increases and therefore the material becomes weaker. Note that this reaction takes place (i) in the presence of water and (ii) at room temperature. At higher temperatures, formation of hydroxyl ions is so fast that the effects cannot be monitored. At very low temperatures, the reaction is so slow that it does not produce any significant reaction.

11.7 FRACTURE MECHANICS APPROACH

In the discussion of fatigue behavior of materials, we have been so far concerned with the stress required to cause fatigue failure in a large number of stress cycles. Samples for such tests need to be smooth and have highly polished surfaces. But, most engineering materials do not have smooth surfaces. Additionally, because the way the materials are processed and fabricated, they almost always contain some preexisting cracks. Thus, even though the above approach is useful, it is not realistic.

Further, fatigue failure involves two stages—crack initiation and crack propagation. In the approach discussed above, we were only determining the total fatigue life of the component and not the individual life in the crack initiation and crack propagation stages. If a specimen contains preexisting cracks, then crack initiation is not required and we need only to consider crack propagation. Thus, fatigue life of specimens with preexisting flaws may be much shorter than one without flaws. Therefore, the fracture mechanics approach has been developed to study the crack growth behavior as a function of the loading cycles and this approach is independent of the geometry of the crack. Using this approach it is possible to predict the fatigue life of a component that contains flaws and under conditions of cyclic loading.

In the fracture mechanics approach, the crack length, a is measured as a function of the number of stress cycles, N. The crack length is measured by the change in electrical potential produced by the crack being further opened and extended by the fatigue action. This measurement will help in determining the crack growth rate, defined as $\mathrm{d}a/\mathrm{d}N$, the amount of crack extension per loading cycle. The actual experimentation involves taking a large number of samples into which a crack of known geometry and size is introduced. These samples are then tested at different stress levels and the crack length is measured as a function of the number of stress cycles. The resulting plot will appear as shown in Figure 11.16. It is clear from this figure that the crack growth rate, $\mathrm{d}a/\mathrm{d}N$ (measured in m cycle^{-1} as the slope of the curve at designated crack length) increases with increasing crack length. Thus, it is clear from Figure 11.16 that $\mathrm{d}a/\mathrm{d}N$ is higher at the crack length a_j than at a_i, since $a_\mathrm{j} > a_\mathrm{i}$. Note also that $\mathrm{d}a/\mathrm{d}N$ increases with increasing stress applied.

It has been experimentally shown in the 1960s by Paul Paris that fatigue crack growth rate is a function of the stress intensity factor, K (mode I). Any geometry for which the load applied is normal to the crack surface is referred to as mode I. The stress intensity factor, K, in turn, is a function of the stress and crack length and may be written down as

$$K = YS\sqrt{\pi a} \tag{11.7}$$

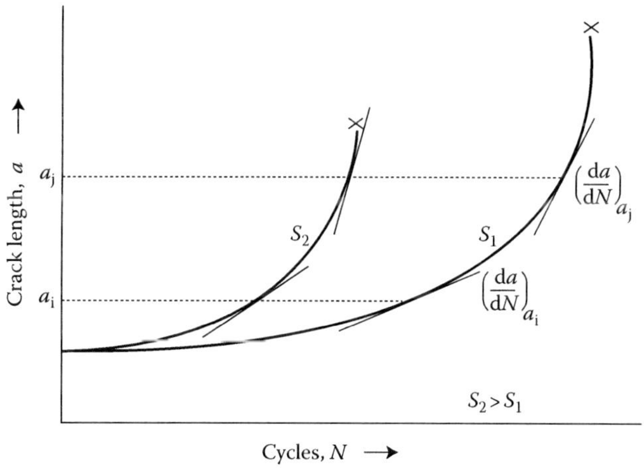

FIGURE 11.16 Schematic representation of crack length, a vs. number of stress cycles, N at two different stress levels, S_1 and S_2. Note that $S_2 > S_1$. Note that the crack growth rate, da/dN increases with increasing crack length and increasing stress.

where S = applied nominal stress, a = edge crack length or half the length of an internal crack, and Y = dimensionless geometric constant that varies depending on the geometry of the crack, but usually taken as 1.

Thus, we can consider the stress intensity factor corresponding to S_{max} and S_{min} as K_{max} and K_{min}, respectively and the difference between these two values is taken as the stress intensity factor range, that is,

$$\Delta K = K_{max} - K_{min} \tag{11.8}$$

From Equation 11.7,

$$K_{max} = YS_{max}\sqrt{\pi a} \tag{11.9}$$

and

$$K_{min} = YS_{min}\sqrt{\pi a} \tag{11.10}$$

and therefore,

$$\begin{aligned}\Delta K &= K_{max} - K_{min} \\ &= YS_{max}\sqrt{\pi a} - YS_{min}\sqrt{\pi a} \\ &= Y\sqrt{\pi a}\,(S_{max} - S_{min}) \\ &= Y\sqrt{\pi a}\,\Delta S\end{aligned} \tag{11.11}$$

Recall that if S_{min} is compressive in nature, its value is taken as zero. A typical plot of da/dN versus ΔK is shown in Figure 11.17. Notice that the plot is on a log–log

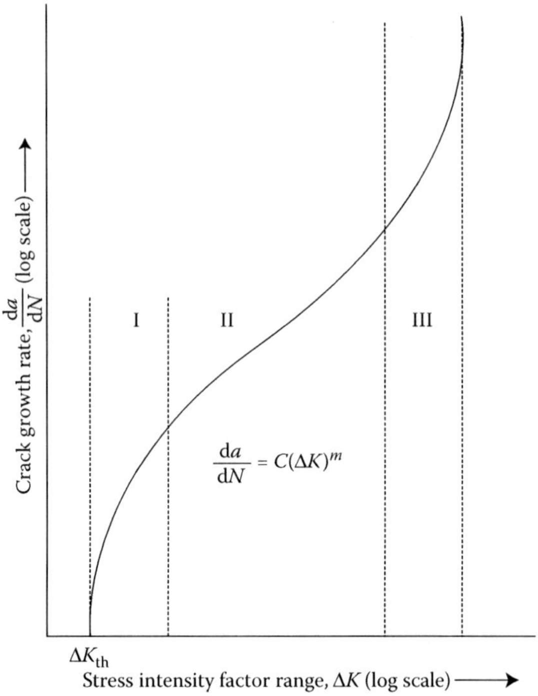

FIGURE 11.17 Schematic representation of fatigue crack growth rate, da/dN as a function of the stress intensity factor range, ΔK.

scale, that is, both da/dN and ΔK are plotted on a logarithmic scale because of the large range of these values. Also note that there is a threshold value of the stress intensity factor range, ΔK below which crack growth does not occur.

Examination of Figure 11.17 indicates that the curve can be divided into three regions. In region I, crack growth rate is slow. In region II, referred to as the Paris regime, crack growth rate obeys the relationship

$$\frac{da}{dN} = C(\Delta K)^m \tag{11.12}$$

where C and m are material constants. Taking logarithms on both sides of the above equation, we get

$$\log\frac{da}{dN} = m\log\Delta K + \log C \tag{11.13}$$

Therefore, in the Paris regime, the log da/dN versus log ΔK plot is a straight line with a slope of m. The value of m usually varies between 2.5 and 6 in this region. It should be realized that this is the most significant region of fatigue crack growth behavior in engineering applications. In region III, crack growth rate is very rapid and the material fails.

Figure 11.18 presents the fatigue crack growth behavior of the ASTM A533 B1 steel sample tested in air at 24°C. The stress ratio employed was $R = 0.1$.

The fatigue crack growth rate can be combined with the fracture toughness data to predict the fatigue life of a material, that is, the number of stress cycles a material can withstand under the given conditions of stress and crack size. Such an estimate will be useful in designing new engineering components using a specific material. The fatigue life can be calculated in the following manner.

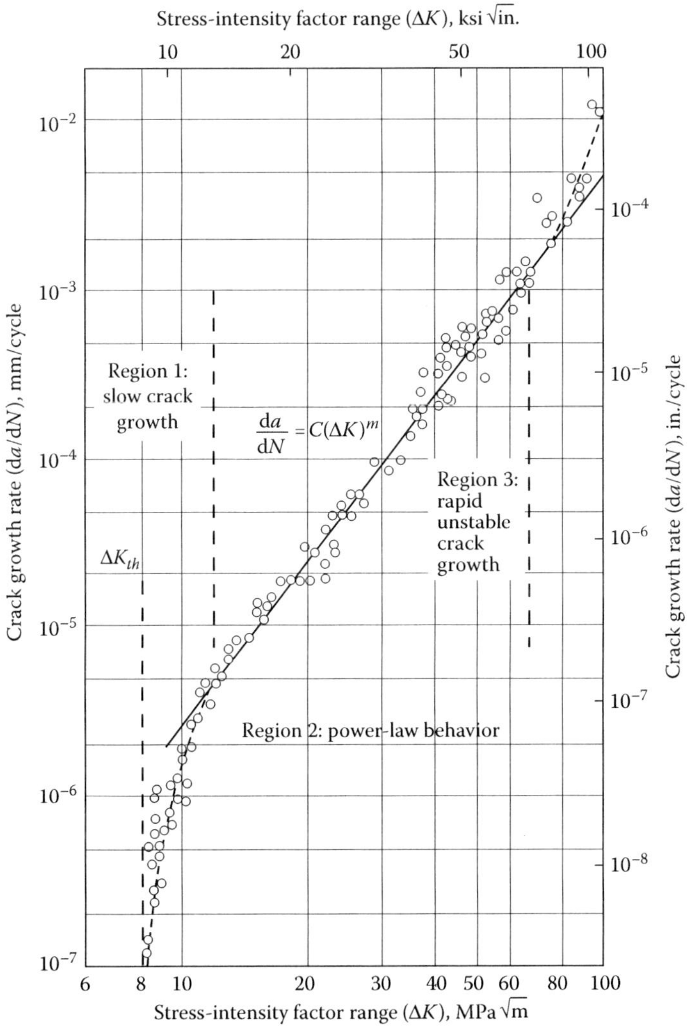

FIGURE 11.18 Fatigue crack growth behavior of ASTM A533 B1 steel tested in air at 24°C at a stress amplitude, $R = 0.1$.

Since it is known that in the Paris region (region II)

$$\frac{da}{dN} = C(\Delta K)^m \tag{11.12}$$

and that $\Delta K = Y\sqrt{\pi a}\,\Delta S$, we can combine the above two equations and write

$$\frac{da}{dN} = C(\Delta K)^m = C\left(Y^m \Delta S^m \pi^{m/2} a^{m/2}\right) \tag{11.14}$$

Equation 11.14 can be rearranged and then we integrate the crack size from the initial value a_0 to the final value at failure, a_f and the number of stress cycles from 0 to the value at failure, N_f. Thus,

$$\int_{a_0}^{a_f} da = C\,Y^m\,\Delta S^m\,\pi^{m/2}\,a^{m/2}\int_0^{N_f} dN \tag{11.15}$$

or

$$\int_0^{N_f} dN = \frac{1}{C\,Y^m\,\Delta S^m\,\pi^{m/2}}\int_{a_0}^{a_f}\frac{da}{a^{m/2}} \tag{11.16}$$

Therefore,

$$N_f = \frac{a_f^{1-(m/2)} - a_0^{1-(m/2)}}{C\,Y^m\,\Delta S^m\,\pi^{m/2}\left[1-(m/2)\right]} \tag{11.17}$$

The above equation may also be expressed as

$$N_f = \frac{2\left[a_f^{(2-m)/2} - a_0^{(2-m)/2}\right]}{(2-m)C\,Y^m\,\Delta S^m\,\pi^{m/2}} \tag{11.18}$$

The above equation assumes that $m \neq 2$ and that Y is independent of crack length, which is not usually the case. For a more accurate prediction of the fatigue life, the variation of Y with crack length (and also ΔK) must be taken into account. In other words, the value of N must be calculated for small successive changes of crack length.

Example Problem 11.2

A high-strength steel plate is subjected to fatigue with a maximum tensile stress of 120 MPa and a minimum compressive stress of 40 MPa. Assume that (i) $Y = 1$ in the equation for the stress intensity factor, (ii) the edge crack length is 5 mm, and (iii) the steel plate has a plane strain fracture toughness of 70 MPa√m. Using the equation da/dN (m cycle^{-1}) = $1.5 \times 10^{-9}\ \Delta K^3$ (MPa m$^{1/2}$)^3, estimate the fatigue life of the component.

Solution 11.2

From the data provided, $\Delta S = S_{max} - S_{min} = 120$ MPa – 0 = 120 MPa. (Since the minimum stress is compressive in nature and the cracks do not propagate during compression, S_{min} is taken as zero.)

The final crack size, a_f is determined from the fracture toughness Equation 11.7 (remember that $Y = 1$).

$$a_f = \frac{1}{\pi}\left(\frac{K_{IC}}{S_{max}}\right)^2 = \frac{1}{\pi}\left(\frac{70\,\text{MPa}\sqrt{\text{m}}}{120\ \text{MPa}}\right)^2 = 0.1083\ \text{m}$$

It is also given that

$$\frac{da}{dN}\ (\text{m cycle}^{-1}) = 1.5 \times 10^{-9}\left(\Delta K\right)^3$$

and therefore, in Equation 11.12, $C = 1.5 \times 10^{-9}$ and $m = 3$. Substituting all the respective values into Equation 11.17 we get

$$N_f = \frac{a_f^{1-(m/2)} - a_o^{1-(m/2)}}{C\,Y^m\,\Delta S^m\,\pi^{m/2}\left[1-(m/2)\right]}$$

$$N_f = \frac{(0.1083\ \text{m})^{1-(3/2)} - (5 \times 10^{-3}\,\text{m})^{1-(3/2)}}{1.5 \times 10^{-9} \times (1)^3 \times (120\,\text{MPa})^3\,\pi^{3/2}\left[1-(3/2)\right]}$$

$$= 1{,}540\ \text{cycles.}$$

Example Problem 11.3

Assume that a steel specimen contains a crack with an initial size of 0.5 mm. (a) Calculate the fatigue life of the specimen if the stress range used is 200 MPa and if the specimen fails when the crack reaches a size of 5 mm. Assume for calculations purposes that $C = 3 \times 10^{-12}$, $Y = 1$, and $m = 4$. (b) Calculate the lifetime if the specimen were to fail when the final crack size is 3 mm, and all other parameters remain the same as in (a).

Solution 11.3

a. Since all the required parameters are given, we can use Equation 11.17 to solve for N_f. Therefore,

$$N_f = \frac{a_f^{1-(m/2)} - a_o^{1-(m/2)}}{C\, Y^m\, \Delta S^m\, \pi^{m/2}\left[1-(m/2)\right]}$$

$$N_f = \frac{\left(5\times10^{-3}\text{m}\right)^{1-(4/2)} - (0.5\times10^{-3}\,\text{m})^{1-(4/2)}}{3\times10^{-12}\times(1)^4\times\left(200\ \text{MPa}\right)^4\pi^{4/2}\left[1-(4/2)\right]}$$

$$= \frac{\left(5\times10^{-3}\text{m}\right)^{-1} - (0.5\times10^{-3}\,\text{m})^{-1}}{3\times10^{-12}\times(1)^4\times\left(200\ \text{MPa}\right)^4\pi^{2}\left[1-2\right]}$$

$$= 3.8\times10^4\text{cycles}$$

b. Since all other parameters remain constant, but the final crack size is now only 3 mm, the above equation can be written down as

$$N_f = \frac{\left(3\times10^{-3}\,\text{m}\right)^{-1} - (0.5\times10^{-3}\,\text{m})^{-1}}{3\times10^{-12}\times(1)^4\times\left(200\ \text{MPa}\right)^4\text{p}^{2}\left[1-2\right]}$$

$$= 3.52\times\ 10^3\text{cycles}$$

From the above result, it can be clearly seen that the specimen would fail at less number of cycles if the crack size at which it would fail were to decrease.

Example Problem 11.4

Nondestructive examination of a steel component that is subjected to a repeated stress of 300 MPa every 2 h revealed the presence of 3-mm-long edge crack. The plane strain fracture toughness of the steel sample is given as 80 MPa√m. If the crack propagation rate is expressed as

$$\frac{da}{dN}\text{m cycle}^{-1} = 4\times10^{-13}\left(\Delta K\right)^4$$

(a) Calculate the remaining lifetime of the steel part. (b) Is it necessary to replace the component if the next inspection procedure is scheduled after 6 months' time?

Solution 11.4

a. The plane strain fracture toughness value is given as 80 MPa√m. Therefore, the crack size for failure can be calculated from Equation 11.7 as

$$a_f = \frac{1}{\pi}\left(\frac{K_{IC}}{S_{max}}\right)^2 = \frac{1}{\pi}\left(\frac{80\,\text{MPa}\sqrt{\text{m}}}{300\,\text{MPa}}\right)^2 = 0.023\text{m}$$

The fatigue life can be calculated from Equation 11.17 as

$$N_f = \frac{a_f^{1-(m/2)} - a_o^{1-(m/2)}}{C\,Y^m\,\Delta S^m \pi^{m/2}\left[1-(m/2)\right]}$$

$$N_f = \frac{(2.3\times10^{-2}\ \text{m})^{1-(4/2)} - (3\times10^{-3}\ \text{m})^{1-(4/2)}}{4\times10^{-13}\times(1)^4\times(300\ \text{MPa})^4 p^{4/2}[1-(4/2)]}$$

$$= \frac{(2.3\times10^{-2}\ \text{m})^{-1} - (3\times10^{-3}\ \text{m})^{-1}}{4\times10^{-13}\times(1)^4\times(300\ \text{MPa})^4\,\pi^2\,[1-2]}$$

$$= 9064 \text{ cycles}$$

b. Since the next inspection procedure is scheduled in 6-months' time, the number of cycles the part would experience in 6 months (6 × 30 days × 24 h = 4320 h) is 4320/2 (since the stress cycle is 2 h) = 2160 cycles. Since the fatigue life calculated in (a) is 9064 cycles, it is not necessary for the part to be replaced.

11.8 CORRELATIONS BETWEEN FATIGUE STRENGTH AND OTHER MECHANICAL PROPERTIES

The fatigue behavior of materials can only be obtained from a large number of test results conducted on a number of specimens and at different stress levels. In addition, fatigue tests are more complicated than tensile or hardness tests and require more expensive and specialized equipment. Further, these tests are time consuming (the tests require several hours to weeks) and the fatigue data require statistical treatment because of reproducibility difficulties. Consequently, attempts have been made to relate the fatigue properties of materials to other more easily measurable mechanical properties such as tensile strength and surface hardness.

It is generally noted that the fatigue strength increases with increasing tensile strength or surface hardness of the materials. A comparison of the fatigue strength and tensile strength of different materials is shown in Table 11.1. It is noted that the fatigue strength of materials usually lies between one-third and three-fifth of the tensile strength (Figure 11.19). This ratio of endurance limit to tensile strength is known as *fatigue ratio* or *endurance ratio*. Thus,

$$0.35 < \frac{\text{Endurance limit, } S_e}{\text{Tensile strength, } \sigma_{ts}} < 0.60 \tag{11.19}$$

By knowing the endurance ratio of materials, it becomes possible to choose a safe stress to operate the component.

Example Problem 11.5

The tensile strength of a structural steel alloy is reported as 600 MPa. Determine the minimum and maximum permissible service stress levels during fatigue testing such that the specimen will not fail.

TABLE 11.1
Tensile Strength, Fatigue Strength, and Endurance Ratios of Some Selected Engineering Alloys

Material	Condition	Yield Strength, σ_{ys} (MPa)	Tensile Strength, σ_{ts} (MPa)	Fatigue Strength, σ_f (MPa)	Endurance Ratio, σ_f/σ_{ts}
	Steels (Endurance Limit Based on 10^7 Cycles)				
1015	Cold drawn–0%	275	455	240	0.527
1015	Cold drawn–60%	605	710	350	0.493
1040	Cold drawn–0%	405	670	345	0.515
1040	Cold drawn–50%	855	965	410	0.425
4340	Annealed	475	745	340	0.456
4340	Quenched and Tempered (204°C)	1640	1950	480	0.246
4340	Quenched and Tempered (427°C)	1380	1530	470	0.307
4340	Quenched and Tempered (538°C)	1170	1260	670	0.532
	Aluminum Alloys (Endurance Limit Based on 5 × 10^8 Cycles)				
1100–0		34	90	34	0.378
2014–T6		414	483	124	0.257
2024–T3		345	483	138	0.286
6061–T6		276	310	97	0.313
7075–T6		503	572	159	0.278
	Titanium Alloys (Endurance Limit Based on 10^7 Cycles)				
Ti–6Al–4V		885	1035	515	0.498
Ti–6Al–2Sn–4Zr–2Mo		825	895	485	0.542
Ti–5Al–2Sn–2Zr–4Mo–4Cr		1130	1185	675	0.570
	Copper Alloys (Endurance Limit Based on 10^8 Cycles)				
70Cu–30Zn brass	Hard	435	524	145	0.277
90Cu–10Zn	Hard	370	420	160	0.381
	Magnesium Alloys (Endurance Limit Based on 10^8 Cycles)				
HK31A–T6		110	215	62–83	0.288–0.386
AZ91A		160	235	69–96	0.294–0.409

Note: All compositions are expressed in weight percentage.

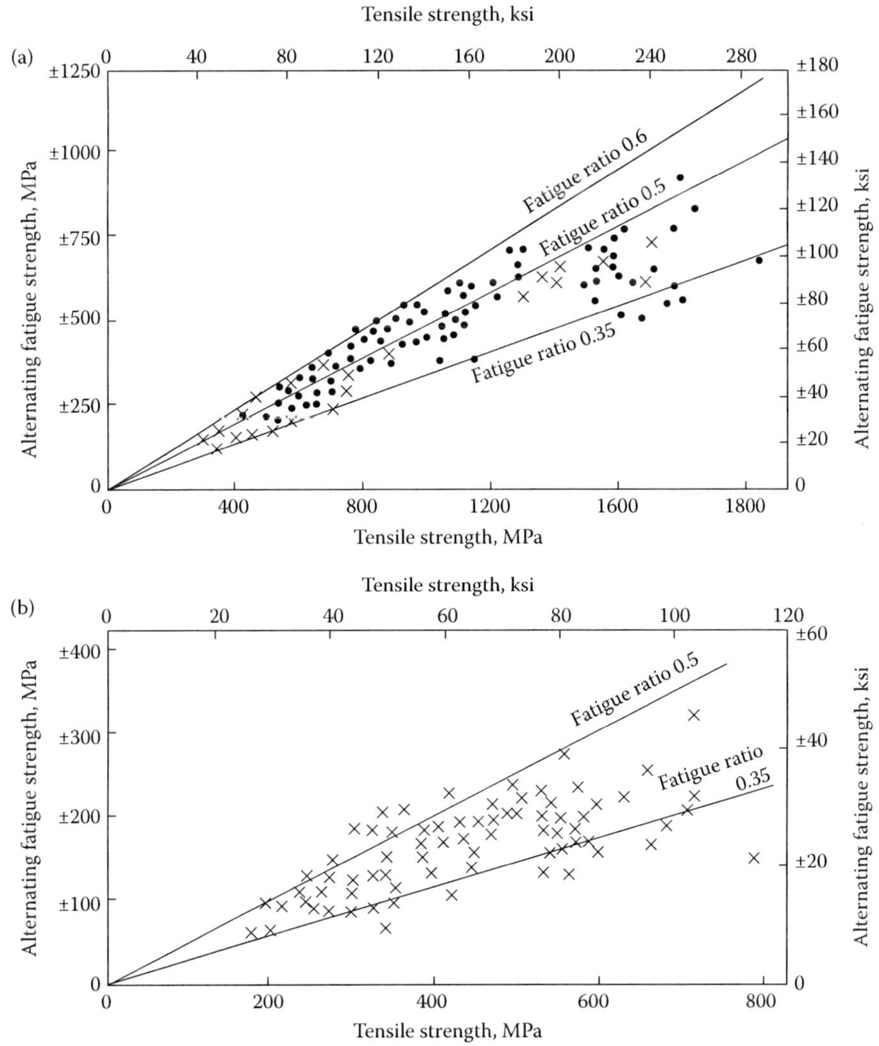

FIGURE 11.19 Relationship between fatigue strength and tensile strength for (a) alloy (•) and carbon (×) steels and (b) wrought copper alloys. Note that the fatigue ratio (ratio of fatigue strength to tensile strength) usually lies between 0.35 and 0.6. (After Hertzberg, R. W. 1996. *Deformation and Fracture Mechanics of Engineering Materials*, 4th edition. New York, NY: John Wiley & Sons, Inc. With permission.)

Solution 11.5

Knowing that the endurance ratio of materials usually lies between 0.35 and 0.6, we can calculate the conservative endurance limit (or fatigue strength) as

$$\text{Fatigue strength} = 0.35 \times 600\ \text{MPa} = 210\ \text{MPa}$$

On the other hand, the highest value of service stress that could be applied is

$$0.6 \times 600 \text{ MPa} = 360 \text{ MPa}$$

Note: Generally, a safety factor is used to account the statistical variation of the fatigue strength and also to take into account the approximate nature of the correlation between fatigue strength and tensile strength of materials. This safety factor is normally 2. Thus, in the above case, the stresses that are permitted without specimen failure are a minimum of 210/2 = 105 MPa and a maximum of 360/2 = 180 MPa.

11.9 PRECAUTIONS

Determination of reliable fatigue strength data requires use of specimens with highly polished surfaces that are free of scratches, pits, notches, and other surface imperfections.

The *S–N* data that are obtained from laboratory tests is useful. But, it must be noted that the specimens in laboratory tests are unconstrained. On the other hand, in actual practice, the engineering components are constrained, particularly at stress concentrations. In such a case, it is observed that the fatigue performance is controlled by the ductility or toughness of a material at short fatigue lives and the strength of the material at longer lives.

It may be appreciated that fatigue experiments require long time for the results to be obtained. Therefore, these experiments are accelerated by increasing the stress levels. Consequently, the specimens fail in shorter times. For example, in low-cycle fatigue (LCF), the stress levels are higher and therefore, the fatigue life is of the order of 10^4 cycles. To distinguish this from the regular experiments, the latter are referred to as *high-cycle fatigue* (HCF) experiments in which the fatigue life is of the order of 10^7 cycles.

11.10 EXPERIMENTAL PROCEDURE

Fatigue testing machines are not frequently available in all the university laboratories, unless research investigations are being conducted in the area of fatigue behavior of materials. If you have a fatigue-testing machine in your laboratory, get familiar with the type of equipment you have. Check whether the specimen is loaded in single cantilever bending or otherwise. In the cantilever-type loading, the specimen is supported at one end and loaded at the other through self-aligning ball bearings which permit free rotation of the specimen under load without any torsional effects. Take a large number of specimens, prepared according to the ASTM standards (a sketch of which was shown in Figure 11.6), and ensure that the specimen surface is polished to a smooth finish. In other words, the specimens should not have any stress concentrators (pits, notches, scratches, inclusions, etc.) on their surfaces. Load the specimen into the machine and choose the desired fatigue cycle (the maximum and minimum stresses, the cycle duration, etc.). Conduct the test and determine the number of cycles required for the specimen to fail under these conditions.

Repeat the test with different stress amplitudes and determine the number of cycles to failure in each case. (Alternately, different batches of students could do a similar test at different stress levels and the results of all the batches can be pooled together.) Using the data obtained, plot the *S–N* curve and determine the endurance limit of the sample. Recall that not all types of materials exhibit a well-defined endurance limit. If the specimen does not exhibit a clear endurance limit, the endurance limit is taken as the stress value at which the specimen fails in 10^7 cycles.

Note: The fatigue experiments require long time to be completed and therefore, these experiments may not be possible to be completed in one regular laboratory session.

11.11 ADDITIONAL EXPERIMENTS

Depending on the time available, similar tests could be conducted on other samples. For example, one could take different steel and nonferrous alloy samples, or polymeric samples, or the recently developed bulk metallic glassy samples.

At an advanced level, one could also determine the fatigue crack growth rate as a function of the stress intensity factor range and determine the constants C and m in Equation 11.12.

A useful exercise would be to look at the fractured surface of the specimen in an SEM and identify the features. Of particular interest would be the location of the point of origin of fracture.

EXERCISES

11.1 Define fatigue.

Define the following terms:

11.2 Range of stress

11.3 Mean stress

11.4 Stress amplitude

11.5 Stress ratio

11.6 Fatigue cycle

11.7 What is an *S–N* curve?

11.8 What is fatigue strength?

11.9 What is fatigue life?

11.10 What is endurance limit?

11.11 Draw a typical *S–N* curve showing a well-defined endurance limit. Label the axes.

11.12 Differentiate between the *S–N* curves of steel and aluminum alloy samples.

11.13 Where do fatigue failures usually originate on a metal section?

11.14 What are the distinctive features of the fracture surface of a specimen that has failed by fatigue?

11.15 How does fatigue failure occur in metals?

11.16 Why is it necessary to have a smooth surface on a fatigue testing sample?

11.17 What is the relation between fatigue strength and tensile strength?

11.18 What are the factors on which fatigue strength depends?
11.19 How can one increase the fatigue life of a component?
11.20 What is the effect of stress concentration on the fatigue strength of a material?
11.21 What is corrosion fatigue?
11.22 What is thermal fatigue?
11.23 What will happen to the fatigue life if the surface of a steel specimen is case carburized?
11.24 What will happen to the fatigue life if the surface of a steel specimen is decarburized?
11.25 How can one predict the fatigue life of a material?
11.26 What is the relation between fatigue strength and endurance ratio?
11.27 What is the ratio of the fatigue strength to the tensile strength of a metal?

FURTHER READING

Callister, Jr., W. D. and D. G. Rethwisch. 2010. *Materials Science and Engineering, An Introduction*, 8th edition. New York, NY: John Wiley & Sons, Inc.

Courtney, T. H. 2000. *Mechanical Behavior of Materials*, 2nd edition. New York, NY: McGraw-Hill.

Dowling, N. E. 2007. *Mechanical Behavior of Materials*, 3rd edition. Upper Saddle River, NJ: Pearson: Prentice-Hall.

Hayden, H. W., W. G. Moffatt, and J. Wulff, 1965. *The Structure and Properties of Materials, Vol. III: Mechanical Behavior*, New York, NY: John Wiley & Sons, Inc.

Hertzberg, R. W. 1996. *Deformation and Fracture Mechanics of Engineering Materials*, 4th edition. New York, NY: John Wiley & Sons, Inc.

McEvily, A. J. 2002. *Metal Failures: Mechanisms, Analysis, Prevention*. New York, NY: John Wiley & Sons, Inc.

Suresh, S. 1998. *Fatigue of Materials*, 2nd edition. Cambridge, UK: Cambridge University Press.

12 Creep Testing

12.1 INTRODUCTION

We had earlier considered elastic and plastic deformation of specimens during tensile testing. It was noted that the specimen experiences strain on subjecting it to a stress and that the strain is proportional to stress in the elastic portion of the stress–strain curve. Permanent or irrecoverable strain is experienced by the specimen when it is subjected to stresses beyond the yield point. In either case, the strain experienced by the specimen is assumed to be independent of time, that is, the magnitude of strain remains constant and does not change with time. However, depending on the material of the specimen and the temperature at which the specimen is subjected to deformation, a situation will arise when the strain induced in the specimen, at a constant stress (or load), changes (usually it increases) with time. This time-dependent plastic (permanent) deformation at constant stresses (or loads) occurring at high temperatures is called *creep*. Because creep deformation is significant at elevated temperatures, it is frequently described as a *thermally activated process*, that is, temperature plays an important role.

It is important to realize that creep can occur at any temperature. But, the full range of effects of creep is visible only at elevated temperatures. At lower temperatures, creep is characterized by an ever-decreasing strain rate, while at elevated temperatures, it proceeds through three distinct stages and ultimately results in specimen failure.

It is appropriate to state that creep occurs in metallic specimens at high temperatures. But, the concept of "high" temperatures is relative and depends on the material and temperature combination. It is common to describe the use temperature in terms of the "*homologous temperature*," defined as the ratio of the use temperature to the melting point (both expressed in degrees Kelvin), that is, T/T_m. Creep generally occurs at temperatures above $0.4T_m$. Thus, if a specimen of lead (melting point is 327°C or 600 K) is subjected to deformation at room temperature (say 300 K), this temperature is $0.5T_m$ and so creep occurs in lead specimens at room temperature. On the other hand, creep does not occur in a Ni-based superalloy at room temperature since this temperature is much less than $0.2T_m$. To put this in the proper perspective, it is better to say that this time-dependent deformation does not occur in high-melting point materials (to any measurable level) at room temperature.

Creep occurs in all types of materials—metallic, polymeric, and ceramic. While creep is important in metallic materials at moderate temperatures and stresses, it occurs at relatively low stresses and temperatures in polymeric materials and at high temperatures and stresses in ceramic materials.

The creep behavior of metals and alloys is very important for some of the engineering applications, especially when the components are used at elevated temperatures and under loads. For example, materials used in turbine blades of a gas turbine engine experience very high temperatures and so creep is of great concern. The creep behavior of materials is usually evaluated in terms of the *creep rate*, that is, the change in creep strain with time ($d\varepsilon/dt = \dot{\varepsilon}$). Since creep rate corresponds to the strain rate, from now onward, these two terms will be used interchangeably. The creep rate is an important design criterion in selecting materials for high-temperature applications. That is, the creep rate, at a given stress, determines the maximum temperature up to which the material may safely be used. But, luckily, the creep behavior of a material can be modified through an intelligent design and judicious choice of the alloy composition, processing techniques, and microstructure, and consequently the use temperature can be increased.

12.2 THE CREEP TEST

In a conventional creep test (ASTM Standard E 139: Standard Practice for Conducting Creep, Creep Rupture, and Stress Rupture Tests of Metallic Materials), a specimen is subjected to a constant load or stress, while maintaining the temperature constant. Most engineering tests are conducted at constant loads. Although the method is simple, recall that since the cross-sectional area in the gauge length portion changes during deformation, the stress also varies with strain (time). But, to get a better understanding of the creep mechanisms, creep tests are conducted under constant stress conditions. The resultant plastic (or permanent) strain is measured and plotted as a function of time.

The specimen used to determine the creep behavior of materials in tension is exactly the same as the one used in conventional tensile testing (see Chapter 9: Tensile Testing, for details of specimen dimensions). However, for brittle materials it may be more appropriate to conduct the creep tests under compression. The specimens employed for compression creep testing are usually right cylinders with an aspect ratio (length to diameter) ranging from 2 to 4 (similar to compression testing). A typical setup for creep testing is shown in Figure 12.1.

A number of creep tests are conducted at a given temperature but varying stress levels or at different temperatures for a given stress level. All the information is gathered together to obtain the creep behavior of the material.

12.3 THE CREEP CURVE

Let us assume that we wish to determine the creep behavior of a polycrystalline specimen. We will then take a tensile specimen in a well-annealed condition and subject it to a constant load or constant stress (sufficient to cause creep deformation) while maintaining the temperature constant at $>0.4T_m$. The strain in the specimen is plotted as a function of time. Figure 12.2 represents schematic of a typical creep curve obtained.

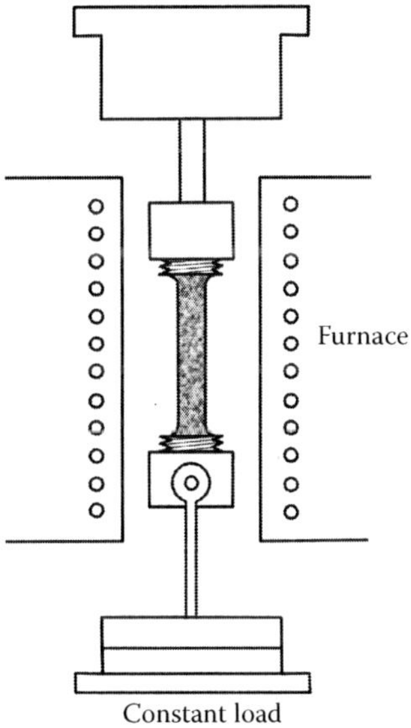

FIGURE 12.1 A typical setup for creep testing.

The creep curve (Figure 12.2) can be divided into three different stages of creep, each with its own characteristic strain versus time behavior. Note that since the specimen has been subjected to stress, it shows an instantaneous and rapid elongation and consequently, there is an instantaneous elastic strain, ε_o in the specimen at time $t = 0$. Following this, the specimen exhibits three distinctive stages of creep.

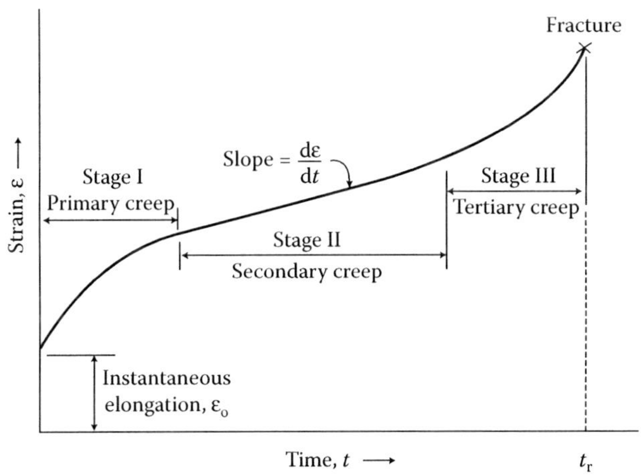

FIGURE 12.2 A typical creep curve. It represents the strain as a function of time when the test is conducted at a constant temperature and constant load or stress.

12.3.1 Primary or Stage I Creep

The first stage, referred to as the *primary creep* or *Stage I creep*, is characterized by a continuously decreasing slope of the creep curve, that is, the creep rate, $\dot{\varepsilon}$, (= $d\varepsilon/dt$) decreases with time but the creep strain, ε still continues to increase.

During Stage I of creep, the material work hardens and therefore more deformation is required to obtain the same amount of strain. Thus, it can be mentioned that, during Stage I, the creep resistance increases with time. The creep strain in Stage I can be described using the equation

$$\varepsilon = At^{1/3} \tag{12.1}$$

where ε is the creep strain, t is the time, and A is a material constant. This type of behavior can continue indefinitely at low temperatures. At higher temperatures, however, the Stage I regime leads to conditions where the rate of deformation becomes independent of time and strain. When this occurs, creep is in Stage II or secondary or steady-state regime.

The reason why the strain rate decreases with time in Stage I is because of rearrangement of dislocations in the specimen to maintain equilibrium with the applied load. Such a dislocation rearrangement occurs through a process called *dislocation climb*, illustrated in Figure 12.3. In this process, the dislocation "climbs" or moves up one atomic distance. The mechanism by which this occurs can be described as follows. Let us consider the edge dislocation, indicated by the symbol ⊥ in Figure 12.3a. Recall that an edge dislocation is characterized by an extra half-plane of atoms, indicated by the solid circles. In the climb process (to be more accurate, we should refer to this as *positive climb*), atoms are removed from the extra half-plane of atoms so that the row of atoms constituting the bottom of the extra half-plane moves up by one atomic spacing. This can occur when vacancies in the lattice, indicated by the cube symbol □ diffuse to the dislocation line from somewhere within the lattice as indicated in Figure 12.3b. Under these circumstances, the vacancies move into the atom positions and the atoms at the bottom of the extra half plane of the dislocation move into the vacant lattice sites in the lattice, as shown in Figure 12.3c. Consequently, we can say that the dislocation has moved one atom distance upward. On the other hand, if the dislocation moves downward, or the row of atoms moves upward, then this will be referred to as *negative climb*. Realize that it is highly unlikely that complete rows of atoms are removed from the extra half plane of atoms. The more common situation is for a small number of vacancies to diffuse to the extra half-plane of atoms and therefore climb occurs over a short segment of the dislocation line.

12.3.2 Secondary or Stage II Creep

In this stage, referred to as *Secondary Creep* or *Stage II Creep*, the strain rate is constant, that is, the creep strain increases linearly with time. For this reason, this stage of creep is also referred to as *steady-state creep*. In this stage, equilibrium is established between the recovery processes involving highly mobile dislocations and

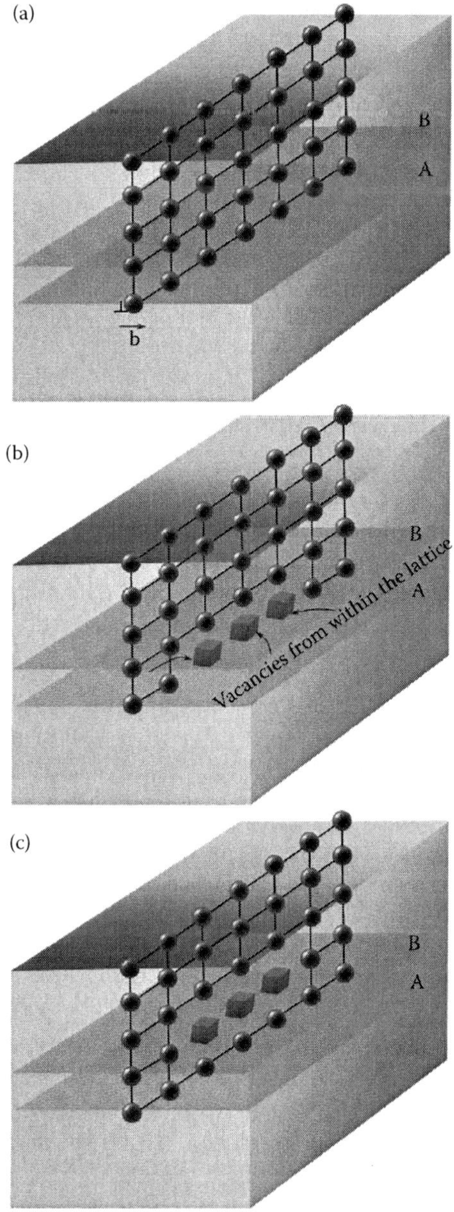

FIGURE 12.3 Mechanism of dislocation climb. The edge dislocation is characterized by an extra half-plane of atoms indicated by the solid circles. The position of the dislocation is indicated by the symbol ⊥ in (a). Vacancies, indicated by the cube symbol □, can diffuse to the bottom row of atoms constituting the edge dislocation as shown in (b). The vacancies then migrate to the atoms constituting the bottom row of the extra half-plane and the atoms constituting the bottom row of the edge dislocation diffuse into the lattice. This results in the upward movement of the dislocation by one atomic spacing (c).

strain hardening of the specimen due to the applied load. Therefore, the creep rate is constant and also minimum, and it is this minimum creep rate that is used in design considerations. Consequently, the creep resistance of the material is the highest in this stage. Further, because the creep rate is minimum in this stage, the secondary creep stage lasts much longer than the other stages. Since the creep strain varies linearly with time during the secondary stage, the creep strain in this stage is given by the equation

$$\varepsilon = \varepsilon_0 + \dot{\varepsilon}_s t \tag{12.2}$$

where $\dot{\varepsilon}_s$, the steady-state creep rate is a material constant. The value of $\dot{\varepsilon}_s$ is most important and is used in evaluating the useful life of engineering components that are expected to be operated for a long period of time (several years), for example, nuclear power plant components.

Example Problem 12.1

A specimen has been subjected to creep. If the strain measured in Stage II is 0.003 at 600 h and 0.002 at 200 h, calculate the steady-state creep rate of the specimen.

Solution 12.1

Since the steady-state creep rate is the slope of the strain versus time straight line, the steady-state creep rate can be calculated as

$$\text{Steady-state creep rate} = \frac{d\varepsilon}{dt} = \frac{0.003 - 0.002}{600\,\text{h} - 200\,\text{h}} = \frac{0.001}{400\,\text{h}} = 0.0000025\,\text{h}^{-1}$$

$$\text{or,} \quad \dot{\varepsilon}_s = 2.5 \times 10^{-6}\,\text{h}^{-1}$$

12.3.3 Tertiary or Stage III Creep

By the time this stage, referred to as *tertiary creep*, is reached, the specimen has experienced considerable elongation and consequently the specimen experiences necking resulting in a decreased cross-sectional area. Recall that we discussed about the phenomenon of necking earlier (in Chapter 9: Tensile Testing). Consequently, the strain rate increases due to an increase in the true stress value of the specimen. Further, voids and cracks also begin to form inside the specimen, particularly along the grain boundaries. As a result of all these effects, the strain rate in the specimen in this stage increases exponentially with time until failure occurs. This failure is frequently referred to as rupture. The time at which rupture occurs is called *rupture lifetime*, t_r.

The variation of strain with time during Stage III is expressed by the relation

$$\varepsilon = B + C\exp(\gamma t) \tag{12.3}$$

where B, C, and γ are material constants. It is during Stage III that the maximum amount of strain is accumulated in the specimen. However, since this strain is accumulated in a short time, this is not useful.

While $\dot{\varepsilon}_s$ is the parameter used to calculate the useful life of components used for long-life applications, the rupture lifetime, t_r is the design parameter used to evaluate the short-life applications, for example, for turbine blades in military aircraft and rocket motor nozzles. If a creep test is conducted until the specimen ruptures, then that type of test is referred to as *creep-rupture test*. Compared to conventional creep testing, the stress levels employed in a creep-rupture test are higher. Thus, the availability of both the parameters $\dot{\varepsilon}_s$ and t_r will enable a proper design of the component for the desired application.

Even though three distinct stages of creep have been identified in Figure 12.2, it is not necessary that a specimen will always exhibit all the three stages. In some cases, fracture occurs in the secondary stage and in that case, the final stage (Stage III) will not be observed. Further, the relative durations of the three stages could be different depending on the material, the stress that is applied and the temperature at which creep is occurring.

12.4 EFFECT OF STRESS AND TEMPERATURE

Both the magnitude of stress applied and the temperature at which the creep test is conducted affect the creep behavior of the sample. At low stresses and low temperatures, metals predominantly exhibit primary creep and negligible amount of secondary creep. This is because the temperature is too low for recovery processes to occur and therefore, the creep strain decreases with increasing time. But, if the stress employed is higher, the metal will deform like in a conventional tensile test and therefore the creep strain increases. At higher temperatures, the recovery processes are much faster and so the specimen will have Stage II and Stage III of creep and may exhibit minimum amount of Stage I creep. Thus, it is easy to visualize that the creep strain increases with increasing stress and temperature, as indicated in Figure 12.4.

Since creep is a thermally activated process, the creep rate can be expressed in terms of an Arrhenius-type equation as

$$\dot{\varepsilon}_s = C\exp^{-(Q/RT)} \tag{12.4}$$

where C is the preexponential constant, Q is the activation energy for creep, R is the universal gas constant, and T is the temperature expressed in degrees Kelvin. By taking natural logarithms on both sides of the above equation, we get

$$\ln \dot{\varepsilon}_s = \ln \mathrm{C} - \left(\frac{Q}{RT}\right) \tag{12.5}$$

Thus, by plotting ln $\dot{\varepsilon}_s$ against $1/T$, a straight line is obtained (Figure 12.5) with a slope of $-Q/R$. Since the value of R (the gas constant) is known, it is possible to calculate the activation energy for creep. The value of the activation energy, Q enables one to predict the mechanism by which creep occurs.

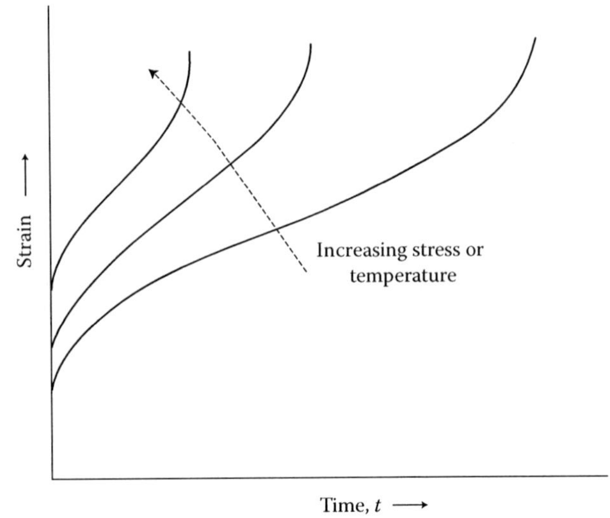

FIGURE 12.4 Effect of stress and temperature on the creep strain of a metallic specimen. The creep strain increases with increasing stress and increasing temperature. Also note that the creep rate increases with increasing stress.

It was mentioned earlier that the creep strain rate is higher at higher temperatures and/or higher stresses. That is, the times required for creep failure are shorter at high temperatures. Therefore, by extrapolating the high-temperature short-time data to lower temperatures, it should be possible to predict the long-term creep behavior at lower temperatures. This extrapolation, of course, is valid only when the same creep mechanism operates in the whole temperature range considered.

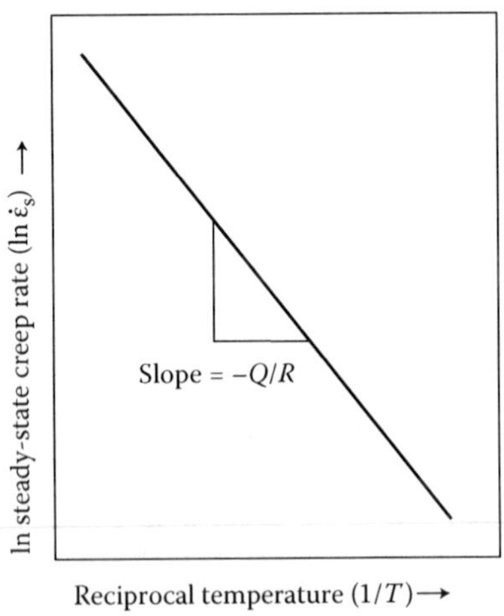

FIGURE 12.5 Plot of ln $\dot{\varepsilon}_s$ vs. $1/T$, where $\dot{\varepsilon}_s$ is the steady-state creep rate and T is the temperature expressed in degrees Kelvin. A straight line is obtained, whose slope is $-Q/R$, where Q is the activation energy for creep.

Example Problem 12.2

A metallic alloy exhibits a steady-state creep rate of 2.5×10^{-4} h^{-1} at a temperature of 550°C and 3.2×10^{-3} h^{-1} at 650°C. Calculate the activation energy for creep.

Solution 12.2

Since the steady-state creep rate is given at two different temperatures, and we wish to calculate the activation energy, we can use Equation 12.4. The temperatures need to be expressed in degrees K. Therefore, 550°C = 550 + 273 = 823 K and 650°C = 650 + 273 = 923 K. Substituting the appropriate values into this equation at these two temperatures, we obtain

$$\dot{\varepsilon}_s = C\exp^{-(Q/RT)}$$

$$3.2 \times 10^{-3}\,\text{h}^{-1} = C\exp\left(-\frac{Q}{8.314\ \text{J mol}^{-1}\,\text{K}^{-1} \times (923\,\text{K})}\right)$$

and

$$2.5 \times 10^{-4}\,\text{h}^{-1} = C\exp\left(-\frac{Q}{8.314\ \text{J mol}^{-1}\,\text{K}^{-1} \times (823\,\text{K})}\right)$$

Dividing the above two equations with one another, we get

$$\frac{3.2 \times 10^{-3}\,\text{h}^{-1}}{2.5 \times 10^{-4}\,\text{h}^{-1}} = \exp\left[-\frac{Q}{8.314\ \text{J mol}^{-1}\,\text{K}^{-1}}\left(\frac{1}{923\,\text{K}} - \frac{1}{823\,\text{K}}\right)\right]$$

$$1.28 \times 10\text{h}^{-1} = \exp\left[-\frac{Q}{8.314\ \text{J mol}^{-1}\,\text{K}^{-1}}(-1.316 \times 10^{-4})\right]$$

Taking natural logarithms on both sides,

$$\ln(1.28 \times 10) = \left(\frac{Q}{8.314}\right)(1.316 \times 10^{-4})$$

or,

$$2.55 = Q \times 1.583 \times 10^{-5}$$

$$Q = 1.611 \times 10^{5}\ \text{J mol}^{-1},\ \text{that is, } 161\ \text{kJ mol}^{-1}$$

12.5 CREEP-RUPTURE TEST

As mentioned earlier, a creep-rupture test (also commonly referred to as a *stress-rupture test*) is conducted to obtain data on rupture time when the specimen is subjected to creep till failure at a constant stress and at a constant temperature. If the creep test is conducted at different stress levels and a constant temperature, and the rupture lifetime measured, then it will be possible to plot the stress required for failure against rupture lifetime at a constant temperature. Similar tests can also be conducted at different temperatures. All these data can be combined to produce a plot such as the one shown in Figure 12.6. Note that both the stress and rupture lifetime are plotted on a logarithmic scale. Figure 12.6 shows the stress-rupture data at five different temperatures for a mechanically alloyed oxide-dispersion-strengthened (ODS) superalloy MA 6000. MA6000 is a nickel-based superalloy with the composition 15 Cr, 2.0 Mo, 4.0 W, 4.5 Al, 2.5 Ti, 2.0 Ta, and 1.1 Y_2O_3 (all in weight percent). The plot shows that a linear relationship exists between the logarithm of applied stress and logarithm of rupture life at each temperature. It may be noted that the rupture lifetime decreases as the applied stress and/or temperature are increased. But, nonlinearity or slope changes in the curves have been noted in some alloys over relatively large stress/temperature ranges. These are caused by factors such as oxidation, corrosion, recrystallization, or phase changes at different temperatures or stresses. The data available from plots such as those in Figure 12.6 are of great help to design engineers. For example, knowing the temperature of operation, the engineer can limit the maximum stress to which the component will be subjected to, without failure occurring. Alternatively, if the maximum stress is known, then the highest temperature can be limited. Thus, these plots provide a convenient data set for design engineers.

Based on the available data, empirical relationships have been developed between the steady-state creep rate and the applied stress and temperature. The stress dependence of the steady-state creep rate $\dot{\varepsilon}_s$ can be expressed as

$$\dot{\varepsilon}_s = A_1\sigma^n \tag{12.6}$$

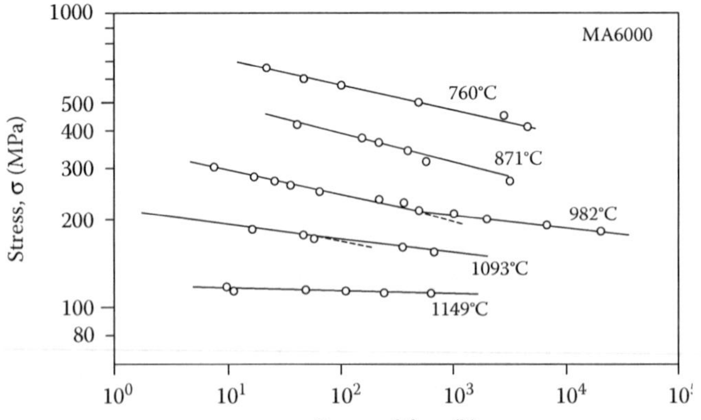

FIGURE 12.6 Variation of logarithm of stress with rupture lifetime for a mechanically alloyed ODS nickel-based superalloy MA6000 at five different temperatures.

where A_1 and n are material constants. However, if both stress and temperature effects are considered together, then the steady-state creep rate $\dot{\varepsilon}_s$ is expressed as

$$\dot{\varepsilon}_s = A_2 \sigma^n \exp\left(-\frac{Q_c}{RT}\right) \tag{12.7}$$

The above equation is known as the power-law equation.

This activation energy, Q_c for steady-state creep is often found to be essentially equal to the activation energy for self-diffusion. That this is true is amply proven by the correlation represented in Figure 12.7 for different types of materials. This correlation also confirms that vacancy motion is essential for the creep process in materials.

Example Problem 12.3

A superalloy sample is subjected to creep studies at a stress level of 60 MPa and the stress exponent, n is 6.5. If the steady-state creep rate was measured as 5.6×10^{-4} h^{-1} at a temperature of 800°C and 7.8×10^{-3} h^{-1} at a temperature of 900°C, calculate (a) the activation energy for the creep process, and (b) the steady-state creep rate at a temperature of 1000°C. You may assume that the creep mechanism is the same at all these temperatures and that n is independent of temperature.

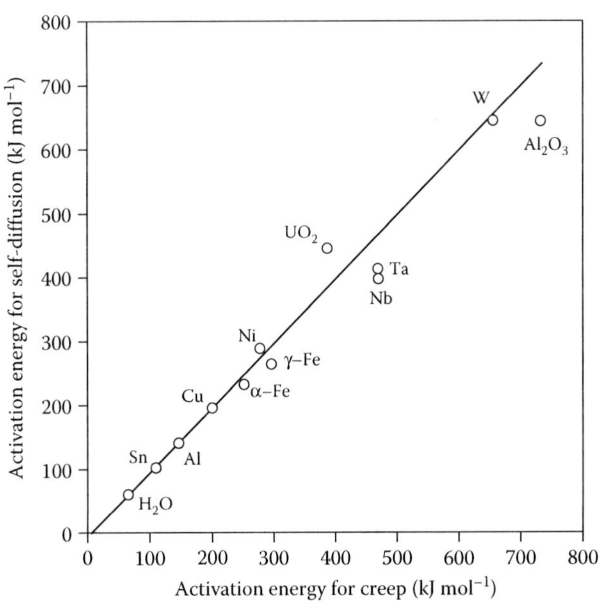

FIGURE 12.7 Relationship between the activation energy for high-temperature creep and activation energy for self-diffusion in different materials.

Solution 12.3

We can use the power-law Equation 12.7 to solve this problem. Since the temperatures have to be expressed in K, the temperatures are 800°C = 800 + 273 = 1073 K; 900°C = 900 + 273 = 1173 K; and 1000°C = 1000 + 273 = 1273 K. Substituting the steady-state creep rate, stress exponent, and stress values into Equation 12.7, we get

$$\dot{\varepsilon}_s = A_2\sigma^n \exp\left(-\frac{Q_c}{RT}\right)$$

$$7.8 \times 10^{-3}\,\mathrm{h}^{-1} = \mathrm{A}_2(60\,\mathrm{MPa})^{6.5} \exp\left(-\frac{Q_c}{8.314\,\mathrm{J\,mol^{-1}\,K^{-1}} \times 1173\mathrm{K}}\right)$$

and

$$5.6 \times 10^{-4}\,\mathrm{h}^{-1} = \mathrm{A}_2(60\,\mathrm{MPa})^{6.5} \exp\left(-\frac{Q_c}{8.314\,\mathrm{J\,mol^{-1}\,K^{-1}} \times 1073\mathrm{K}}\right)$$

a. Dividing the above two equations one with the other, we get

$$\frac{7.8 \times 10^{-3}\,\mathrm{h}^{-1}}{5.6 \times 10^{-4}\,\mathrm{h}^{-1}} = \exp\left[-\frac{Q_c}{8.314\,\mathrm{J\,mol^{-1}\,K^{-1}}}\left(\frac{1}{1173\mathrm{K}} - \frac{1}{1073\mathrm{K}}\right)\right]$$

$$1.39 \times 10\mathrm{h}^{-1} = \exp\left[-\frac{Q_c}{8.314\,\mathrm{J\,mol^{-1}\,K^{-1}}}(-7.945 \times 10^{-5})\right]$$

Taking natural logarithms on both sides of the equation, we get

$$\ln(1.39 \times 10) = \left(\frac{Q}{8.314}\right)(7.945 \times 10^{-5})$$

$$2.632 = Q \times 9.556 \times 10^{-6}$$

or,

$$Q = 2.75 \times 10^5\,\mathrm{J\,mol^{-1}}, \text{ that is, } 275\mathrm{kJ\,mol^{-1}}.$$

b. Since we know the activation energy for creep as 275 kJ/mol, we can now calculate the steady-state creep rate at 1000°C (= 1273 K), using Equation 12.7. But, before we can calculate the steady-state creep rate, we also need to know the value of the preexponential constant A_2. The value of A_2 can be calculated from any one of the above equations, since the activation energy is now known. Thus, at 800°C (= 1073 K),

$$5.6 \times 10^{-4}\,\mathrm{h}^{-1} = \mathrm{A}_2(60\,\mathrm{MPa})^{6.5} \exp\left(-\frac{275{,}000\,\mathrm{J\,mol^{-1}}}{8.314\,\mathrm{J\,mol^{-1}K^{-1}} \times 1073\mathrm{K}}\right)$$

$$5.6 \times 10^{-4}\,\mathrm{h}^{-1} = \mathrm{A}_2(60\,\mathrm{MPa})^{6.5} \exp(-30.83)$$

$$\mathrm{A}_2 = \frac{5.6 \times 10^{-4}}{(60\,\mathrm{MPa})^{6.5} \exp(-30.83)} = 3.8 \times 10^{-2}$$

Substituting this value of A_2, activation energy, Q_c and the temperature T into Equation 12.7, we now get

$$\dot{\varepsilon}_s = 3.8 \times 10^{-2} \times (60\ \text{MPa})^{6.5} \exp\left(-\frac{275{,}000\ \text{J mol}^{-1}}{8.314\ \text{J mol}^{-1}\text{K}^{-1} \times 1273\ \text{K}}\right)$$

$$= 3.8 \times 10^{-2}(60\ \text{MPa})^{6.5} \exp(-25.98) = 7.15 \times 10^{-2}\ \text{h}^{-1}$$

12.6 CREEP RESISTANCE CRITERIA

The main purpose of conducting the creep tests is to evaluate the creep resistance of the material, that is, to determine the maximum stress and temperature up to which the material can be exposed for a long time. As we saw earlier, the steady-state creep rate represents the minimum creep rate experienced by the material. Therefore, in practice, the minimum creep rate (i.e., the steady-state creep rate) is determined at a constant temperature and at different stress levels. A plot is then made with the stress plotted as the ordinate (Y-axis) and the minimum creep rate as the abscissa (X-axis) at a constant temperature. Plots like this are obtained at different temperatures. The relationship between stress and the minimum creep rate is linear, as shown in Figure 12.8. Once a plot like this is available, we can always determine the minimum creep rate at any given temperature and/or stress level. The stress required to produce a minimum creep rate of 10^{-5}% per hour at a given temperature is a common way of designating the creep strength of a material.

12.7 LARSON–MILLER PARAMETER

One of the most convenient and common ways of representing creep data is to use the *Larson–Miller parameter.* Larson and Miller proposed use of a parameter P which combines the effects of both temperature and time. The Larson–Miller

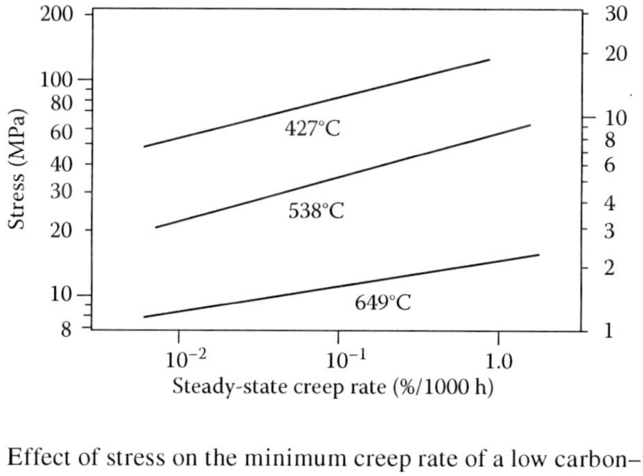

FIGURE 12.8 Effect of stress on the minimum creep rate of a low carbon–nickel alloy at three different temperatures.

parameter P, widely used to correlate stress versus rupture lifetime data over a wide range of temperatures, is defined as

$$P = T\left(C + \log t_{\mathrm{r}}\right) \tag{12.8}$$

where T is the temperature expressed in degrees Kelvin, t_{r} is the rupture lifetime, and C is a constant. (The "constant" C is usually expressed in units of time, hours.) Thus, the units of P are K-h. The above equation suggests that the log time to stress rupture plus a constant multiplied by the temperature in degrees Kelvin is a constant at a given stress level. That is to say, the rupture lifetime of a sample at a given stress level will vary with test temperature in such a way that the Larson–Miller parameter remains unchanged.

Sometimes, the temperature is expressed in degrees Rankine (°R). The conversion between °F and °R is: °R = °F + 460.

Originally, the constant C was set equal to 20. In practice, it varies between 17 and 23 for different types of materials. However, C is currently permitted to assume a value that best describes the data. The value of C can be experimentally determined from a minimum of two sets of time and temperature data. Assuming P to be a constant, Equation 12.8 can be written for two different temperatures as

$$P = T_1\left(C + \log t_{\mathrm{r}1}\right) \tag{12.9a}$$

and

$$P = T_2\left(C + \log t_{\mathrm{r}2}\right) \tag{12.9b}$$

Rearranging the above two equations, we get

$$C = \frac{T_2 \log t_{\mathrm{r}2} - T_1 \log t_{\mathrm{r}1}}{T_1 - T_2} \tag{12.10}$$

C can also be determined graphically by rearranging Equation 12.8 as

$$\log t_{\mathrm{r}} = -C + \frac{P}{T} \tag{12.11}$$

From the above Equation 12.11, it is clear that when the experimental creep rupture data, t_{r} is plotted against reciprocal of temperature, the intersection of the curves for different stress values at $1/T = 0$ defines the value of C.

Thus, when the log stress required to rupture is plotted against the Larson–Miller parameter, P, a single curve is obtained. The advantage of a plot of this type, called the *Larson–Miller diagram*, is that the Larson–Miller parameter allows one to combine rupture lifetime data from several different temperatures and stresses into one curve. This avoids looking at a vast amount of data plotted in different curves. Figure 12.9 shows the Larson–Miller diagram for an iron-based alloy (S-590).

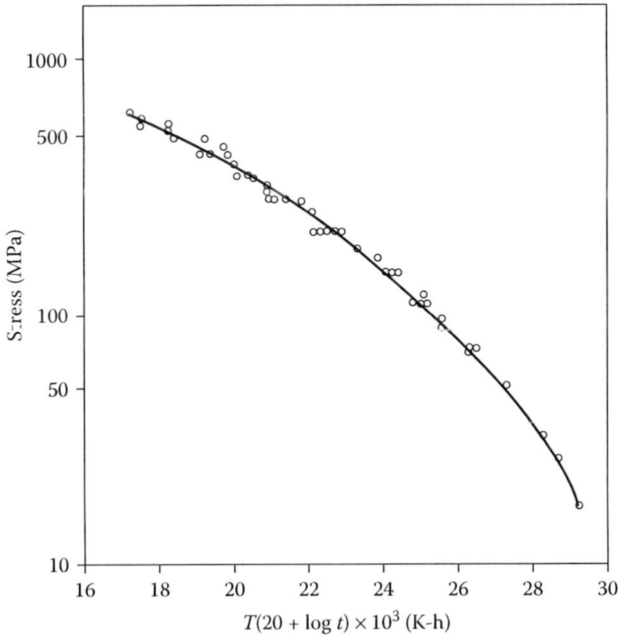

FIGURE 12.9 Larson–Miller diagram for an iron-based alloy (S-590).

The main advantage of the Larson–Miller diagram is that among the three parameters—time to rupture, temperature at which creep is occurring and the stress applied—if two of them are known, the third can be determined.

Example Problem 12.4

Using the Larson–Miller diagram represented in Figure 12.9, (a) determine the time to stress rupture at 650°C for the S-590 alloy at a stress level of 200 MPa, and (b) estimate the maximum service temperature if a rupture life of 1000 h is required.

Solution 12.4

From Figure 12.9 the value of the Larson–Miller parameter corresponding to a stress level of 200 MPa is 22.8×10^3 K-h. Since the value of the constant C was chosen as 20 in Figure 12.9, and the temperature at which rupture is expected is 650°C (= 650 + 273 = 923 K), Equation 12.8, namely, $P = T$ (K) $(C + \log t_r)$ can be used.

a. The time to rupture can be calculated from the above equation by substituting the values of P, T, and C. Thus,

$$22.8 \times 10^3\,\text{K-h} = 923\,\text{K}\left(20 + \log t_r\right)$$

or,

$$\log t_r = \frac{22.8 \times 10^3 \text{ K-h}}{923\text{K}} - 20\text{h}$$
$$= 24.70\text{h} - 20\text{h}$$
$$= 4.70$$

Therefore, $t_r = 50{,}119$ h or 5.7 years.

b. If we now wish to calculate the maximum service temperature for a rupture life of 1000 h, we can use the above data again. Thus,

$$22.8 \times 10^3 \text{ K-h} = T(\text{K})(20 \text{ h} + \log(1000 \text{ h}))$$
$$= T \text{ (K) } (20 \text{ h} + 3)$$

or,

$$T(\text{K}) = \left[\frac{(22.8 \times 10^3 \text{ K-h})}{23 \text{ h}}\right] = 991\text{K} = 718°\text{C}$$

You will notice, from the above example, that a small increase of 68°C is decreasing the rupture life from 5.7 years to only 1000 h (about 41 days). This is a clear demonstration of the fact that a small change in temperature can have a large effect on the rupture life of specimens.

There are other ways of representing the Larson–Miller plots. A set amount of strain, for example, 0.2% produced by variable stresses at different times and temperatures are also used to compare the high-temperature creep properties of materials.

12.8 CREEP IN CERAMIC MATERIALS

Since ceramic materials are used at elevated temperatures, the phenomenon of creep plays an important role in them also. However, the phenomenon is more complex in ceramics than in metallic specimens. This is because, ceramics contain ions of different charges, and it is always difficult to maintain charge neutrality. Further, the cations and anions in the ceramic may have different diffusivities. Because of this (instead of dislocation climb in metallic materials) grain boundary sliding assumes importance in the high-temperature deformation of ceramic materials. Consequently, it is the grain boundary diffusion that is more important in ceramic materials than lattice diffusion. In fact, Coble creep, involving grain boundary diffusion is the predominant mechanism of creep in ceramic materials and also the novel nanostructured materials (with a grain size of ≤100 nm).

The temperatures at which significant creep occurs in ceramics are much higher than those in the case of metals.

12.9 CREEP IN POLYMERIC MATERIALS

In spite of the relatively low temperatures at which polymeric materials are used in industry, creep is a significant design factor. In polymeric materials also, the strain increases with time at a constant stress. The related phenomenon of *stress relaxation* is another important design criterion for polymers. In the case of metallic materials, we have noted that the strain increases with time at a constant stress. On the other hand, in polymers, the stress required to maintain a particular strain decreases with time. This process is known as stress relaxation.

12.10 EXPERIMENTAL PROCEDURE

Creep-testing machines consist of a loading arrangement that produces a constant load or a constant stress on the specimen. The machines also have the provision to

- Heat specimens to the desired temperature.
- Accurately control the temperature.
- Accurately measure extremely small deformations of the order of 0.0001 cm, using strain-measuring devices.

Note that the forms of the strain versus time plots are the same under constant load and constant stress conditions. However, the time spent by the specimen in the first two stages of creep (Stage I and Stage II) can be much longer under constant stress than under constant load. Hence, rupture life is longer under constant stress conditions.

The actual experiment will consist of loading the specimen into the machine and then heating it to the desired temperature. Strain measurement is usually carried out by an optical extensometer which, by means of a light beam focused on tilting mirrors attached to the gage length of the specimen, magnifies the elongation by three orders of magnitude. This leads to accurate measurements of 0.0001 cm. The strain is automatically plotted as a function of time.

Since most metallic specimens require a long time and high temperatures to obtain a significant creep curve, it is convenient to take a lead metal sample and conduct the test at room temperature and then at a slightly elevated temperature, may be up to about 50–60°C. By measuring the steady-state creep rate at a minimum of two different temperatures, one can calculate the activation energy for the creep process.

Alternately, the students can be divided into small batches of three students each, and each batch can do one experiment (at one temperature and one stress level). But, the difficulty of not having many creep machines in the Laboratory could be a big constraint for this. Realize that creep testing requires a really large number of specimens. In the industry, normally one uses a number of machines simultaneously (battery of machines) to get meaningful data and also to understand the mechanisms by which creep occurs.

EXERCISES

12.1 Define creep.

12.2 What is the temperature range in which creep occurs?

12.3 Draw a typical creep curve for a metal under constant load and at a relatively high temperature, and indicate on it all the three stages of creep.

12.4 How does one define the minimum creep rate with respect to the creep curve?
12.5 Does the shape of the creep curve depend on stress and temperature? If so, what is the effect?
12.6 What is a stress rupture test?
12.7 How can one predict the creep behavior of a material at low temperatures and/or stresses?
12.8 Does creep occur in ceramic samples?
12.9 What parameters determine the shape of the creep curve?
12.10 Explain how one can predict the creep behavior of a material at low temperatures?

FURTHER READING

Callister, Jr., W. D. and D. G. Rethwisch, 2010. *Materials Science and Engineering, An Introduction*, 8th edition. New York, NY: John Wiley & Sons, Inc.

Courtney, T. H. 2000. *Mechanical Behavior of Materials*, 2nd edition. New York, NY: McGraw-Hill.

Dieter, G. E. 1986. *Mechanical Metallurgy*, 3rd edition. Boston, MA: McGraw-Hill.

Dowling, N. E. 2007. *Mechanical Behavior of Materials*, 3rd edition. Upper Saddle River, NJ: Pearson: Prentice-Hall.

Hertzberg, R. W. 1996. *Deformation and Fracture Mechanics of Engineering Materials*, 4th edition. New York, NY: John Wiley & Sons, Inc.

Meyers, M. A. and K. K. Chawla 2008. *Mechanical Behavior of Materials*, 2nd edition. Cambridge, UK: Cambridge University Press.

Index

Note: n = Footnote.

C

D

E

F

G

H

I

M

N

O

P

U

V

W

X

Y

Z